A Problem-Solving Approach to Single-Variable Calculus

Farzin Asadi

A Problem-Solving Approach to Single-Variable Calculus

Volume I

Springer

Farzin Asadi
Department of Computer Engineering
Ostim Technical University
Ankara, Türkiye

ISBN 978-3-032-13504-9 ISBN 978-3-032-13505-6 (eBook)
https://doi.org/10.1007/978-3-032-13505-6

This Springer imprint is published by the registered company Springer Nature Switzerland AG
The registered company address is: Gewerbestrasse 11, 6330 Cham, Switzerland

Preface

Single-variable calculus is a fundamental mathematical framework essential for students across diverse fields—including the basic sciences, engineering, and economics—because it enables the precise modeling and understanding of change. It introduces two indispensable tools: the derivative, which quantifies instantaneous rates of change, and the integral, which measures accumulation over time or space. These concepts form the foundation of nearly all quantitative disciplines. For example, engineers rely on calculus to analyze motion, forces, and optimize designs; scientists apply it to describe phenomena ranging from planetary motion in physics and reaction kinetics in chemistry to population growth and drug concentration in biology; and economists use it to determine marginal cost, marginal revenue, and optimal production levels.

To truly grasp the concepts and techniques of single-variable calculus, practice is paramount. This book is your comprehensive companion—a two-volume set filled with meticulously solved, step-by-step problems designed to build your skills and deepen your understanding. With these resources, you will gain the confidence to approach any single-variable calculus problem and transform challenges into achievements. A solid foundation in precalculus, algebra, and trigonometry is the only prerequisite.

This book is designed as a valuable supplement to any standard single-variable calculus textbook. Each chapter begins with a concise review of key concepts, followed by a collection of fully worked examples presented in a clear, step-by-step format. You are encouraged to attempt each problem on your own before consulting the detailed solutions.

Because computational methods are central to modern science and engineering, this book adopts a dual approach to problem-solving. It first guides you through traditional hand calculations and then demonstrates how to apply the same concepts using MATLAB®, bridging theoretical understanding with practical computational skills.

I hope this book will serve as a reliable companion throughout your studies. By embracing this problem-solving approach, you will not only build a strong understanding of single-variable calculus but also develop the confidence to tackle new and challenging problems. I wish you a productive and rewarding journey as you master this fascinating subject.

Ankara, Turkey Farzin Asadi

Competing Interests The author has no competing interests to declare that are relevant to the content of this manuscript.

Contents

Notation Used in This Book[1]

$\log(x)$	$= \log_{10}(x)$.
$\sin^{-1}(x)$	The arcsine of x.
$\cos^{-1}(x)$	The arccosine of x.
$\tan^{-1}(x)$	The arctangent of x.
$\cot^{-1}(x)$	The arccotangent of x.
$\sec^{-1}(x)$	The arcsecant of x.
$\csc^{-1}(x)$	The arccosecant of x.
$\sec(x)$	The secant of x ($\sec(x) = \frac{1}{\cos(x)} = (\cos(x))^{-1}$).
$\csc(x)$	The cosecant of x ($\csc(x) = \frac{1}{\sin(x)} = (\sin(x))^{-1}$).
$\sin^n(x)$	$= (\sin(x))^n$.
$\cos^n(x)$	$= (\cos(x))^n$.
$[x]$	The greatest integer that is less than or equal to x. For example, $[2.3] = 2$ and $[-2.3] = -3$.
i and j	The imaginary unit, where $i = j = \sqrt{-1}$.
$\cong$	Denotes that two values are approximately equal.

[1]Note that, for example, $\cos^{-1}(x)$ denotes the inverse cosine function (also written as $\arccos(x)$, meaning the function that undoes cosine), while $(\cos(x))^{-1}$ represents the reciprocal of cosine, which equals $\frac{1}{\cos(x)}$.

Chapter 1
Mathematical Prerequisites

1.1 Introduction

This chapter provides a concise yet essential review of the mathematical foundation you established during high school, serving as the launching pad for advanced study.

1.2 Set of Numbers

Numbers, the fundamental units of mathematics, have evolved over centuries, shaping our understanding of quantity, measurement, and abstract concepts. From the simple counting of objects to the complex calculations of modern science, numbers underpin our world. Important number sets are:

1.3 Basic Number Sets

Natural Numbers ($\mathbb{N}$): Positive integers starting from 1: $\{1, 2, 3, \ldots\}$

Whole Numbers ($\mathbb{W}$): Natural numbers including 0: $\{0, 1, 2, 3, \ldots\}$

Integers ($\mathbb{Z}$): Positive and negative whole numbers, including zero: $\{\ldots, -3, -2, -1, 0, 1, 2, 3, \ldots\}$

Rational Numbers ($\mathbb{Q}$): Numbers that can be expressed as a fraction of two integers: $\left\{\frac{p}{q} \mid p, q \in Z, q \neq 0\right\}$

Irrational Numbers ($\mathbb{I}$): Numbers that cannot be expressed as a simple fraction, often represented by non-repeating, non-terminating decimals: $\left\{\sqrt{2}, \pi, e, \ldots\right\}$

F. Asadi, *A Problem-Solving Approach to Single-Variable Calculus*,
https://doi.org/10.1007/978-3-032-13505-6_1

Real Numbers ($\mathbb{R}$): All rational and irrational numbers: This set encompasses all numbers on the number line.

Complex Numbers ($\mathbb{C}$): Numbers of the form $a + bi$, where a and b are real numbers, and i is the imaginary unit ($i = \sqrt{-1}$): $\{a + bi | a, b \in \mathbb{R}\}$. In electrical engineering, i denotes current. To avoid confusion, electrical engineers prefer the symbol j for the imaginary unit. Note that both i and j may be used for the imaginary unit in this book.

1.4 Functions

A function in mathematics is a rule that assigns to each input exactly one output. Key components of a function:

Domain: The set of all possible input values.

Codomain: The set of all possible output values. It's like the entire menu at a restaurant, even if not all dishes are currently available.

Range: The subset of the codomain that consists of all actual output values. It's like the dishes that are actually served, based on the ingredients and the chef's choices.

Let's study some numeric examples.

Example 1.1 Value of $f(x) = x^2 + \sin(x)$ at $x = 0.5$ at $x = 0.5$ equals to $f(0.5) = 0.5^2 + \sin(0.5) = 0.7294$.

Example 1.2 Value of $z = f(x, y) = x^2 + y^2 + 3x^2y + \sin(y)$ at point $(1, 2)$ equals to $f(1, 2) = 1^2 + 2^2 + 3 \times 1^2 \times 2 + \sin(2) = 11.9093$.

Example 1.3 Domain of $f(x) = \frac{x+2}{x+3}$ is $\mathbb{R} - \{x : x + 3 = 0\} = \mathbb{R} - \{-3\}$.

Example 1.4 Domain of $f(x) = \ln\left(\frac{x+2}{x+3}\right)$ is $\mathbb{R} - [-3, -2] = (-\infty, -3) \cup (-2, \infty)$ (Fig. 1.1).

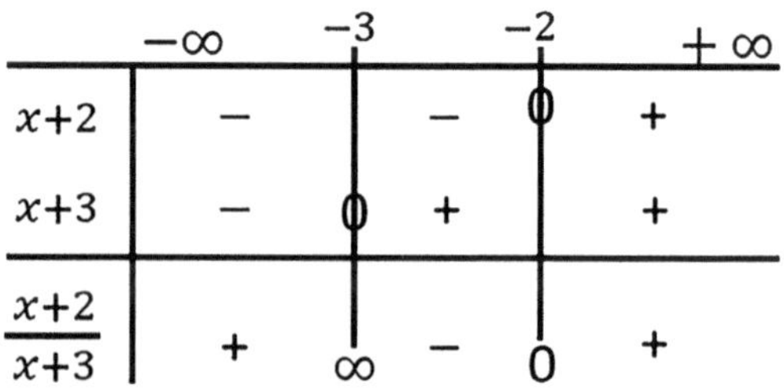

Fig. 1.1 Sign of $\frac{x+2}{x+3}$ for different values of x

Exercise 1.1 Determine the domain of $f(x) = \sqrt{16 - x^2}$.

Exercise 1.2 Determine the domain of $f(x) = \sqrt[3]{16 - x^2}$.

Exercise 1.3 Determine the domain of $f(x) = \frac{1}{\sqrt{x^2 - x - 6}}$.

Exercise 1.4 Determine the domain of $f(x) = \frac{\sqrt{x}}{x - 4}$.

Exercise 1.5 Determine the domain of $f(x) = \log_2(x^2 - 4)$.

1.5 Line Equation

Consider the line that passes through points $A = \begin{bmatrix} x_1 \\ y_1 \end{bmatrix}$ and $B = \begin{bmatrix} x_2 \\ y_2 \end{bmatrix}$ (Fig. 1.2).

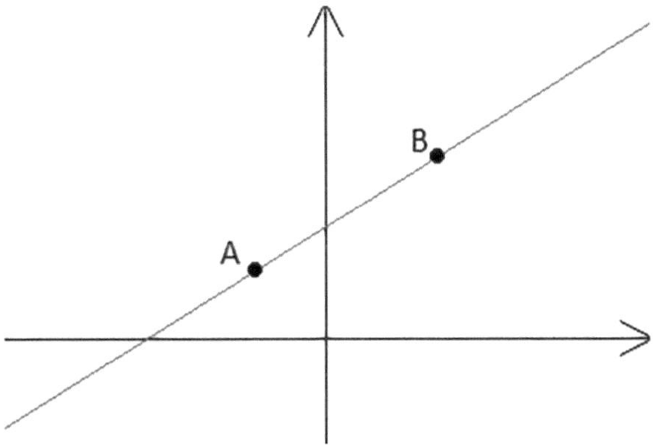

Fig. 1.2 Line passes through *A* and *B*

Equation of the line that passes through points $A = \begin{bmatrix} x_1 \\ y_1 \end{bmatrix}$ and $B = \begin{bmatrix} x_2 \\ y_2 \end{bmatrix}$ is:

$$m = \frac{y_2 - y_1}{x_2 - x_1} = \frac{y_1 - y_2}{x_1 - x_2}$$

$$y - y_1 = m(x - x_1) \Rightarrow y = m(x - x_1) + y_1 \Rightarrow y = mx + (y_1 - mx_1)$$

or

$$y - y_2 = m(x - x_2) \Rightarrow y = m(x - x_2) + y_2 \Rightarrow y = mx + (y_2 - mx_2)$$

The equation for a line perpendicular to $y = mx + b$ is: $y = -\frac{1}{m}x + c$. The variable c represents the new y-intercept. This value can be any real number unless the perpendicular line is required to pass through a specific point.

Example 1.5 Find the equation of the line passing through points $A = \begin{bmatrix} -1 \\ 2 \end{bmatrix}$ and $B = \begin{bmatrix} 3 \\ 4 \end{bmatrix}$.

Solution In this case, $m = \frac{4-2}{3-(-1)} = \frac{2-4}{-1-3} = \frac{2}{4} = 0.5$

$$y = 0.5(x - (-1)) + 2 = 0.5x + 2.5$$

$$y = 0.5(x - 3) + 4 = 0.5x + 2.5$$

Example 1.6 Find the equation of the line passing through point $A = \begin{bmatrix} -1 \\ 2 \end{bmatrix}$ with a slope of −5.

Solution $y - y_1 = m(x - x_1) \Rightarrow y = -5(x - (-1)) + 2 = -5x - 3$

Example 1.7 Find the equation of the line shown in Fig. 1.3.

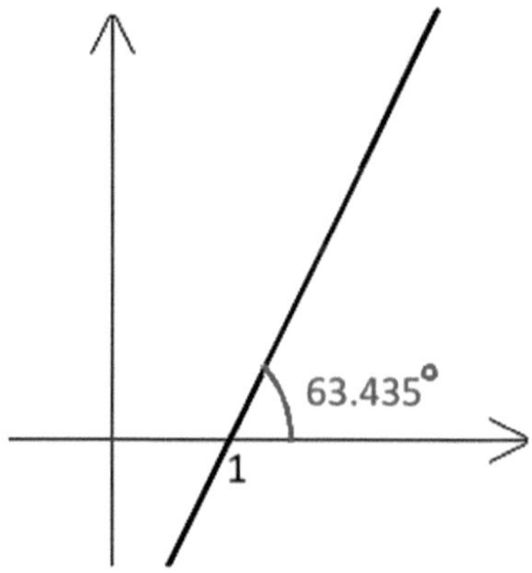

Fig. 1.3 Example line for equation determination

Solution $m = \tan(63.435°) = 2 \Rightarrow y - y_1 = m(x - x_1) \Rightarrow y - 0 = 2(x - 1) \Rightarrow y = 2x - 2$

Example 1.8 Find the equation of the line shown in Fig. 1.4.

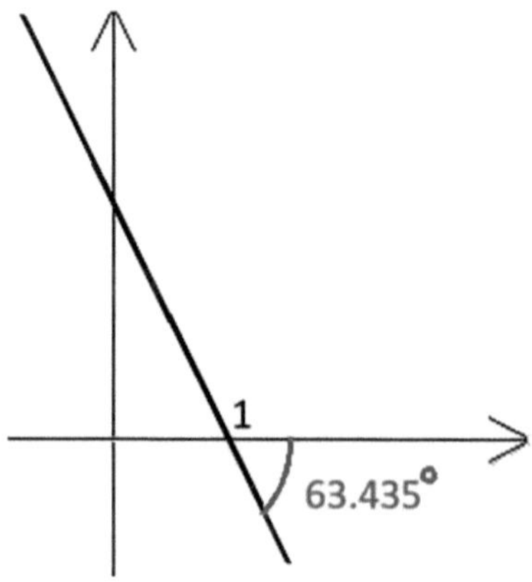

Fig. 1.4 Example line for equation determination

Solution $m = -\tan(63.435°) = -2 \Rightarrow y - y_1 = m(x - x_1) \Rightarrow y - 0 = -2(x - 1) \Rightarrow y = -2x + 2$

1.6 Angle Between Two Lines

The angle between two lines, $y_1 = m_1x + b_1$ and $y_2 = m_2x + b_2$ is $\alpha = \tan^{-1}\left|\frac{m_2 - m_1}{1+m_2m_1}\right|$.

Example 1.9 Angle between $y_1 = 2x$ and $y_2 = x$ is $\alpha = \tan^{-1}\left|\frac{1-2}{1+1\times 2}\right| = \tan^{-1}\left(\frac{1}{3}\right) = 0.3218\,Rad = 18.435°$ (Fig. 1.5).

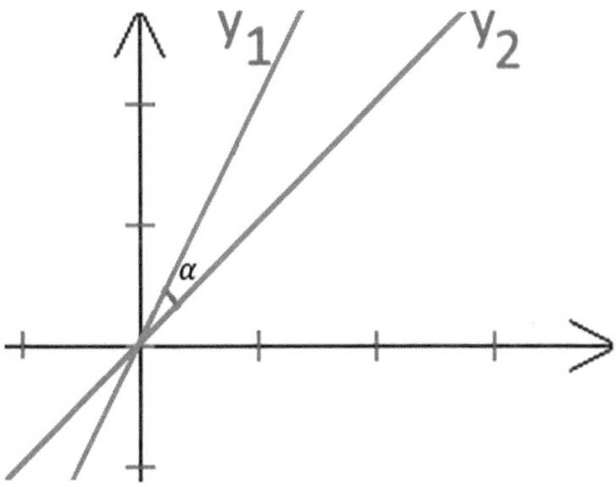

Fig. 1.5 Angle between lines y_1 and y_2

1.7 Distance Between Two Points

The distance between two points $A = \begin{bmatrix} x_1 \\ y_1 \end{bmatrix}$ and $B = \begin{bmatrix} x_2 \\ y_2 \end{bmatrix}$ is $d = \sqrt{(x_1 - x_2)^2 + (y_1 - y_2)^2}$. The distance between two points $A = \begin{bmatrix} x_1 \\ y_1 \\ z_1 \end{bmatrix}$ and $B = \begin{bmatrix} x_2 \\ y_2 \\ z_2 \end{bmatrix}$ is $d = \sqrt{(x_1 - x_2)^2 + (y_1 - y_2)^2 + (z_1 - z_2)^2}$.

Example 1.10 The distance between points $A = \begin{bmatrix} 1 \\ 2 \end{bmatrix}$ and $B = \begin{bmatrix} -1 \\ -3 \end{bmatrix}$ (Fig. 1.6) is $d = \sqrt{(1-(-1))^2 + (2-(-3))^2} = \sqrt{29} = 5.38$.

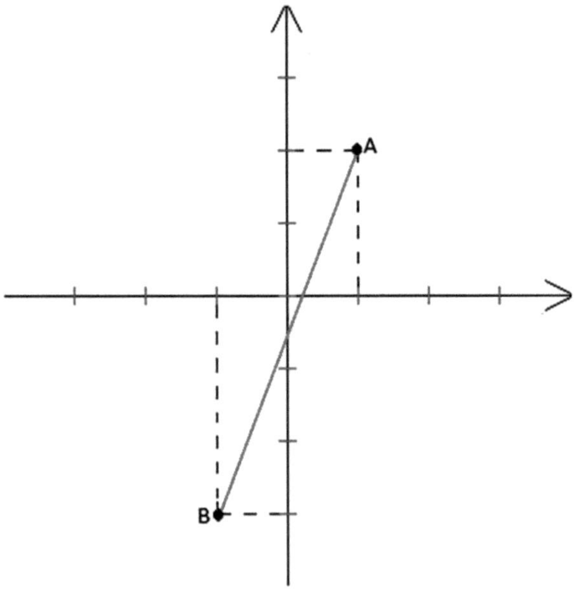

Fig. 1.6 Distance between points A and B

The distance between point $A = \begin{bmatrix} x_1 \\ y_1 \end{bmatrix}$ and origin $O = \begin{bmatrix} 0 \\ 0 \end{bmatrix}$ is: $d = \sqrt{(x_1 - 0)^2 + (y_1 - 0)^2} = \sqrt{{x_1}^2 + {y_1}^2}$. For instance, the distance between point $A = \begin{bmatrix} 1 \\ 2 \end{bmatrix}$ and the origin (Fig. 1.7) is: $d = \sqrt{1^2 + 2^2} = \sqrt{5} = 2.24$.

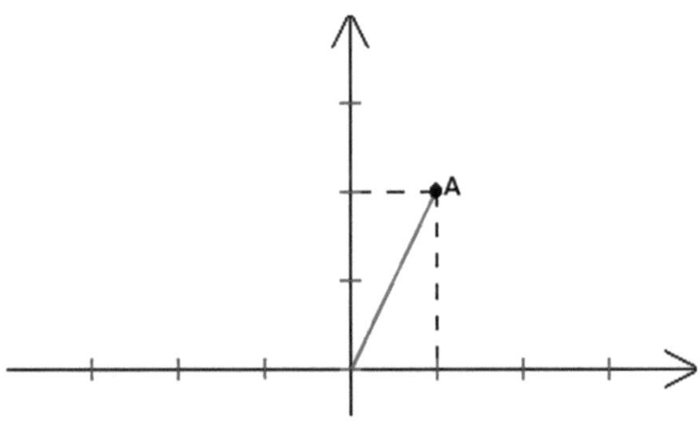

Fig. 1.7 Distance between point A and origin

1.8 Distance Between a Point and a Plane

Given a plane defined by the equation $Ax + By + Cz + D = 0$ and a point $P(x_0, y_0, z_0)$, the distance d between the point and the plane is given by the formula $d = \frac{|Ax_0 + By_0 + Cz_0 + D|}{\sqrt{A^2 + B^2 + C^2}}$.

Example 1.11 Calculate the distance between point $\begin{bmatrix} 1 \\ 0 \\ -2 \end{bmatrix}$ and $x + 2y + z = 4$.

Solution $d = \frac{|Ax_0 + By_0 + Cz_0 + D|}{\sqrt{A^2 + B^2 + C^2}} \Rightarrow d = \frac{|1 + 2 \times 0 - 2 - 4|}{\sqrt{1^2 + 2^2 + 1^2}} = \frac{5}{\sqrt{6}} = 2.0412$

1.9 Exponential Function

An exponential function is a mathematical function of the form $f(x) = a^x$, where x is a variable and a is a constant called the base, and it must be positive and not equal to 1. When the base of function is $e \approx 2.7182$, the function is called "natural exponential function." Important properties of the natural exponential function are studied in this section.

Consider $f(t) = e^{-pt}$ where $p > 0$ is a real numbers. $\int_0^{\infty} e^{-pt} dt$ equals to:

$$\int_0^{\infty} e^{-pt} dt = \frac{e^{-pt}}{-p}\bigg|_0^{\infty} = \frac{e^{-p\infty}}{-p} - \frac{e^{-p0}}{-p} = 0 - \frac{1}{-p} = \frac{1}{p}$$

$f(t) = e^{-pt}$ decays with the rate shown in Table 1.1.

Table 1.1 Rate of decay for $f(t) = e^{-pt}$

pt	e^{-pt}
0	1
1	0.37
2	0.13
3	0.05
4	0.02
5	0.006

For instance, graph of $f(t) = e^{-2t}$ is shown in Fig. 1.8. $f(t)$ is almost zero for $t \geq \frac{5}{2} = 2.5$.

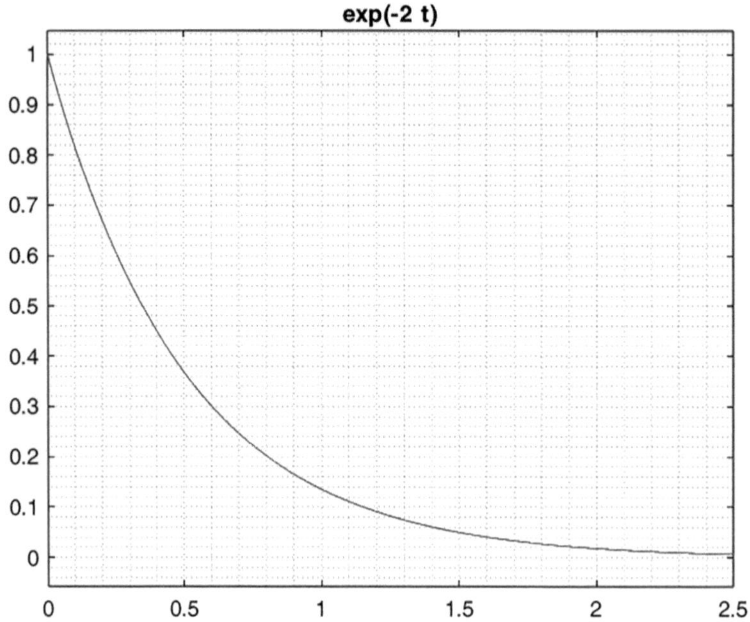

Fig. 1.8 Plot of e^{-2t}

If $a < 0$, then $\lim\limits_{x \to \infty} e^{(a+jb)x} = 0$ (note that j shows the imaginary unit). For instance, $\lim\limits_{x \to \infty} e^{(-3+j4)x}$, $\lim\limits_{x \to \infty} e^{(-3-j4)x}$, or $\lim\limits_{x \to \infty} e^{-6x}$ equals to zero.

If $a > 0$, then $\lim\limits_{x \to -\infty} e^{(a+jb)x} = 0$ (note that j shows the imaginary unit). For instance, $\lim\limits_{x \to -\infty} e^{(7+j2)x}$, $\lim\limits_{x \to -\infty} e^{(7-j2)x}$, or $\lim\limits_{x \to -\infty} e^{12x}$ equals to zero.

Let's study another important function: $f(t) = (I - F)e^{-pt} + F$. I and F are two real constants. Note that:

$$\lim_{t \to 0} f(t) = \lim_{t \to 0} (I - F)e^{-pt} + F = I$$

$$\lim_{t \to \infty} f(t) = \lim_{t \to \infty} (I - F)e^{-pt} + F = F$$

The time needed to travel from I to F is approximately $\frac{5}{p}$.

For instance, for $I = 3$, $F = 5$, and $p = 2$, $f(t) = (I - F)e^{-pt} + F = (3 - 5)e^{-2t} + 5 = 5 - 2e^{-2t}$. Note that $\lim_{t\to 0} f(t) = \lim_{t\to 0} 5 - 2e^{-2t} = 3$ and $\lim_{t\to\infty} f(t) = \lim_{t\to\infty} 5 - 2e^{-2t} = 5$. Graph of $f(t) = 5 - 2e^{-2t}$ is shown in Fig. 1.9. Travel from 3 to 5 takes around $\frac{5}{2} = 2.5\ s$.

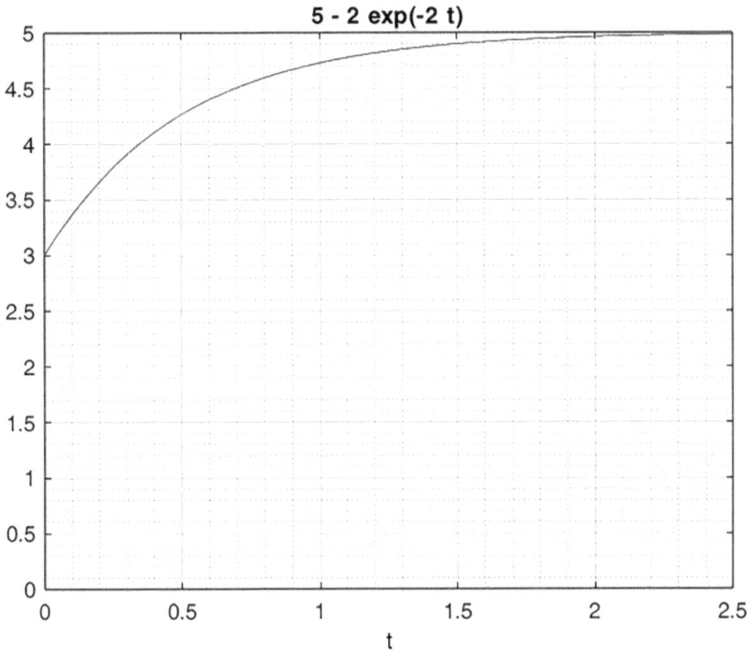

Fig. 1.9 Plot of $5 - 2e^{-2t}$

Let's study another numeric example. For $I = 5$, $F = 3$, and $p = 2$, $f(t) = (I - F)e^{-pt} + F = (5 - 3)e^{-2t} + 3 = 3 + 2e^{-2t}$. Note that $lim_{t\to 0} f(t) = lim_{t\to 0} 3 + 2e^{-2t} = 5$ and $lim_{t\to\infty} f(t) = lim_{t\to\infty} 3 + 2e^{-2t} = 3$. Graph of $f(t) = 3 + 2e^{-2t}$ is shown in Fig. 1.10. Travel from 5 to 3 takes around $\frac{5}{2} = 2.5\ s$.

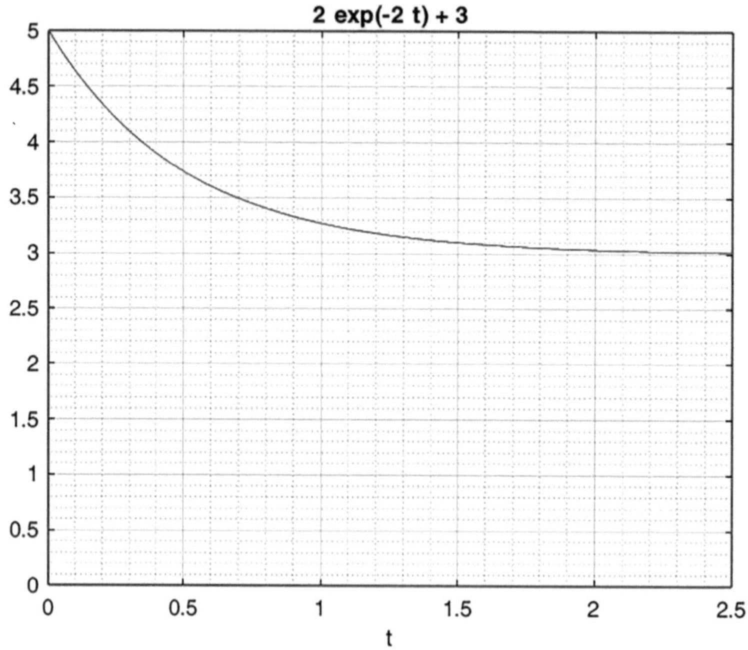

Fig. 1.10 Plot of $3 + 2e^{-2t}$

1.10 Logarithmic Function

A logarithm is essentially the inverse operation of exponentiation. It's a way to determine the exponent to which a base must be raised to produce a given number. For instance, $3^4 = 81 \Rightarrow \log_3 81 = 4$.

When no base is explicitly written for a logarithm, it is understood to be base 10. Thus, $\log(x)$ is the same as $\log_{10}(x)$.

The following are the most important properties of logarithmic functions:

$$\log_c(a \times b) = \log_c(a) + \log_c(b)$$

$$\log_c\left(\frac{a}{b}\right) = \log_c(a) - \log_c(b)$$

$$\log_c(a^n) = n \times \log_c(a)$$

$$a^{\log_a(x)} = x$$

$$\log_c(a) = \frac{\log_d(a)}{\log_d(c)}$$

Let's study some numeric example.

Example 1.12 Find solutions of $\log_2 x + \log_2(x - 3) = 2$.

Solution

$$\log_2 x + \log_2(x-3) = 2 \Rightarrow \log_2 x(x-3) = 2 \Rightarrow x(x-3) = 2^2 \Rightarrow x^2 - 3x$$
$$= 4 \Rightarrow x^2 - 3x - 4 = 0 \Rightarrow x_1 = 4, x_2 = -1$$

Logarithms are only defined for positive arguments. The arguments in the original equation are x and $(x - 3)$, so we must have $x > 0$ and $x - 3 > 0$ (which means $x > 3$). $-1 \ngtr 0$. Therefore, the only solution to the equation is: $x = 4$.

Example 1.13 Find solutions of $\log(x^2 - 11x + 130) = 2$.

Solution

$$x^2 - 11x + 130 = 10^2 \Rightarrow x^2 - 11x + 30 = 0 \Rightarrow x_1 = 5, x_2 = 6$$

For a logarithmic solution to be valid, the argument of the logarithm must be positive $(x^2 - 11x + 130 > 0)$.

$x^2 - 11x + 130 \xRightarrow{x=5} 5^2 - 11 \times 5 + 130 = 100 > 0 \Rightarrow$ Since $100 > 0$, $x = 5$ is a valid solution.

$x^2 - 11x + 130 \xRightarrow{x=6} 6^2 - 11 \times 6 + 130 = 100 > 0 \Rightarrow$ Since $100 > 0$, $x = 6$ is a valid solution.

Example 1.14 Calculate the value of $e^{3 \ln(4)}$, $10^{2 \log(4)}$, and $\log_7(14)$.

Solution

$$e^{3\ln(4)} = e^{\ln(4^3)} = 4^3 = 64$$

$$10^{2\log(4)} = 10^{\log(4^2)} = 4^2 = 16$$

$$\log_7(14) = \frac{\log_{10}(14)}{\log_{10}(7)} = \frac{1.1461}{0.8451} = 1.3562 \text{ or } \log_7 14 = \frac{\ln(14)}{\ln(7)} = \frac{2.6391}{1.9459} = 1.3562.$$

1.11 Even and Odd Functions

A function $f(x)$ is even if $f(-x) = f(x)$ for all x in the domain. For instance, $f(x) = x^2$, $f(x) = \cos(x)$, or $f(x) = |x|$ are even function. The graph of an even function is symmetric about the y-axis. This means that if you reflect the graph across the y-axis, it will look exactly the same (Fig. 1.11).

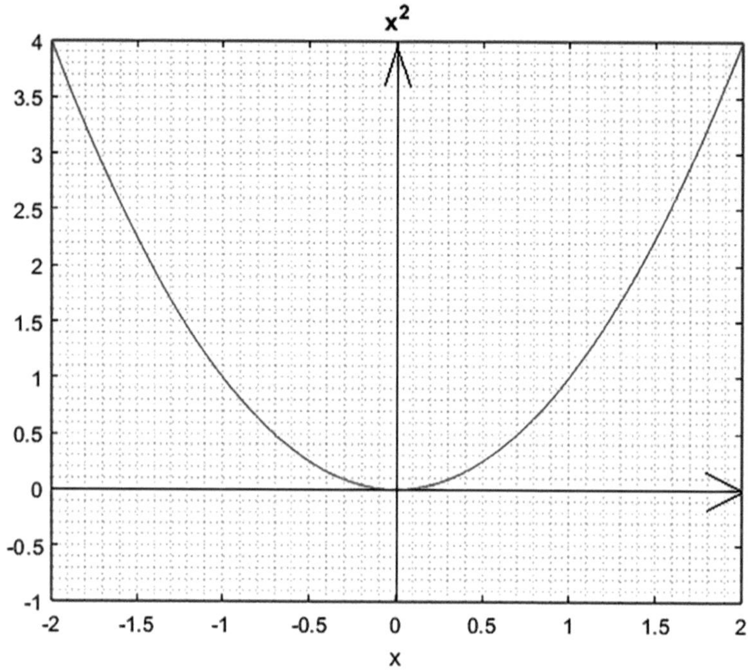

Fig. 1.11 Plot of $f(x) = x^2$

A function $f(x)$ is odd if $f(-x) = -f(x)$ for all x in the domain. For instance, $f(x) = x^3$, $f(x) = \sin(x)$, or $f(x) = x^3 - x$ are odd functions. The graph of an odd function is symmetric about the origin. If you draw a line from any point on the graph of an odd function to the origin, and then extend that line an equal distance on the other side of the origin, you will reach another point on the graph (Fig. 1.12).

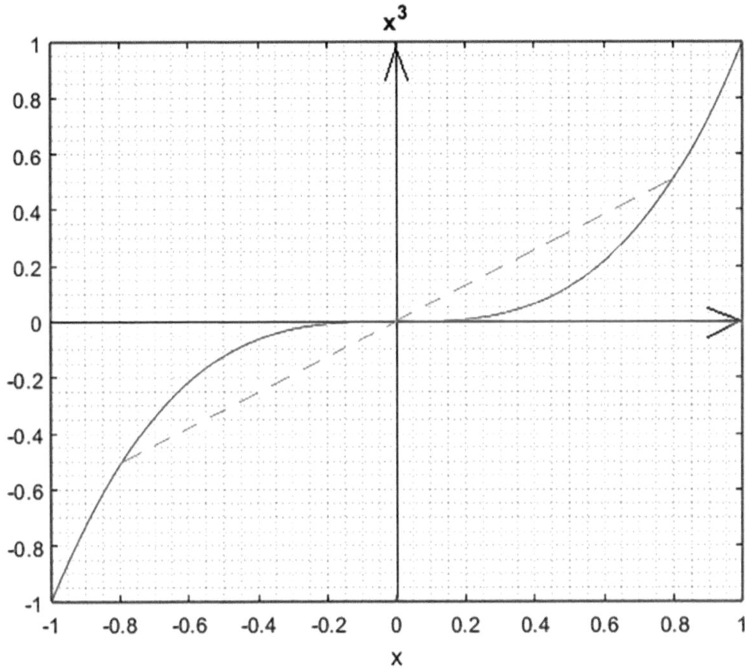

Fig. 1.12 Plot of $f(x) = x^3$

Not all functions are either even or odd. Many functions fall into the category of "neither even nor odd." The sum of two even functions is even. The sum of two odd functions is odd. The product of two even functions is even. The product of two odd functions is even.

For an even function $f(x)$ integrated over a symmetric interval $[-a, a]$, the integral can be simplified as $\int_{-a}^{a} f(x)dx = 2\int_{0}^{a} f(x)dx$.

For an odd function $f(x)$ integrated over a symmetric interval $[-a, a]$, the integral can be simplified as $\int_{-a}^{a} f(x)dx = 0$.

Exercise 1.6 Show that tan(kx) and cotan(kx) are odd functions.

1.12 Rotation

The rotation matrix $R(\theta) = \begin{bmatrix} \cos(\theta) & -\sin(\theta) \\ \sin(\theta) & \cos(\theta) \end{bmatrix}$ rotates a point by θ radians when multiplied to its coordinates.

Example 1.15 Rotate point $A = \begin{bmatrix} 1 \\ 2 \end{bmatrix}$ by 90° in the counterclockwise direction.

Solution Rotation is counterclockwise; therefore, $\theta = +90°$.

$$\begin{bmatrix} \cos(\theta) & -\sin(\theta) \\ \sin(\theta) & \cos(\theta) \end{bmatrix} \times \begin{bmatrix} x \\ y \end{bmatrix} = \begin{bmatrix} \cos(\theta)x - \sin(\theta)y \\ \sin(\theta)x + \cos(\theta)y \end{bmatrix}$$

$$B = \begin{bmatrix} \cos(90°) & -\sin(90°) \\ \sin(90°) & \cos(90°) \end{bmatrix} \times \begin{bmatrix} 1 \\ 2 \end{bmatrix} = \begin{bmatrix} 0 & -1 \\ 1 & 0 \end{bmatrix} \times \begin{bmatrix} 1 \\ 2 \end{bmatrix} = \begin{bmatrix} -2 \\ 1 \end{bmatrix}$$

Figure 1.13 shows point A and the result of the rotation (point B).

Fig. 1.13 Point rotation

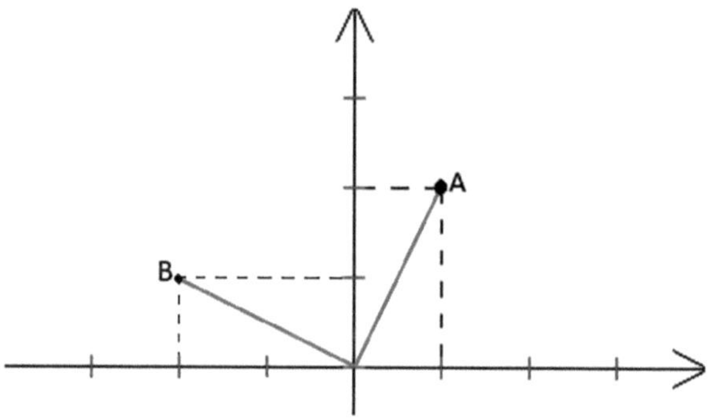

This problem can be solved with the aid of complex numbers as well.

$$A = \begin{bmatrix} 1 \\ 2 \end{bmatrix} = 1 + 2j \Rightarrow B = (1 + 2j) \times e^{j\frac{\pi}{2}} = -2 + j \Rightarrow B = \begin{bmatrix} -2 \\ 1 \end{bmatrix}$$

Example 1.16 Rotate point $A = \begin{bmatrix} 1 \\ 2 \end{bmatrix}$ by 63.435° in the clockwise direction.

Solution Rotation is clockwise; therefore, $\theta = -63.435°$.

$$C = \begin{bmatrix} \cos(-63.435°) & -\sin(-63.435°) \\ \sin(-63.435°) & \cos(-63.435°) \end{bmatrix} \times \begin{bmatrix} 1 \\ 2 \end{bmatrix} = \begin{bmatrix} 0.4473 & 0.8944 \\ -0.8944 & 0.4473 \end{bmatrix}$$
$$\times \begin{bmatrix} 1 \\ 2 \end{bmatrix} = \begin{bmatrix} 2.2361 \\ 0 \end{bmatrix}$$

Figure 1.14 shows point A and the result of the rotation (point C).

Fig. 1.14 Point rotation

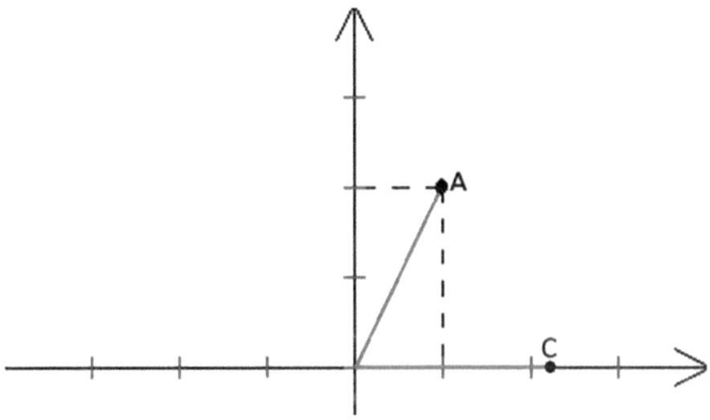

This problem can be solved with the aid of complex numbers as well (−63.435 ° = − 1.1071 *rad*).

$$A = \begin{bmatrix} 1 \\ 2 \end{bmatrix} = 1 + 2j \Rightarrow C = (1 + 2j) \times e^{-j1.1071} = 2.2361 \Rightarrow C = \begin{bmatrix} 2.2361 \\ 0 \end{bmatrix}$$

1.13 Scalar Product of Two Vectors

The dot product or scalar product of two vectors, $\vec{A} = x_1 i + y_1 j$ and $\vec{B} = x_2 i + y_2 j$, is equal to the product of their magnitudes and the cosine of the angle α between them. The dot product of two vectors, $\vec{A}$ and $\vec{B}$, is denoted by $\vec{A}.\vec{B}$.

$$\vec{A} = x_1 i + y_1 j \Rightarrow \left|\vec{A}\right| = \sqrt{x_1^2 + y_1^2}$$

$$\vec{B} = x_2 i + y_2 j \Rightarrow \left|\vec{B}\right| = \sqrt{x_2^2 + y_2^2}$$

$$\vec{A}.\vec{B} = \left|\vec{A}\right| \times \left|\vec{B}\right| \times \cos(\alpha) = x_1 x_2 + y_1 y_2$$

$$\cos(\alpha) = \frac{x_1 x_2 + y_1 y_2}{\sqrt{x_1^2 + y_1^2} \times \sqrt{x_2^2 + y_2^2}} \Rightarrow \alpha = \cos^{-1}\left(\frac{x_1 x_2 + y_1 y_2}{\sqrt{x_1^2 + y_1^2} \times \sqrt{x_2^2 + y_2^2}}\right)$$

Note that here *i* and *j* represent unit vectors along the x and y axes, respectively, in a two-dimensional Cartesian coordinate system.

Example 1.17 Angle between $\vec{A} = i + 2j$ and $\vec{B} = i + j$ is $\alpha = \cos^{-1}\left(\frac{1+2}{\sqrt{2} \times \sqrt{5}}\right) = 18.435\,°$ (Fig. 1.15).

Fig. 1.15 Vectors $\vec{A}$ and $\vec{B}$

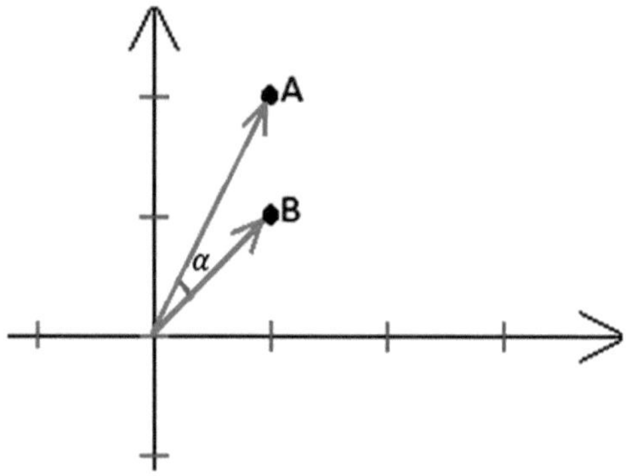

You can solve this problem with the aid of complex numbers as well:

$$\vec{A} = 1 + 2j = \sqrt{1^2 + 2^2} e^{j \tan^{-1}\left(\frac{2}{1}\right)} = \sqrt{5} e^{j1.1071}$$

$$\vec{B} = 1 + j = \sqrt{1^2 + 1^2} e^{j \tan^{-1}\left(\frac{1}{1}\right)} = \sqrt{2} e^{j\frac{\pi}{4}} = \sqrt{2} e^{j0.7854}$$

$$\alpha = 1.1071 - 0.7854 = 0.3217\ rad = 18.4321\ ^\circ$$

Very small deviations between obtained results come from rounding the numbers.

1.14 Equation of Circle in Cartesian Coordinates

Equation of a circle with radius of R and center of $\begin{bmatrix} x_0 \\ y_0 \end{bmatrix}$ is: $(x - x_0)^2 + (y - y_0)^2 = R^2$. For instance, equation of circle shown in Fig. 1.16 is: $(x - 2)^2 + (y - 1)^2 = 1^2$.

Fig. 1.16 Plot of $(x - 2)^2 + (y - 1)^2 = 1^2$

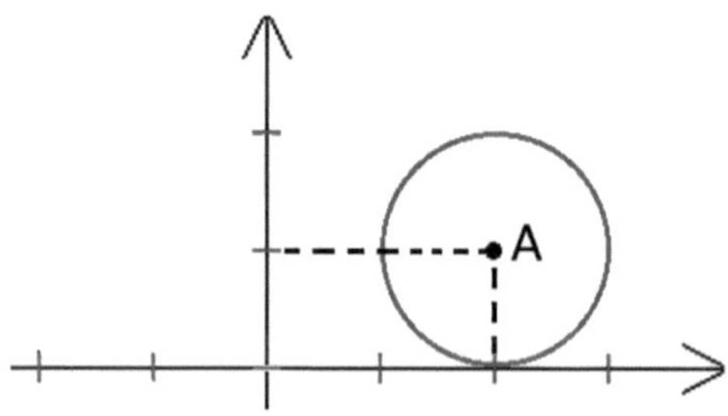

$(x - x_0)^2 + (y - y_0)^2 < R^2$ represents the points that fall inside a circle with a radius of R and a center at the point $\begin{bmatrix} x_0 \\ y_0 \end{bmatrix}$ but excluding the circle itself. Such a region is called an opened disk.

$(x - x_0)^2 + (y - y_0)^2 \leq R^2$ describes the interior of a circle with a radius of R and a center at the point $\begin{bmatrix} x_0 \\ y_0 \end{bmatrix}$ and the circle itself. Such a region is called a closed disk.

For instance, equation of closed disk shown in Fig. 1.17 is: $(x - 2)^2 + (y - 1)^2 \leq 1^2$.

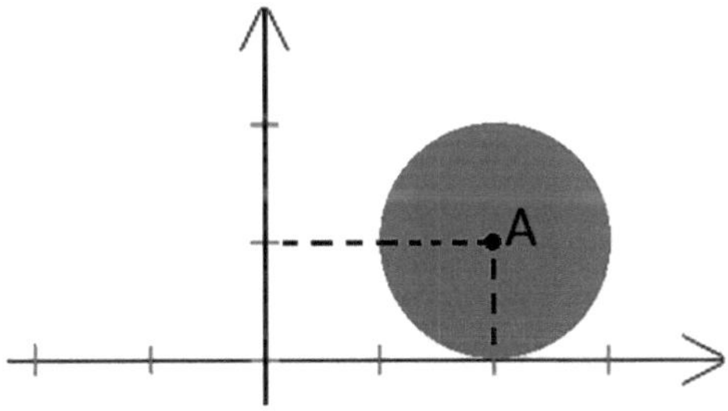

Fig. 1.17 Plot of $(x - 2)^2 + (y - 1)^2 \leq 1^2$

$(x - x_0)^2 + (y - y_0)^2 > R^2$ represents the points that fall outside a circle with a radius of R and a center at the point $\begin{bmatrix} x_0 \\ y_0 \end{bmatrix}$ but excluding the circle itself.

$(x - x_0)^2 + (y - y_0)^2 \geq R^2$ describes the exterior of a circle with a radius of R and a center at the point $\begin{bmatrix} x_0 \\ y_0 \end{bmatrix}$ and the circle itself.

For instance, Fig. 1.18 shows the points that satisfy the inequality $(x - 2)^2 + (y - 1)^2 > 1^2$.

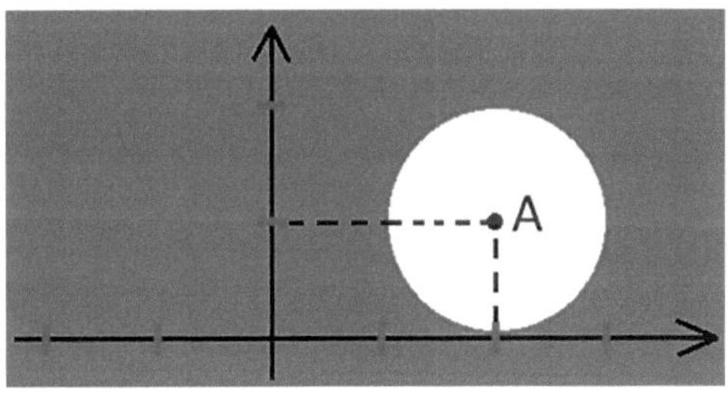

Fig. 1.18 Plot of $(x - 2)^2 + (y - 1)^2 > 1^2$

1.15 Equation of Ellipse in Cartesian Coordinates

An ellipse is a plane curve surrounding two focal points, such that for all points on the curve, the sum of the two distances to the focal points is a constant (Fig. 1.19). It generalizes a circle, which is the special type of ellipse in which the two focal points are the same.

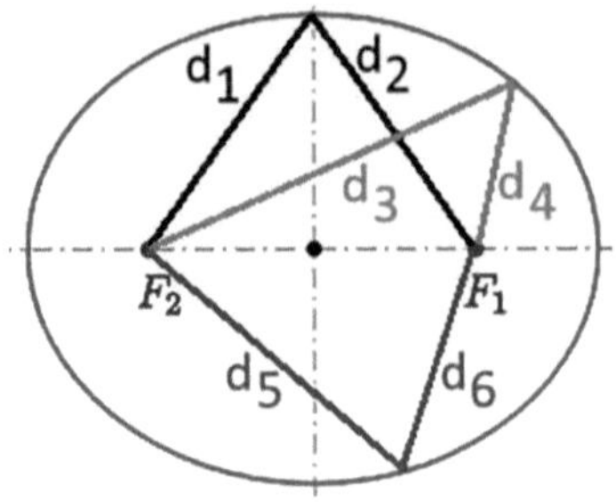

Fig. 1.19 An ellipse ($d_1 + d_2 = d_3 + d_4 = d_5 + d_6$)

Equation of the ellipse shown in Fig. 1.20 is $\frac{(x-x_0)^2}{a^2} + \frac{(y-y_0)^2}{b^2} = 1$. Note that $a > b$.

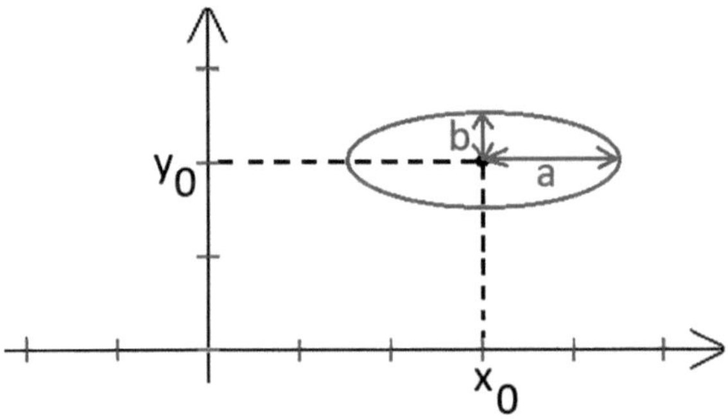

Fig. 1.20 A horizontal ellipse ($a > b$)

Equation of the ellipse shown in Fig. 1.21 is $\frac{(x-x_0)^2}{a^2} + \frac{(y-y_0)^2}{b^2} = 1$. Note that $b > a$.

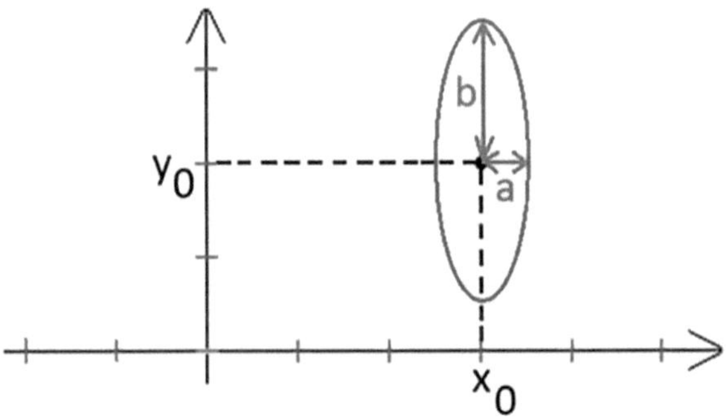

Fig. 1.21 A vertical ellipse ($b > a$)

Example 1.18 Equation of ellipse shown in Fig. 1.22 is $\frac{(x-0)^2}{2^2} + \frac{(y-2)^2}{1^2} = 1 \Rightarrow \frac{x^2}{4} + (y-2)^2 = 1$.

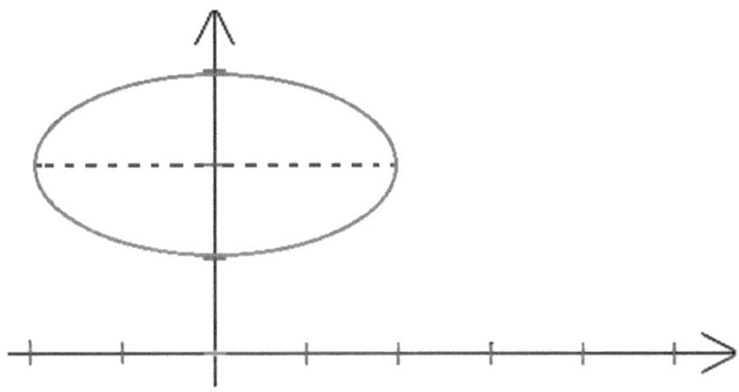

Fig. 1.22 Plot of $\frac{x^2}{4} + (y-2)^2 = 1$

1.16 Graphical Approach to Function Transformations

The function $f(t) + c$ represents a vertical shift of the function $f(t)$. When you add a constant c to the function, you shift the entire graph of $f(t)$ upward by c units if c is positive, and downward by $|c|$ units if c is negative.

For $t_0 > 0$, the graph of $f(t - t_0)$ is a horizontal translation of the graph of $f(t)$ by t_0 units to the right. The graph of $f(t + t_0)$ is a horizontal translation of the graph of $f(t)$ by t_0 units to the left.

The function $f(-t)$ represents a reflection of the function $f(t)$ about the y-axis. This means that the graph of $f(-t)$ is the mirror image of the graph of $f(t)$ across the y-axis.

The function $-f(t)$ represents a reflection of the function $f(t)$ about the x-axis. This means that the graph of $-f(t)$ is the mirror image of the graph of $f(t)$ across the x-axis.

The function $f(at)$ represents a horizontal scaling of the function $f(t)$. If $a > 1$, the graph of $f(at)$ is compressed horizontally. If $0 < a < 1$, the graph of $f(at)$ is stretched horizontally.

Example 1.19 Consider the function $f(t)$ shown in Fig. 1.23.

Fig. 1.23 Plot of $f(t)$

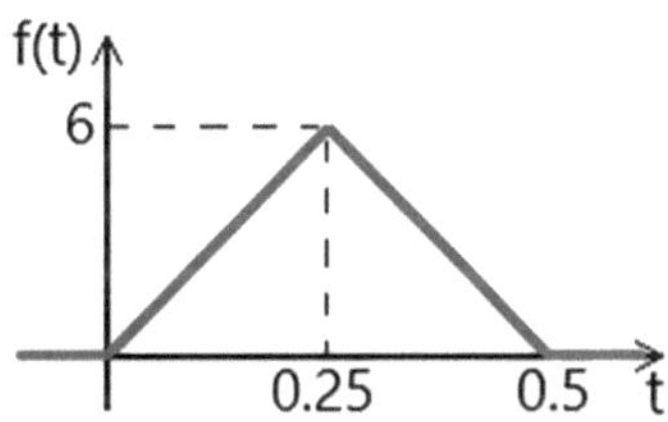

Graph of $f(t) + 5$ is shown in Fig. 1.24.

Fig. 1.24 Plot of $f(t) + 5$

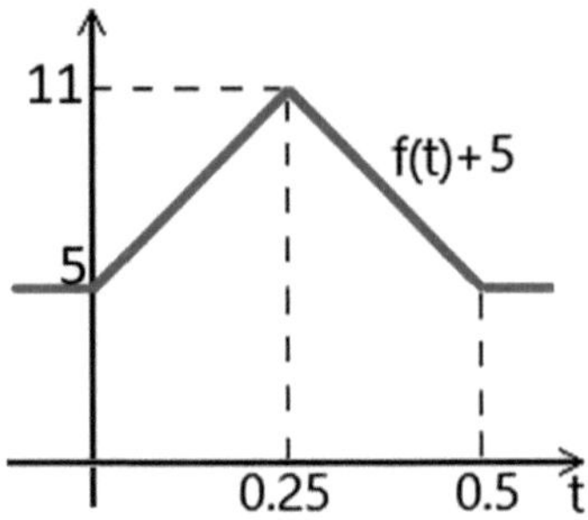

Graph of $f(-t)$ is shown in Fig. 1.25.

Fig. 1.25 Plot of $f(-t)$

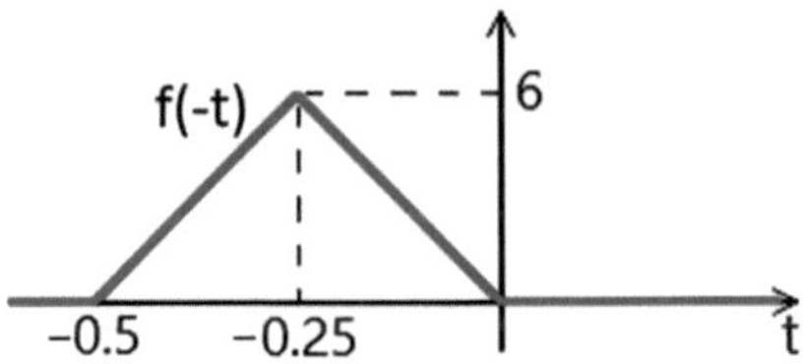

Graph of $f(t - 1)$ is shown in Fig. 1.26.

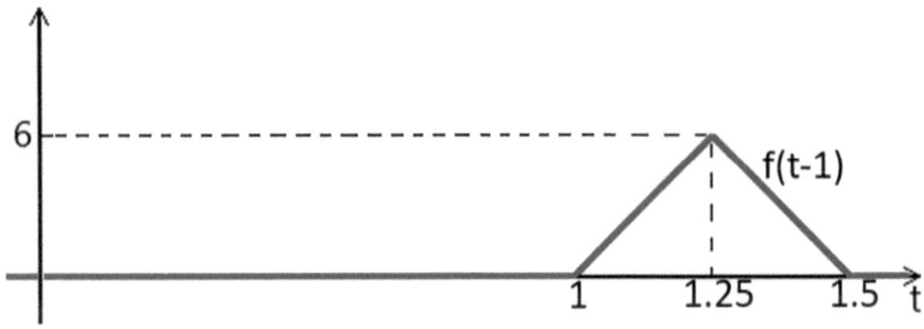

Fig. 1.26 Plot of $f(t - 1)$

Graph of $f(t + 1)$ is shown in Fig. 1.27.

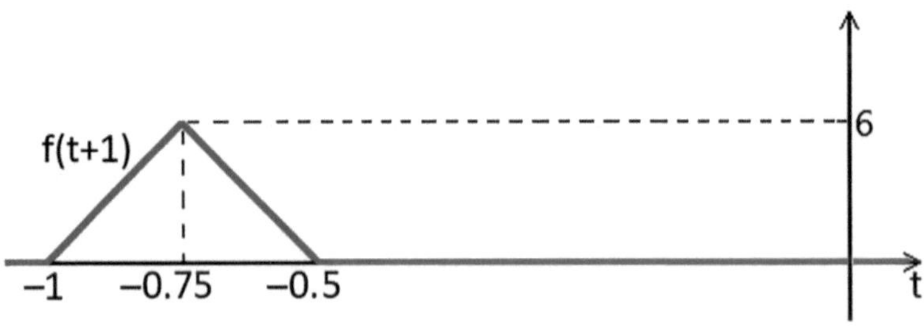

Fig. 1.27 Plot of $f(t + 1)$

Graph of $f(2t)$ is shown in Fig. 1.28.

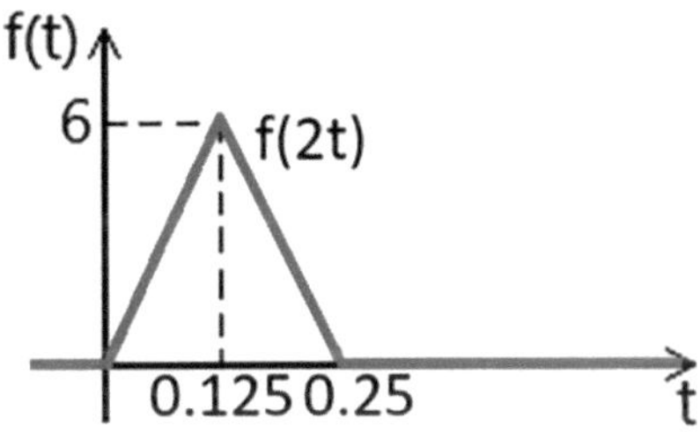

Fig. 1.28 Plot of $f(2t)$

Graph of $f(0.5t)$ is shown in Fig. 1.29.

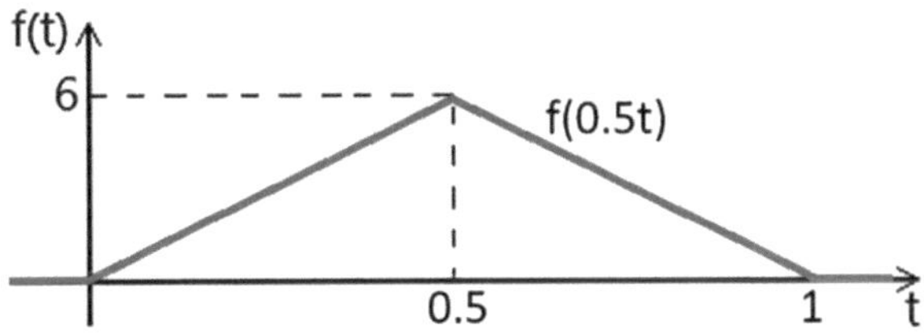

Fig. 1.29 Plot of $f(0.5t)$

Exercise 1.7 Draw the graph of $-f(-t + 2)$.

1.17 Determining the Sign of Expressions

Figure 1.30 shows how to determine the sign of $f(x) = ax + b$.

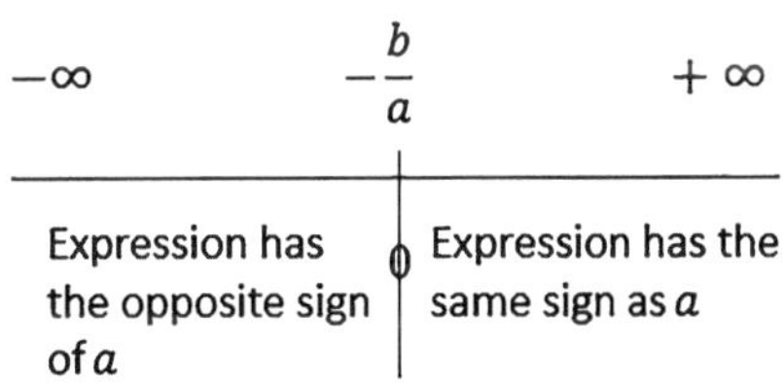

Fig. 1.30 Sign of $f(x) = ax + b$

Figure 1.31 shows how to determine the sign of $f(x) = ax^2 + bx + c$ when $ax^2 + bx + c$ has two real roots x_1 and x_2.

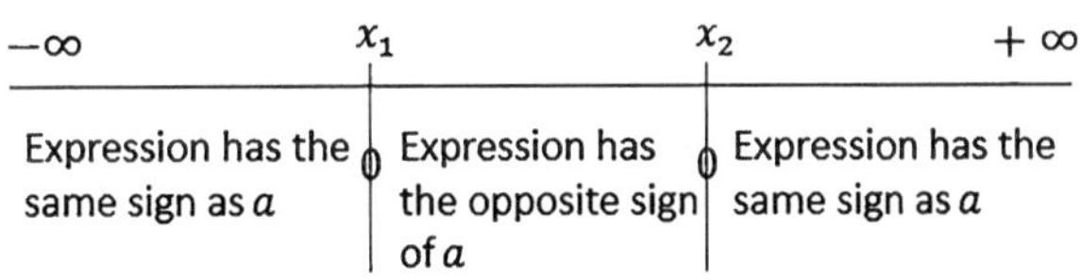

Fig. 1.31 Sign of $f(x) = ax^2 + bx + c$ (roots are real and different)

Figure 1.32 shows how to determine the sign of $f(x) = ax^2 + bx + c$ when $ax^2 + bx + c$ has repeated real roots.

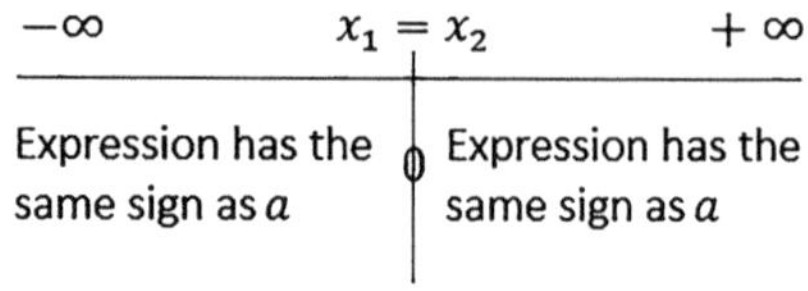

Fig. 1.32 Sign of $f(x) = ax^2 + bx + c$ (roots are real and repeated)

Figure 1.33 shows how to determine the sign of $f(x) = ax^2 + bx + c$ when $ax^2 + bx + c$ has no real roots.

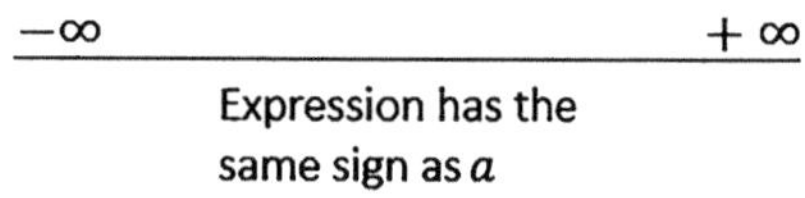

Fig. 1.33 Sign of $f(x) = ax^2 + bx + c$ (roots are complex)

Example 1.20 Determine the sign of $f(x) = x + 12$.

Solution Sign of $f(x) = x + 12$ is shown in Fig. 1.34.

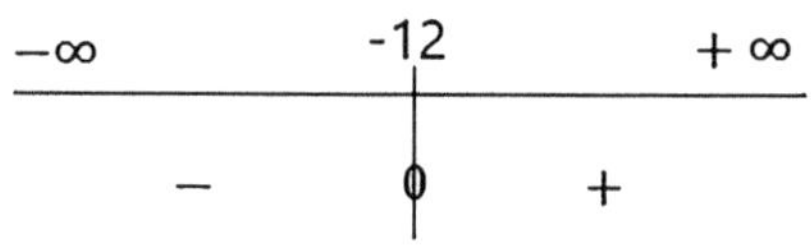

Fig. 1.34 Sign of $f(x) = x + 12$

Example 1.21 Determine the sign of $f(x) = \frac{x+1}{x+7}$.

Solution Sign of $f(x) = \frac{x+1}{x+7}$ is shown in Fig. 1.35.

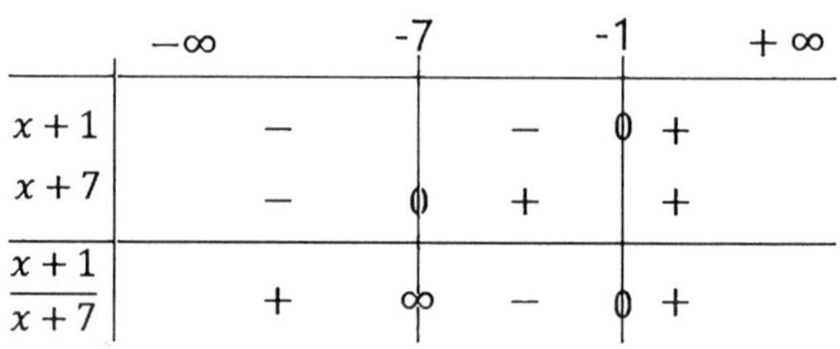

Fig. 1.35 Sign of $f(x) = \frac{x+1}{x+7}$

Graph of $f(x) = \frac{x+1}{x+7}$ is shown in Fig. 1.36, as well. Use this graph to verify the result shown in Fig. 1.35.

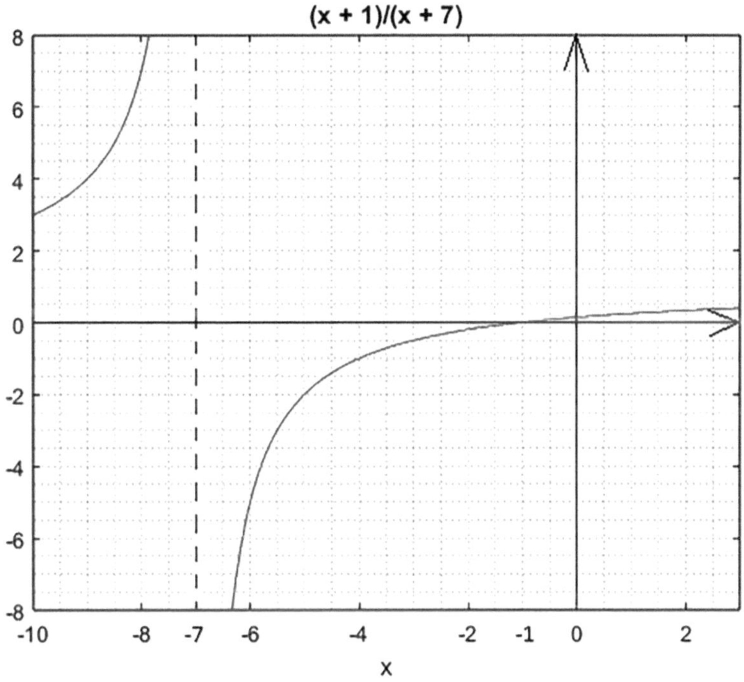

Fig. 1.36 Plot of $f(x) = \frac{x+1}{x+7}$

Example 1.22 Determine the sign of $f(x) = -x^2 + 4x - 10$.

Solution $-x^2 + 4x - 10 = 0 \Rightarrow \Delta = 4^2 - 4(-1)(-10) = 16 - 40 = -24$. Therefore, this equation has no real roots. Sign of this expression is always negative since $a = -1 < 0$. Note that $f(x) = -x^2 + 4x - 10$ can be written as $f(x) = -(x - 2)^2 - 6$. This expression clearly shows that $f(x)$ is always negative.

Example 1.23 Determine the sign of $f(x) = 2x^2 + 4x + 10$.

Solution $2x^2 + 4x + 10 = 0 \Rightarrow \Delta = 4^2 - 4(2)(10) = 16 - 80 = -64$. Therefore, this equation has no real roots. Sign of this expression is: always positive since $a = 2 > 0$. Note that $f(x) = 2x^2 + 4x + 10$ can be written as $f(x) = 2(x + 1)^2 + 8$. This expression clearly shows that $f(x)$ is always positive.

Example 1.24 Determine the sign of $f(x) = \frac{x+1}{x^2+5x+6}$.

Solution Sign of $f(x) = \frac{x+1}{x^2+5x+6} = \frac{x+1}{(x+2)(x+3)}$ is shown in Fig. 1.37.

	$-\infty$	-3		-2		-1	$+\infty$
$x+1$	$-$		$-$		$-$	0	$+$
$x+2$	$-$		$-$	0	$+$		$+$
$x+3$	$-$	0	$+$		$+$		$+$
$\frac{x+1}{(x+2)(x+3)}$	$-$	∞	$+$	∞	$-$	0	$+$

Fig. 1.37 Sign of $f(x) = \frac{x+1}{x^2+5x+6} = \frac{x+1}{(x+2)(x+3)}$

Graph of $f(x) = \frac{x+1}{x^2+5x+6}$ is shown in Fig. 1.38, as well. Use this graph to verify the result shown in Fig. 1.37.

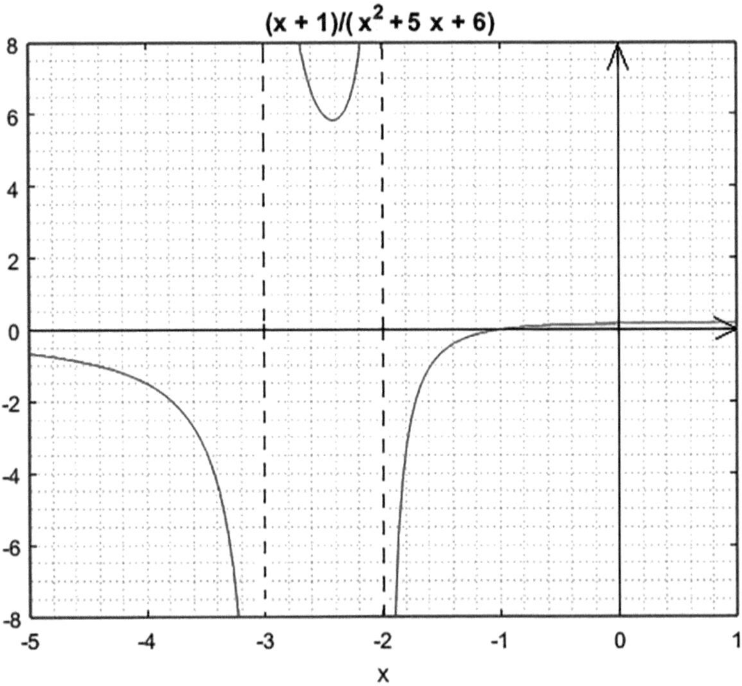

Fig. 1.38 Plot of $f(x) = \frac{x+1}{x^2+5x+6}$

1.18 Optimizing Single-Variable Functions

Single-variable optimization is a mathematical technique used to find the maximum or minimum value of a function of one variable within a specified domain.

To find the optimum point of a function of one variable, we typically calculate the derivative of the function, set it equal to zero, and solve for the critical points. Critical points are points where the derivative of the function is zero or undefined.

To classify these critical points as local maxima, minima, or points of inflection, we typically employ the second derivative test. By evaluating the second derivative at each critical point, we can determine the concavity of the function at that point. A positive second derivative indicates a local minimum, while a negative second derivative signifies a local maximum.

Example 1.25 Find the optimum point of $f(x) = x^2 + 3x + 5$.

Solution

$$f(x) = x^2 + 3x + 5 \Rightarrow f'(x) = 2x + 3 \Rightarrow 2x + 3 = 0 \Rightarrow x = -1.5$$

$$f''(x) = 2 > 0 \Rightarrow f''(-1.5) = 2 > 0$$

Therefore, the critical point $x = -1.5$ is a minimum (Fig. 1.39). Value of function at $x = -1.5$ is $f(-1.5) = (-1.5)^2 + 3 \times -1.5 + 5 = 2.7500$.

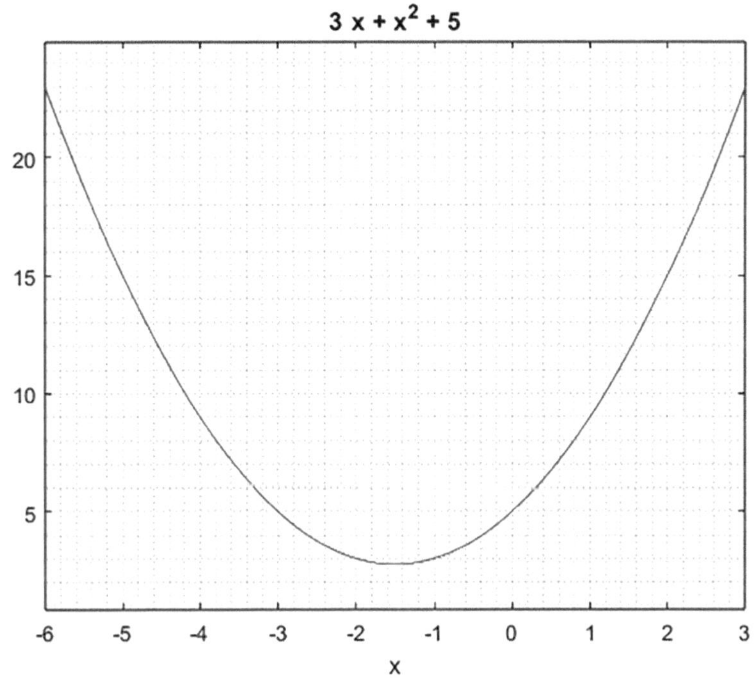

Fig. 1.39 Graph of $f(x) = x^2 + 3x + 5$

Example 1.26 Given the constraint $x + y = 7$, determine the maximum value of the $x \times y$.

Solution

$$f = x.y = x.(7 - x) = -x^2 + 7x \Rightarrow f'(x) = -2x + 7 = 0 \Rightarrow x = 3.5 \Rightarrow y = 3.5$$

$$x.y = 3.5 \times 3.5 = 12.25$$

Example 1.27 The circuit in Fig. 1.40, has a fixed resistor R_1 and a fixed voltage source V. Determine the optimal value of resistor R_2 to maximize the power dissipated across it.

Fig. 1.40 A simple circuit

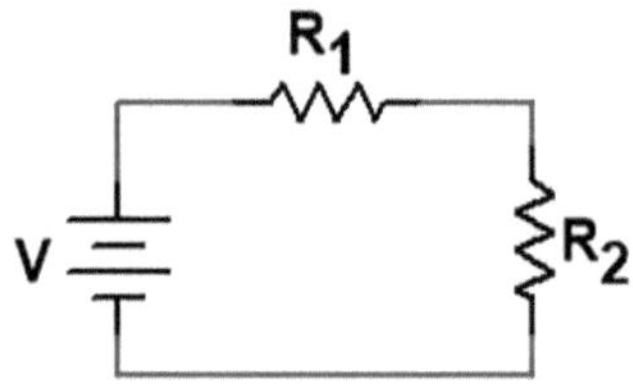

Solution According to the principles of physics, power dissipation in the resistor R_2 is given by:

$$P_{R_2} = V_{R_2} \times I_{R_2} = \frac{R_2}{R_1 + R_2} V \times \frac{1}{R_1 + R_2} V = \frac{V^2 R_2}{(R_1 + R_2)^2}$$

We want to maximize the power dissipated in the resistor R_2.

$$\frac{dP_{R_2}}{dR_2} = 0 \Rightarrow \frac{V^2 (R_1 + R_2)^2 - 2(R_1 + R_2)(V^2 R_2)}{(R_1 + R_2)^4} = 0 \Rightarrow V^2 (R_1 + R_2)^2 - 2(R_1 + R_2)(V^2 R_2) = 0 \Rightarrow V^2 (R_1 + R_2) - 2V^2 R_2 = 0 \Rightarrow R_2 = R_1$$

$$R_2 = R_1 \Rightarrow P_{max} = \frac{V^2 R_1}{(R_1 + R_1)^2} = \frac{V^2}{4R_1}$$

For $V = 10\ V$ and $R_1 = 2.5\ \Omega$, $P_{R_2} = \frac{100 R_2}{(2.5 + R_2)^2}$ (Fig. 1.41). When $R_2 = R_1 = 2.5\ \Omega$, $P_{max} = \frac{V^2}{4R_1} = \frac{100}{4 \times 2.5} = 10\ W$ (Fig. 1.42).

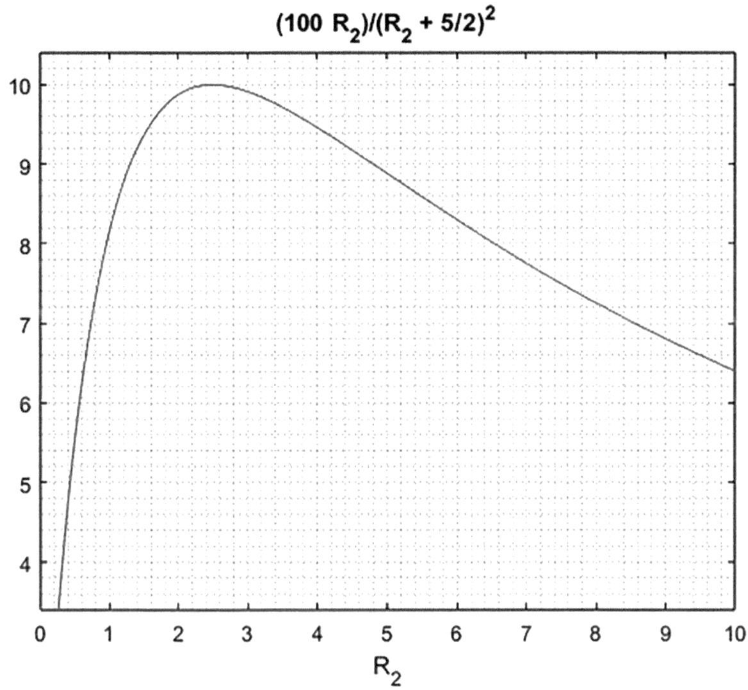

Fig. 1.41 Graph of $P_{R_2} = \frac{100R_2}{(2.5+R_2)^2}$

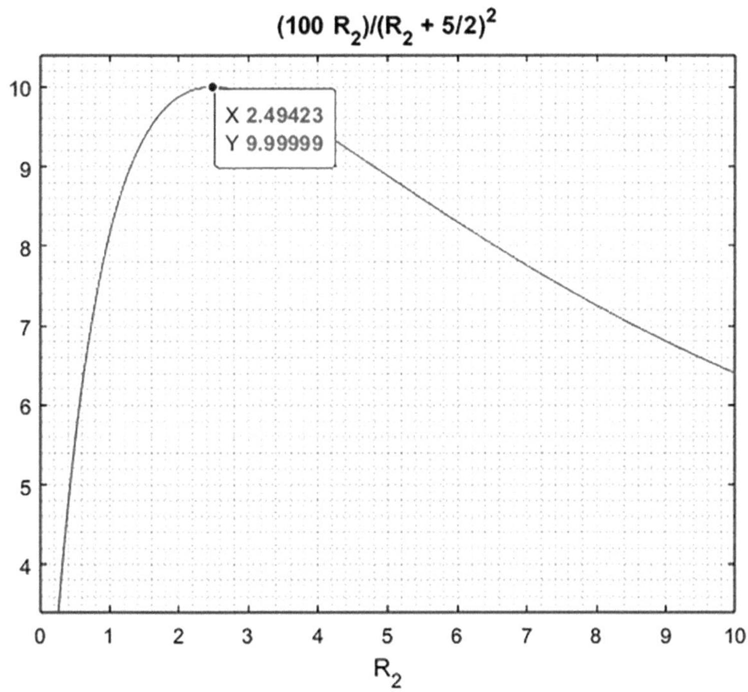

Fig. 1.42 The function $P_{R_2} = \frac{100R_2}{(2.5+R_2)^2}$ reaches its peak value of 10 around the point 2.5 on the x-axis

Example 1.28 A manufacturer needs to design a cylindrical tin can that must hold a fixed volume of 100 cm^3. Determine the radius (r) and height (h) of the can that will minimize the total surface area (A), thereby minimizing the material used.

Solution The volume (V) is fixed at 100 cm^3:

$$V = \pi r^2 h \Rightarrow \pi r^2 h = 100$$

The total surface area (A) is the sum of the two circular ends and the side wall:

$$A = 2\left(\pi r^2\right) + 2\pi r h$$

Solve the constraint for h and substitute it into the area equation to make A a function of r only:

$$h = \frac{100}{\pi r^2}$$

$$A(r) = 2\pi r^2 + 2\pi r\left(\frac{100}{\pi r^2}\right) \Rightarrow 2\pi r^2 + \frac{200}{r}$$

Take the first derivative of $A(r)$ and set it to zero:

$$A(r) = 2\pi r^2 + 2\pi r\left(\frac{200}{r}\right) \Rightarrow \frac{dA(r)}{dr} = 4\pi r - \frac{200}{r^2}$$

$$\frac{dA(r)}{dr} = 0 \Rightarrow 4\pi r - \frac{200}{r^2} = 0 \Rightarrow \frac{4\pi r^3 - 200}{r^2} = 0 \Rightarrow 4\pi r^3 - 200 = 0 \Rightarrow r = \sqrt[3]{\frac{50}{\pi}} \approx 2.515$$

Use the second derivative test to confirm it's a minimum:
$\frac{d^2A(r)}{dr^2} = 4\pi + \frac{400}{r^3} \Rightarrow \left.\frac{d^2A(r)}{dr^2}\right|_{r=2.515} = +37.6991 \Rightarrow$ The critical point corresponds to a minimum.

Substitute the optimal radius r back into the height constraint equation:

$$h = \frac{100}{\pi r^2} \Rightarrow h = \frac{100}{\pi (2.515)^2} \Rightarrow h \approx 5.0324$$

The minimum surface area is achieved when the height of the can is equal to its diameter: $h = 2r$.

Optimal radius: $r \approx 2.515$ cm
Optimal height: $h \approx 5.0324$ cm

1.19 GeoGebra

GeoGebra (https://www.geogebra.org) is a dynamic mathematics software application that integrates geometry, algebra, spreadsheets, graphing, statistics, and calculus into a single, interactive platform. It is designed for teaching and learning mathematics and science from primary school through university level. Its main purpose is to help students and teachers visualize mathematical concepts, allowing for the dynamic manipulation of points, vectors, functions, and equations so that users can immediately see the effect of changes in both the algebraic and graphical representations.

The software is available as free apps (like the calculator suite, graphing calculator, and 3d calculator) for web, desktop, and mobile devices, and it hosts a vast online repository of interactive learning resources created by its global community.

1.20 Problem Solving with MATLAB®

This section offers MATLAB examples relevant to the problems in this chapter. If you are new to MATLAB, begin with Chap. 11.

Example 1.29 Employ MATLAB to calculate $-\frac{16}{2025}-\frac{104}{3375}$.

Solution The required code is shown in Fig. 1.43.

Fig. 1.43 MATLAB code

```
Command Window
>> sym(-16/2025-104/3375)

ans =

-392/10125

>> eval(ans)

ans =

   -0.0387

fx >>
```

According to Fig. 1.43, $-\frac{16}{2025}-\frac{104}{3375}=\frac{-392}{10125}=-0.0387$.

Example 1.30 Employ MATLAB to evaluate the expression $-3x^2+7x+x\cos(x)$ at $x=\frac{\pi}{6}$.

Solution The required code is shown in Fig. 1.44.

Fig. 1.44 MATLAB code

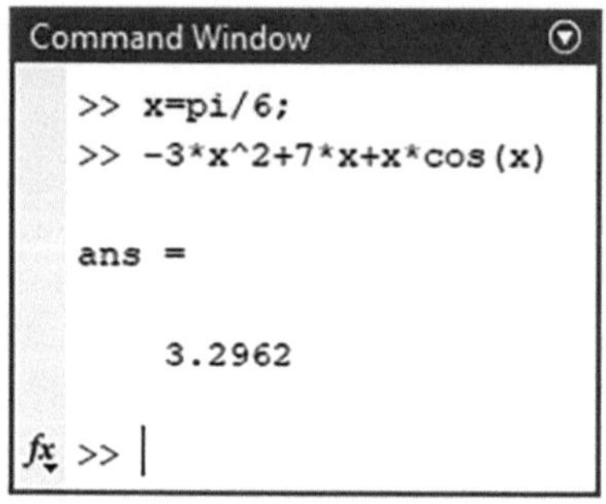

Example 1.31 Employ MATLAB to evaluate the expression $-\log(x^2) + 4\ln(x)$ at $x = 2.5$.

Solution The required code is shown in Fig. 1.45.

Fig. 1.45 MATLAB code

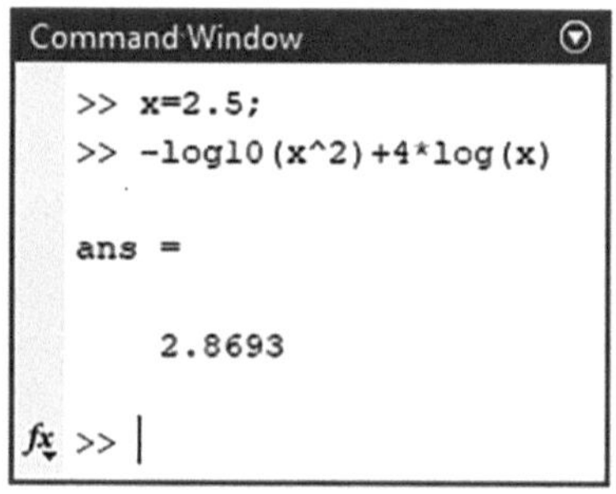

Example 1.32 Employ MATLAB to calculate the value of $\log_6 24$.

Solution Recall that $\log_b a = \frac{\log_c a}{\log_c b}$. The required code is shown in Fig. 1.46.

Fig. 1.46 MATLAB code

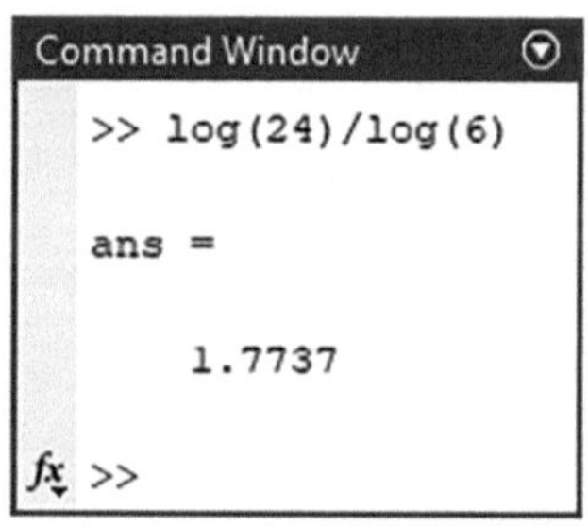

Example 1.33 Employ MATLAB to evaluate the expression $e^{-x2} + 3^x + x^6$ at $x = -0.5$.

Solution The required code is shown in Fig. 1.47.

Fig. 1.47 MATLAB code

```
Command Window
>> x=-0.5;
>> exp(-x^2)+3^x+x^6

ans =

    1.3718

fx >>
```

Example 1.34 Employ MATLAB to evaluate the expression $|x^2 - 8x + 12|$ at $x = 3$.

Solution The required code is shown in Fig. 1.48.

Fig. 1.48 MATLAB code

```
Command Window
>> x=3;
>> abs(x^2-8*x+12)

ans =

     3

fx >>
```

Example 1.35 Employ MATLAB to find the roots of $x^2 - 8x + 12$.

Solution The required code is shown in Fig. 1.49.

Fig. 1.49 MATLAB code

```
Command Window
>> roots([1 -8 12])

ans =

     6
     2

fx >>
```

Example 1.36 Employ MATLAB to find the roots of $x^2 - 8x$.

Solution The required code is shown in Fig. 1.50.

Fig. 1.50 MATLAB code

```
Command Window
>> roots([1 -8 0])

ans =

     0
     8

>>
```

Example 1.37 Employ MATLAB to find the roots of $x^2 - 8$.

Solution The required code is shown in Fig. 1.51.

Fig. 1.51 MATLAB code

```
Command Window
>> roots([1 0 -8])

ans =

    2.8284
   -2.8284
```

Example 1.38 Employ MATLAB to find the irreducible polynomial factors of the symbolic expression $x^6 - 1$.

Solution The required code is shown in Fig. 1.52. According to the result shown in Fig. 1.52, $x^6 - 1 = (x - 1)(x + 1)(x^2 + x + 1)(x^2 - x + 1)$.

Fig. 1.52 MATLAB code

```
Command Window
>> syms x
>> p=x^6-1;
>> F=factor(p)

F =

[- 1 + x, 1 + x, 1 + x + x^2, 1 - x + x^2]

>>
```

Example 1.39 Employ MATLAB to find the prime factorization of the integer 200.

Solution The required code is shown in Fig. 1.53. According to the result shown in Fig. 1.53, $200 = 2^3 \times 5^2$.

Fig. 1.53 MATLAB code

```
Command Window
>> f=factor(200)

f =

     2     2     2     5     5

fx >>
```

Example 1.40 Employ MATLAB to write the expression $\frac{1}{x-3} + \frac{3}{x}$ as a single rational fraction, combining the terms using a common denominator.

Solution The required code is shown in Fig. 1.54.

Fig. 1.54 MATLAB code

```
Command Window
>> syms x
>> simplifyFraction(1/(x-3)+3/x)

ans =

(- 9 + 4*x)/(x*(- 3 + x))

>> pretty(ans)
 - 9 + 4 x
-----------
x (- 3 + x)

fx >>
```

Example 1.41 Use MATLAB to plot the function $y = \cos(x) + x \sin(x)$ over the interval $[-\pi, \pi]$.

Solution Using the ezplot command is the simplest method for solving this example (Fig. 1.55). The output of the code provided in Fig. 1.55 can be seen in Fig. 1.56.

Fig. 1.55 MATLAB code

```
Command Window
>> syms x
>> ezplot(cos(x)+x*sin(x),[-pi,pi])
>> grid on
fx >>
```

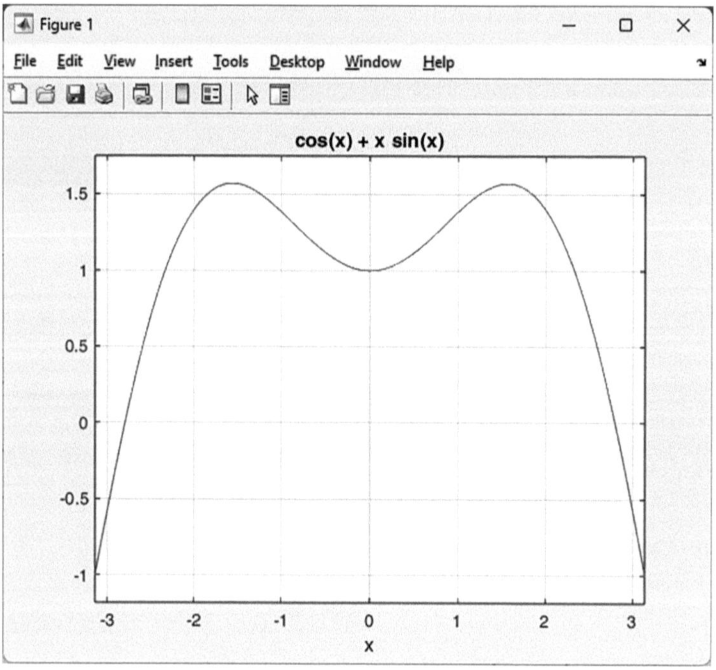

Fig. 1.56 Output of the code in Fig. 1.55

You can view the coordinates of any point by clicking on it (Fig. 1.57).

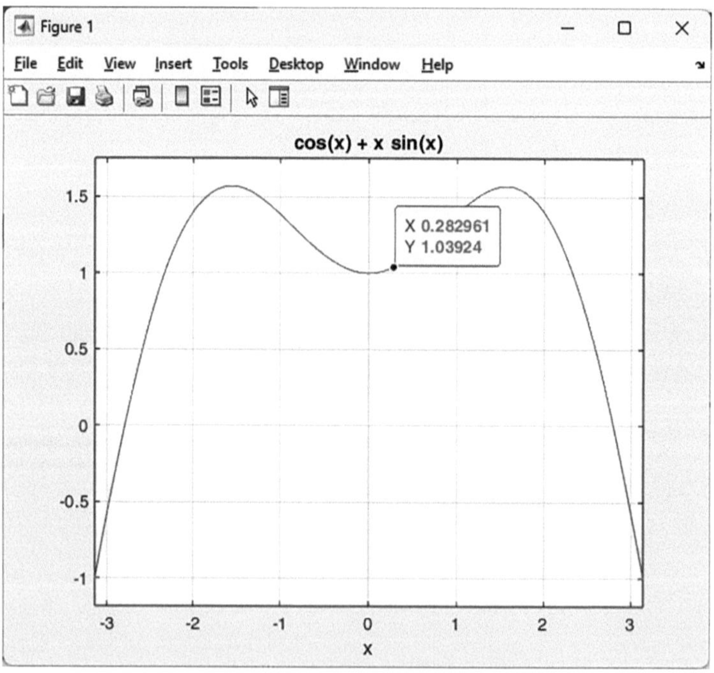

Fig. 1.57 The coordinates of a point are shown upon clicking it

Example 1.42 Use MATLAB to plot the function $y = \cos(x) + x\sin(x)$ over the interval $[-\pi, \pi]$.

Solution While this problem was addressed in the preceding example, Fig. 1.58 presents a different method for its solution (Fig. 1.59).

Fig. 1.58 MATLAB code

```
Command Window
>> x= linspace(-pi, pi, 500);
>> y= cos(x)+x.*sin(x);
>> plot(x,y)
>> grid on
>> xlabel('x');
>> ylabel('y');
>> title('plot of cos(x)+x*sin(x)');
fx >>
```

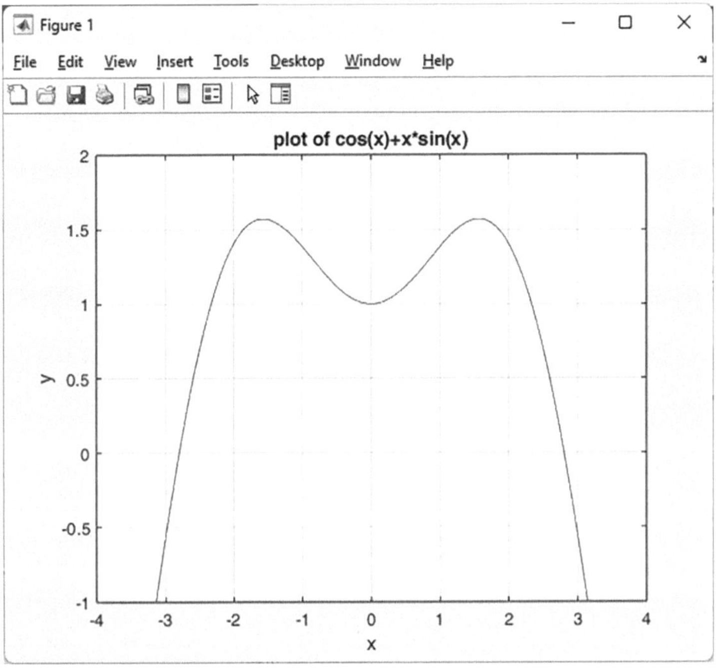

Fig. 1.59 Output of the code in Fig. 1.58

Example 1.43 Calculate the factorial of five (5!) using MATLAB.

Solution The required code is shown in Fig. 1.60.

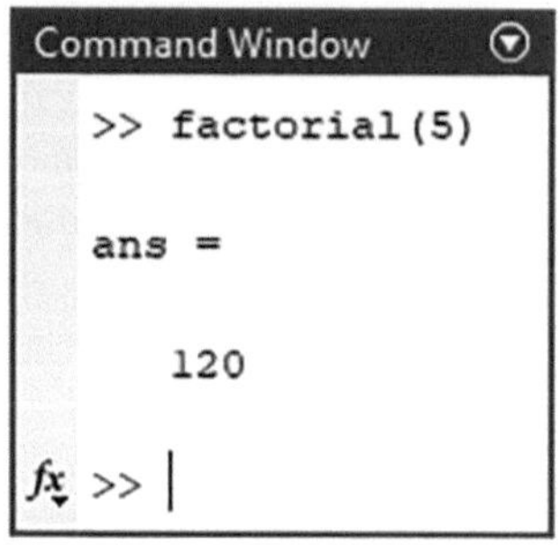

Fig. 1.60 MATLAB code

References for Further Study

1. Engineering Mathematics with MATLAB® and Simulink®, Farzin Asadi, Springer, 2025.
2. Applied Numerical Analysis with MATLAB®/Simulink®: For Engineers and Scientists, Farzin Asadi, Springer, 2023.
3. Advanced Engineering Mathematics, Erwin Kreyszig, Wiley, 2020.
4. A Problem-Solving Approach to Single-Variable Calculus (Volume II), Farzin Asadi, Springer, 2026.
5. A Problem-Solving Approach to Multi-Variable Calculus (Volume I), Farzin Asadi, Springer, 2026.
6. A Problem-Solving Approach to Multi-Variable Calculus (Volume II), Farzin Asadi, Springer, 2026.
7. A Problem-Solving Approach to Multi-Variable Calculus (Volume III), Farzin Asadi, Springer, 2026.
8. A Problem-Solving Approach to Ordinary Differential Equations (Volume I), Farzin Asadi, Springer, 2026.
9. A Problem-Solving Approach to Ordinary Differential Equations (Volume II), Farzin Asadi, Springer, 2026.
10. A Problem-Solving Approach to Ordinary Differential Equations (Volume III), Farzin Asadi, Springer, 2026.
11. A Problem-Solving Approach to Ordinary Differential Equations (Volume IV), Farzin Asadi, Springer, 2026.

Chapter 2
Fundamentals of Trigonometry

2.1 Introduction

Trigonometry is a branch of mathematics that deals with the study of triangles, particularly right-angled triangles. It explores the relationships between the sides and angles of these triangles. This chapter covers essential trigonometric formulas needed in subsequent chapters.

2.2 Degree, Radian, and Grad

Degree, radian, and grad are units of measurement for angles:

Degree: A full circle is divided into 360°.

Radian: A radian is the angle subtended at the center of a circle by an arc equal in length to the radius of the circle. There are 2π radians in a full circle.

Grad: A full circle is divided into 400 grads.

Most of the scientific calculations are done with radian.

Example 2.1 Convert 75° to radians.

Solution 1 radian equals to $\frac{180^\circ}{\pi} \cong 57.30^\circ$. Therefore, $75^\circ = \frac{75}{57.30} rad = 1.309\ rad.$

Example 2.2 Convert 2.5 radians to degree.

Solution

$$2.5 \times 57.30^\circ = 143.25^\circ.$$

Example 2.3 A shaft is rotating at a speed of 1490 revolutions per minute. Convert this rotational speed to radians per second.

F. Asadi, *A Problem-Solving Approach to Single-Variable Calculus*,
https://doi.org/10.1007/978-3-032-13505-6_2

Solution

$$1490\ rpm = 1490 \times 2\pi\ \frac{rad}{min} = 1490 \times 2\pi\ \frac{rad}{60\ s} = \frac{1490 \times 2\pi}{60}\ \frac{rad}{s} = 156.0324\ \frac{rad}{s}$$

2.3 Trigonometric Functions

Trigonometric functions for the right triangle *ABC* (Fig. 2.1) are defined as shown in Table 2.1.

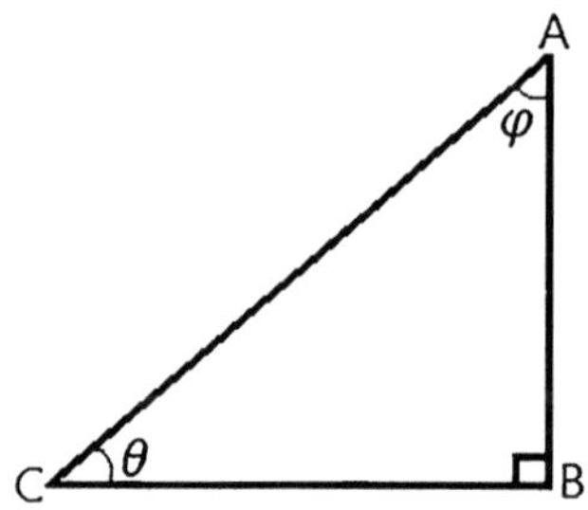

Fig. 2.1 A right triangle

Table 2.1 Definition of different trigonometric functions for the right triangle ABC (Fig. 2.1)

For angle θ	For angle φ
$\sin(\theta) = \frac{AB}{AC}$ $\cos(\theta) = \frac{BC}{AC}$ $\tan(\theta) = \frac{\sin(\theta)}{\cos(\theta)} = \frac{AB}{BC}$ $\cot(\theta) = \frac{\cos(\theta)}{\sin(\theta)} = \frac{BC}{AB}$	$\sin(\varphi) = \frac{BC}{AC}$ $\cos(\varphi) = \frac{AB}{AC}$ $\tan(\varphi) = \frac{\sin(\varphi)}{\cos(\varphi)} = \frac{BC}{AB}$ $\cot(\varphi) = \frac{\cos(\varphi)}{\sin(\varphi)} = \frac{AB}{BC}$

It is recommended to memorize the sine and cosine values for the angles shown in Table 2.2. Secant ($\sec(\theta)$) is the reciprocal of the cosine: $\sec(\theta) = \frac{1}{\cos(\theta)}$. Cosecant ($\csc(\theta)$ or $\mathrm{cosec}(\theta)$) is the reciprocal of the sine: $\csc(\theta) = \mathrm{cosec}(\theta) = \frac{1}{\sin(\theta)}$.

Table 2.2 Sine and cosine values for common angles

Angle	Sine	Cosine
0°	0	1
15°	0.26	0.97
30°	$\frac{1}{2}$	$\frac{\sqrt{3}}{2}$
45°	$\frac{\sqrt{2}}{2}$	$\frac{\sqrt{2}}{2}$
60°	$\frac{\sqrt{3}}{2}$	$\frac{1}{2}$
90°	1	0
180°	0	−1

Note that $\sin(x)$ is an odd function since: $\sin(-x) = -\sin(x)$. $\cos(x)$ is an even function since $\cos(-x) = \cos(x)$.

2.4 The Law of Sines and Cosines

For a triangle with sides a, b, and c, and their corresponding opposite angles α, β, and γ (Fig. 2.2), the law of sines states:

$$\frac{\sin(\alpha)}{a} = \frac{\sin(\beta)}{b} = \frac{\sin(\gamma)}{c}$$

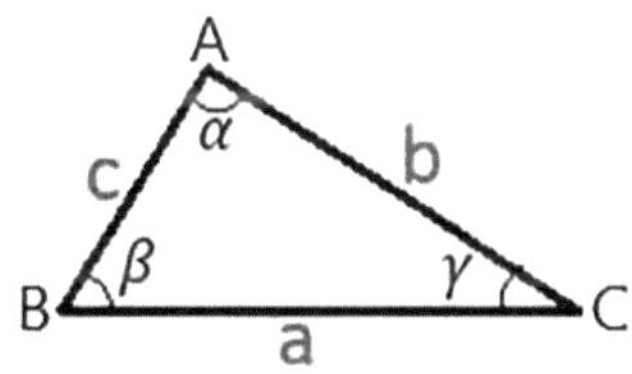

Fig. 2.2 A general triangle

Consider the triangle shown in Fig. 2.2. The Law of Cosines states that:

$$a^2 = b^2 + c^2 - 2 \times b \times c \times \cos(\alpha)$$
$$b^2 = a^2 + c^2 - 2 \times a \times c \times \cos(\beta)$$
$$c^2 = a^2 + b^2 - 2 \times a \times b \times \cos(\gamma)$$

2.5 Isosceles Triangle

In the isosceles triangle ABC depicted in Fig. 2.3, we have: $BC = 2 \times l \times \sin\left(\frac{\theta}{2}\right)$.

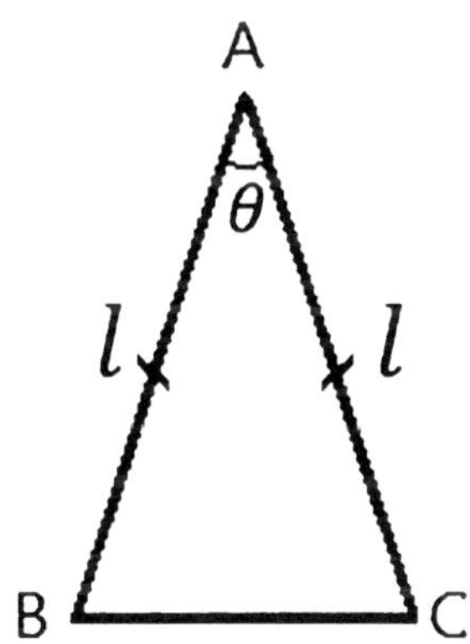

Fig. 2.3 Isosceles triangle ($AB = AC = l$)

Area of the isosceles triangle depicted in Fig. 2.3 is: $S = \frac{1}{2} l^2 \sin(\theta)$.

2.6 Area of Triangle

Heron's Formula is used to calculate the area of a triangle when all three side lengths are known. Given a triangle with side lengths a, b, and c (Fig. 2.4), the area (S) can be calculated using the formula $S = \sqrt{p(p-a)(p-b)(p-c)}$. Note that p is the semi-perimeter of the triangle, i.e., $p = \frac{a+b+c}{2}$.

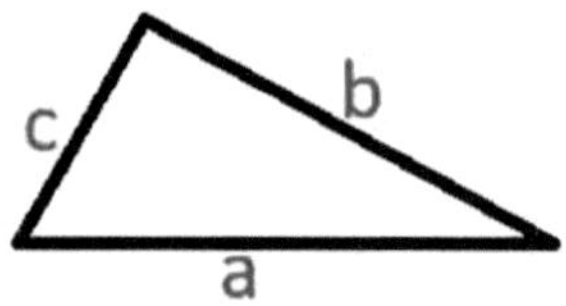

Fig. 2.4 A general triangle with a, b, and c sides

Figure 2.5 provides formulas for calculating the areas of various geometric shapes.

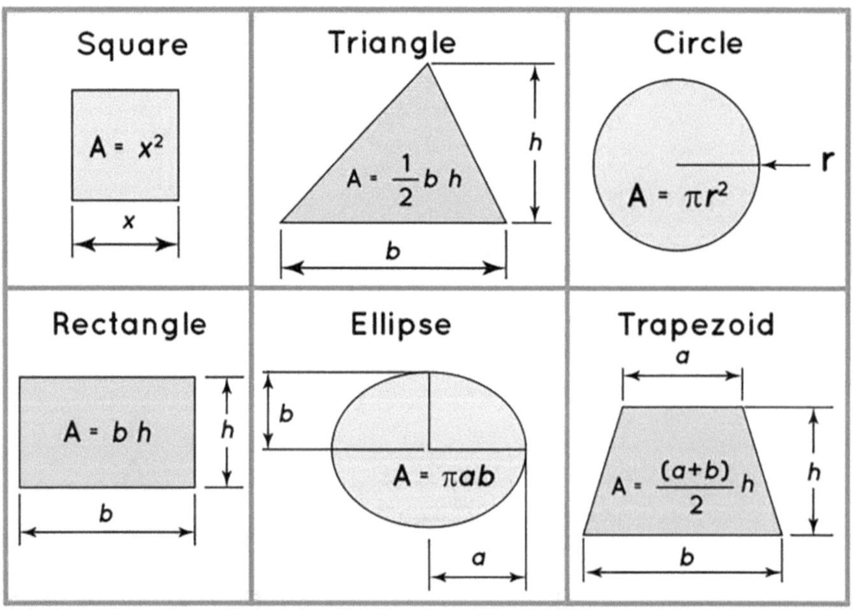

Fig. 2.5 Formulas for calculation of area for different shapes

2.7 Important Trigonometric Relationships

The following are some of the important trigonometric relationships (note that sign

$$(x) = \begin{cases} +1, & x > 1 \\ 0, & x = 0 \\ -1, & x < 0 \end{cases}):$$

1.

$$(\sin(x))^2 + (\cos(x))^2 = 1 \Rightarrow \begin{cases} \sin(x) = \sqrt{1 - (\cos(x))^2} \\ \cos(x) = \sqrt{1 - (\sin(x))^2} \end{cases}$$

2.

$$a \times \sin(\omega t) + b \times \cos(\omega t) = \operatorname{sign}(a).\sqrt{a^2 + b^2} \sin\left(\omega t + \tan^{-1}\left(\frac{b}{a}\right)\right)$$

3.

$$\sin^2(\theta) = \frac{1 - \cos(2\theta)}{2}$$

4.

$$\cos^2(\theta) = \frac{1 + \cos(2\theta)}{2}$$

5.

$$\sin(2\theta) = 2\sin(\theta)\cos(\theta)$$

6.

$$\cos(2\theta) = \cos^2(\theta) - \sin^2(\theta)$$

7.

$$\sin(\theta \pm \varphi) = \sin(\theta)\cos(\varphi) \pm \cos(\theta)\sin(\varphi)$$

8.

$$\cos(\theta \pm \varphi) = \cos(\theta)\cos(\varphi) \mp \sin(\theta)\sin(\varphi)$$

9.

$$\tan(\theta \pm \varphi) = \frac{\tan(\theta) \pm \tan(\varphi)}{1 \mp \tan(\theta)\tan(\varphi)}$$

10.

$$x + y = \frac{\pi}{2} \Rightarrow \sin(x) = \cos(y)$$

11.
$$x + y = \frac{\pi}{2} \Rightarrow \tan(x) = \text{cotan}(y)$$

12.
$$\sin\left(x + \frac{\pi}{2}\right) = \cos(x)$$

13.
$$\cos\left(x + \frac{\pi}{2}\right) = -\sin(x)$$

14.
$$\tan\left(x + \frac{\pi}{2}\right) = -\text{cotan}(x)$$

15.
$$\sin(x \pm \pi) = -\sin(x)$$

16.
$$\cos(x \pm \pi) = -\cos(x)$$

17.
$$\tan(x \pm \pi) = \tan(x)$$

18.
$$\sin^{-1}(x) + \cos^{-1}(x) = \frac{\pi}{2}$$

19.
$$\tan^{-1}(x) + \text{cotan}^{-1}(x) = \frac{\pi}{2}$$

20.
$$\begin{cases} x \in \left[0, \frac{\pi}{2}\right) \Rightarrow \sin(x) \leq x \leq \tan(x) \\ x \in \left(0, \frac{\pi}{2}\right) \Rightarrow \sin(x) < x < \tan(x) \end{cases}$$

21.
$$\begin{cases} x \in \left(-\frac{\pi}{2}, 0\right] \Rightarrow \tan(x) \leq x \leq \sin(x) \\ x \in \left(-\frac{\pi}{2}, 0\right) \Rightarrow \tan(x) < x < \sin(x) \end{cases}$$

Note that $\sin^{-1}(x)$, $\cos^{-1}(x)$, $\tan^{-1}(x)$, and $\text{cotan}^{-1}(x)$ denote the inverse trigonometric functions arcsine, arccosine, arctangent, and arc cotangent, respectively.

Example 2.4 Express the function 5 sin (θ) + 7 cos (θ) as a single sine function.

Solution

$$a\times \sin(\omega t)+b\times \cos(\omega t)=\text{sign}(a).\sqrt{a^2+b^2}\sin\left(\omega t+\tan^{-1}\left(\frac{b}{a}\right)\right)$$

$$5\sin(\theta)+7\cos(\theta)=\text{sign}(5).\sqrt{5^2+7^2}\sin\left(\theta+\tan^{-1}\left(\frac{7}{5}\right)\right)=+8.60\sin(\theta+54.46°)$$

Example 2.5 Express the function –5 sin (θ) + 7 cos (θ) as a single sine function.

Solution

$$a\times \sin(\omega t)+b\times \cos(\omega t)=\text{sign}(a).\sqrt{a^2+b^2}\sin\left(\omega t+\tan^{-1}\left(\frac{b}{a}\right)\right)$$

$$-5\sin(\theta)+7\cos(\theta)=\text{sign}(-5)\sqrt{5^2+7^2}\sin\left(\theta+\tan^{-1}\left(\frac{7}{-5}\right)\right)=$$

$$-8.60\sin(\theta-54.46°)=8.60\sin(\theta-54.46°+180°)$$
$$=8.60\sin(\theta+125.54°)$$

Example 2.6 Express the function 5 sin (θ) – 7 cos (θ) as a single sine function.

Solution

$$a\times \sin(\omega t)+b\times \cos(\omega t)=\text{sign}(a).\sqrt{a^2+b^2}\sin\left(\omega t+\tan^{-1}\left(\frac{b}{a}\right)\right)$$

$$5\sin(\theta)-7\cos(\theta)=\text{sign}(5)\sqrt{5^2+7^2}\sin\left(\theta+\tan^{-1}\left(\frac{-7}{5}\right)\right)=+8.60\sin(\theta-54.46°)$$

Example 2.7 Express the function –5 sin (θ) – 7 cos (θ) as a single sine function.

Solution

$$a\times \sin(\omega t)+b\times \cos(\omega t)=\text{sign}(a).\sqrt{a^2+b^2}\sin\left(\omega t+\tan^{-1}\left(\frac{b}{a}\right)\right)$$

$$-5\sin(\theta)-7\cos(\theta)=\text{sign}(-5)\sqrt{5^2+7^2}\sin\left(\theta+\tan^{-1}\left(\frac{-7}{-5}\right)\right)=$$

$$-8.60\sin(\theta+54.46°)=8.60\sin(\theta+54.46°+180°)$$
$$=8.60\sin(\theta+234.46°)$$

Example 2.8 Calculate the cos(150°) and sin(120°).

Solution

$$\cos(150°)=\sin(-60°)=-\sin(60°)=-\frac{\sqrt{3}}{2}$$

$$\sin(120°)=\cos(-30°)=\cos(30°)=\frac{\sqrt{3}}{2}$$

2.8 Period of Trigonometric Functions

$\sin(kx + \varphi_0)$ or $\cos(kx + \varphi_0)$ has period of $T = \frac{2\pi}{k}$. $\tan(kx + \varphi_0)$ or $\cot(kx + \varphi_0)$ has period of $T = \frac{\pi}{k}$.

Example 2.9 Calculate the period of $\sin\left(3x + \frac{\pi}{6}\right)$ and $\tan\left(6x + \frac{\pi}{3}\right)$.

Solution $\sin\left(3x + \frac{\pi}{6}\right) \Rightarrow T = \frac{2\pi}{3}$ and $\tan\left(6x + \frac{\pi}{3}\right) \Rightarrow T = \frac{\pi}{6}$.

Example 2.10 Calculate the period of $\sin\left(2x + \frac{\pi}{3}\right) + \cos\left(6x + \frac{\pi}{4}\right)$.

Solution

$$\sin\left(2x + \frac{\pi}{3}\right) \Rightarrow T_1 = \frac{2\pi}{2} = \pi = 1\pi$$

$$\cos\left(6x + \frac{\pi}{4}\right) \Rightarrow T_2 = \frac{2\pi}{6} = \frac{\pi}{3} = \frac{1}{3}\pi$$

We need to find the least common multiple (LCM) of T_1 and T_2. LCM of T_1 and T_2 is π. Graph of $\sin\left(2x + \frac{\pi}{3}\right) + \cos\left(6x + \frac{\pi}{4}\right)$ is shown in Fig. 2.6. You can use Fig. 2.6 to verify that period is π.

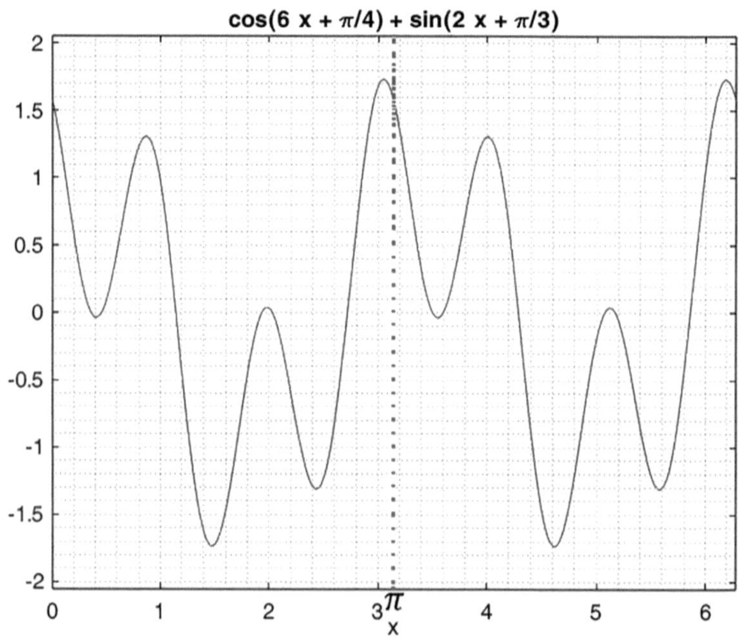

Fig. 2.6 Plot of $\sin\left(2x + \frac{\pi}{3}\right) + \cos\left(6x + \frac{\pi}{4}\right)$ over $[0, 2\pi]$ interval

2.9 Hyperbolic Functions

Hyperbolic functions are defined as:

1.

$$\sinh(x) = \frac{e^x - e^{-x}}{2}$$

2.

$$\cosh(x) = \frac{e^x + e^{-x}}{2}$$

3.

$$\tanh(x) = \frac{\sinh(x)}{\cosh(x)} = \frac{e^x - e^{-x}}{e^x + e^{-x}}$$

4.

$$\coth(x) = \frac{e^x + e^{-x}}{e^x - e^{-x}}$$

5.

$$\text{sech}(x) = \frac{1}{\cosh(x)} = \frac{2}{e^x + e^{-x}}$$

6.

$$\text{csch}(x) = \frac{1}{\sinh(x)} = \frac{2}{e^x - e^{-x}}$$

2.10 Phase Difference

Phase difference is the difference in the phase angle of two sinusoidal waves. It's essentially the difference in the starting point of two waves with the same frequency.

Example 2.11 Determine the phase difference between $y_1(t) = 5\sin\left(314t + \frac{\pi}{3}\right)$ and $y_2(t) = 3\sin\left(314t + \frac{\pi}{2}\right)$.

Solution

$$\Delta\varphi = \varphi_{y_1} - \varphi_{y_2} = 314t + \frac{\pi}{3} - \left(314t + \frac{\pi}{2}\right) = \frac{\pi}{3} - \frac{\pi}{2} = -\frac{\pi}{6}$$

$y_1(t)$ lags $y_2(t)$ by $\frac{\pi}{6}$ radians. In other words, $y_2(t)$ leads $y_1(t)$ by $\frac{\pi}{6}$ radians. The graphs of $y_1(t)$ and $y_2(t)$ are shown in Fig. 2.7.

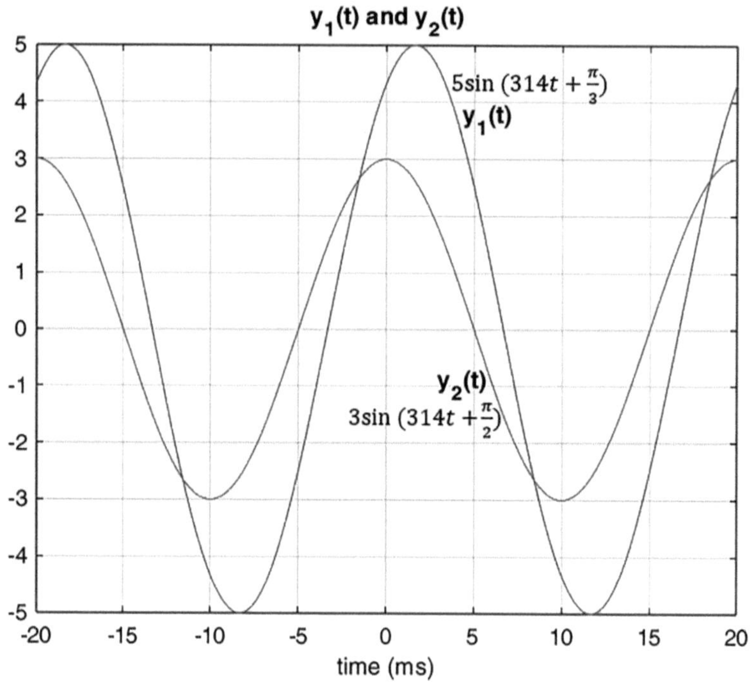

Fig. 2.7 Plot of $y_1(t) = 5\sin\left(314t + \frac{\pi}{3}\right)$ and $y_2(t) = 3\sin\left(314t + \frac{\pi}{2}\right)$

Example 2.12 Determine the phase difference between $y_1(t) = 5\cos\left(314t + \frac{\pi}{6}\right)$ and $y_2(t) = 3\cos(314t + \pi)$.

Solution

$$\Delta\varphi = \varphi_{y_1} - \varphi_{y_2} = 314t + \frac{\pi}{6} - (314t + \pi) = \frac{\pi}{6} - \pi = -\frac{5\pi}{6}$$

$y_1(t)$ lags $y_2(t)$ by $\frac{5\pi}{6}$ radians. In other words, $y_2(t)$ leads $y_1(t)$ by $\frac{5\pi}{6}$ radians. The graphs of $y_1(t)$ and $y_2(t)$ are shown in Fig. 2.8.

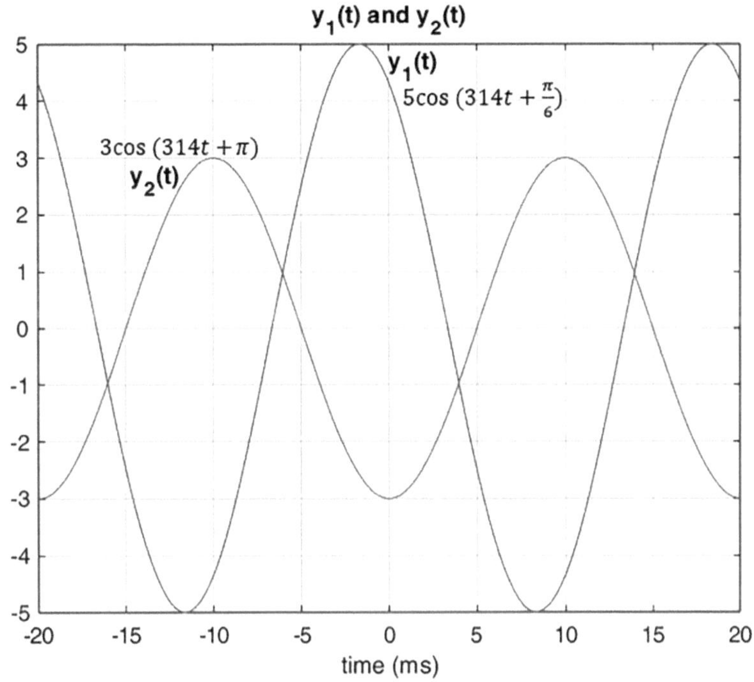

Fig. 2.8 Plot of $y_1(t) = 5\cos\left(314t + \frac{\pi}{6}\right)$ and $y_2(t) = 3\cos(314t + \pi)$

Example 2.13 Determine the phase difference between $y_1(t) = 5\sin(314t)$ and $y_2(t) = 3\cos\left(314t + \frac{\pi}{4}\right)$.

Solution

$$y_1(t) = 5\sin(314t)$$

$$y_2(t) = 3\cos\left(314t + \frac{\pi}{4}\right) = 3\sin\left(314t + \frac{\pi}{4} + \frac{\pi}{2}\right) = 3\sin\left(314t + \frac{3\pi}{4}\right)$$

$$\Delta\varphi = \varphi_{y_1} - \varphi_{y_2} = 314t - \left(314t + \frac{3\pi}{4}\right) = -\frac{3\pi}{4}$$

Or

$$y_1(t) = 5\sin(314t) = 5\cos\left(314t - \frac{\pi}{2}\right)$$

$$y_2(t) = 3\cos\left(314t + \frac{\pi}{4}\right)$$

$$\Delta\varphi = \varphi_{y_1} - \varphi_{y_2} = \left(314t - \frac{\pi}{2}\right) - \left(314t + \frac{\pi}{4}\right) = -\frac{3\pi}{4}$$

$y_1(t)$ lags $y_2(t)$ by $\frac{3\pi}{4}$ radians. In other words, $y_2(t)$ leads $y_1(t)$ by $\frac{3\pi}{4}$ radians. The graphs of $y_1(t)$ and $y_2(t)$ are shown in Fig. 2.9.

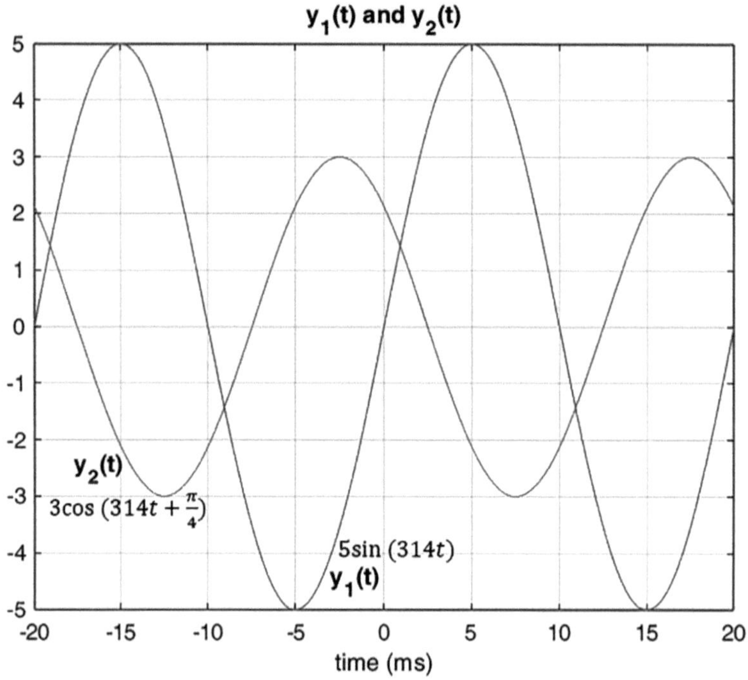

Fig. 2.9 Plot of $y_1(t) = 5\ \sin\ (314t)$ and $y_2(t) = 3\cos\left(314t + \frac{\pi}{4}\right)$

2.11 Graphical Phase Difference Measurement

This section demonstrates graphical phase difference measurement. Consider the waveforms shown in Fig. 2.10.

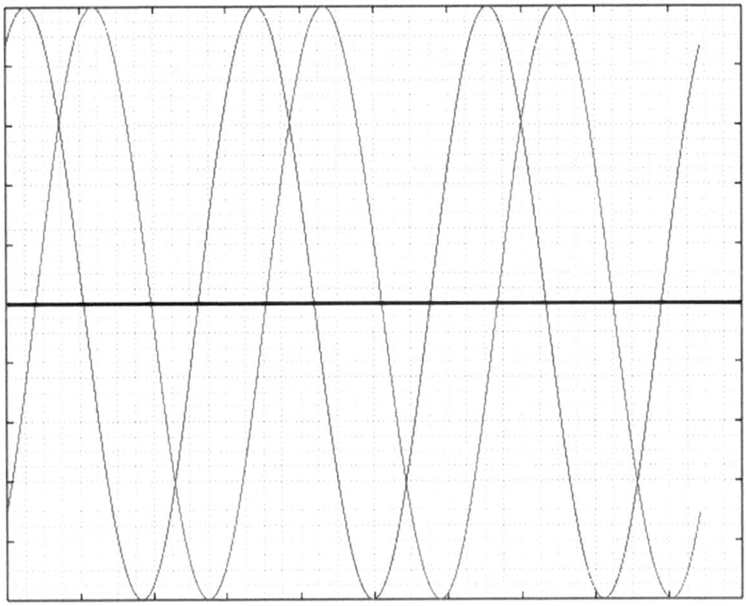

Fig. 2.10 Two sinusoidal signals

We need to measure the period first (Fig. 2.11).

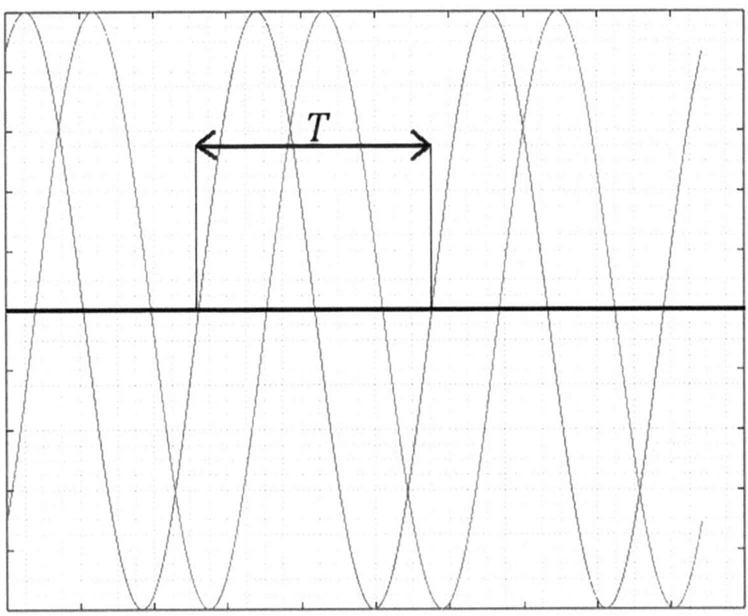

Fig. 2.11 Measurement of period

Next, we determine the time difference between corresponding zero-crossings (or peak points) on the two waveforms (Fig. 2.12). The phase difference can be calculated using $\Delta\varphi = \frac{\Delta t}{T} \times 360^\circ$ formula.

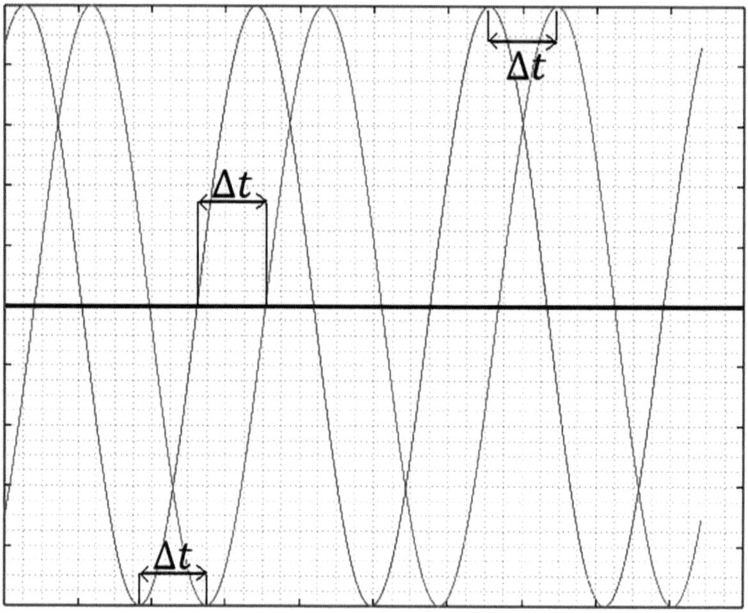

Fig. 2.12 Measurement of time shift

The proposed method does not explicitly identify the leading or lagging waveform. To determine this, we designate one waveform as a reference and identify a zero-crossing point on this waveform where it transitions from negative to positive values.

Next, we examine the value of the second waveform at the selected zero-crossing point. If the value is positive, the second waveform leads the reference. Conversely, if the value is negative, the second waveform lags the reference.

For instance, according to Fig. 2.13, the red waveform lags the blue waveform.

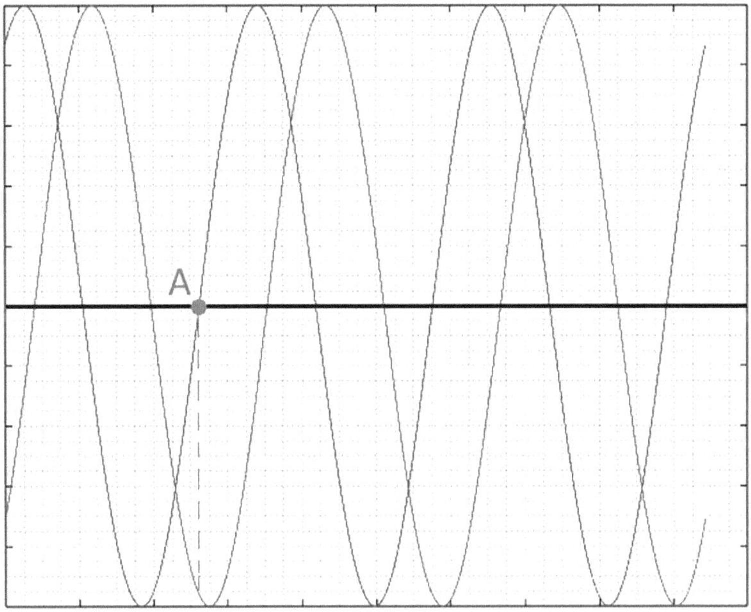

Fig. 2.13 The blue waveform starts to rise at point *A*

For instance, according to Fig. 2.14, the blue waveform leads the red waveform.

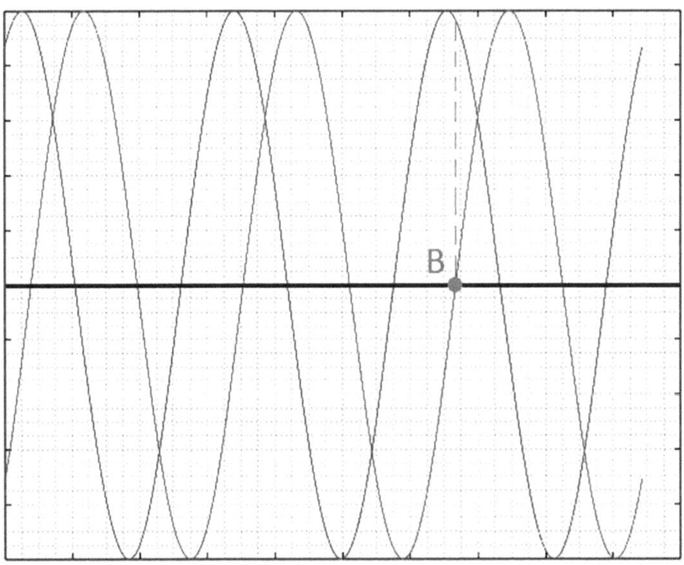

Fig. 2.14 The red waveform starts to rise at point *B*

2.12 Problem Solving with MATLAB®

This section offers MATLAB examples relevant to the problems in this chapter. If you are new to MATLAB, begin with Chap. 11.

Example 2.14 The code in Fig. 2.15 converts $\frac{\pi}{6}$ radians to degrees.

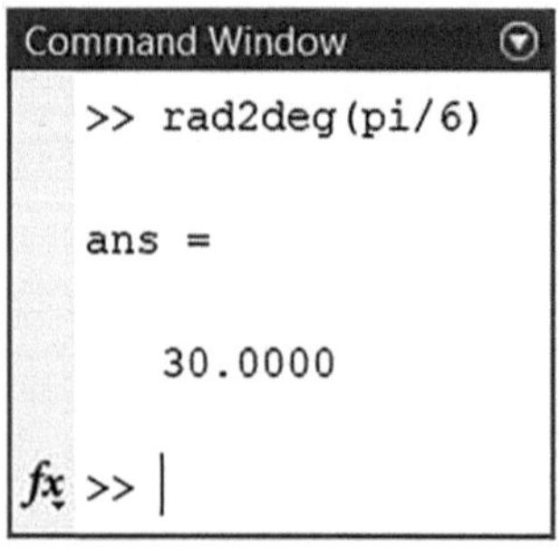

Fig. 2.15 Conversion of $\frac{\pi}{6}$ radians to degrees

Example 2.15 The code in Fig. 2.16 converts 45° to radians.

Fig. 2.16 Conversion of 45° to radians

Example 2.16 The code in Fig. 2.17 computes the sine, cosine, tangent, cotangent, secant, and cosecant of $\frac{\pi}{6}$ radians.

```
Command Window
>> x=pi/6;
>> sin(x)

ans =

    0.5000

>> cos(x)

ans =

    0.8660

>> tan(x)

ans =

    0.5774

>> cot(x)

ans =

    1.7321

>> sec(x)

ans =

    1.1547

>> csc(x)

ans =

    2.0000

fx >>
```

Fig. 2.17 Calculation of trigonometric ratios for $\frac{\pi}{6}$ radians

Example 2.17 The code in Fig. 2.18 computes the sine, cosine, tangent, cotangent, secant, and cosecant of 30°.

Fig. 2.18 Calculation of trigonometric ratios for 30°

```
Command Window
>> x=30;
>> sind(x)

ans =

    0.5000

>> cosd(x)

ans =

    0.8660

>> tand(x)

ans =

    0.5774

>> cotd(x)

ans =

    1.7321

>> secd(x)

ans =

    1.1547

>> cscd(x)

ans =

     2

fx >>
```

Example 2.18 The code in Fig. 2.19 computes the arcsine, arccosine, arctangent, arc cotangent, arc secant, and arc cosecant of 0.5. Results are in radians.

Fig. 2.19 Calculation of the inverse trigonometric functions for $x = 0.5$ (Results are in radians)

```
Command Window
>> x=0.5;
>> asin(x)

ans =

    0.5236

>> acos(x)

ans =

    1.0472

>> atan(x)

ans =

    0.4636

>> acot(x)

ans =

    1.1071

fx >>
```

Example 2.19 The code in Fig. 2.20 computes the arcsine, arccosine, arctangent, arc cotangent, arc secant, and arc cosecant of 0.5. Results are in degrees.

Fig. 2.20 Calculation of the inverse trigonometric functions for $x = 0.5$ (Results are in Degrees)

```
Command Window
>> x=0.5;
>> asind(x)

ans =

   30.0000

>> acosd(x)

ans =

   60.0000

>> atand(x)

ans =

   26.5651

>> acotd(x)

ans =

   63.4349

fx >>
```

Example 2.20 The code in Fig. 2.21 computes the hyperbolic sine, hyperbolic cosine, hyperbolic tangent, hyperbolic cotangent, hyperbolic secant, and hyperbolic cosecant of 1.2.

```
Command Window
>> x=1.2;
>> sinh(x)

ans =

    1.5095

>> cosh(x)

ans =

    1.8107

>> tanh(x)

ans =

    0.8337

>> coth(x)

ans =

    1.1995

>> sech(x)

ans =

    0.5523

>> csch(x)

ans =

    0.6625

fx >>
```

Fig. 2.21 Calculation of hyperbolic ratios for $x = 1.2$

Example 2.21 The code in Fig. 2.22 computes the hyperbolic arcsine, hyperbolic arccosine, hyperbolic arctangent, hyperbolic arc cotangent, hyperbolic arc secant, and hyperbolic arc cosecant of 0.5.

Fig. 2.22 Calculation of inverse hyperbolic functions for $x = 0.5$

```
Command Window
>> x=0.5;
>> asinh(x)

ans =

    0.4812

>> acosh(x)

ans =

   0.0000 + 1.0472i

>> atanh(x)

ans =

    0.5493

>> acoth(x)

ans =

   0.5493 + 1.5708i

fx >>
```

Example 2.22 The code in Fig. 2.23 expresses the function 5 sin (x) + 7 cos (x) as a single cosine function.

```
Command Window
>> syms x
>> y=5*sin(x)+7*cos(x);
>> simplify(vpa(y))

ans =

8.6023252670426267717294735350497*cos(x - 0.62024948598282147133653255617593)

fx >>
```

Fig. 2.23 Conversion of 5 sin (x) + 7 cos (x) into a single cosine function

The number of decimal places in Fig. 2.23 is excessive. Use the digits(5) command to reduce the number of significant digits to 5 (Fig. 2.24).

Fig. 2.24 The digits (5) command decreases the number of significant digits to 5

```
Command Window
>> syms x
>> y=5*sin(x)+7*cos(x);
>> digits(5)
>> simplify(vpa(y))

ans =

8.6023*cos(- 0.62025 + x)

fx >>
```

From Fig. 2.24 we deduce that $5\sin(x)+7\cos(x)=8.6023\cos(x-0.62025)=8.6023\sin\left(x-0.62025+\frac{\pi}{2}\right)=8.6023\sin(x-0.62025+1.5708)=8.6023\sin(x+0.9505)$. Recall that $\cos(x)=\sin\left(x+\frac{\pi}{2}\right)$.

Example 2.23 The code in Fig. 2.25 expresses the function -5 sin (x) + 7 cos (x) as a single cosine function.

Fig. 2.25 Conversion of $-5\sin(x) + 7\cos(x)$ into a single cosine function

```
Command Window
>> syms x
>> y=-5*sin(x)+7*cos(x);
>> digits(5)
>> simplify(vpa(y))

ans =

8.6023*cos(0.62025 + x)

fx >>
```

From Fig. 2.25 we deduce that $-5\sin(x)+7\cos(x)=8.6023\cos(x+0.62025)=8.6023\sin\left(x+0.62025+\frac{\pi}{2}\right)=8.6023\sin(x+0.62025+1.5708)=8.6023\sin(x+2.1910)$.

Example 2.24 The code in Fig. 2.26 expresses the function $5\sin(x) - 7\cos(x)$ as a single cosine function.

Fig. 2.26 Conversion of $5\sin(x) - 7\cos(x)$ into a single cosine function

```
Command Window
>> syms x
>> y=5*sin(x)-7*cos(x);
>> digits(5)
>> simplify(vpa(y))

ans =

-8.6023*cos(0.62025 + x)

fx >>
```

From Fig. 2.26 we deduce that $5\sin(x)-7\cos(x)=-8.6023\cos(x+0.62025)=8.6023\cos(x+0.62025+\pi)=8.6023\sin\left(x+0.62025+\pi+\frac{\pi}{2}\right)=8.6023\sin(x+5.3326)=8.6023\sin(x-0.9506)$. Recall that $-\cos(x) = \cos(x+\pi)$ and $\cos(x)=\sin\left(x+\frac{\pi}{2}\right)$.

Example 2.25 The code in Fig. 2.27 expresses the function $-5\sin(x) - 7\cos(x)$ as a single cosine function.

```
Command Window
>> syms x
>> y=-5*sin(x)-7*cos(x);
>> digits(5)
>> simplify(vpa(y))

ans =

-8.6023*cos(- 0.62025 + x)

fx >>
```

Fig. 2.27 Conversion of $-5\ \sin\ (x) - 7\ \cos\ (x)$ into a single cosine function

From Fig. 2.27 we deduce that $-5\sin(x)-7\cos(x)=-8.6023\cos(x-0.62025)=8.6023\cos(x-0.62025+\pi)=8.6023\sin\left(x-0.62025+\pi+\frac{\pi}{2}\right)=8.6023\sin(x+4.0921)=8.6023\sin(x+4.0921-2\pi)=8.6023\sin(x-2.1911)$. Recall that $-\cos\ (x) = \cos\ (x + \pi)$ and $\cos(x)=\sin\left(x+\frac{\pi}{2}\right)$.

Example 2.26 The code in Fig. 2.28 plots the function $f(x)=\sin\left(2x+\frac{\pi}{3}\right)+\cos\left(6x+\frac{\pi}{4}\right)$ on $[0, 2\pi]$ interval. Output of this code is shown in Fig. 2.29.

```
Command Window
>> syms x
>> ezplot(sin(3*x+pi/6)+cos(6*x+pi/4),[0,2*pi])
>> grid minor
fx >>
```

Fig. 2.28 Plotting the $f(x)=\sin\left(2x+\frac{\pi}{3}\right)+\cos\left(6x+\frac{\pi}{4}\right)$ on $[0, 2\pi]$ interval using the ezplot command

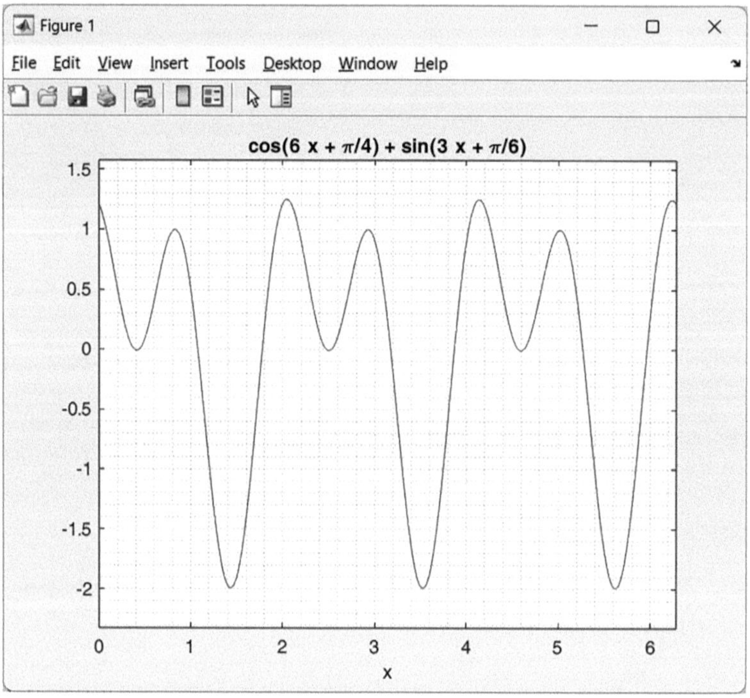

Fig. 2.29 Output of the code shown in Fig. 2.28

According to Fig. 2.30, the period of this function is around 4.13558 − 2.04603 = 2.0896.

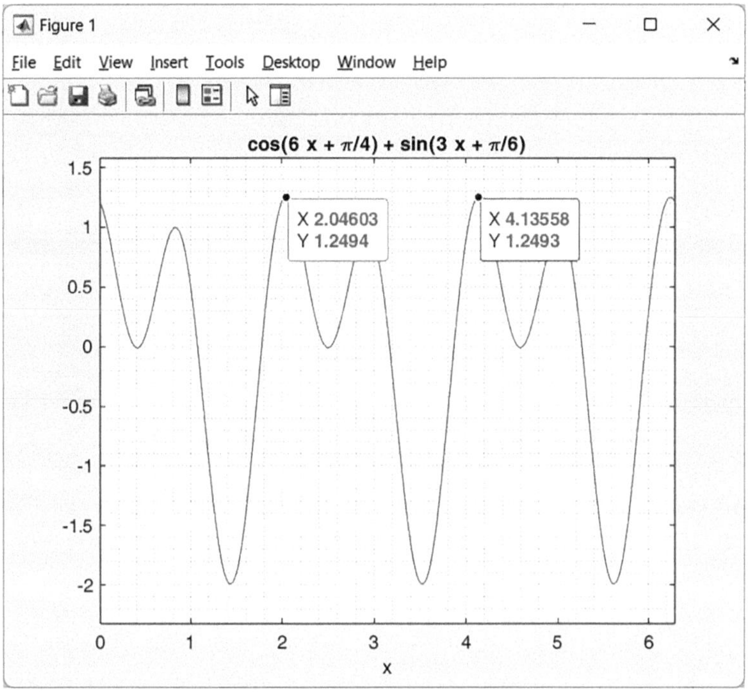

Fig. 2.30 Measuring coordinates of two consecutive maximums

Example 2.27 The previous example used the ezplot command to plot the function $f(x) = \sin\left(2x + \frac{\pi}{3}\right) + cos\left(6x + \frac{\pi}{4}\right)$. Although ezplot is still active, MathWorks® recommends using fplot. The code required to plot $f(x) = \sin\left(2x + \frac{\pi}{3}\right) + \cos\left(6x + \frac{\pi}{4}\right)$ over the interval $[0, 2\pi]$ is shown in Fig. 2.31. The resulting output plot is shown in Fig. 2.32.

```
Command Window
>> syms x
>> fplot(sin(2*x+pi/3)+cos(6*x+pi/4),[0,2*pi]);
>> title('Plot of f(x) = sin(2x + \pi/3) + cos(6x + \pi/4)')
>> xlabel('x');
>> ylabel('y');
>> grid on
fx >>
```

```
Command Window
>> f=@(x) sin(2*x + pi/3) + cos(6*x + pi/4);
>> fplot(f,[0,2*pi])
>> title('Plot of f(x) = sin(2x + \pi/3) + cos(6x + \pi/4)')
>> xlabel('x');
>> ylabel('y');
>> grid on;
fx >>
```

Fig. 2.31 Plotting the $f(x) = \sin\left(2x + \frac{\pi}{3}\right) + \cos\left(6x + \frac{\pi}{4}\right)$ on $[0, 2\pi]$ interval using the fplot command

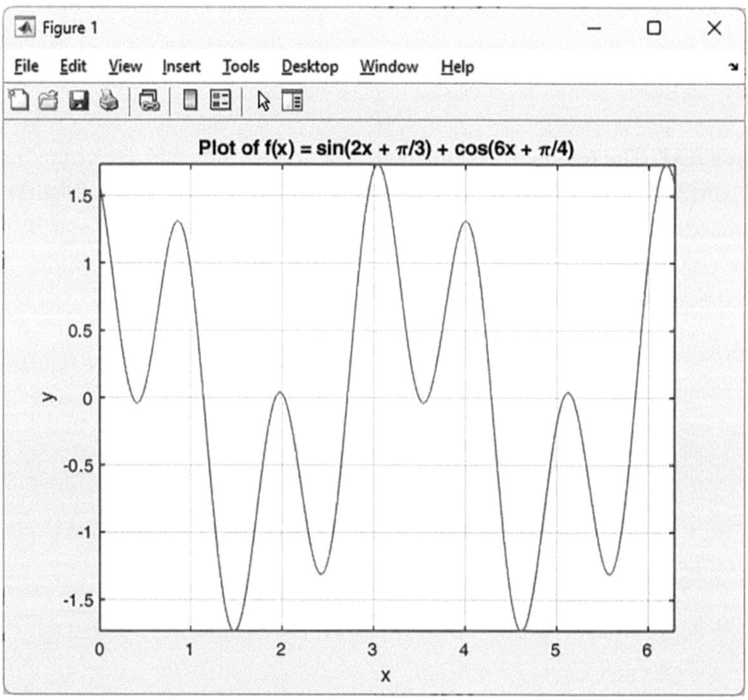

Fig. 2.32 Output of the code shown in Fig. 2.31

Example 2.28 The commands shown in Fig. 2.33 draw the graph of $f(x) = x^2 - 2x$ on the interval $[-4, 4]$. The output of this code is shown in Fig. 2.34.

```
Command Window
>> syms x
>> fplot(x^2-2*x,[-4,4]);
>> title('Plot of f(x) = x^2-2x')
>> xlabel('x');
>> ylabel('y');
>> grid on
fx >>
```

```
Command Window
>> f=@(x) x.^2-2.*x;
>> fplot(f,[-4,4]);
>> title('Plot of f(x) = x^2-2x')
>> xlabel('x');
>> ylabel('y');
>> grid on
fx >>
```

Fig. 2.33 Plotting the $f(x) = x^2 - 2x$ on $[-4, 4]$ interval using the fplot command

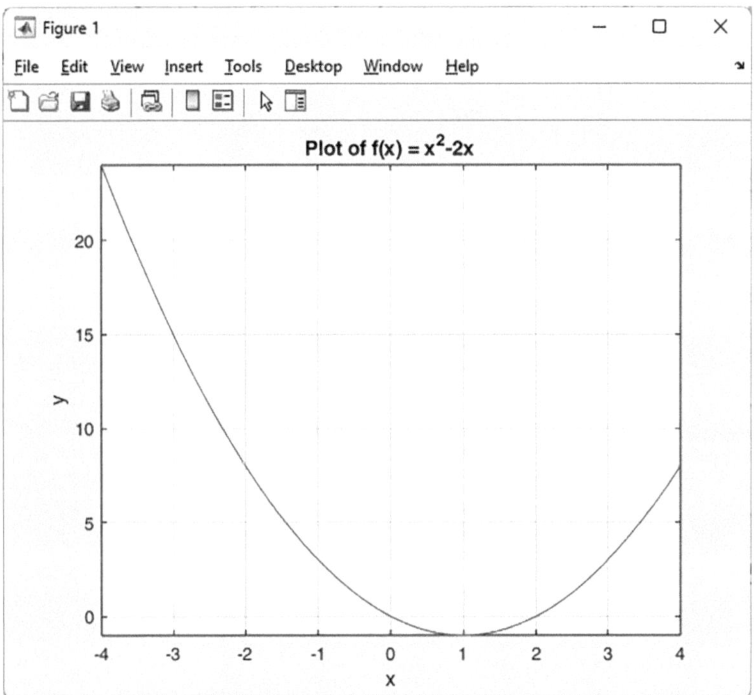

Fig. 2.34 Output of the code shown in Fig. 2.33

Example 2.29 The commands shown in Fig. 2.35 draw the graph of $f(x) = x^3 \sin(5x^2 + 7x)$ on the interval $[0, 2]$. The output of this code is shown in Fig. 2.36.

```
>> syms x
>> fplot(x^3*sin(5*x^2+7*x),[0,2]);
>> title('Plot of f(x) = x^3sin(5x^2+7x)')
>> xlabel('x');
>> ylabel('y');
>> grid on
>>
```

```
>> f = @(x) x.^3 .* sin(5.*x.^2 + 7.*x);
>> fplot(f,[0,2]);
>> title('Plot of f(x) = x^3sin(5x^2+7x)')
>> xlabel('x');
>> ylabel('y');
>> grid on
>>
```

Fig. 2.35 Plotting the $f(x) = x^3 \sin(5x^2 + 7x)$ on $[0, 2]$ interval using the fplot command

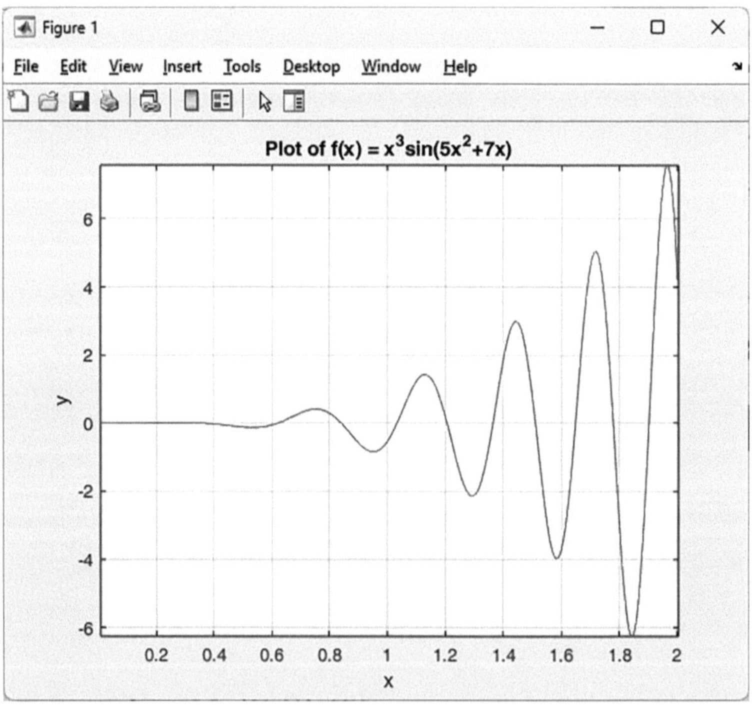

Fig. 2.36 Output of the code shown in Fig. 2.35

Example 2.30 The commands shown in Fig. 2.37 simultaneously plot the graphs of $f(x) = x^2 \sin(7x)$ and $g(x) = e^{-3x}$ on a single set of axes over the interval $[0, 2]$. Fig. 2.38 shows the resulting output of this code.

```
Command Window
>> syms x
>> f=x^2*sin(7*x);
>> g=exp(-3*x);
>> fplot(f,[0,2]);
>> hold on
>> fplot(g,[0,2]);
>> title('Plots of f(x) = x^2sin(7x) and g(x)=exp(-3x)');
>> xlabel('x');
>> ylabel('y');
>> grid on
fx >>
```

```
Command Window
>> f = @(x) x.^2 .* sin(7.*x);
>> g = @(x) exp(-3.*x);
>> fplot(f,[0,2]);
>> hold on
>> fplot(g,[0,2]);
>> title('Plots of f(x) = x^2sin(7x) and g(x)=exp(-3x)');
>> xlabel('x');
>> ylabel('y');
>> grid on
fx >>
```

Fig. 2.37 Plotting the graphs of $f(x) = x^2 \sin(7x)$ and $g(x) = e^{-3x}$ simultaneously on a single set of axes

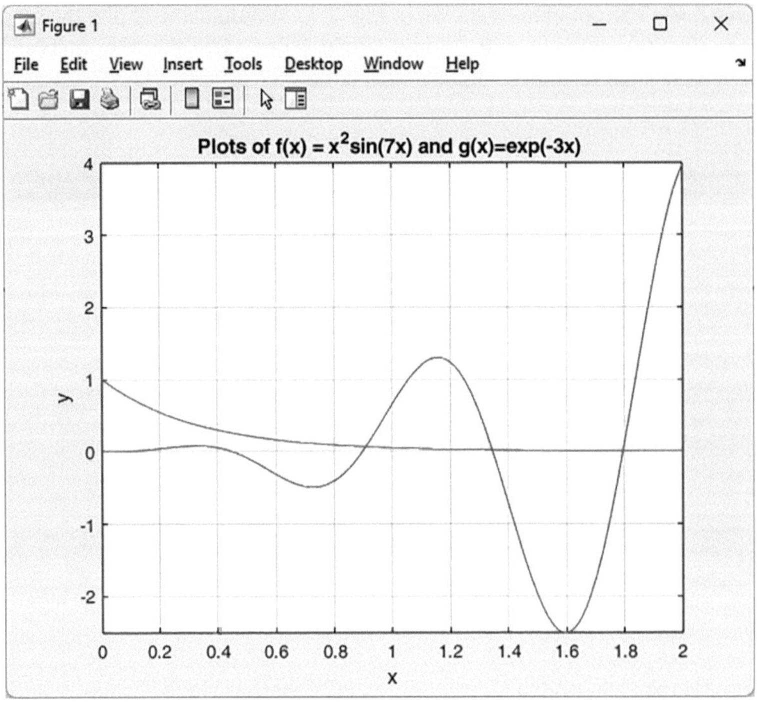

Fig. 2.38 Output of the code shown in Fig. 2.37

References for Further Study

1. Engineering Mathematics with MATLAB® and Simulink®, Farzin Asadi, Springer, 2025.
2. Applied Numerical Analysis with MATLAB®/Simulink®: For Engineers and Scientists, Farzin Asadi, Springer, 2023.
3. Advanced Engineering Mathematics, Erwin Kreyszig, Wiley, 2020.
4. A Problem-Solving Approach to Single-Variable Calculus (Volume II), Farzin Asadi, Springer, 2026.
5. A Problem-Solving Approach to Multi-Variable Calculus (Volume I), Farzin Asadi, Springer, 2026.
6. A Problem-Solving Approach to Multi-Variable Calculus (Volume II), Farzin Asadi, Springer, 2026.
7. A Problem-Solving Approach to Multi-Variable Calculus (Volume III), Farzin Asadi, Springer, 2026.
8. A Problem-Solving Approach to Ordinary Differential Equations (Volume I), Farzin Asadi, Springer, 2026.
9. A Problem-Solving Approach to Ordinary Differential Equations (Volume II), Farzin Asadi, Springer, 2026.
10. A Problem-Solving Approach to Ordinary Differential Equations (Volume III), Farzin Asadi, Springer, 2026.
11. A Problem-Solving Approach to Ordinary Differential Equations (Volume IV), Farzin Asadi, Springer, 2026.

Chapter 3
A Concise Review of Matrices and Linear Algebra

3.1 Introduction

Matrices serve as the foundational building blocks of linear algebra, offering a concise and powerful means to represent linear transformations, systems of equations, and complex data structures. This chapter explores the key concepts and relationships within matrix algebra.

3.2 Matrix Addition

The following example provides a review of matrix addition.

Example 3.1 Calculate the $\begin{bmatrix} 1 & 2 & 3 \\ 4 & 5 & 6 \end{bmatrix} + \begin{bmatrix} 7 & 8 & 9 \\ 10 & 11 & 13 \end{bmatrix}$.

Solution

$$\begin{bmatrix} 1 & 2 & 3 \\ 4 & 5 & 6 \end{bmatrix} + \begin{bmatrix} 7 & 8 & 9 \\ 10 & 11 & 13 \end{bmatrix} = \begin{bmatrix} 8 & 10 & 12 \\ 14 & 16 & 19 \end{bmatrix}$$

3.3 Matrix Multiplication

The following examples provide a review of matrix multiplication.

Example 3.2 Calculate the $7 \times \begin{bmatrix} 1 & 2 & 3 \\ 4 & 5 & 6 \end{bmatrix}$.

F. Asadi, *A Problem-Solving Approach to Single-Variable Calculus*,
https://doi.org/10.1007/978-3-032-13505-6_3

Solution

$$7 \times \begin{bmatrix} 1 & 2 & 3 \\ 4 & 5 & 6 \end{bmatrix} = \begin{bmatrix} 7 & 14 & 21 \\ 28 & 35 & 42 \end{bmatrix}$$

Example 3.3 Calculate the $\begin{bmatrix} 1 & 2 & 3 \\ 4 & 5 & 6 \end{bmatrix} \times \begin{bmatrix} 7 & 8 & 9 \\ 0 & 1 & 2 \\ 3 & 5 & 5 \end{bmatrix}$.

Solution

$$\begin{bmatrix} 1 & 2 & 3 \\ 4 & 5 & 6 \end{bmatrix} \times \begin{bmatrix} 7 & 8 & 9 \\ 0 & 1 & 2 \\ 3 & 5 & 5 \end{bmatrix}$$
$$= \begin{bmatrix} 1\times7+2\times0+3\times3 & 1\times8+2\times1+3\times5 & 1\times9+2\times2+3\times5 \\ 4\times7+5\times0+6\times3 & 4\times8+5\times1+6\times5 & 4\times9+5\times2+6\times5 \end{bmatrix} = \begin{bmatrix} 16 & 25 & 28 \\ 46 & 67 & 76 \end{bmatrix}$$

Example 3.4 Calculate the $\begin{bmatrix} 7 & 0 & 0 \\ 0 & 6 & 0 \\ 0 & 0 & 5 \end{bmatrix} \times \begin{bmatrix} 3 & 0 & 0 \\ 0 & 4 & 0 \\ 0 & 0 & -2 \end{bmatrix}$.

Solution

$$\begin{bmatrix} 7 & 0 & 0 \\ 0 & 6 & 0 \\ 0 & 0 & 5 \end{bmatrix} \times \begin{bmatrix} 3 & 0 & 0 \\ 0 & 4 & 0 \\ 0 & 0 & -2 \end{bmatrix} = \begin{bmatrix} 7\times3 & 0 & 0 \\ 0 & 6\times4 & 0 \\ 0 & 0 & 5\times-2 \end{bmatrix} = \begin{bmatrix} 21 & 0 & 0 \\ 0 & 24 & 0 \\ 0 & 0 & -10 \end{bmatrix}$$

For any positive integer n, the n-th power of a diagonal matrix A, denoted as A^n, is also a diagonal matrix. The diagonal elements of A^n are simply the n-th powers of the corresponding diagonal elements of A. For instance, $\begin{bmatrix} a & 0 & 0 \\ 0 & b & 0 \\ 0 & 0 & c \end{bmatrix}^n = \begin{bmatrix} a^n & 0 & 0 \\ 0 & b^n & 0 \\ 0 & 0 & c^n \end{bmatrix}$.

3.4 Transpose of a Matrix

If A is an $m \times n$ matrix, then its transpose, denoted as A^T, is an $n \times m$ matrix, where ${A^T}_{ij} = A_{ji}$.

Example 3.5 Calculate the transpose of $A = \begin{bmatrix} 1 & 2 & 3 \\ 4 & 5 & 6 \end{bmatrix}$.

Solution $A^T = \begin{bmatrix} 1 & 4 \\ 2 & 5 \\ 3 & 6 \end{bmatrix}$.

3.5 Minors of a Matrix

The minor of the element a_{ij} (in the i-th row, j-th column) which is shown by M_{ij} is the determinant of the sub-matrix obtained by deleting the i-th row and j-th column. For instance, for $A = \begin{bmatrix} a_{11} & a_{12} & a_{13} \\ a_{21} & a_{22} & a_{23} \\ a_{31} & a_{32} & a_{33} \end{bmatrix}$.

$$M_{11} = \begin{vmatrix} a_{22} & a_{23} \\ a_{32} & a_{33} \end{vmatrix}, M_{12} = \begin{vmatrix} a_{21} & a_{23} \\ a_{31} & a_{33} \end{vmatrix}, M_{13} = \begin{vmatrix} a_{21} & a_{22} \\ a_{31} & a_{32} \end{vmatrix}$$

$$M_{21} = \begin{vmatrix} a_{12} & a_{13} \\ a_{32} & a_{33} \end{vmatrix}, M_{22} = \begin{vmatrix} a_{11} & a_{13} \\ a_{31} & a_{33} \end{vmatrix}, M_{23} = \begin{vmatrix} a_{11} & a_{12} \\ a_{31} & a_{32} \end{vmatrix}$$

$$M_{31} = \begin{vmatrix} a_{12} & a_{13} \\ a_{22} & a_{23} \end{vmatrix}, M_{32} = \begin{vmatrix} a_{11} & a_{13} \\ a_{21} & a_{23} \end{vmatrix}, M_{33} = \begin{vmatrix} a_{11} & a_{12} \\ a_{21} & a_{22} \end{vmatrix}$$

Example 3.6 For $A = \begin{bmatrix} 1 & 1 & 2 \\ 5 & 0 & -1 \\ 7 & 2 & 3 \end{bmatrix}$, $M_{21} = \begin{vmatrix} 1 & 2 \\ 2 & 3 \end{vmatrix} = 1 \times 3 - 2 \times 2 = -1$.

3.6 Cofactors of a Matrix

The cofactor of the element a_{ij} (in the i-th row, j-th column) which is shown by C_{ij} equals to $(-1)^{i+j} M_{ij}$. For instance, for $A = \begin{bmatrix} a_{11} & a_{12} & a_{13} \\ a_{21} & a_{22} & a_{23} \\ a_{31} & a_{32} & a_{33} \end{bmatrix}$,

$$C_{11} = + \begin{vmatrix} a_{22} & a_{23} \\ a_{32} & a_{33} \end{vmatrix}, C_{12} = - \begin{vmatrix} a_{21} & a_{23} \\ a_{31} & a_{33} \end{vmatrix}, C_{13} = + \begin{vmatrix} a_{21} & a_{22} \\ a_{31} & a_{32} \end{vmatrix}$$

$$C_{21} = - \begin{vmatrix} a_{12} & a_{13} \\ a_{32} & a_{33} \end{vmatrix}, C_{22} = + \begin{vmatrix} a_{11} & a_{13} \\ a_{31} & a_{33} \end{vmatrix}, C_{23} = - \begin{vmatrix} a_{11} & a_{12} \\ a_{31} & a_{32} \end{vmatrix}$$

$$C_{31} = + \begin{vmatrix} a_{12} & a_{13} \\ a_{22} & a_{23} \end{vmatrix}, C_{32} = - \begin{vmatrix} a_{11} & a_{13} \\ a_{21} & a_{23} \end{vmatrix}, C_{33} = + \begin{vmatrix} a_{11} & a_{12} \\ a_{21} & a_{22} \end{vmatrix}$$

Example **3.7** For $A = \begin{bmatrix} 1 & 1 & 2 \\ 5 & 0 & -1 \\ 7 & 2 & 3 \end{bmatrix}$, $C_{23} = (-1)^{2+3} \begin{vmatrix} 1 & 1 \\ 7 & 2 \end{vmatrix} = -\begin{vmatrix} 1 & 1 \\ 7 & 2 \end{vmatrix} = -(1 \times 2 - 1 \times 7) = 5.$

3.7 Determinant of a Matrix

The determinant of a matrix is a scalar value that characterizes certain properties of the matrix and the linear transformation it represents. Determinants are defined only for square matrices (matrices with the same number of rows and columns).

Determinant of $\begin{bmatrix} a & b \\ c & d \end{bmatrix}$ equals to $a \times d - b \times c$. Determinant of 3×3 matrix $\begin{bmatrix} a & b & c \\ d & e & f \\ g & h & i \end{bmatrix}$ can be calculated using one of the following three methods:

$$det\left(\begin{bmatrix} a & b & c \\ d & e & f \\ g & h & i \end{bmatrix}\right) = a\begin{vmatrix} e & f \\ h & i \end{vmatrix} - b\begin{vmatrix} d & f \\ g & i \end{vmatrix} + c\begin{vmatrix} d & e \\ g & h \end{vmatrix}$$
$$= a \times (e \times i - f \times h) - b \times (d \times i - f \times g) + c \times (d \times h - e \times g)$$

$$det\left(\begin{bmatrix} a & b & c \\ d & e & f \\ g & h & i \end{bmatrix}\right) = -d\begin{vmatrix} b & c \\ h & i \end{vmatrix} + e\begin{vmatrix} a & c \\ g & i \end{vmatrix} - f\begin{vmatrix} a & b \\ g & h \end{vmatrix}$$
$$= -d \times (b \times i - c \times h) + e \times (a \times i - c \times g) - f \times (a \times h - b \times g)$$

$$det\left(\begin{bmatrix} a & b & c \\ d & e & f \\ g & h & i \end{bmatrix}\right) = g\begin{vmatrix} b & c \\ e & f \end{vmatrix} - h\begin{vmatrix} a & c \\ d & f \end{vmatrix} + i\begin{vmatrix} a & b \\ d & e \end{vmatrix}$$
$$= g \times (b \times f - c \times e) - h \times (a \times f - c \times d) + i \times (a \times e - b \times d)$$

The Sarrus method is a quick and efficient way to calculate the determinant of a 3x3 matrix. The calculation of the determinant for matrix $\begin{bmatrix} a_{11} & a_{12} & a_{13} \\ a_{21} & a_{22} & a_{23} \\ a_{31} & a_{32} & a_{33} \end{bmatrix}$ using the Sarrus method is shown in Fig. 3.1.

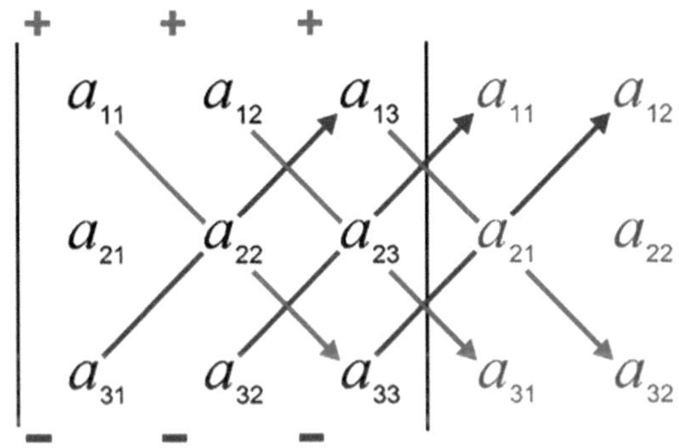

$$det = a_{11}a_{22}a_{33} + a_{12}a_{23}a_{31} + a_{13}a_{21}a_{32} - a_{31}a_{22}a_{13} - a_{32}a_{23}a_{11} - a_{33}a_{21}a_{12}$$

Fig. 3.1 Calculation of determinant using Sarrus method

Example 3.8 Use the Sarrus method to calculate the determinant of $\begin{bmatrix} 2 & 3 & 5 \\ -1 & 4 & 6 \\ 3 & -2 & 7 \end{bmatrix}$.

Solution The solution is shown in Fig. 3.2.

Fig. 3.2 Calculation of the determinant of $\begin{bmatrix} 2 & 3 & 5 \\ -1 & 4 & 6 \\ 3 & -2 & 7 \end{bmatrix}$ using Sarrus method

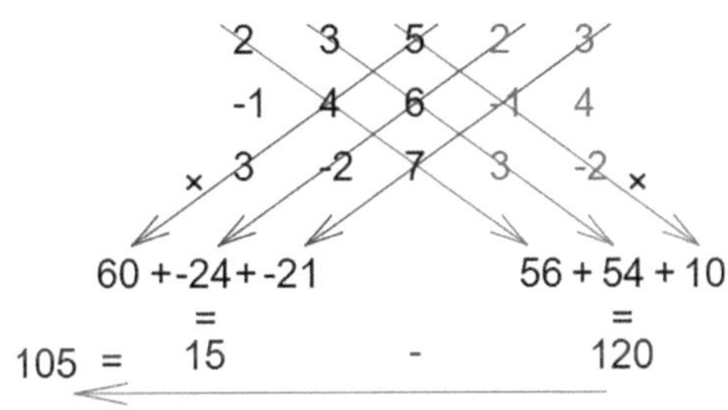

The following are some important properties of determinants.

A. The determinant of a diagonal matrix is simply the product of its diagonal elements. For instance, determinant of $A = diag(7, 6, -2) = \begin{bmatrix} 7 & 0 & 0 \\ 0 & 6 & 0 \\ 0 & 0 & -2 \end{bmatrix}$ equals to $7 \times 6 \times -2 = -84$.
B. The determinant of the product of two square matrices A and B of the same order equals to the product of their individual determinants, i.e., $|A \times B| = |A| \times |B|$.
C. The determinant of a matrix and its transpose are equal. That is, $|A| = |A^T|$.

D. The determinant of a triangular matrix (upper triangular or lower triangular) is the product of its diagonal elements. For instance, $\begin{vmatrix} a & b & c \\ 0 & e & f \\ 0 & 0 & g \end{vmatrix} = a.e.g$ or $\begin{vmatrix} a & 0 & 0 \\ b & c & 0 \\ d & e & f \end{vmatrix} = a.c.f.$

E. A square matrix A is invertible if and only if $|A| \neq 0$.

F. If a row or column of a matrix is multiplied by a scalar k, the determinant is multiplied by k. For instance,

$$A = \begin{bmatrix} 8 & -2 & 1 \\ 2 & -1 & 3 \\ 1 & 1 & 4 \end{bmatrix} \Rightarrow |A| = \begin{vmatrix} 8 & -2 & 1 \\ 2 & -1 & 3 \\ 1 & 1 & 4 \end{vmatrix} = -43$$

$$B = \begin{bmatrix} 8 & -2 & 1 \\ 2\times 2 & -1\times 2 & 3\times 2 \\ 1 & 1 & 4 \end{bmatrix} = \begin{bmatrix} 8 & -2 & 1 \\ 4 & -2 & 6 \\ 1 & 1 & 4 \end{bmatrix} \Rightarrow |B| = \begin{vmatrix} 8 & -2 & 1 \\ 4 & -2 & 6 \\ 1 & 1 & 4 \end{vmatrix} = 2$$

$$\times \begin{vmatrix} 8 & -2 & 1 \\ 2 & -1 & 3 \\ 1 & 1 & 4 \end{vmatrix} = 2 \times -43 = -86$$

$$C = \begin{bmatrix} 8 & -2 & 1 \\ 2\times 2 & -1\times 2 & 3\times 2 \\ 1\times 5 & 1\times 5 & 4\times 5 \end{bmatrix} = \begin{bmatrix} 8 & -2 & 1 \\ 4 & -2 & 6 \\ 5 & 5 & 20 \end{bmatrix} \Rightarrow |C| = \begin{vmatrix} 8 & -2 & 1 \\ 4 & -2 & 6 \\ 5 & 5 & 20 \end{vmatrix} = 2$$

$$\times \begin{vmatrix} 8 & -2 & 1 \\ 2 & -1 & 3 \\ 5 & 5 & 20 \end{vmatrix} = 2\times 5 \times \begin{vmatrix} 8 & -2 & 1 \\ 2 & -1 & 3 \\ 1 & 1 & 4 \end{vmatrix} = 2\times 5\times -43 = -430$$

$$D = \begin{bmatrix} 8\times 4 & -2\times 4 & 1\times 4 \\ 2\times 4 & -1\times 4 & 3\times 4 \\ 1\times 4 & 1\times 4 & 4\times 4 \end{bmatrix} = \begin{bmatrix} 32 & -8 & 4 \\ 8 & -4 & 12 \\ 4 & 4 & 16 \end{bmatrix} \Rightarrow |D| = \begin{vmatrix} 32 & -8 & 4 \\ 8 & -4 & 12 \\ 4 & 4 & 16 \end{vmatrix}$$

$$= 4^3 \times \begin{vmatrix} 8 & -2 & 1 \\ 2 & -1 & 3 \\ 1 & 1 & 4 \end{vmatrix} = 64 \times -43 = -2752$$

G. If two rows or columns of a matrix are interchanged, the sign of the determinant changes. For instance,

$$A = \begin{bmatrix} 8 & -2 & 1 \\ 2 & -1 & 3 \\ 1 & 1 & 4 \end{bmatrix} \Rightarrow |A| = \begin{vmatrix} 8 & -2 & 1 \\ 2 & -1 & 3 \\ 1 & 1 & 4 \end{vmatrix} = -43$$

$$B=\begin{bmatrix}8 & -2 & 1\\ 1 & 1 & 4\\ 2 & -1 & 3\end{bmatrix}\Rightarrow|B|=\begin{vmatrix}8 & -2 & 1\\ 1 & 1 & 4\\ 2 & -1 & 3\end{vmatrix}=43$$

$$C=\begin{bmatrix}1 & -2 & 8\\ 3 & -1 & 2\\ 4 & 1 & 1\end{bmatrix}\Rightarrow|C|=\begin{vmatrix}1 & -2 & 8\\ 3 & -1 & 2\\ 4 & 1 & 1\end{vmatrix}=43$$

3.8 Matrix Inversion

Inverse of a 2 × 2 matrix $\begin{bmatrix}a & b\\ c & d\end{bmatrix}$ is:

$$\begin{bmatrix}a & b\\ c & d\end{bmatrix}^{-1}=\frac{1}{a\times d-b\times c}\begin{bmatrix}d & -b\\ -c & a\end{bmatrix}$$

Inverse of a 3 × 3 matrix $\begin{bmatrix}a & b & c\\ d & e & f\\ g & h & i\end{bmatrix}$ is:

$$\begin{bmatrix}a & b & c\\ d & e & f\\ g & h & i\end{bmatrix}^{-1}=\frac{1}{det\left(\begin{bmatrix}a & b & c\\ d & e & f\\ g & h & i\end{bmatrix}\right)}adj\left(\begin{bmatrix}a & b & c\\ d & e & f\\ g & h & i\end{bmatrix}\right)$$

$$=\frac{1}{det\left(\begin{bmatrix}a & b & c\\ d & e & f\\ g & h & i\end{bmatrix}\right)}\begin{bmatrix}C_{11} & C_{12} & C_{13}\\ C_{21} & C_{22} & C_{23}\\ C_{31} & C_{32} & C_{33}\end{bmatrix}^{T}$$

$$C_{11}=+\begin{vmatrix}e & f\\ h & i\end{vmatrix}, C_{12}=-\begin{vmatrix}d & f\\ g & i\end{vmatrix}, C_{13}=+\begin{vmatrix}d & e\\ g & h\end{vmatrix}$$

$$C_{21}=-\begin{vmatrix}b & c\\ h & i\end{vmatrix}, C_{22}=+\begin{vmatrix}a & c\\ g & i\end{vmatrix}, C_{23}=-\begin{vmatrix}a & b\\ g & h\end{vmatrix}$$

$$C_{31}=+\begin{vmatrix}b & c\\ e & f\end{vmatrix}, C_{32}=-\begin{vmatrix}a & c\\ d & f\end{vmatrix}, C_{33}=+\begin{vmatrix}a & b\\ d & e\end{vmatrix}$$

Let’s study some numeric examples.

Example 3.9 Calculate the inverse of $\begin{bmatrix} 7 & 5 \\ 1 & 2 \end{bmatrix}$.

Solution

$$\begin{bmatrix} 7 & 5 \\ 1 & 2 \end{bmatrix}^{-1} = \frac{1}{7\times 2 - 5\times 1}\begin{bmatrix} 2 & -5 \\ -1 & 7 \end{bmatrix} = \frac{1}{9}\begin{bmatrix} 2 & -5 \\ -1 & 7 \end{bmatrix} = \begin{bmatrix} \frac{2}{9} & \frac{-5}{9} \\ \frac{-1}{9} & \frac{7}{9} \end{bmatrix}$$

Example 3.10 Calculate the inverse of $\begin{bmatrix} 0 & 0 & 1 \\ 2 & -1 & 3 \\ 1 & 1 & 4 \end{bmatrix}$.

Solution

$$\det\left(\begin{bmatrix} 0 & 0 & 1 \\ 2 & -1 & 3 \\ 1 & 1 & 4 \end{bmatrix}\right) = 0\begin{vmatrix} -1 & 3 \\ 1 & 4 \end{vmatrix} - 0\begin{vmatrix} 2 & 3 \\ 1 & 4 \end{vmatrix} + 1\begin{vmatrix} 2 & -1 \\ 1 & 1 \end{vmatrix} = 0 - 0 + 1(2+1) = 3$$

$$\begin{bmatrix} 0 & 0 & 1 \\ 2 & -1 & 3 \\ 1 & 1 & 4 \end{bmatrix}^{-1} = \frac{1}{3}\begin{bmatrix} +\begin{vmatrix} -1 & 3 \\ 1 & 4 \end{vmatrix} & -\begin{vmatrix} 2 & 3 \\ 1 & 4 \end{vmatrix} & +\begin{vmatrix} 2 & -1 \\ 1 & 1 \end{vmatrix} \\ -\begin{vmatrix} 0 & 1 \\ 1 & 4 \end{vmatrix} & +\begin{vmatrix} 0 & 1 \\ 1 & 4 \end{vmatrix} & -\begin{vmatrix} 0 & 0 \\ 1 & 1 \end{vmatrix} \\ +\begin{vmatrix} 0 & 1 \\ -1 & 3 \end{vmatrix} & -\begin{vmatrix} 0 & 1 \\ 2 & 3 \end{vmatrix} & +\begin{vmatrix} 0 & 0 \\ 2 & -1 \end{vmatrix} \end{bmatrix}^T$$

$$= \frac{1}{3}\begin{bmatrix} -7 & -5 & 3 \\ 1 & -1 & 0 \\ 1 & 2 & 0 \end{bmatrix}^T = \begin{bmatrix} \frac{-7}{3} & \frac{1}{3} & \frac{1}{3} \\ \frac{-5}{3} & \frac{-1}{3} & \frac{2}{3} \\ 1 & 0 & 0 \end{bmatrix}$$

Example 3.11 Calculate the inverse of $\begin{bmatrix} -4 & -6 & 2 \\ 5 & -1 & 3 \\ -2 & 4 & -3 \end{bmatrix}$.

Solution

$$\det\left(\begin{bmatrix} -4 & -6 & 2 \\ 5 & -1 & 3 \\ -2 & 4 & -3 \end{bmatrix}\right) = -4\begin{vmatrix} -1 & 3 \\ 4 & -3 \end{vmatrix} - (-6)\begin{vmatrix} 5 & 3 \\ -2 & -3 \end{vmatrix} + 2\begin{vmatrix} 5 & -1 \\ -2 & 4 \end{vmatrix} = -4(3-12) + 6(-15+6) + 2(20-2) = 36 - 54 + 36 = 18$$

$$\begin{bmatrix} -4 & -6 & 2 \\ 5 & -1 & 3 \\ -2 & 4 & -3 \end{bmatrix}^{-1} = \frac{1}{18}\begin{bmatrix} +\begin{vmatrix} -1 & 3 \\ 4 & -3 \end{vmatrix} & -\begin{vmatrix} 5 & 3 \\ -2 & -3 \end{vmatrix} & +\begin{vmatrix} 5 & -1 \\ -2 & 4 \end{vmatrix} \\ -\begin{vmatrix} -6 & 2 \\ 4 & -3 \end{vmatrix} & +\begin{vmatrix} -4 & 2 \\ -2 & -3 \end{vmatrix} & -\begin{vmatrix} -4 & -6 \\ -2 & 4 \end{vmatrix} \\ +\begin{vmatrix} -6 & 2 \\ -1 & 3 \end{vmatrix} & -\begin{vmatrix} -4 & 2 \\ 5 & 3 \end{vmatrix} & +\begin{vmatrix} -4 & -6 \\ 5 & -1 \end{vmatrix} \end{bmatrix}^T$$

$$= \frac{1}{18}\begin{bmatrix} -9 & 9 & 18 \\ -10 & -8 & 4 \\ -16 & 22 & 34 \end{bmatrix}^T = \frac{1}{18}\begin{bmatrix} -9 & -10 & -16 \\ 9 & -8 & 22 \\ 18 & 4 & 34 \end{bmatrix} = \begin{bmatrix} \frac{-1}{2} & \frac{-5}{9} & \frac{-8}{9} \\ \frac{1}{2} & \frac{-4}{9} & \frac{11}{9} \\ 1 & \frac{2}{9} & \frac{17}{9} \end{bmatrix}$$

3.9 Solving Systems of Equations Using Matrices

This section shows how matrix method can be used to solve system of linear equations. The following examples show how to solve a system of linear equations using the matrix method.

Example 3.12 Solve $\begin{cases} x + 2y = 3 \\ 4x + 5y = 6 \end{cases}$.

Solution

$$\begin{cases} x+2y=3 \\ 4x+5y=6 \end{cases} \Rightarrow \begin{bmatrix} 1 & 2 \\ 4 & 5 \end{bmatrix}\begin{bmatrix} x \\ y \end{bmatrix} = \begin{bmatrix} 3 \\ 6 \end{bmatrix} \Rightarrow \begin{bmatrix} 1 & 2 \\ 4 & 5 \end{bmatrix}^{-1}\begin{bmatrix} 1 & 2 \\ 4 & 5 \end{bmatrix}\begin{bmatrix} x \\ y \end{bmatrix}$$

$$= \begin{bmatrix} 1 & 2 \\ 4 & 5 \end{bmatrix}^{-1}\begin{bmatrix} 3 \\ 6 \end{bmatrix} \Rightarrow \begin{bmatrix} 1 & 0 \\ 0 & 1 \end{bmatrix}\begin{bmatrix} x \\ y \end{bmatrix} = \begin{bmatrix} 1 & 2 \\ 4 & 5 \end{bmatrix}^{-1}\begin{bmatrix} 3 \\ 6 \end{bmatrix} \Rightarrow \begin{bmatrix} x \\ y \end{bmatrix} = \begin{bmatrix} 1 & 2 \\ 4 & 5 \end{bmatrix}^{-1}\begin{bmatrix} 3 \\ 6 \end{bmatrix}$$

$$= \frac{1}{1\times5-2\times4}\begin{bmatrix} 5 & -2 \\ -4 & 1 \end{bmatrix}\begin{bmatrix} 3 \\ 6 \end{bmatrix} = \frac{1}{-3}\begin{bmatrix} 5\times3-2\times6 \\ -4\times3+1\times6 \end{bmatrix} = \begin{bmatrix} -1 \\ 2 \end{bmatrix} \Rightarrow x = -1, y = 2.$$

You can use the following method as well:

$$\begin{cases} x + 2y = 3 \\ 4x + 5y = 6 \end{cases} \Rightarrow \begin{cases} -4x - 8y = -12 \\ 4x + 5y = 6 \end{cases} \Rightarrow (-4x - 8y) + (4x + 5y) = -12$$

$$+6 \Rightarrow -3y = -6 \Rightarrow y = 2$$

$$x + 2y = 3 \Rightarrow x + 2\times2 = 3 \Rightarrow x = 3 - 4 = -1 \Rightarrow x = -1$$

The following method is another option:

$$\begin{cases} x + 2y = 3 \\ 4x + 5y = 6 \end{cases} \Rightarrow x = 3 - 2y$$

$$4x + 5y = 6 \Rightarrow 4(3 - 2y) + 5y = 6 \Rightarrow 12 - 8y + 5y = 6 \Rightarrow -3y = -6 \Rightarrow y = 2$$

$$x = 3 - 2y = 3 - 2 \times 2 = 3 - 4 = -1 \Rightarrow x = -1.$$

Example 3.13 Solve $\begin{cases} x + 2y + 3z = 10 \\ 4x + 6y + 5z = 12 \\ 7x + 9y + 8z = 13 \end{cases}.$

Solution

$$\begin{cases} x + 2y + 3z = 10 \\ 4x + 6y + 5z = 12 \\ 7x + 9y + 8z = 13 \end{cases} \Rightarrow \begin{bmatrix} 1 & 2 & 3 \\ 4 & 6 & 5 \\ 7 & 9 & 8 \end{bmatrix} \begin{bmatrix} x \\ y \\ z \end{bmatrix}$$

$$= \begin{bmatrix} 10 \\ 12 \\ 13 \end{bmatrix} \Rightarrow \begin{bmatrix} 1 & 2 & 3 \\ 4 & 6 & 5 \\ 7 & 9 & 8 \end{bmatrix}^{-1} \begin{bmatrix} 1 & 2 & 3 \\ 4 & 6 & 5 \\ 7 & 9 & 8 \end{bmatrix} \begin{bmatrix} x \\ y \\ z \end{bmatrix}$$

$$= \begin{bmatrix} 1 & 2 & 3 \\ 4 & 6 & 5 \\ 7 & 9 & 8 \end{bmatrix}^{-1} \begin{bmatrix} 10 \\ 12 \\ 13 \end{bmatrix} \Rightarrow \begin{bmatrix} 1 & 0 & 0 \\ 0 & 1 & 0 \\ 0 & 0 & 1 \end{bmatrix} \begin{bmatrix} x \\ y \\ z \end{bmatrix}$$

$$= \begin{bmatrix} 1 & 2 & 3 \\ 4 & 6 & 5 \\ 7 & 9 & 8 \end{bmatrix}^{-1} \begin{bmatrix} 10 \\ 12 \\ 13 \end{bmatrix} \Rightarrow \begin{bmatrix} x \\ y \\ z \end{bmatrix} = \begin{bmatrix} 1 & 2 & 3 \\ 4 & 6 & 5 \\ 7 & 9 & 8 \end{bmatrix}^{-1} \begin{bmatrix} 10 \\ 12 \\ 13 \end{bmatrix}$$

Let's calculate $\begin{bmatrix} 1 & 2 & 3 \\ 4 & 6 & 5 \\ 7 & 9 & 8 \end{bmatrix}^{-1}$:

$$\begin{bmatrix} 1 & 2 & 3 \\ 4 & 6 & 5 \\ 7 & 9 & 8 \end{bmatrix}^{-1} = \frac{1}{det\left(\begin{bmatrix} 1 & 2 & 3 \\ 4 & 6 & 5 \\ 7 & 9 & 8 \end{bmatrix}\right)} \times adj\left(\begin{bmatrix} 1 & 2 & 3 \\ 4 & 6 & 5 \\ 7 & 9 & 8 \end{bmatrix}\right)$$

$$det\left(\begin{bmatrix} 1 & 2 & 3 \\ 4 & 6 & 5 \\ 7 & 9 & 8 \end{bmatrix}\right) = 1 \times \begin{vmatrix} 6 & 5 \\ 9 & 8 \end{vmatrix} - 2 \times \begin{vmatrix} 4 & 5 \\ 7 & 8 \end{vmatrix} + 3 \times \begin{vmatrix} 4 & 6 \\ 7 & 9 \end{vmatrix} = 1$$

$$\times (6 \times 8 - 5 \times 9) - 2 \times (4 \times 8 - 5 \times 7) + 3 \times (4 \times 9 - 6 \times 7) = 3 + 6 - 18 = -9$$

$$adj\left(\begin{bmatrix} 1 & 2 & 3 \\ 4 & 6 & 5 \\ 7 & 9 & 8 \end{bmatrix}\right) = \begin{bmatrix} +\begin{vmatrix} 6 & 5 \\ 9 & 8 \end{vmatrix} & -\begin{vmatrix} 4 & 5 \\ 7 & 8 \end{vmatrix} & +\begin{vmatrix} 4 & 6 \\ 7 & 9 \end{vmatrix} \\ -\begin{vmatrix} 2 & 3 \\ 9 & 8 \end{vmatrix} & +\begin{vmatrix} 1 & 3 \\ 7 & 8 \end{vmatrix} & -\begin{vmatrix} 1 & 2 \\ 7 & 9 \end{vmatrix} \\ +\begin{vmatrix} 2 & 3 \\ 6 & 5 \end{vmatrix} & -\begin{vmatrix} 1 & 3 \\ 4 & 5 \end{vmatrix} & +\begin{vmatrix} 1 & 2 \\ 4 & 6 \end{vmatrix} \end{bmatrix}^T$$

$$= \begin{bmatrix} 3 & 3 & -6 \\ 11 & -13 & 5 \\ -8 & 7 & -2 \end{bmatrix}^T = \begin{bmatrix} 3 & 11 & -8 \\ 3 & -13 & 7 \\ -6 & 5 & -2 \end{bmatrix}$$

$$\begin{bmatrix} 1 & 2 & 3 \\ 4 & 6 & 5 \\ 7 & 9 & 8 \end{bmatrix}^{-1} = \frac{1}{det\left(\begin{bmatrix} 1 & 2 & 3 \\ 4 & 6 & 5 \\ 7 & 9 & 8 \end{bmatrix}\right)} \times adj\left(\begin{bmatrix} 1 & 2 & 3 \\ 4 & 6 & 5 \\ 7 & 9 & 8 \end{bmatrix}\right)$$

$$= \frac{1}{-9}\begin{bmatrix} 3 & 11 & -8 \\ 3 & -13 & 7 \\ -6 & 5 & -2 \end{bmatrix} = \begin{bmatrix} -\frac{1}{3} & -\frac{11}{9} & \frac{8}{9} \\ -\frac{1}{3} & \frac{13}{9} & -\frac{7}{9} \\ \frac{2}{3} & -\frac{5}{9} & \frac{2}{9} \end{bmatrix}$$

Therefore,

$$\begin{bmatrix} x \\ y \\ z \end{bmatrix} = \begin{bmatrix} 1 & 2 & 3 \\ 4 & 6 & 5 \\ 7 & 9 & 8 \end{bmatrix}^{-1} \begin{bmatrix} 10 \\ 12 \\ 13 \end{bmatrix} = \begin{bmatrix} -\frac{1}{3} & -\frac{11}{9} & \frac{8}{9} \\ -\frac{1}{3} & \frac{13}{9} & -\frac{7}{9} \\ \frac{2}{3} & -\frac{5}{9} & \frac{2}{9} \end{bmatrix} \times \begin{bmatrix} 10 \\ 12 \\ 13 \end{bmatrix}$$

$$= \begin{bmatrix} \frac{-58}{9} \\ \frac{35}{9} \\ \frac{26}{9} \end{bmatrix} \Rightarrow \begin{bmatrix} x \\ y \\ z \end{bmatrix} = \begin{bmatrix} \frac{-58}{9} \\ \frac{35}{9} \\ \frac{26}{9} \end{bmatrix} = \begin{bmatrix} -6.4444 \\ 3.8889 \\ 2.8889 \end{bmatrix}$$

The following method is another option:

$$\begin{cases} x + 2y + 3z = 10 \\ 4x + 6y + 5z = 12 \Rightarrow x + 2y + 3z = 10 \Rightarrow x = 10 - 2y - 3z \\ 7x + 9y + 8z = 13 \end{cases}$$

$$\begin{cases} 4x+6y+5z=12 \\ 7x+9y+8z=13 \end{cases} \Rightarrow \begin{cases} 4(10-2y-3z)+6y+5z=12 \\ 7(10-2y-3z)+9y+8z=13 \end{cases}$$

$$\Rightarrow \begin{cases} 40-8y-12z+6y+5z=12 \\ 70-14y-21z+9y+8z=13 \end{cases} \Rightarrow \begin{cases} -2y-7z=-28 \\ -5y-13z=-57 \end{cases} \Rightarrow -2y-7z$$

$$=-28 \Rightarrow y=\frac{-28+7z}{-2}=14-3.5z$$

$$-5y-13z=-57 \Rightarrow -5(14-3.5z)-13z=-57 \Rightarrow -70+17.5z-13z=$$

$$-57 \Rightarrow 4.5z=13 \Rightarrow z=\frac{13}{4.5}=\frac{26}{9}=2.8889$$

$$y=14-3.5z=14-3.5\times 2.8889=14-10.1111=3.8889$$

$$x=10-2y-3z=10-2\times 3.8889-3\times 2.8889=-6.4444$$

3.10 Underdetermined System of Linear Equations

An underdetermined system of equations is: a system where there are fewer equations than unknowns. This means that there are more variables than constraints, leading to infinitely many solutions or no solutions at all. An underdetermined system cannot have a unique solution.

Row Reduction (Gauss-Jordan elimination) can be used to determine whether an undetermined system has no solution or infinitely many solutions: Write the augmented matrix of the system. Perform row operations to reduce the matrix to row-echelon form or reduced row-echelon form. Analyze the resulting matrix: If you encounter a row of the form [0 0…0 | c] where c is non-zero, the system is inconsistent and has no solution. If you have a row of zeros on the left side and a zero on the right side (a row of the form [0 0…0 | 0]), the system has infinitely many solutions.

Example 3.14 $2x + 3y = 5$ has two unknowns (x and y) but only one equation. In this case, there are infinitely many solutions. Any pair of numbers that adds up to 5 is a valid solution. For example, $(1,4)$, $(2,3)$, $(3,2)$, and so on, are all solutions to this system.

Example 3.15 $\begin{cases} x+y+z=1 \\ 2x+2y+2z=2 \end{cases}$ has three unknowns (x, y, and z) but only two equations. There are infinitely many solutions to this system. Any combination of x, y, and z that satisfies the first equation will also satisfy the second equation. For example, $(1,0,0)$, $(0,1,0)$, and $(0,0,1)$ are all valid solutions.

Example 3.16 $\begin{cases} x+y+z=1 \\ 2x+2y+2z=3 \end{cases}$ has three unknowns (x, y, and z) but only two equations. As you can see, the second equation is: a multiple of the first, but the right-hand side values are inconsistent. This means there's no combination of x, y, and z that can satisfy both equations simultaneously.

Example 3.17 Determine the number of solutions for $\begin{cases} x+2y+3z=1 \\ 11x+12y+3z=2 \end{cases}$.

Solution Let's assume the given underdetermined system has infinitely many solutions.

$$\begin{cases} x+2y+3z=1 \\ 11x+12y+3z=2 \end{cases} \Rightarrow \begin{cases} x+2y=1-3z \\ 11x+12y=2-3z \end{cases} \Rightarrow \begin{bmatrix} 1 & 2 \\ 11 & 12 \end{bmatrix}\begin{bmatrix} x \\ y \end{bmatrix} = \begin{bmatrix} 1-3z \\ 2-3z \end{bmatrix}$$

$$\Rightarrow . \begin{bmatrix} x \\ y \end{bmatrix} = \begin{bmatrix} 1 & 2 \\ 11 & 12 \end{bmatrix}^{-1}\begin{bmatrix} 1-3z \\ 2-3z \end{bmatrix} = \begin{bmatrix} -1.2 & 0.2 \\ 1.1 & -0.1 \end{bmatrix}\begin{bmatrix} 1-3z \\ 2-3z \end{bmatrix}$$

$$= \begin{bmatrix} -1.2 & 0.2 \\ 1.1 & -0.1 \end{bmatrix}\begin{bmatrix} 1-3z \\ 2-3z \end{bmatrix} = \begin{bmatrix} 3z-0.8 \\ 0.9-3z \end{bmatrix} \Rightarrow \begin{cases} x=3z-0.8 \\ y=0.9-3z \end{cases} \Rightarrow \begin{cases} x=3t-0.8 \\ y=0.9-3t \\ z=t \end{cases}$$

Now, we substitute this into the original equation:

$$\begin{cases} x+2y+3z=1 \\ 11x+12y+3z=2 \end{cases} \Rightarrow \begin{cases} 3t-0.8+2(0.9-3t)+3t=1 \\ 11(3t-0.8)+12(0.9-3t)+3t=2 \end{cases} \Rightarrow \begin{cases} 1=1 \\ 2=2 \end{cases}$$

Therefore, the assumption of having infinitely many solutions is correct.

3.11 Overdetermined System of Linear Equations

An overdetermined system of linear equations is: a system where there are more equations than unknowns. This means that there are more constraints than degrees of freedom, making it unlikely that an exact solution exists that satisfies all equations simultaneously.

Let's study some numeric examples.

Example 3.18 Determine the number of solutions for $\begin{cases} x+y=3 \\ 2x+3y=7 \\ x+y=4 \end{cases}$.

Solution This system has three equations and two unknowns. Let's solve a subsystem of two equations and two unknowns. For instance, if we select first and second equations:

$$\begin{cases} x+y=3 \\ 2x+3y=7 \end{cases} \Rightarrow \begin{bmatrix} x \\ y \end{bmatrix} = \begin{bmatrix} 1 & 1 \\ 2 & 3 \end{bmatrix}^{-1} \begin{bmatrix} 3 \\ 7 \end{bmatrix} = \frac{1}{3-2} \begin{bmatrix} 3 & -1 \\ -2 & 1 \end{bmatrix} \begin{bmatrix} 3 \\ 7 \end{bmatrix} = \begin{bmatrix} 2 \\ 1 \end{bmatrix}$$

Substituting the solution into the third equation yields a contradiction since $x + y = 2 + 1 = 3 \neq 4$.

Let's select another subsystem of two equations and two unknowns: $\begin{cases} x+y=3 \\ x+y=4 \end{cases}$. This subsystem has no solution.

The last subsystem of two equations and two unknowns is $\begin{cases} 2x+3y=7 \\ x+y=4 \end{cases}$.

$$\begin{cases} 2x+3y=7 \\ x+y=4 \end{cases} \Rightarrow \begin{bmatrix} x \\ y \end{bmatrix} = \begin{bmatrix} 2 & 3 \\ 1 & 1 \end{bmatrix}^{-1} \begin{bmatrix} 7 \\ 4 \end{bmatrix} = \frac{1}{2-3} \begin{bmatrix} 1 & -3 \\ -1 & 2 \end{bmatrix} \begin{bmatrix} 7 \\ 4 \end{bmatrix} = \begin{bmatrix} 5 \\ -1 \end{bmatrix}$$

The solution fails to satisfy the remaining equation since $x + y = 5 - 1 \neq 3$.

Therefore, the given system does not have an exact solution that satisfies all equations simultaneously. However, we can find an approximate solution that minimizes the sum of the squared errors. The cost function is defined as:

$$f(x,y) = (x+y-3)^2 + (2x+3y-7)^2 + (x+y-4)^2$$

It can be shown that $x = 3.5$ and $y = 0$ minimizes this cost function. Therefore, $x = 3.5$ and $y = 0$ is the approximate solution to the overdetermined system $\begin{cases} x+y=3 \\ 2x+3y=7 \\ x+y=4 \end{cases}$.

For the overdetermined system of linear equations $Ax = b$, the approximate solution that minimizes the sum of squared errors is $x = (A^TA)^{-1}A^Tb$. Note that A^T is the transpose of matrix A.

Example 3.19 Find the approximate solution of $\begin{cases} x+y=3 \\ 2x+3y=7 \\ x+y=4 \end{cases}$.

Solution

$$\begin{cases} x+y=3 \\ 2x+3y=7 \\ x+y=4 \end{cases} \Rightarrow A.\begin{bmatrix} x \\ y \end{bmatrix} = b \text{ where } A = \begin{bmatrix} 1 & 1 \\ 2 & 3 \\ 1 & 1 \end{bmatrix}, b = \begin{bmatrix} 3 \\ 7 \\ 4 \end{bmatrix}.$$

$$\begin{bmatrix} x \\ y \end{bmatrix} = (A^T A)^{-1} A^T b = \left(\begin{bmatrix} 1 & 2 & 1 \\ 1 & 3 & 1 \end{bmatrix} \begin{bmatrix} 1 & 1 \\ 2 & 3 \\ 1 & 1 \end{bmatrix} \right)^{-1} \begin{bmatrix} 1 & 2 & 1 \\ 1 & 3 & 1 \end{bmatrix} \begin{bmatrix} 3 \\ 7 \\ 4 \end{bmatrix}$$

$$= \begin{bmatrix} 6 & 8 \\ 8 & 11 \end{bmatrix}^{-1} \begin{bmatrix} 1 & 2 & 1 \\ 1 & 3 & 1 \end{bmatrix} \begin{bmatrix} 3 \\ 7 \\ 4 \end{bmatrix} = \begin{bmatrix} 6 & 8 \\ 8 & 11 \end{bmatrix}^{-1} \begin{bmatrix} 21 \\ 28 \end{bmatrix}$$

$$= \frac{1}{66 - 64} \begin{bmatrix} 11 & -8 \\ -8 & 6 \end{bmatrix} \begin{bmatrix} 21 \\ 28 \end{bmatrix} = \begin{bmatrix} 3.5 \\ 0 \end{bmatrix}$$

Therefore, approximate solution of this overdetermined system is: $x = 3.5$ and $y = 0$.

3.12 Square System of Equations

A linear system of equations in which the number of variables equals the number of equations. The coefficient matrix of a square system is a square matrix.

A square system of linear equations has a unique solution if and only if the determinant of its coefficient matrix is non-zero.

When the determinant of a system of linear equations is: zero, the system has either no solutions or infinitely many solutions. If the system has no solutions, it is said to be inconsistent. If the system has infinitely many solutions, it is said to be dependent. This occurs when at least one row of coefficient matrix can be expressed as a linear combination of the other rows.

Row Reduction (Gauss-Jordan elimination) can be used to determine whether a system of linear equations with a zero determinant has no solution or infinitely many solutions: Write the augmented matrix of the system. Perform row operations to reduce the matrix to row-echelon form or reduced row-echelon form. Analyze the resulting matrix: If you encounter a row of the form [0 0…0 | c] where c is non-zero, the system is inconsistent and has no solution. If you have a row of zeros on the left side and a zero on the right side (a row of the form [0 0…0 | 0]), the system is dependent and has infinitely many solutions.

Example 3.20 $\begin{cases} x + 2y + 3z = 10 \\ 4x + 6y + 5z = 12 \\ 7x + 9y + 8z = 13 \end{cases}$ can be written as: $\begin{bmatrix} 1 & 2 & 3 \\ 4 & 6 & 5 \\ 7 & 9 & 8 \end{bmatrix} \begin{bmatrix} x \\ y \\ z \end{bmatrix} = \begin{bmatrix} 10 \\ 12 \\ 13 \end{bmatrix}$.

The determinant of the coefficients is -9. Since the determinant is non-zero, a unique solution exists. Let's use the Gauss-Jordan elimination method to find the solution. The augmented matrix is:

$$\begin{bmatrix} 1 & 2 & 3 & 10 \\ 4 & 6 & 5 & 12 \\ 7 & 9 & 8 & 13 \end{bmatrix} \rightarrow \begin{bmatrix} 1 & 2 & 3 & 10 \\ 0 & -2 & -7 & -28 \\ 7 & 9 & 8 & 13 \end{bmatrix} \rightarrow \begin{bmatrix} 1 & 2 & 3 & 10 \\ 0 & -2 & -7 & -28 \\ 0 & -5 & -13 & -57 \end{bmatrix}$$

$$\rightarrow \begin{bmatrix} 1 & 0 & -4 & -18 \\ 0 & -2 & -7 & -28 \\ 0 & -5 & -13 & -57 \end{bmatrix} \rightarrow \begin{bmatrix} 1 & 0 & -4 & -18 \\ 0 & -2 & -7 & -28 \\ 0 & 0 & 4.5 & 13 \end{bmatrix}$$

$$\rightarrow \begin{bmatrix} 1 & 0 & -4 & -18 \\ 0 & -2 & 0 & -7.7772 \\ 0 & 0 & 4.5 & 13 \end{bmatrix} \rightarrow \begin{bmatrix} 1 & 0 & 0 & -6.4443 \\ 0 & -2 & 0 & -7.7772 \\ 0 & 0 & 4.5 & 13 \end{bmatrix}$$

$$\rightarrow \begin{bmatrix} 1 & 0 & 0 & -6.4443 \\ 0 & 1 & 0 & 3.8886 \\ 0 & 0 & 1 & 2.8889 \end{bmatrix}$$

Therefore, solution of the given system is $x = -6.4443$, $y = 3.8886$, and $z = 2.8889$.

Example 3.21 $\begin{cases} 2x + y + z = 1 \\ 4x + 2y + 2z = 3 \\ 6x + 3y + 3z = 5 \end{cases}$ can be written as: $\begin{bmatrix} 2 & 1 & 1 \\ 4 & 2 & 2 \\ 6 & 3 & 3 \end{bmatrix} \begin{bmatrix} x \\ y \\ z \end{bmatrix} = \begin{bmatrix} 1 \\ 3 \\ 5 \end{bmatrix}$.

Determinant of coefficients is zero:

$$\begin{vmatrix} 2 & 1 & 1 \\ 4 & 2 & 2 \\ 6 & 3 & 3 \end{vmatrix} = 2\begin{vmatrix} 2 & 2 \\ 3 & 3 \end{vmatrix} - 1\begin{vmatrix} 4 & 2 \\ 6 & 3 \end{vmatrix} + 1\begin{vmatrix} 4 & 2 \\ 6 & 3 \end{vmatrix} = 0$$

Therefore, the given system may be inconsistent, i.e., have no solution, or it may have infinitely many solutions. The augmented matrix is:

$$\begin{bmatrix} 2 & 1 & 1 & 1 \\ 4 & 2 & 2 & 3 \\ 6 & 3 & 3 & 5 \end{bmatrix}$$

Basic row operations lead to:

$$\begin{bmatrix} 2 & 1 & 1 & 1 \\ 4 & 2 & 2 & 3 \\ 6 & 3 & 3 & 5 \end{bmatrix} \longrightarrow \begin{bmatrix} 2 & 1 & 1 & 1 \\ 0 & 0 & 0 & 1 \\ 6 & 3 & 2 & 5 \end{bmatrix}$$

Since the second has a non-zero entry on the right side, the system is inconsistent and has no solution.

Example 3.22 $\begin{cases} 2x+y+z=1 \\ 4x+2y+2z=2 \\ 6x+3y+3z=3 \end{cases}$ can be written as: $\begin{bmatrix} 2 & 1 & 1 \\ 4 & 2 & 2 \\ 6 & 3 & 3 \end{bmatrix}\begin{bmatrix} x \\ y \\ z \end{bmatrix} = \begin{bmatrix} 1 \\ 2 \\ 3 \end{bmatrix}$.

Determinant of coefficients is zero:

$$\begin{vmatrix} 2 & 1 & 1 \\ 4 & 2 & 2 \\ 6 & 3 & 3 \end{vmatrix} = 2\begin{vmatrix} 2 & 2 \\ 3 & 3 \end{vmatrix} - 1\begin{vmatrix} 4 & 2 \\ 6 & 3 \end{vmatrix} + 1\begin{vmatrix} 4 & 2 \\ 6 & 3 \end{vmatrix} = 0$$

Therefore, the given system may be inconsistent, i.e., have no solution, or it may have infinitely many solutions. The augmented matrix is:

$$\begin{bmatrix} 2 & 1 & 1 & 1 \\ 4 & 2 & 2 & 2 \\ 6 & 3 & 3 & 3 \end{bmatrix}$$

Basic row operations lead to:

$$\begin{bmatrix} 2 & 1 & 1 & 1 \\ 4 & 2 & 2 & 2 \\ 6 & 3 & 3 & 3 \end{bmatrix} \longrightarrow \begin{bmatrix} 2 & 1 & 1 & 1 \\ 0 & 0 & 0 & 0 \\ 6 & 3 & 2 & 5 \end{bmatrix}$$

Since the second row is entirely zeros, the system is dependent and has infinitely many solutions. Let's find the solutions:

$$\begin{bmatrix} 2 & 1 & 1 & 1 \\ 4 & 2 & 2 & 2 \\ 6 & 3 & 3 & 3 \end{bmatrix} \longrightarrow \begin{bmatrix} 2 & 1 & 1 & 1 \\ 0 & 0 & 0 & 0 \\ 6 & 3 & 2 & 5 \end{bmatrix} \longrightarrow \begin{bmatrix} 2 & 1 & 1 & 1 \\ 0 & 0 & 0 & 0 \\ 0 & 0 & 0 & 0 \end{bmatrix} \Rightarrow 2x+y+z$$

$$=1 \Rightarrow \begin{cases} x=t \\ y=u \\ z=1-2t-y \end{cases}$$

For instance, $\begin{bmatrix} 3 \\ 4 \\ -9 \end{bmatrix}$, $\begin{bmatrix} 0 \\ 0 \\ 1 \end{bmatrix}$ or $\begin{bmatrix} -1 \\ 3 \\ 0 \end{bmatrix}$ satisfy the given system.

Example 3.23 $\begin{cases} x+2y+3z=1 \\ 4x+8y+12z=4 \\ 11x+12y+3z=2 \end{cases}$ can be written as: $\begin{bmatrix} 1 & 2 & 3 \\ 4 & 8 & 12 \\ 11 & 12 & 3 \end{bmatrix}\begin{bmatrix} x \\ y \\ z \end{bmatrix} = \begin{bmatrix} 1 \\ 4 \\ 2 \end{bmatrix}$.

Determinant of coefficients is zero:

$$\begin{vmatrix} 1 & 2 & 3 \\ 4 & 8 & 12 \\ 11 & 12 & 3 \end{vmatrix} = 1\begin{vmatrix} 8 & 12 \\ 12 & 3 \end{vmatrix} - 2\begin{vmatrix} 4 & 12 \\ 11 & 3 \end{vmatrix} + 3\begin{vmatrix} 4 & 8 \\ 11 & 12 \end{vmatrix} = 1(24 - 144)$$
$$- 2(12 - 132) + 3(48 - 88) = 0$$

Therefore, the given system may be inconsistent, i.e., have no solution, or it may have infinitely many solutions. The augmented matrix is:

$$\begin{bmatrix} 1 & 2 & 3 & 1 \\ 4 & 8 & 12 & 4 \\ 11 & 12 & 3 & 2 \end{bmatrix}$$

Basic row operations lead to:

$$\begin{bmatrix} 1 & 2 & 3 & 1 \\ 4 & 8 & 12 & 4 \\ 11 & 12 & 3 & 2 \end{bmatrix} \to \begin{bmatrix} 1 & 2 & 3 & 1 \\ 0 & 0 & 0 & 0 \\ 11 & 12 & 3 & 2 \end{bmatrix}$$

Since the second row is entirely zeros, the system is dependent and has infinitely many solutions. Let's find the solutions:

$$\begin{bmatrix} 1 & 2 & 3 & 1 \\ 4 & 8 & 12 & 4 \\ 11 & 12 & 3 & 2 \end{bmatrix} \to \begin{bmatrix} 1 & 2 & 3 & 1 \\ 0 & 0 & 0 & 0 \\ 11 & 12 & 3 & 2 \end{bmatrix} \to \begin{bmatrix} 1 & 2 & 3 & 1 \\ 0 & 0 & 0 & 0 \\ 0 & -10 & -30 & -9 \end{bmatrix}$$
$$\to \begin{bmatrix} 1 & 2 & 3 & 1 \\ 0 & 0 & 0 & 0 \\ 0 & 1 & 3 & 0.9 \end{bmatrix} \to \begin{bmatrix} 1 & 0 & -3 & -0.8 \\ 0 & 0 & 0 & 0 \\ 0 & 1 & 3 & 0.9 \end{bmatrix} \Rightarrow \begin{cases} x - 3z = -0.8 \\ y + 3z = 0.9 \end{cases} \Rightarrow \begin{cases} x = 3t - 0.8 \\ y = -3t + 0.9 \\ z = t \end{cases}$$

For instance, $\begin{bmatrix} -0.8 \\ 0.9 \\ 0 \end{bmatrix}$, $\begin{bmatrix} 2.2 \\ -2.1 \\ 1 \end{bmatrix}$ or $\begin{bmatrix} -6.8 \\ 6.9 \\ -2 \end{bmatrix}$ satisfy the given system.

3.13 Rank of a Matrix

The rank of a matrix represents the maximum number of linearly independent rows or columns in the matrix. The rank of a matrix is not limited to square matrices. The maximum possible rank of an $m \times n$ matrix is the minimum of m and n.

The rank of a matrix may be determined by employing the following methods:

A. Row reduction method: Reduce the matrix to row-echelon form or reduced row-echelon form. The number of non-zero rows in the resulting matrix is the rank.
B. Determinant method: For square matrices, the rank is equal to the size of the largest non-zero minor.

Example 3.24 Calculate the rank of $A = \begin{bmatrix} 1 & 0 & 5 \\ 2 & 2 & -1 \\ 7 & 5 & 5 \end{bmatrix}$.

Solution Given matrix is a square matrix and its determinant is $\begin{vmatrix} 1 & 0 & 5 \\ 2 & 2 & -1 \\ 7 & 5 & 5 \end{vmatrix} = 3.$

Therefore, given matrix is full rank (its rank is 3).

Example 3.25 Calculate the rank of $A = \begin{bmatrix} 1 & 1 & 2 \\ 5 & 0 & -1 \\ 7 & 2 & 3 \end{bmatrix}$.

Solution Given matrix is a square matrix and its determinant is $\begin{vmatrix} 1 & 1 & 2 \\ 5 & 0 & -1 \\ 7 & 2 & 3 \end{vmatrix} = 0.$

So, the given matrix is not full rank. For instance, $M_{11} = \begin{vmatrix} 0 & -1 \\ 2 & 3 \end{vmatrix} = 0 + 2 = 2 \neq 0.$ Therefore, rank of this matrix is 2. We can stop checking other minors since we found a full-rank sub-matrix.

Example 3.26 Calculate the rank of $A = \begin{bmatrix} 1 & 2 & -5 \\ 3 & 6 & -15 \\ 2 & 4 & -10 \end{bmatrix}$.

Solution Given matrix is a square matrix and its determinant is $\begin{vmatrix} 1 & 2 & -5 \\ 3 & 6 & -15 \\ 2 & 4 & -10 \end{vmatrix} = 0.$

So, the given matrix is not full rank. Let's see if there exists at least one full-rank 2×2 sub-matrix. 2×2 sub-matrices are:

$$\begin{bmatrix} 6 & -15 \\ 4 & -10 \end{bmatrix}, \begin{bmatrix} 3 & -15 \\ 2 & -10 \end{bmatrix}, \begin{bmatrix} 3 & 6 \\ 2 & 4 \end{bmatrix}$$

$$\begin{bmatrix} 2 & -5 \\ 4 & -10 \end{bmatrix}, \begin{bmatrix} 1 & -5 \\ 2 & 10 \end{bmatrix}, \begin{bmatrix} 1 & 2 \\ 2 & 4 \end{bmatrix}$$

$$\begin{bmatrix} 2 & -5 \\ 6 & -15 \end{bmatrix}, \begin{bmatrix} 1 & -5 \\ 3 & -15 \end{bmatrix}, \begin{bmatrix} 1 & 2 \\ 3 & 6 \end{bmatrix}$$

However, none of them are full ranked:

$$\begin{vmatrix} 6 & -15 \\ 4 & -10 \end{vmatrix} = 0, \begin{vmatrix} 3 & -15 \\ 2 & -10 \end{vmatrix} = 0, \begin{vmatrix} 3 & 6 \\ 2 & 4 \end{vmatrix} = 0$$

$$\begin{vmatrix} 2 & -5 \\ 4 & -10 \end{vmatrix} = 0, \begin{vmatrix} 1 & -5 \\ 2 & 10 \end{vmatrix} = 0, \begin{vmatrix} 1 & 2 \\ 2 & 4 \end{vmatrix} = 0$$

$$\begin{vmatrix} 2 & -5 \\ 6 & -15 \end{vmatrix} = 0, \begin{vmatrix} 1 & -5 \\ 3 & -15 \end{vmatrix} = 0, \begin{vmatrix} 1 & 2 \\ 3 & 6 \end{vmatrix} = 0$$

Therefore, rank of this matrix is 1.
Alternatively, let's apply the row reduction method to the given examples.

Example 3.27 Calculate the rank of $A = \begin{bmatrix} 1 & 0 & 5 \\ 2 & 2 & -1 \\ 7 & 5 & 5 \end{bmatrix}$ with row reduction method.

Solution

$$\begin{bmatrix} 1 & 0 & 5 \\ 2 & 2 & -1 \\ 7 & 5 & 5 \end{bmatrix} \rightarrow \begin{bmatrix} 1 & 0 & 5 \\ 0 & 2 & -11 \\ 7 & 5 & 5 \end{bmatrix} \rightarrow \begin{bmatrix} 1 & 0 & 5 \\ 0 & 2 & -11 \\ 0 & 5 & -30 \end{bmatrix} \rightarrow \begin{bmatrix} 1 & 0 & 5 \\ 0 & 2 & -11 \\ 0 & 0 & -2.5 \end{bmatrix}$$
$$\rightarrow \begin{bmatrix} 1 & 0 & 0 \\ 0 & 2 & -11 \\ 0 & 0 & -2.5 \end{bmatrix} \rightarrow \begin{bmatrix} 1 & 0 & 0 \\ 0 & 2 & 0 \\ 0 & 0 & -2.5 \end{bmatrix}$$

Number of non-zero rows is 3. Therefore, the rank is 3.

Example 3.28 Calculate the rank of $A = \begin{bmatrix} 1 & 1 & 2 \\ 5 & 0 & -1 \\ 7 & 2 & 3 \end{bmatrix}$ with row reduction method.

Solution

$$\begin{bmatrix} 1 & 1 & 2 \\ 5 & 0 & -1 \\ 7 & 2 & 3 \end{bmatrix} \rightarrow \begin{bmatrix} 1 & 1 & 2 \\ 0 & -5 & -11 \\ 7 & 2 & 3 \end{bmatrix} \rightarrow \begin{bmatrix} 1 & 1 & 2 \\ 0 & -5 & -11 \\ 0 & -5 & -11 \end{bmatrix}$$
$$\rightarrow \begin{bmatrix} 1 & 1 & 2 \\ 0 & -5 & -11 \\ 0 & 0 & 0 \end{bmatrix} \rightarrow \begin{bmatrix} 1 & 0 & \frac{-1}{5} \\ 0 & -5 & -11 \\ 0 & 0 & 0 \end{bmatrix}$$

Number of non-zero rows is 2. Therefore, the rank is 2.

Example 3.29 Calculate the rank of $A=\begin{bmatrix}1 & 2 & -5\\ 3 & 6 & -15\\ 2 & 4 & -10\end{bmatrix}$ with row reduction method.

Solution

$$\begin{bmatrix}1 & 2 & -5\\ 3 & 6 & -15\\ 2 & 4 & -10\end{bmatrix}\to\begin{bmatrix}1 & 2 & -5\\ 0 & 0 & 0\\ 2 & 4 & -10\end{bmatrix}\to\begin{bmatrix}1 & 2 & -5\\ 0 & 0 & 0\\ 0 & 0 & 0\end{bmatrix}$$

Number of non-zero rows is 1. Therefore, the rank is 1.

Example 3.30 Calculate the rank of $A=\begin{bmatrix}1 & 2 & -2 & 1\\ 7 & 6 & 3 & 2\\ 1 & 1 & 4 & 9\end{bmatrix}$ with row reduction method.

Solution

$$\begin{bmatrix}1 & 2 & -2 & 1\\ 7 & 6 & 3 & 2\\ 1 & 1 & 4 & 9\end{bmatrix}\to\begin{bmatrix}1 & 2 & -2 & 1\\ 0 & -8 & 17 & -5\\ 1 & 1 & 4 & 9\end{bmatrix}$$

$$\to\begin{bmatrix}1 & 2 & -2 & 1\\ 0 & -8 & 17 & -5\\ 0 & -1 & 6 & 8\end{bmatrix}\to\begin{bmatrix}1 & 2 & -2 & 1\\ 0 & -8 & 17 & -5\\ 0 & 0 & 31/8 & 69/8\end{bmatrix}$$

$$\to\begin{bmatrix}1 & 2 & -2 & 1\\ 0 & -8 & 17 & -5\\ 0 & 0 & 1 & 69/31\end{bmatrix}\to\begin{bmatrix}1 & 0 & 9/4 & -1/4\\ 0 & -8 & 17 & -5\\ 0 & 0 & 1 & 69/31\end{bmatrix}$$

$$\to\begin{bmatrix}1 & 0 & 0 & -652/31\\ 0 & -8 & 17 & -5\\ 0 & 0 & 1 & 69/31\end{bmatrix}\to\begin{bmatrix}1 & 0 & 0 & -652/124\\ 0 & -8 & 0 & -1328/31\\ 0 & 0 & 1 & 69/31\end{bmatrix}$$

$$\to\begin{bmatrix}1 & 0 & 0 & -652/124\\ 0 & 1 & 0 & 1328/248\\ 0 & 0 & 1 & 69/31\end{bmatrix}=\begin{bmatrix}1 & 0 & 0 & -5.2581\\ 0 & 1 & 0 & 5.3548\\ 0 & 0 & 1 & 2.2258\end{bmatrix}$$

Number of non-zero rows is 3. Therefore, the rank is 3.

3.14 Problem Solving with MATLAB®

This section offers MATLAB examples relevant to the problems in this chapter. If you are new to MATLAB, begin with Chap. 11.

Example 3.31 The code in Fig. 3.3 calculates $\begin{bmatrix} 1 & 2 & 0 & -1 \\ 7 & 5 & 1 & 0 \\ 2 & 2 & 3 & 4 \end{bmatrix} \times \begin{bmatrix} 1 & 2 \\ 0 & -1 \\ 6 & 3 \\ -9 & 1 \end{bmatrix}$.

Fig. 3.3 Matrix multiplication

```
Command Window
>> A=[1 2 0 -1;7 5 1 0;2 2 3 4];
>> B=[1 2;0 -1;6 3;-9 1];
>> A*B

ans =

    10    -1
    13    12
   -16    15

fx >>
```

Example 3.32 The code in Fig. 3.4 demonstrates how to extract the first and second columns of the matrix $A = \begin{bmatrix} 1 & 2 & 0 & -1 \\ 7 & 5 & 1 & 0 \\ 2 & 2 & 3 & 4 \end{bmatrix}$ and assign them to the variables V_1 and V_2, respectively.

Fig. 3.4 Extracting the first and second columns of the matrix *A*

```
Command Window
>> A=[1 2 0 -1;7 5 1 0;2 2 3 4];
>> V1=A(:,1)

V1 =

     1
     7
     2

>> V2=A(:,2)

V2 =

     2
     5
     2

fx >>
```

Example 3.33 The code in Fig. 3.5 demonstrates how to extract the first and second rows of the matrix $A = \begin{bmatrix} 1 & 2 & 0 & -1 \\ 7 & 5 & 1 & 0 \\ 2 & 2 & 3 & 4 \end{bmatrix}$ and assign them to the variables W_1 and W_2, respectively.

Fig. 3.5 Extracting the first and second rows of the matrix *A*

```
Command Window
>> A=[1 2 0 -1;7 5 1 0;2 2 3 4];
>> W1=A(1,:)

W1 =

     1     2     0    -1

>> W2=A(2,:)

W2 =

     7     5     1     0

fx >>
```

Example 3.34 Figure 3.6 shows MATLAB code for generating a 3 × 2 random matrix.

Fig. 3.6 Generating a random 3 × 2 matrix

```
Command Window
>> rand(3,2)

ans =

    0.8147    0.9134
    0.9058    0.6324
    0.1270    0.0975

fx >>
```

Example 3.35 Figure 3.7 shows how to create a 2 × 3 matrix of ones in MATLAB.

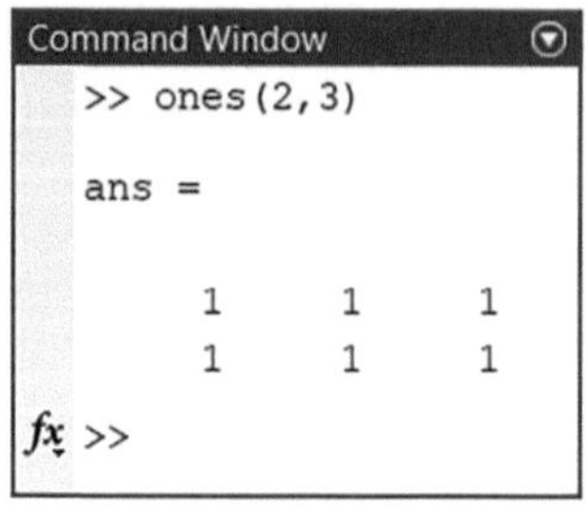

Fig. 3.7 Generating a 2 × 3 ones matrix

Example 3.36 Figure 3.8 shows how to create a 3 × 2 matrix of zeros in MATLAB.

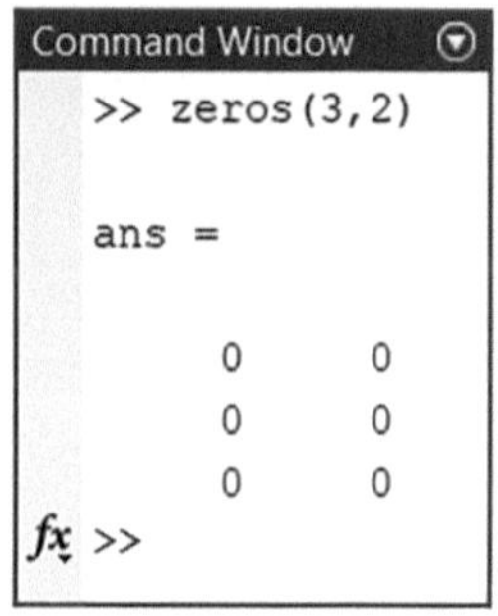

Fig. 3.8 Generating a 3 × 2 zeros matrix

Example 3.37 Figure 3.9 shows how to create a 3 × 3 identity matrix in MATLAB.

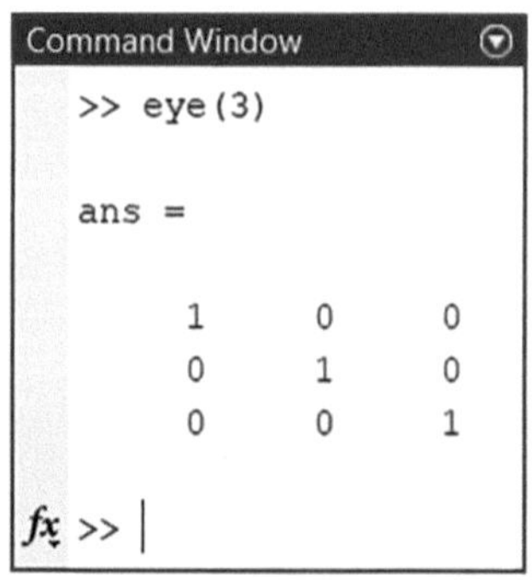

Fig. 3.9 Generating a 3 × 3 identity matrix

Example 3.38 MATLAB provides two types of transpose operation: Conjugate transpose and non-conjugate transpose. Table 3.1 compares these two operators.

Table 3.1 Transpose operators in MATLAB

Operator	Name	Behavior
'	Conjugate transpose	Transposes and conjugates complex elements ($\overline{A^T}$)
.'	Non-conjugate transpose	Transposes without conjugating complex elements (A^T)

The code in Fig. 3.10 calculates the conjugate and non-conjugate transpose of the matrix $A = \begin{bmatrix} 1 & 1+3i \\ 9-2i & -1+2i \end{bmatrix}$.

Fig. 3.10 Calculating the conjugate and non-conjugate transpose of matrix *A*

```
Command Window
>> A=[1 1+3i;9-2i -1+2i];
>> A'

ans =

   1.0000 + 0.0000i   9.0000 + 2.0000i
   1.0000 - 3.0000i  -1.0000 - 2.0000i

>> A.'

ans =

   1.0000 + 0.0000i   9.0000 - 2.0000i
   1.0000 + 3.0000i  -1.0000 + 2.0000i

>> transpose(A)

ans =

   1.0000 + 0.0000i   9.0000 - 2.0000i
   1.0000 + 3.0000i  -1.0000 + 2.0000i

fx >>
```

Example 3.39 The code in Fig. 3.11 calculates the determinant of $A = \begin{bmatrix} -4 & -6 & 2 \\ 5 & -1 & 3 \\ -2 & 4 & -3 \end{bmatrix}$.

Fig. 3.11 Calculating the determinant of the matrix *A*

```
Command Window
>> A=[-4 -6 2;5 -1 3;-2 4 -3];
>> det(A)

ans =

   18.0000

fx >>
```

Example 3.40 The code in Figs. 3.12 and 3.13 calculates the inverse of $A=\begin{bmatrix} 8 & -2 & 1 \\ 2 & -1 & 3 \\ 1 & 1 & 4 \end{bmatrix}$.

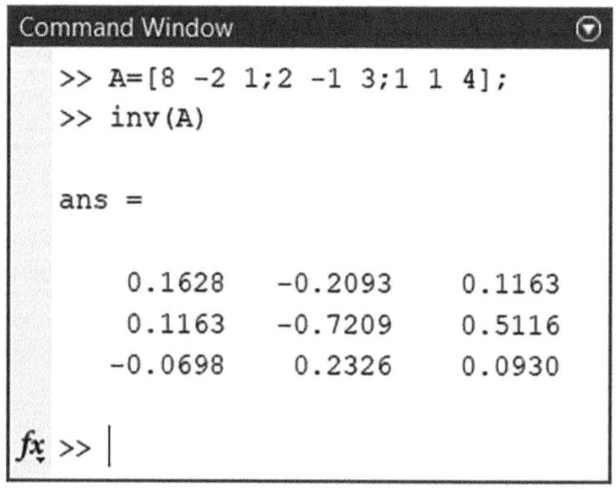

Fig. 3.12 Calculating the inverse of matrix A

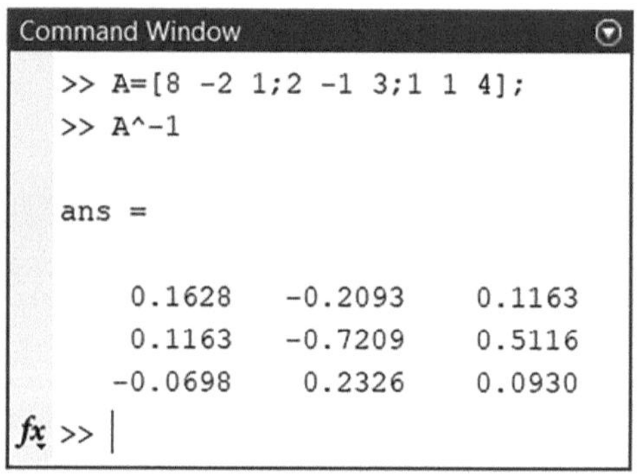

Fig. 3.13 Calculating the inverse of matrix A

Example 3.41 $\begin{cases} x+2y+3z=10 \\ 4x+6y+5z=12 \\ 7x+9y+8z=13 \end{cases}$ can be written as: $\begin{bmatrix} 1 & 2 & 3 \\ 4 & 6 & 5 \\ 7 & 9 & 8 \end{bmatrix}\begin{bmatrix} x \\ y \\ z \end{bmatrix}=\begin{bmatrix} 10 \\ 12 \\ 13 \end{bmatrix}$.
The determinant of the coefficient matrix is non-zero, so a unique solution exists. The code presented in Figs. 3.14, 3.15, 3.16, and 3.17 demonstrates various approaches for finding the solution.

Fig. 3.14 Solving the system of equations in Example 3.41

```
Command Window
>> A=[1 2 3;4 6 5;7 9 8];
>> b=[10;12;13];
>> A\b

ans =

   -6.4444
    3.8889
    2.8889

fx >>
```

Fig. 3.15 Solving the system of equations in Example 3.41

```
Command Window
>> A=[1 2 3;4 6 5;7 9 8];
>> b=[10;12;13];
>> linsolve(A,b)

ans =

   -6.4444
    3.8889
    2.8889

fx >>
```

Fig. 3.16 Solving the system of equations in Example 3.41

```
Command Window
>> A=[1 2 3;4 6 5;7 9 8];
>> b=[10;12;13];
>> sol=inv(A)*b

sol =

   -6.4444
    3.8889
    2.8889

fx >>
```

```
Command Window
>> A=[1 2 3;4 6 5;7 9 8];
>> b=[10;12;13];
>> (A^-1)*b

ans =

   -6.4444
    3.8889
    2.8889

fx >>
```

Fig. 3.17 Solving the system of equations in Example 3.41

Example 3.42 The code in Fig. 3.18 calculates the rank of $A = \begin{bmatrix} 1 & 2 & -2 & 1 \\ 7 & 6 & 3 & 2 \\ 1 & 1 & 4 & 9 \end{bmatrix}$.

```
Command Window
>> A=[1 2 -2 1;7 6 3 2;1 1 4 9];
>> rank(A)

ans =

     3

fx >>
```

Fig. 3.18 Calculation of rank for matrix A

Example 3.43 The code in Fig. 3.19 calculates the reduced row-echelon form of $A = \begin{bmatrix} -5 & 1 & 1 & 0 \\ 2 & -4 & 2 & 0 \\ 3 & 3 & -3 & 0 \end{bmatrix}$.

```
Command Window
>> A=[-5 1 1 0;2 -4 2 0;3 3 -3 0];
>> rref(A)

ans =

    1.0000         0   -0.3333         0
         0    1.0000   -0.6667         0
         0         0         0         0

fx >>
```

Fig. 3.19 Calculating the reduced row-echelon form of matrix A

Exercise 3.1 Calculate the height of the table illustrated in Fig. 3.20. (Ans. 150 cm)

Fig. 3.20 Figure for Exercise 3.1

References for Further Study

1. Engineering Mathematics with MATLAB® and Simulink®, Farzin Asadi, Springer, 2025.
2. Applied Numerical Analysis with MATLAB®/Simulink®: For Engineers and Scientists, Farzin Asadi, Springer, 2023.
3. Advanced Engineering Mathematics, Erwin Kreyszig, Wiley, 2020.
4. A Problem-Solving Approach to Single-Variable Calculus (Volume II), Farzin Asadi, Springer, 2026.
5. A Problem-Solving Approach to Multi-Variable Calculus (Volume I), Farzin Asadi, Springer, 2026.
6. A Problem-Solving Approach to Multi-Variable Calculus (Volume II), Farzin Asadi, Springer, 2026.
7. A Problem-Solving Approach to Multi-Variable Calculus (Volume III), Farzin Asadi, Springer, 2026.
8. A Problem-Solving Approach to Ordinary Differential Equations (Volume I), Farzin Asadi, Springer, 2026.
9. A Problem-Solving Approach to Ordinary Differential Equations (Volume II), Farzin Asadi, Springer, 2026.
10. A Problem-Solving Approach to Ordinary Differential Equations (Volume III), Farzin Asadi, Springer, 2026.
11. A Problem-Solving Approach to Ordinary Differential Equations (Volume IV), Farzin Asadi, Springer, 2026.

Chapter 4
Techniques for Solving Equations

4.1 Introduction

This chapter is dedicated to reviewing the critical equation-solving methods.

4.2 First-Order Algebraic Equations

Linear equations can be solved as: $ax + b = 0 \Rightarrow x = -\frac{b}{a}$. For instance, $3x + 5 = 0 \Rightarrow x = \frac{-5}{3}$ or $3x + 5 = 6 \Rightarrow x = \frac{6-5}{3} = \frac{1}{3}$.

4.3 Second-Order Algebraic Equations

$ax^2 + bx + c = 0$ can be solved as: $ax^2 + bx + c = 0 \Rightarrow x_{1,2} = \frac{-b \pm \sqrt{\Delta}}{2a}$, where $\Delta = b^2 - 4 \times a \times c$. Let's study some numeric examples.

Example 4.1 Solve $x^2 - 3x + 2 = 0$, $x^2 + 6x + 9 = 0$, and $x^2 + 2x + 10 = 0$.

Solution

$$x^2 - 3x + 2 = 0 \Rightarrow \Delta = (-3)^2 - 4(1)(2) = 1 \Rightarrow x_{1,2} = \frac{3 \pm \sqrt{1}}{2} \Rightarrow x_1 = 1, x_2 = 2$$

$$x^2 + 6x + 9 = 0 \Rightarrow \Delta = (6)^2 - 4(1)(9) = 0 \Rightarrow x_{1,2} = \frac{-6 \pm \sqrt{0}}{2} \Rightarrow x_{1,2} = -3, x_1 = x_2 = -3$$

$$x^2 + 2x + 10 = 0 \Rightarrow \Delta = (2)^2 - 4(1)(10) = -36 \Rightarrow x_{1,2}$$

$$= \frac{-2 \pm \sqrt{-36}}{2} = \frac{-2 \pm 6i}{2} = -1 \pm 3i \Rightarrow x_1 = -1 + 3i, x_2 = -1 - 3i$$

F. Asadi, *A Problem-Solving Approach to Single-Variable Calculus*,
https://doi.org/10.1007/978-3-032-13505-6_4

4.3.1 Special Case I

When b is an even number, i.e., $b = 2b'$, you can use the following technique:

$$ax^2 + bx + c = 0 \Rightarrow ax^2 + 2b'x + c = 0 \Rightarrow x_{1,2} = \frac{-b' \pm \sqrt{\Delta'}}{a}$$

where $\Delta' = b'^2 - a \times c$.

Working with smaller numbers is the advantage of this technique. Use of this technique decreases the chance of making mistake and is recommended for exams. Let's study a numeric example:

$$x^2 - 10x + 21 = 0 \Rightarrow \frac{5 \pm \sqrt{(-5)^2 - (1)(21)}}{1} = \frac{5 \pm \sqrt{25 - 21}}{1} = \frac{5 \pm 2}{1} \Rightarrow x_1 = 3, x_2 = 7$$

Compare the above solution with the normal solution shown below:

$$x^2 - 10x + 21 = 0 \Rightarrow \frac{10 \pm \sqrt{(-10)^2 - 4(1)(21)}}{2} = \frac{10 \pm \sqrt{100 - 84}}{2} = \frac{10 \pm 4}{2} \Rightarrow x_1 = 3, x_2 = 7$$

4.3.2 Special Case II

If $b = a + c$, then roots of equation $ax^2 + bx + c = 0$ are $x_1 = -1$ and $x_2 = -\frac{c}{a}$. For instance, $6x^2 + 11x + 5 = 0 \Rightarrow x_1 = -1, x_2 = -\frac{5}{6}$.

4.3.3 Special Case III

If $a + b + c = 0$, then roots of equation $ax^2 + bx + c = 0$ are $x_1 = 1$ and $x_2 = \frac{c}{a}$. For instance, $3x^2 - 7x + 4 = 0 \Rightarrow x_1 = 1, x_2 = \frac{4}{3}$.

In general, if the sum of the coefficients of a polynomial is zero, then one of its roots is unity. For instance, consider $x^3 - 9x^2 + 23x - 15 = 0$. Sum of coefficients is zero: $1 - 9 + 23 - 15 = 0$; therefore, one of the roots is 1. If we divide the given polynomial equation by $x - 1$, we obtain:

$$\frac{x^3 - 9x^2 + 23x - 15}{x - 1} = \left(x^2 - 8x + 15\right)$$

The other two roots are 3 and 5 since $\left(x^2 - 8x + 15\right) = 0 \Rightarrow x_{1,2} = \frac{4 \pm \sqrt{4^2 - 1 \times 15}}{1} \Rightarrow x_1 = 3, \Rightarrow x_2 = 5$.

4.4 Constructing a Polynomial with Specified Zeros

A polynomial with r_1, r_2, r_3, …, r_n roots can be formed using the $p(x) = \prod_{i=1}^{n} \times (x - r_i) = (x - r_1)(x - r_2)(x - r_3)\ldots(x - r_n)$ formula. For instance, a polynomial with roots of 1, 3, and −5 is: $p(x) = (x - 1)(x - 3)(x - (-5)) = (x - 1)(x - 3)(x + 5) = (x - 1)(x^2 + 2x - 15) = x^3 + x^2 - 17x + 15$.

When number of given roots is two, you can use the $p(x) = x^2 - Sx + P$ formula. S and P shows the sum and product of given roots. For instance, for $r_1 = 3$ and $r_2 = 5$, $S = 3 + 5 = 8$ and $P = 3 \times 5 = 15$. Therefore, $p(x) = x^2 - 8x + 15$. For $r_1 = -2 + 3j$ and $r_2 = -2 - 3j$, $S = -2 + 3j - 2 - 3j = -4$ and $P = (-2 + 3j) \times (-2 - 3j) = (-2)^2 + 3^2 = 13$. Therefore, $p(x) = x^2 + 4x + 13$.

4.5 Exponential Equations

This section shows how to solve equations where the variable appears in the exponent.

Example 4.2 Solve $2^{x+1} = 3^x$.

Solution Isolate the x terms using logarithms:

$$2^{x+1} = 3^x \Rightarrow \ln\left(2^{x+1}\right) = \ln(3^x) \Rightarrow (x+1)\ln(2) = x\ln(3) \Rightarrow \frac{x+1}{x} = \frac{\ln(3)}{\ln(2)} \Rightarrow \frac{x+1}{x} = \frac{1.0986}{0.6931} \Rightarrow 1 + \frac{1}{x} = 1.5850 \Rightarrow \frac{1}{x} = 0.5850 \Rightarrow x = \frac{1}{0.5850} \Rightarrow x = 1.7094$$

Example 4.3 Solve $e^{2x} = 7$.

Solution

$$e^{2x} = 7 \Rightarrow \ln\left(e^{2x}\right) = \ln(7) \Rightarrow 2x = 1.9459 \Rightarrow x = 0.9730$$

Example 4.4 Solve $4^x = 100$.

Solution

$$4^x = 100 \Rightarrow \ln(4^x) = \ln(100) \Rightarrow x\ln(4) = \ln(100) \Rightarrow x = \frac{\ln(100)}{\ln(4)} \Rightarrow x = \frac{4.6052}{1.3863} \Rightarrow x = 3.3219$$

Example 4.5 Solve $\sqrt{(x+2)^2}=10$.

Solution

$$\sqrt{(x+2)^2}=10 \Rightarrow |x+2|=10 \Rightarrow x+2=\pm 10 \Rightarrow \begin{cases} x+2=10 \Rightarrow x=8 \\ x+2=-10 \Rightarrow x=-12 \end{cases}$$

Example 4.6 Solve $(|x|-4)^2=9$.

Solution

$$(|x|-4)^2=9 \Rightarrow (|x|-4)^2=3^2 \Rightarrow \sqrt{(|x|-4)^2}=\sqrt{3^2} \Rightarrow ||x|-4|$$
$$=|3| \Rightarrow ||x|-4|=3 \Rightarrow |x|-4=$$
$$\pm 3 \Rightarrow \begin{cases} |x|-4=3 \\ |x|-4=-3 \end{cases} \Rightarrow \begin{cases} |x|=7 \\ |x|=1 \end{cases} \Rightarrow \begin{cases} x=-7,7 \\ x=-1,1 \end{cases}$$

Therefore, the equation $(|x|-4)^2=9$ has four distinct solutions: $x=\{-7,7,-1,1\}$.

Example 4.7 Solve $8^{2x-1}=16^{x+3}$.

Solution

$$8^{2x-1}=16^{x+3} \Rightarrow (2^3)^{2x-1}=(2^4)^{x+3} \Rightarrow 2^{6x-3}=2^{4x+12} \Rightarrow 6x-3=4x+12 \Rightarrow 2x$$
$$=15 \Rightarrow x=\frac{15}{2} \Rightarrow x=7.5$$

Exercise 4.1 Solve $9^x-3^x-6=0$. (**Hint:** Let $u=3^x$)

4.6 Logarithmic Equations

This section demonstrates how to solve equations in which the variable appears within a logarithm.

Example 4.8 Solve $\log_5(x^2-4)=1$.

Solution For the logarithm to be defined, the argument must be greater than zero:

$$x^2-4>0 \Rightarrow x^2>4 \Rightarrow x \in (-\infty, -2) \cup (2, +\infty)$$

We must check our final answers against this condition.

$$\log_5(x^2-4)=1 \Rightarrow x^2-4=5^1 \Rightarrow x^2=9 \Rightarrow x^2=3^2 \Rightarrow \sqrt{x^2}=\sqrt{3^2} \Rightarrow |x|=3 \Rightarrow x=\pm 3$$

Both solutions satisfy the domain condition $x^2-4>0$:

$$-3 \in (-\infty, -2) \cup (2, +\infty)$$
$$+3 \in (-\infty, -2) \cup (2, +\infty)$$

Example 4.9 Solve $\log_2(3x - 1) = \log_2(x + 5)$.

Solution For the logarithm to be defined, the arguments must be greater than zero:

$$3x - 1 > 0 \Rightarrow x > \frac{1}{3}$$
$$x + 5 > 0 \Rightarrow x > -5$$

For both conditions to be satisfied, the solution must be $x > \frac{1}{3}$. Since the bases of the logarithms are the same (base 2), we can use the one-to-one property which states that if $\log_b A = \log_b B$, then $A = B$.

$$\log_2(3x - 1) = \log_2(x + 5) \Rightarrow 3x - 1 = x + 5 \Rightarrow 2x = 5 + 1 \Rightarrow x = 3$$

We must check if the solution $x = 3$ satisfies the domain condition, $> \frac{1}{3}$. Since $3 > \frac{1}{3}$, the solution is valid.

Exercise 4.2 Solve $\ln(x^2) = \ln(3x - 2)$.

Example 4.10 Solve $\log_4 x + \log_4(x - 3) = 1$.

Solution For the logarithms to be defined, the arguments must be strictly positive:

$$x > 0$$
$$x - 3 > 0 \Rightarrow x > 3$$

For both conditions to be satisfied, the solution must be $x > 3$.

$$\log_4 x + \log_4(x - 3) = 1 \Rightarrow \log_4(x(x - 3)) = 1 \Rightarrow x(x - 3) = 4^1 \Rightarrow x^2 - 3x$$
$$= 4 \Rightarrow x^2 - 3x - 4 = 0 \Rightarrow \begin{cases} x_1 = -1 \\ x_2 = 4 \end{cases}$$

We must check the potential solutions against the domain condition, $x > 3$.

For $x = 4$: Since $4 > 3$, $x = 4$ is a valid solution.

For $x = -1$: Since -1 is not greater than 3, $x = -1$ is an extraneous solution and must be rejected.

Thus, the only valid solution to the equation $\log_4 x + \log_4(x - 3) = 1$ is $x = 4$.

Example 4.11 Solve $\log_4 x - \log_4(x - 3) = 1$.

Solution For the logarithms to be defined, the arguments must be strictly positive:

$$x > 0$$

$$x - 3 > 0 \Rightarrow x > 3$$

For both conditions to be satisfied, the solution must be $x > 3$.

$$\log_4 x - \log_4(x-3) = 1 \Rightarrow \log_4\left(\frac{x}{x-3}\right) = 1 \Rightarrow \frac{x}{x-3} = 4^1 \Rightarrow x = 4x - 12 \Rightarrow x = 4$$

We must check the potential solutions against the domain condition, $x > 3$. Since $4 > 3$, $x = 4$ is a valid solution.

Example 4.12 Solve $\log_4(x) + \log_2(x) = 6$.

Solution For the logarithms to be defined, both arguments must be strictly positive: $x > 0$.

We use the property $\log_{a^n}(b) = \frac{1}{n}\log_a(b)$.

$$\log_4(x) + \log_2(x) = 6 \Rightarrow \log_{2^2}(x) + \log_2(x) = 6 \Rightarrow \frac{1}{2}\log_2(x) + \log_2(x)$$
$$= 6 \Rightarrow \frac{3}{2}\log_2(x) = 6 \Rightarrow \log_2(x) = 4 \Rightarrow x = 2^4 \Rightarrow x = 16$$

We must check the potential solutions against the domain condition, $x > 0$. Since $16 > 0$, $x = 16$ is a valid solution.

Exercise 4.3 Solve $\log_9(x) + \log_3(x^2) = 10$.

Exercise 4.4 Solve $\log_{\sqrt{5}}(x) + \log_5(25x) = 1$.

Example 4.13 Solve $\log_2(4x) = \log_8(x^2)$.

Solution For the logarithms to be defined, both arguments must be strictly positive:

$$4x > 0 \Rightarrow x > 0$$

$$x^2 > 0 \Rightarrow x \in \mathbb{R} - \{0\}$$

For both conditions to be satisfied, the solution must be $x > 0$.

$$\log_2(4x) = \log_8(x^2) \Rightarrow \log_2(4x) = \log_{2^3}(x^2) \Rightarrow \log_2(4x) = \frac{1}{3}\log_2(x^2)$$
$$\Rightarrow 3\log_2(4x) = \log_2(x^2) \Rightarrow \log_2\left((4x)^3\right) = \log_2(x^2) \Rightarrow (4x)^3 = x^2 \Rightarrow 64x^3$$
$$= x^2 \Rightarrow 64x^3 - x^2 = 0 \Rightarrow x^2(64x - 1) = 0 \Rightarrow \begin{cases} x = 0 \\ x = \dfrac{1}{64} \end{cases}$$

We must check the potential solutions against the domain condition, $x > 0$. Thus, the only valid solution to the equation $\log_2(4x) = \log_8(x^2)$ is $x = \frac{1}{64}$.

Example 4.14 Solve $\log_2(4x) = \log_8(x^2)$.

Solution This problem was addressed in the previous example. We will now employ an alternative method. For the logarithms to be defined, both arguments must be strictly positive:

$$4x > 0 \Rightarrow x > 0$$

$$x^2 > 0 \Rightarrow x \in \mathbb{R} - \{0\}$$

For both conditions to be satisfied, the solution must be $x > 0$.

Let's use change of base formula $\log_a b = \frac{\log_c b}{\log_c a}$ to convert both logarithms to a common base (like base e for the natural logarithm ln, or base 10).

Using the natural logarithm (ln):

$$\begin{aligned}&\log_2(4x) = \log_8(x^2) \Rightarrow \frac{\ln(4x)}{\ln(2)} = \frac{\ln(x^2)}{\ln(8)} \Rightarrow \ln(8)\ln(4x)\\&= \ln(2)\ln(x^2) \Rightarrow \ln(2^3)\ln(4x) = \ln(2)\ln(x^2) \Rightarrow 3\ln(2)\ln(4x)\\&= \ln(2)\ln(x^2) \Rightarrow 3\ln(4x) = \ln(x^2) \Rightarrow \ln\left((4x)^3\right) = \ln(x^2) \Rightarrow (4x)^3\\&= x^2 \Rightarrow 64x^3 = x^2 \Rightarrow 64x^3 - x^2 = 0 \Rightarrow x^2(64x - 1) = 0 \Rightarrow \Rightarrow \begin{cases} x = 0 \\ x = \dfrac{1}{64} \end{cases}\end{aligned}$$

We must check the potential solutions against the domain condition, $x > 0$. Thus, the only valid solution to the equation $\log_2(4x) = \log_8(x^2)$ is $x = 1/64$.

4.7 Radical Equations

This section demonstrates how to solve equations in which the variable appears within a radical (root). Solving a radical equation typically involves isolating the radical and then squaring both sides to eliminate it.

Example 4.15 Solve $\sqrt{2x+1} = x - 1$.

Solution In this case, the radical ($\sqrt{2x+1}$) is already isolated on the left side.

$$\begin{aligned}&\sqrt{2x+1} = x - 1 \Rightarrow 2x + 1 = (x-1)^2 \Rightarrow 2x + 1 = x^2 - 2x + 1 \Rightarrow x^2 - 4x\\&= 0 \Rightarrow x(x-4) = 0 \Rightarrow \begin{cases} x = 0 \\ x = 4 \end{cases}\end{aligned}$$

Crucially, we must check all potential solutions in the original radical equation, as squaring both sides can introduce extraneous solutions.

Check $x = 0$: $\sqrt{2\times 0+1}=0-1\Rightarrow 1=-1$ (!)

Since $1 \neq -1$ is an extraneous solution and must be rejected.

Check $x = 4$: $\sqrt{2\times 4+1}=4-1\Rightarrow 3=3$

Since $3 = 3$, $x = 4$ is a valid solution. The only solution to the radical equation is $x = 4$.

Example 4.16 Solve $\sqrt{x+7}-\sqrt{x}=1$.

Solution

$$\sqrt{x+7}-\sqrt{x}=1\Rightarrow\sqrt{x+7}=1+\sqrt{x}\Rightarrow\left(\sqrt{x+7}\right)^2=\left(1+\sqrt{x}\right)^2\Rightarrow x+7=x+2\sqrt{x}+1\Rightarrow 2\sqrt{x}=6\Rightarrow\sqrt{x}=3\Rightarrow x=9$$

Check for extraneous solutions:

$$\sqrt{x+7}-\sqrt{x}=1\overset{x=9}{\Rightarrow}\sqrt{9+7}-\sqrt{9}=1\Rightarrow\sqrt{16}-\sqrt{9}=1\Rightarrow 4-3=1\Rightarrow 1=1$$

Since the check holds true, the solution is $x = 9$.

Example 4.17 Solve $\sqrt{3^x}=9^{x-3}$.

Solution

$$\sqrt{3^x}=9^{x-3}\Rightarrow 3^{\frac{x}{2}}=\left(3^2\right)^{x-3}\Rightarrow 3^{\frac{x}{2}}=3^{2x-6}\Rightarrow\frac{x}{2}=2x-6\Rightarrow x=4x-12\Rightarrow 3x=12\Rightarrow x=4$$

Check for extraneous solutions:

$$\sqrt{3^x}=9^{x-3}\overset{x=4}{\Rightarrow}\sqrt{3^4}=9^{4-3}\Rightarrow\sqrt{81}=9^1\Rightarrow 9=9$$

Since the check holds true, the solution is $x = 4$.

Exercise 4.5 Solve $4+\sqrt{x-3}=x$.

Exercise 4.6 Solve $\sqrt{5x-4}=x-2$.

4.8 Absolute Value Equations

This section shows how equations containing absolute values can be solved.

Example 4.18 Solve $|x + 4| = 10$.

Solution

$$|x+4|=10\Rightarrow x+4=\pm 10\Rightarrow\begin{cases}x+4=10\Rightarrow x=6\\x+4=-10\Rightarrow x=-14\end{cases}$$

Example 4.19 Solve $|x + 4| = |x - 1|$.

Solution

$$|x+4|=|x-1|\Rightarrow x+4=\pm(x-1)\Rightarrow\begin{cases} x+4=x-1\Rightarrow 4=-1\,(!) \\ x+4=-(x-1)\Rightarrow 2x=-3\Rightarrow x=-1.5 \end{cases}$$

Example 4.20 Solve $|x^2 + 7x - 3| = |x^2 + 3x + 1|$.

Solution

$$|x^2+7x-3|=|x^2+3x+1|\Rightarrow x^2+7x-3=$$

$$\pm(x^2+3x+1)\Rightarrow\begin{cases} x^2+7x-3=+(x^2+3x+1)\Rightarrow x=1 \\ x^2+7x-3=-(x^2+3x+1)\Rightarrow x^2+5x-1=0\Rightarrow x=\dfrac{-5\pm\sqrt{29}}{2} \end{cases}$$

Example 4.21 Solve $|3x + 1| = x + 5$.

Solution

$$|3x+1|=x+5\Rightarrow\begin{cases} x\geq-\dfrac{1}{3}\Rightarrow 3x+1=x+5\Rightarrow 2x=4\Rightarrow x=2 \\ x<-\dfrac{1}{3}\Rightarrow-(3x+1)=x+5\Rightarrow-3x-1=x+5\Rightarrow 4x=-6\Rightarrow x=-1.5 \end{cases}$$

The solutions to the given equation are $x = 2$ and $= - 1.5$.

Exercise 4.7 Solve $4 - 3|5x - 2| = - 14$.

Exercise 4.8 Solve $|x^2 - 5x| = 6$.

4.9 Trigonometric Equations

A trigonometric equation is an equation involving one or more trigonometric ratios of unknown angles. This section shows how this type of equations can be solved.

Trigonometric functions are periodic. A function $f(x)$ is said to be periodic with period T if, for every x in the domain of f, $f(x + T) = f(x)$. $\sin(\omega_0 t + \varphi_0)$ and $\cos(\omega_0 t + \varphi_0)$ are periodic functions with period of $T = \frac{2\pi}{\omega_0}$. $\tan(\omega_0 t + \varphi_0)$ and $\cot(\omega_0 t + \varphi_0)$ are periodic functions with period of $T = \frac{\pi}{\omega_0}$. For instance, period of $\sin(3x)$ is $T = \frac{2\pi}{3}$ and period of $\tan\left(2x + \frac{\pi}{6}\right)$ is $T = \frac{\pi}{2}$.

Inverse trigonometric functions are the inverse functions of the basic trigonometric functions. They are used to find the angle whose trigonometric function is a given value. The ranges of the important inverse trigonometric functions are as follows:

1.

$$-\frac{\pi}{2} \leq \sin^{-1}(x) \leq \frac{\pi}{2}$$

2.

$$0 \leq \cos^{-1}(x) \leq \pi$$

3.

$$-\frac{\pi}{2} < \tan^{-1}(x) < \frac{\pi}{2}$$

4.

$$0 < \cot^{-1}(x) < \pi$$

The formulas presented below enable the rapid solution of trigonometric equation:

1. The general solution for $\sin(\theta) = C$, where $-1 \leq C \leq 1$, is given by $\theta = \begin{cases} \sin^{-1}(C) + 2k\pi \\ \pi - \sin^{-1}(C) + 2k\pi \end{cases}$, where k is any integer ($k \in \mathbb{Z}$ or $k = 0, \pm 1, \pm 2, \pm 3, \ldots$).

   The general solution for $\sin(\theta) = C$ can be unified by using the formula: $\theta = (-1)^k \sin^{-1}(C) + \pi k$ where k is any integer ($k \in \mathbb{Z}$ or $k = 0, \pm 1, \pm 2, \pm 3, \ldots$).
2. The general solution for $\cos(\theta) = C$, where $-1 \leq C \leq 1$, is given by $\theta = \pm \cos^{-1}(C) + 2k\pi$, where k is any integer ($k \in \mathbb{Z}$ or $k = 0, \pm 1, \pm 2, \pm 3, \ldots$).
3. The general solution for $\tan(\theta) = C$ is given by $\theta = \tan^{-1}(C) + \pi k$, where k is any integer ($k \in \mathbb{Z}$ or $k = 0, \pm 1, \pm 2, \pm 3, \ldots$).
4. The general solution for $\cot(\theta) = C$ is given by $\theta = \cot^{-1}(C) + \pi k$, where k is any integer ($k \in \mathbb{Z}$ or $k = 0, \pm 1, \pm 2, \pm 3, \ldots$).

Example 4.22 Solve $\sin(x) = 0.7$.

Solution Using a calculator $\sin^{-1}(0.7) = 0.7754$ *rad* (Fig. 4.1).

Fig. 4.1 Sine of 0.7754 *rad* is 0.7

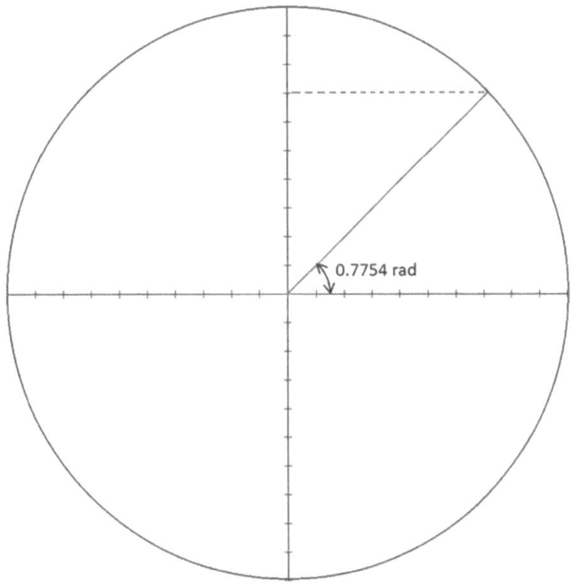

However, there exists another angle whose sine is 0.7 (Fig. 4.2). That angle is $\pi - 0.7754 = 2.3662$ *rad* or $-(\pi + 0.7754) = -3.917$ *rad* (Fig. 4.3).

Fig. 4.2 Sine of 2.3662 *rad* or −3.917 *rad* is 0.7

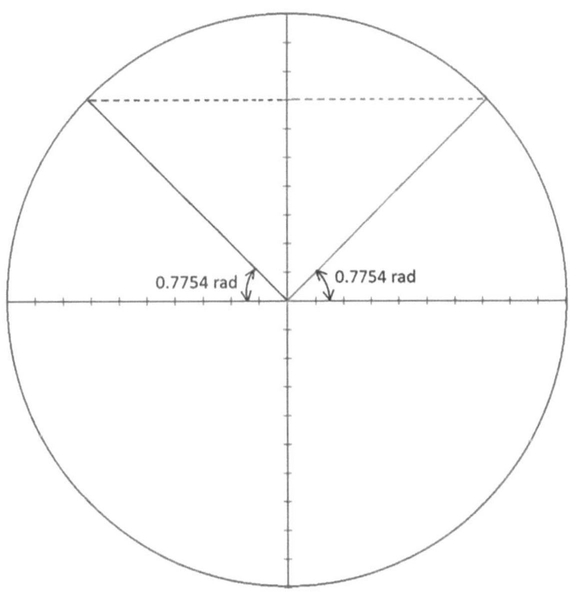

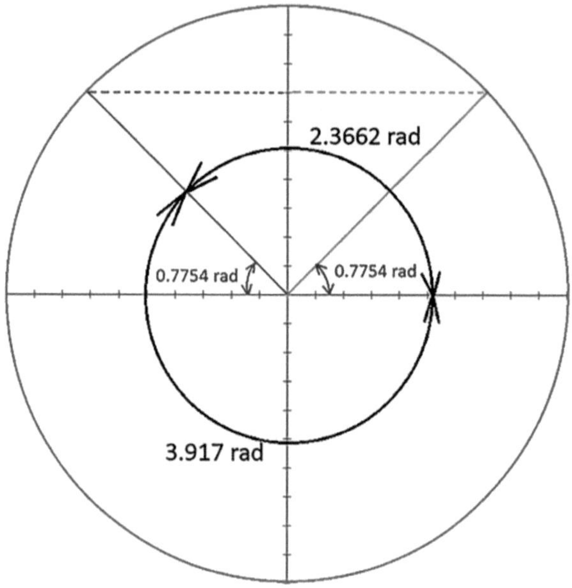

Fig. 4.3 Different ways to measure an angle

$$\sin(x) = 0.7 \Rightarrow \begin{cases} x = 0.7754 + 2k\pi & k = 0, \pm 1, \pm 2, \pm 3, \ldots \\ x = \pi - 0.7754 + 2k\pi = 2.3662 + 2k\pi & k = 0, \pm 1, \pm 2, \pm 3, \ldots \end{cases}$$

or

$$\sin(x) = 0.7 \Rightarrow \begin{cases} x = 0.7754 + 2k\pi & k = 0, \pm 1, \pm 2, \pm 3, \ldots \\ x = -(\pi + 0.7754) + 2k\pi = -3.917 + 2k\pi & k = 0, \pm 1, \pm 2, \pm 3, \ldots \end{cases}$$

Example 4.23 Solve $\sin(x) = -0.7$.

Solution Using a calculator $\sin^{-1}(-0.7) = -0.7754$ *rad* (Fig. 4.4).

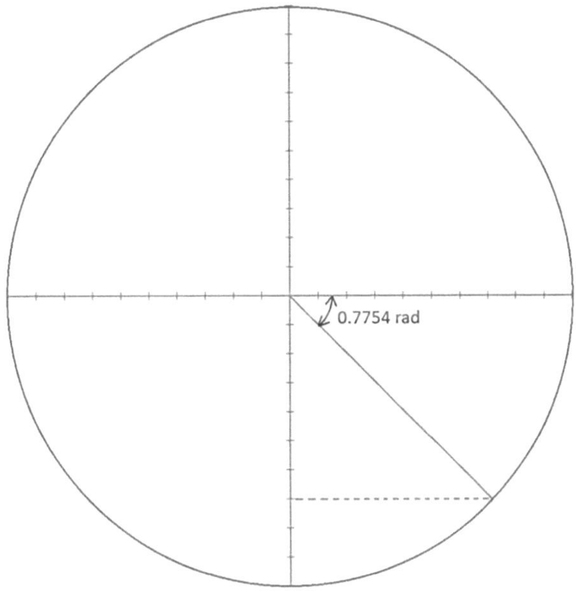

Fig. 4.4 Sine of -0.7754 *rad* is -0.7

However, there exists another angle whose sine is -0.7 (Fig. 4.5). That angle is $-(\pi - 0.7754) = -2.3662$ *rad* or $(\pi + 0.7754) = +3.917$ *rad* (Fig. 4.6).

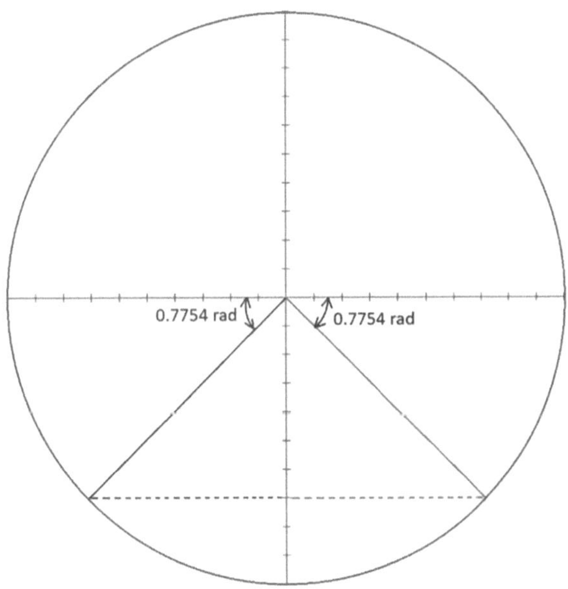

Fig. 4.5 Sine of -2.3662 *rad* or 3.917 *rad* is -0.7

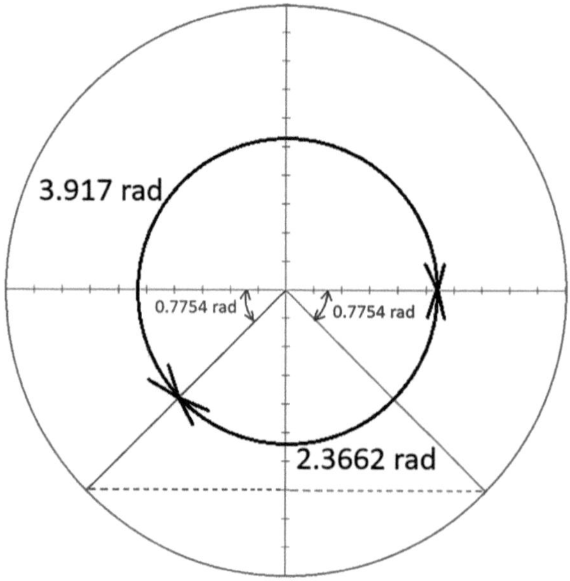

Fig. 4.6 Different ways to measure an angle

$$\sin(x) = -0.7 \Rightarrow \begin{cases} x = -0.7754 + 2k\pi & k = 0, \pm 1, \pm 2, \pm 3, \ldots \\ x = \pi - (-0.7754) + 2k\pi = 3.9170 + 2k\pi & k = 0, \pm 1, \pm 2, \pm 3, \ldots \end{cases}$$

or

$$\sin(x) = -0.7 \Rightarrow \begin{cases} x = -0.7754 + 2k\pi & k = 0, \pm 1, \pm 2, \pm 3, \ldots \\ x = -(\pi - 0.7754) + 2k\pi = -2.3662 + 2k\pi & k = 0, \pm 1, \pm 2, \pm 3, \ldots \end{cases}$$

Example 4.24 Solve $\sin(3x) = 0.7$.

Solution Using a calculator $\sin^{-1}(0.7) = 0.7754\ rad$.

$$\sin(3x) = 0.7 \Rightarrow \begin{cases} 3x = 0.7754 + 2k\pi & k = 0, \pm 1, \pm 2, \pm 3, \ldots \\ 3x = \pi - 0.7754 + 2k\pi = 2.3662 + 2k\pi & k = 0, \pm 1, \pm 2, \pm 3, \ldots \end{cases}$$
$$\Rightarrow \begin{cases} x = \dfrac{0.7754 + 2k\pi}{3} & k = 0, \pm 1, \pm 2, \pm 3, \ldots \\ x = \dfrac{\pi - 0.7754 + 2k\pi}{3} = \dfrac{2.3662 + 2k\pi}{3} & k = 0, \pm 1, \pm 2, \pm 3, \ldots \end{cases}$$

or

$$\sin(3x)=0.7\Rightarrow\begin{cases}3x=0.7754+2k\pi & k=0,\pm1,\pm2,\pm3,\ldots\\ 3x=-(\pi+0.7754)+2k\pi=-3.917+2k\pi & k=0,\pm1,\pm2,\pm3,\ldots\end{cases}$$
$$\Rightarrow\begin{cases}x=\dfrac{0.7754+2k\pi}{3} & k=0,\pm1,\pm2,\pm3,\ldots\\ x=\dfrac{-(\pi+0.7754)+2k\pi}{3}=\dfrac{-3.917+2k\pi}{3} & k=0,\pm1,\pm2,\pm3,\ldots\end{cases}$$

Example 4.25 Solve $y=\sin(x)+\sqrt{2}\cos(x)=1$.

Solution Recall that:

$$a\times\sin(\omega t)+b\times\cos(\omega t)=\text{sign}(a).\sqrt{a^2+b^2}\sin\left(\omega t+\tan^{-1}\left(\frac{b}{a}\right)\right),\text{where sign}(x)$$
$$=\begin{cases}1 & x>0\\ 0 & x=0\\ -1 & x<0\end{cases}.$$

$$\sin(x)+\sqrt{2}\cos(x)=1\Rightarrow\sqrt{3}\sin\left(x+\tan^{-1}\left(\frac{\sqrt{2}}{1}\right)\right)=1\Rightarrow\sin(x+0.9553)=\frac{1}{\sqrt{3}}$$

Using a calculator $\sin^{-1}\left(\frac{1}{\sqrt{3}}\right)=0.6155\ rad$. Therefore,

$$\sin(x+0.9553)$$
$$=\frac{1}{\sqrt{3}}\Rightarrow\begin{cases}x+0.9553=0.6155+2k\pi\Rightarrow x=-0.3398+2k\pi & k=0,\pm1,\pm2,\pm3,\ldots\\ x+0.9553=\pi-0.6155+2k\pi=2.5261+2k\pi\Rightarrow x=+1.5708+2k\pi & k=0,\pm1,\pm2,\pm3,\ldots\end{cases}$$

Graph of $y=1$ and $y=\sin(x)+\sqrt{2}\cos(x)$ for $[-2\pi,2\pi]$ interval are shown in Fig. 4.7. Intersection points are shown in Fig. 4.8. You can use Fig. 4.8 to verify that obtained solution is correct.

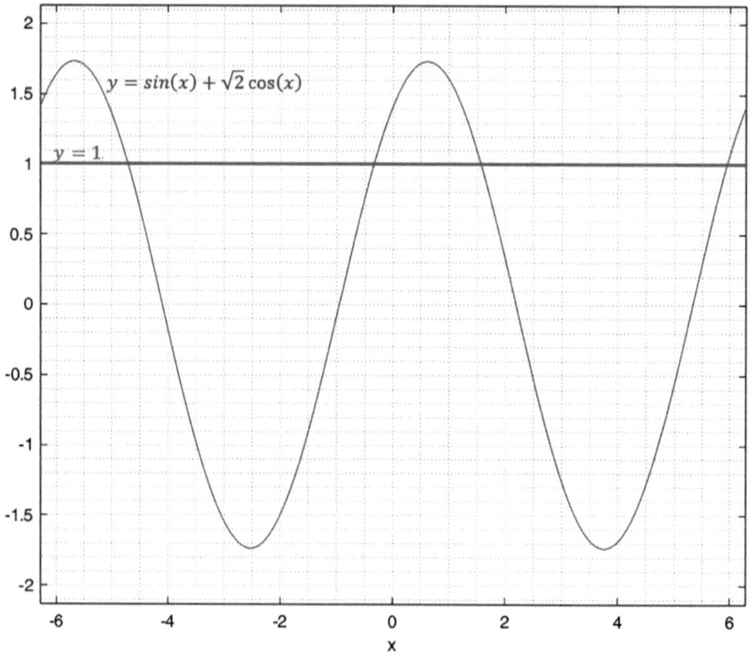

Fig. 4.7 Intersections of $y(x) = \sin(x) + \sqrt{2}\cos(x)$ and $y(x) = 1$

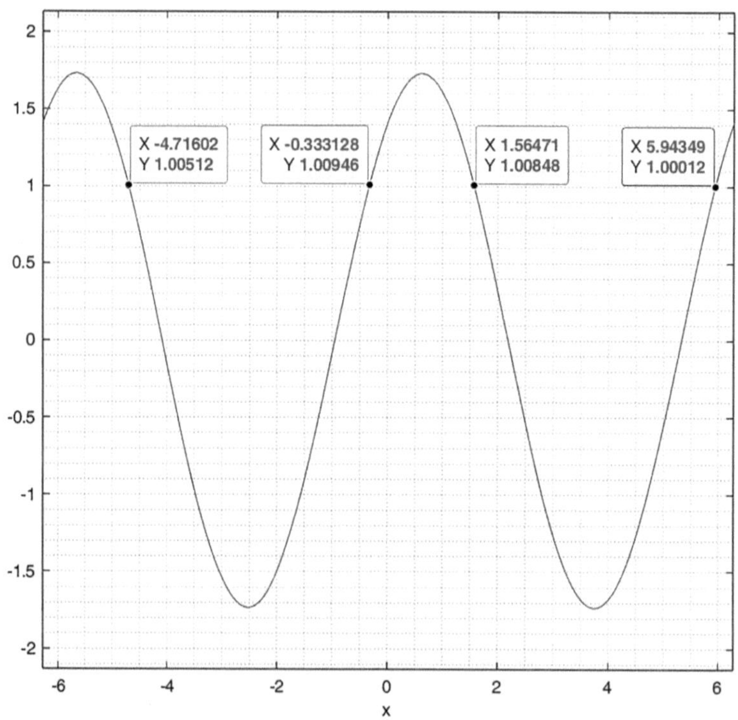

Fig. 4.8 Intersection coordinates

Exercise 4.9 Solve $y = \cos(x) - \sqrt{3}\sin(x) = 2$.

Example 4.26 Solve $\cos(x) = 0.7$.

Solution Using a calculator $\cos^{-1}(0.7) = 0.7954$ *rad* (Fig. 4.9).

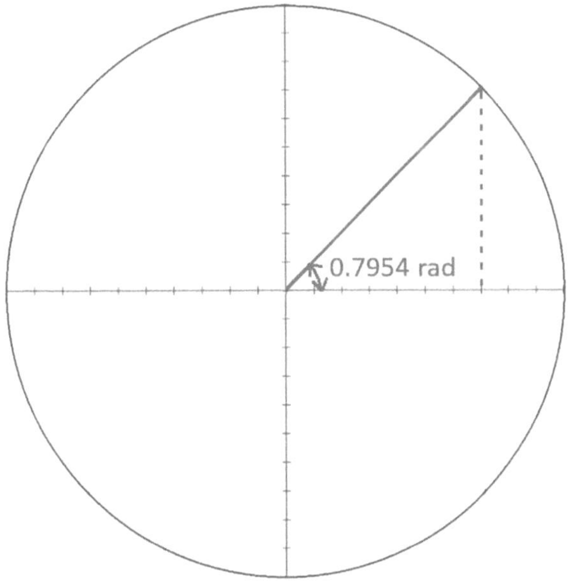

Fig. 4.9 Cosine of 0.7954 *rad* is 0.7

However, there exists another angle whose cosine is 0.7. That angle is -0.7954 *rad* or $2\pi - 0.7954 = +5.4878$ *rad* (Fig. 4.10).

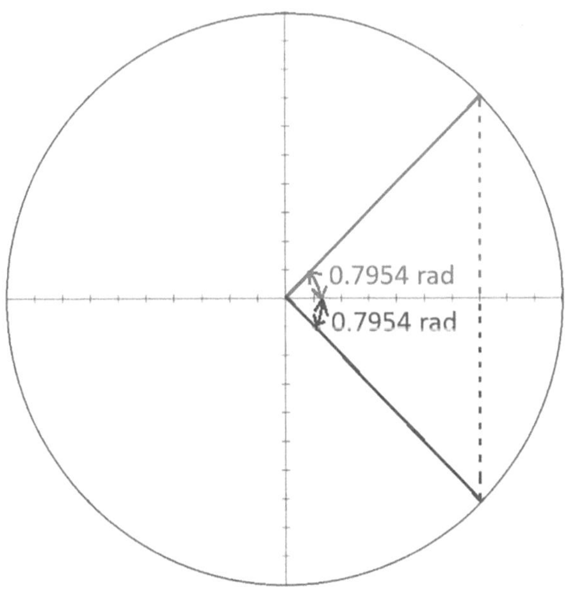

Fig. 4.10 Cosine of 0.7954 *rad* and -0.7954 *rad* is 0.7

$$\cos(x) = 0.7 \Rightarrow \begin{cases} x = 0.7954 + 2k\pi \\ x = -0.7954 + 2k\pi \end{cases} \qquad k = 0, \pm 1, \pm 2, \pm 3, \ldots$$

or

$$\cos(x) = 0.7 \Rightarrow \begin{cases} x = 0.7954 + 2k\pi \\ x = 5.4878 + 2k\pi \end{cases} \qquad k = 0, \pm 1, \pm 2, \pm 3, \ldots$$

Example 4.27 Solve $\cos(3x) = 0.7$.

Solution Using a calculator $\cos^{-1}(0.7) = 0.7954$ *rad* (Fig. 4.10).

$$\cos(3x) = 0.7 \Rightarrow \begin{cases} 3x = 0.7954 + 2k\pi \\ 3x = -0.7954 + 2k\pi \end{cases} \Rightarrow \begin{cases} x = \dfrac{0.7954 + 2k\pi}{3} \\ x = \dfrac{-0.7954 + 2k\pi}{3} \end{cases}$$

$$\Rightarrow \begin{cases} x = 0.2651 + \dfrac{2k\pi}{3} \\ x = -0.2651 + \dfrac{2k\pi}{3} \end{cases} \qquad k = 0, \pm 1, \pm 2, \pm 3, \ldots$$

or

$$\cos(3x) = 0.7 \Rightarrow \begin{cases} 3x = 0.7954 + 2k\pi \\ 3x = 5.4878 + 2k\pi \end{cases} \Rightarrow \begin{cases} x = \dfrac{0.7954 + 2k\pi}{3} \\ x = \dfrac{5.4878 + 2k\pi}{3} \end{cases}$$

$$\Rightarrow \begin{cases} x = 0.2651 + \dfrac{2k\pi}{3} \\ x = 1.8293 + \dfrac{2k\pi}{3} \end{cases} \qquad k = 0, \pm 1, \pm 2, \pm 3, \ldots$$

Example 4.28 Solve $\tan(x) = 0.7$.

Solution Using a calculator $\tan^{-1}(0.7) = 0.6107$ *rad* (Fig. 4.11).

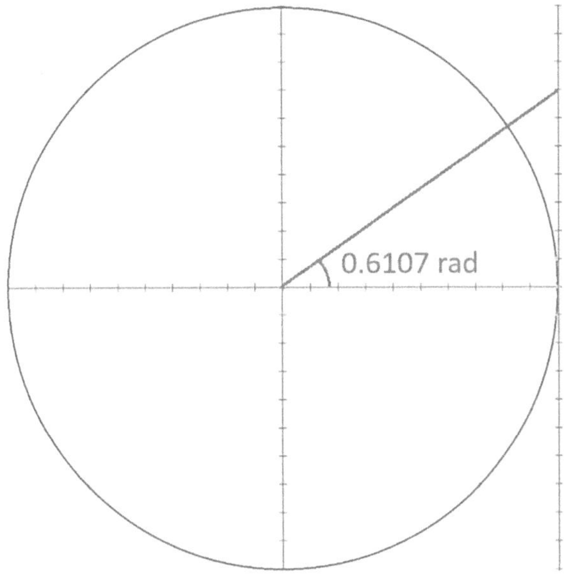

Fig. 4.11 Tangent of 0.6107 *rad* is 0.7

However, there exists another angle whose tangent is 0.7 (Fig. 4.12). That angle is $\pi + 0.6107 = 3.7523$ *rad* or $-(\pi - 0.6107) = -2.5309$ *rad* (Fig. 4.13).

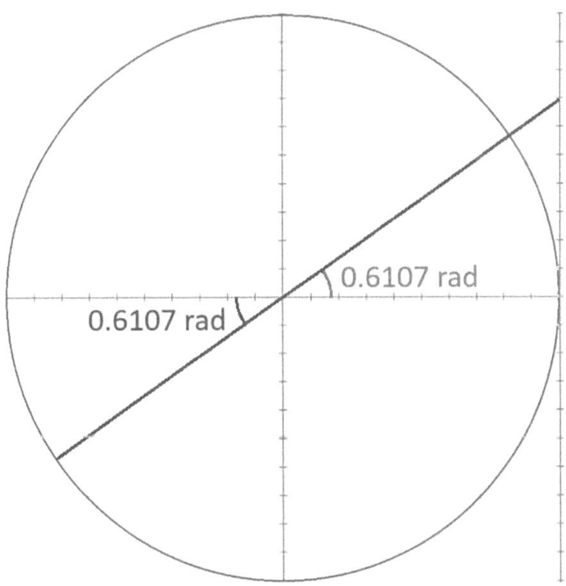

Fig. 4.12 Tangent of 3.7523 *rad* or −2.5309 *rad* is 0.7

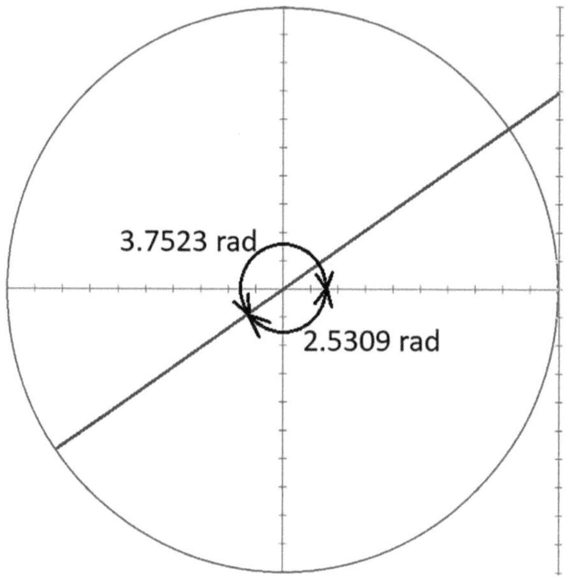

Fig. 4.13 Different ways to measure an angle

Therefore,

$$\tan(x)=0.7 \Rightarrow \begin{cases} x=0.6107+k\pi & k=0,\pm1,\pm2,\pm3,\dots \\ x=\pi+0.6107+k\pi & k=0,\pm1,\pm2,\pm3,\dots \end{cases}$$

$$\Rightarrow \begin{cases} x=0.6107+k\pi & k=0,\pm1,\pm2,\pm3,\dots \\ x=0.6107+(k+1)\pi & k=0,\pm1,\pm2,\pm3,\dots \end{cases} \quad k' = k$$

$$+1 \; k'=k+1 \begin{cases} x=0.6107+k\pi & k=0,\pm1,\pm2,\pm3,\dots \\ x=0.6107+k'\pi & k'=\pm1,\pm2,\pm3,\dots \end{cases}$$

The first equation encompasses the second one, so we only need to consider the first.

$$x=0.6107+k\pi \qquad k=0,\pm1,\pm2,\pm3,\dots$$

Example 4.29 Solve $\tan(3x) = 0.7$.

Solution Using a calculator $\tan^{-1}(0.7) = 0.6107$ *rad*. Therefore,

$$\tan(3x)=0.7 \Rightarrow \begin{cases} 3x=0.6107+k\pi & k=0,\pm1,\pm2,\pm3,\dots \\ 3x=\pi+0.6107+k\pi=3.7523+k\pi & k=0,\pm1,\pm2,\pm3,\dots \end{cases}$$

$$\Rightarrow \begin{cases} 3x=0.6107+k\pi & k=0,\pm1,\pm2,\pm3,\dots \\ 3x=0.6107+(k+1)\pi & k=0,\pm1,\pm2,\pm3,\dots \end{cases} \quad k' \quad = k$$

$$+1 \; k'=k+1 \begin{cases} 3x=0.6107+k\pi & k=0,\pm1,\pm2,\pm3,\dots \\ 3x=0.6107+k'\pi & k'=\pm1,\pm2,\pm3,\pm4,\dots \end{cases}$$

The first equation encompasses the second one, so we only need to consider the first.

$$3x=0.6107+k\pi \quad k=0,\pm1,\pm2,\pm3,\ldots \Rightarrow x=0.2036+k\frac{\pi}{3} \quad k=0,\pm1,\pm2,\pm3,\ldots$$

Example 4.30 Solve $\tan\left(3x+\frac{\pi}{4}\right)=0.7$.

Solution Using a calculator $\tan^{-1}(0.7)=0.6107$. Therefore,

$$\tan\left(3x+\frac{\pi}{4}\right)=0.7\Rightarrow\begin{cases}3x+\frac{\pi}{4}=0.6107+k\pi & k\in\mathbb{Z}\\ 3x+\frac{\pi}{4}=\pi+0.6107+k\pi & k\in\mathbb{Z}\end{cases}$$

$$\Rightarrow\begin{cases}3x+\frac{\pi}{4}=0.6107+k\pi & k\in\mathbb{Z}\\ 3x+\frac{\pi}{4}=0.6107+(k+1)\pi & k\in\mathbb{Z}\end{cases}\overset{k'=k+1}{\Rightarrow}\begin{cases}3x+\frac{\pi}{4}=0.6107+k\pi & k\in\mathbb{Z}\\ 3x+\frac{\pi}{4}=0.6107+k'\pi & k'\in\mathbb{Z}\end{cases}$$

The first equation encompasses the second one, so we only need to consider the first.

$$3x+\frac{\pi}{4}=0.6107+k\pi \quad k\in\mathbb{Z}\Rightarrow 3x=0.6107-\frac{\pi}{4}+k\pi\Rightarrow x=\frac{0.6107}{3}-\frac{\pi}{12}+k\frac{\pi}{3}$$
$$\Rightarrow x=-0.0582+k\frac{\pi}{3} \quad k\in\mathbb{Z}$$

Exercise 4.10 Graph the examples discussed to visually verify the obtained results.

Exercise 4.11 Solve $\cot\left(3x+\frac{\pi}{4}\right)=0.5$. (**Hint:** $\cot\left(3x+\frac{\pi}{4}\right)=0.5\Rightarrow\tan\left(3x+\frac{\pi}{4}\right)=\frac{1}{0.5}=2$)

Example 4.31 The code in Fig. 4.14 solves the equation $2x + 7 = 0$.

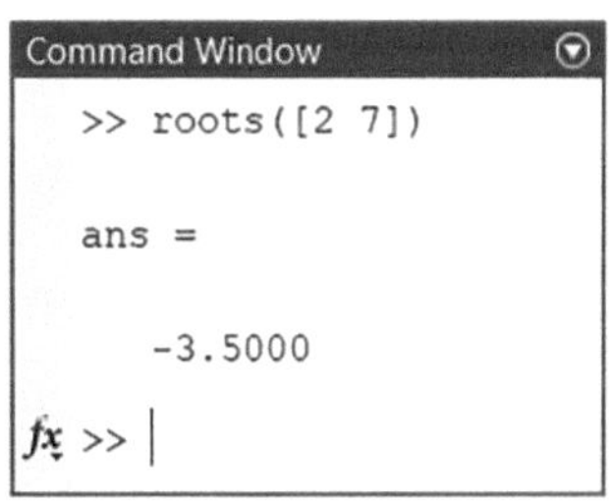

Fig. 4.14 Calculation of the roots of $2x + 7$

Example 4.32 The code in Fig. 4.15 solves the equation $x^2 + 2x + 3 = 0$.

Fig. 4.15 Calculation of the roots of $x^2 + 2x + 3$ and $x^2 + 5x + 6$

```
Command Window
>> roots([1 2 3])

ans =

  -1.0000 + 1.4142i
  -1.0000 - 1.4142i

>> roots([1 5 6])

ans =

   -3.0000
   -2.0000

fx >>
```

Example 4.33 The code in Fig. 4.16 solves the equation $x^3 + 2x^2 + 9x + 7 = 0$.

Fig. 4.16 Calculation of the roots of $x^3 + 2x^2 + 9x + 7$

```
Command Window
>> roots([1 2 9 7])

ans =

  -0.5634 + 2.7746i
  -0.5634 - 2.7746i
  -0.8732 + 0.0000i

fx >>
```

Example 4.34 The code in Fig. 4.17 calculates the coefficients of the polynomial whose roots are 1, $3 + 7j$, and $3 - 7j$. According to Fig. 4.17, the resulting polynomial is $x^3 - 7x^2 + 64x - 58 = 0$.

Fig. 4.17 Calculation of coefficients of the polynomial with roots 1, $3 + 7j$, and $3 - 7j$

```
Command Window
>> poly([1 3+7j 3-7j])

ans =

     1    -7    64   -58

fx >>
```

Example 4.35 To solve the nonlinear equation $e^{-5x}\sin(x) - x^2 + 1 = 0$ in MATLAB, you can use the built-in function fzero. The fzero function finds the root of a single-variable function near a specified starting point or within an interval.

Since this is a transcendental equation, it likely has multiple roots, and you will need to choose an initial guess that is close to the root you are seeking. The commands in Fig. 4.18 generate the plot of the function $e^{-5x}\sin(x) - x^2 + 1$ over the interval $[-1.5, 1.5]$. The resulting graph is shown in Fig. 4.19.

Fig. 4.18 Plotting the function $e^{-5x}\sin(x) - x^2 + 1$ over the interval $[-1.5, 1.5]$

```
Command Window
>> f = @(x) exp(-5*x).*sin(x)-x.^2+1;
>> fplot(f,[-1.5 1.5])
>> grid on
fx >>
```

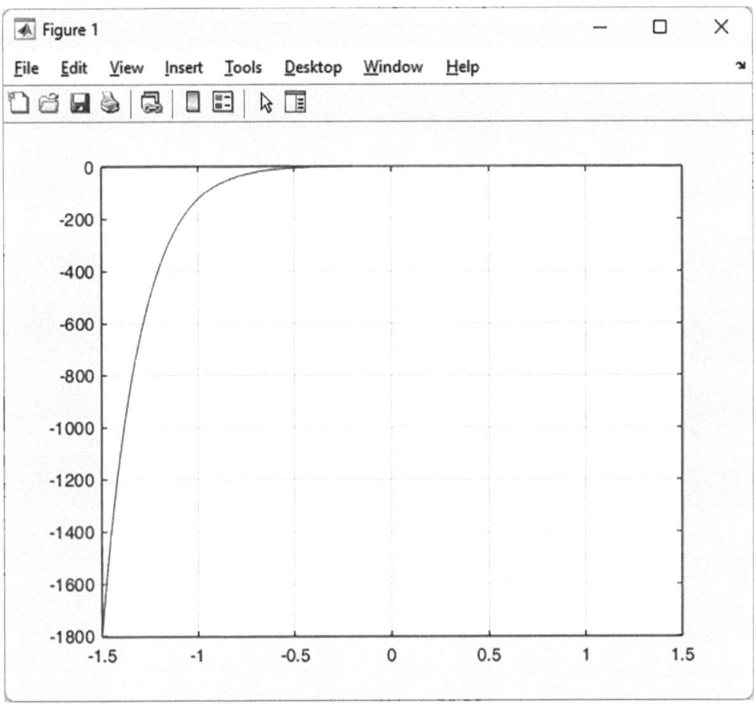

Fig. 4.19 Output of the code shown in Fig. 4.18

Zoom in on the graph to get a better view of the roots (Fig. 4.20). From Fig. 4.21, we can see that the roots are located around -0.26 and 1.

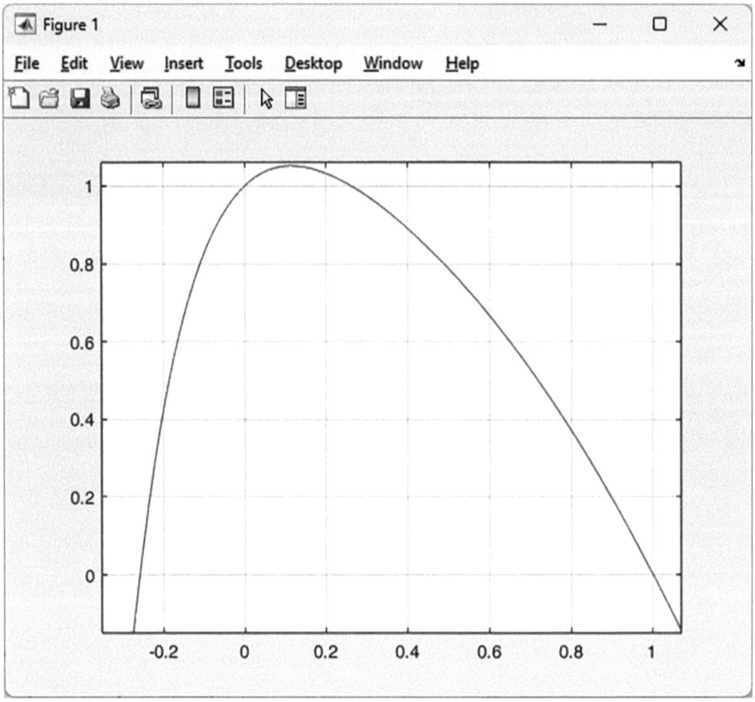

Fig. 4.20 Magnified view of the plot

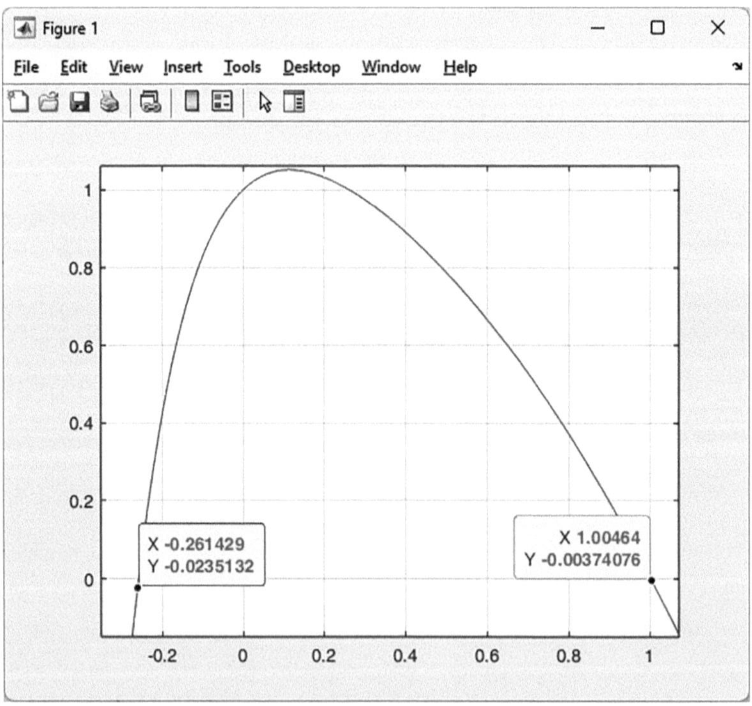

Fig. 4.21 The roots are located near the values −0.25 and 1

Using Fig. 4.21, we can determine a good initial guess for the fzero command. According to Figs. 4.22 and 4.23, the exact roots are −0.2588 and 1.0028.

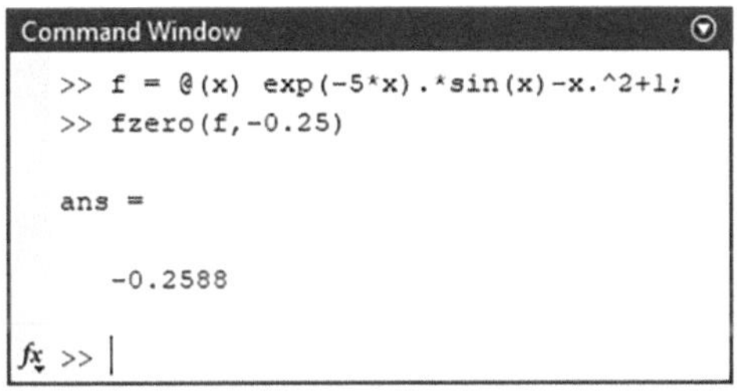

Fig. 4.22 The fzero command converges to −0.2588 when initial guess is −0.25

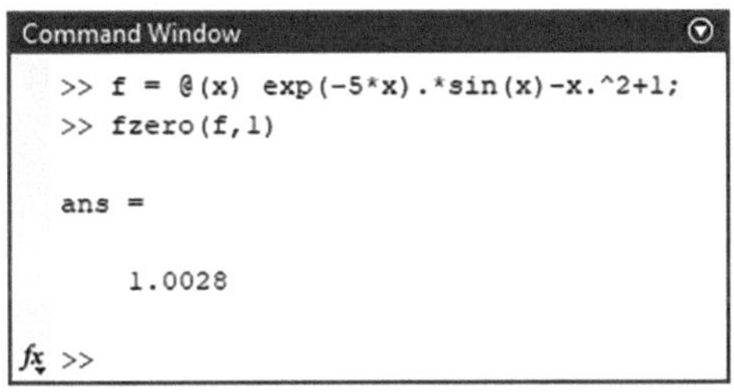

Fig. 4.23 The fzero command converges to −1.0028 when initial guess is 1

Example 4.36 Use MATLAB to solve $\tan\left(3x+\frac{\pi}{4}\right)=0.7$.

Solution Open the MATLAB editor using the edit command (Fig. 4.24).

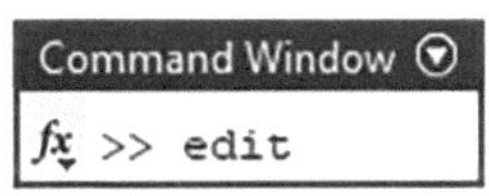

Fig. 4.24 Edit command

After entering the code displayed in Fig. 4.25 into the MATLAB editor, save the file as "SolvingTrigEq" by using the Ctrl+s shortcut.

```
% 1. define symbolic variables and pi
syms x k
p = sym(pi);

% 2. define the equation
eqn = tan(3*x+p/4)==0.7;

% 3. solve for x and return the general solution structure
sol = solve(eqn, x, 'returnconditions', true);

% 4. display the general solution (sol.x) using disp
disp('the general solution for x is:');
pretty(sol.x)

% 5. vpa finds the numerical value of the expression
disp(' ')
disp('the principal numerical solution is:');
disp(vpa(sol.x, 5))
```

Fig. 4.25 Code for solving $\tan\left(3x+\frac{\pi}{4}\right)=0.7$

To execute the code, press F5. Should the prompt displayed in Fig. 4.26 appear regarding the path, select Add to Path to proceed.

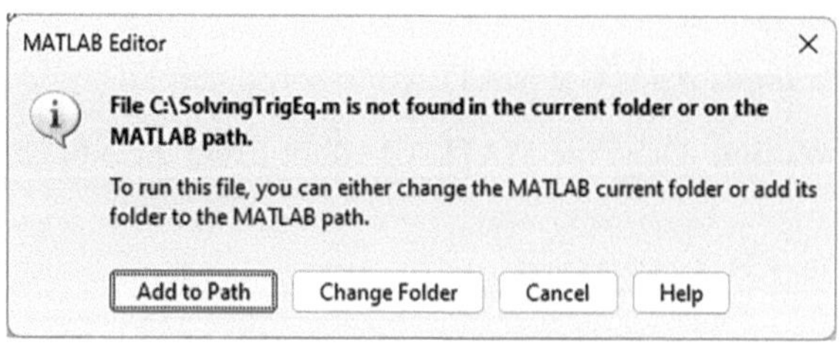

Fig. 4.26 MATLAB Editor window

Figure 4.27 shows the output generated by the code in Fig. 4.25.

```
the general solution for x is:
                / 7 \
           atan| -- |
pi              \ 10 /   pi k
-- + ---------- + ----
 4          3          3


the principal numerical solution is:
0.98897 + 1.0472*k

>>
```

Fig. 4.27 Output of the code in Fig. 4.25

From Fig. 4.27, we get:

$$\tan\left(3x+\frac{\pi}{4}\right)=0.7 \Rightarrow x=\frac{\pi}{4}+\frac{\tan^{-1}\left(\frac{7}{10}\right)}{3}+\frac{k\pi}{3} \Rightarrow x=0.98897+1.0472k$$

which implies:

$$x=\frac{\pi}{4}+\frac{\tan^{-1}\left(\frac{7}{10}\right)}{3}+\frac{k\pi}{3} \Rightarrow x=0.7854+0.2036+\frac{k\pi}{3} \Rightarrow x=0.9890+\frac{k\pi}{3}$$

While this last result ($x=0.9890+\frac{k\pi}{3}$) looks different from the result in Example 4.30 ($x=-0.0582+\frac{k\pi}{3}$), a simple proof shows that they are, in fact, equivalent. The calculations below confirm that both formulas satisfy the original equation:

$$\begin{aligned}\tan\left(3x+\frac{\pi}{4}\right) &= \tan\left(3\left(0.9890+\frac{k\pi}{3}\right)+\frac{\pi}{4}\right)=\tan\left(2.9670+k\pi+\frac{\pi}{4}\right)\\ &= \tan(3.7524+k\pi)=\tan(3.7524)=0.7\end{aligned}$$

$$\begin{aligned}\tan\left(3x+\frac{\pi}{4}\right) &= \tan\left(3\left(-0.0582+\frac{k\pi}{3}\right)+\frac{\pi}{4}\right)=\tan\left(-0.1746+k\pi+\frac{\pi}{4}\right)\\ &= \tan(0.6108+k\pi)=\tan(0.6108)=0.7\end{aligned}$$

As Tables 4.1 and 4.2 illustrate, the values produced by the two formulas are the same. However, the equivalence is shifted, meaning the identical solutions are generated using different integer values of k in each respective equation.

Table 4.1 The values generated by the formula $x=0.9890+\frac{k\pi}{3}$ across the range $-3\leq k\leq 3$

k	−3	−2	−1	0	1	2	3
$x=0.9890+\frac{k\pi}{3}$	−2.1526	−1.1054	−0.0582	0.9890	2.0362	3.0834	4.1306

Table 4.2 The values generated by the formula $x=-0.0582+\frac{k\pi}{3}$ across the range $-3\leq k\leq 3$

k	−3	−2	−1	0	1	2	3
$x=-0.0582+\frac{k\pi}{3}$	−3.1998	−2.1526	−1.1054	−0.0582	0.9890	2.0362	3.0834

Example 4.37 Use MATLAB to solve $\cos\left(3x+\frac{\pi}{4}\right)=0.7$.

Solution This problem can be solved by modifying the code from Example 4.36 (Fig. 4.28).

```
Editor - C:\SolvingTrigEq.m
SolvingTrigEq.m
% 1. define symbolic variables and pi
syms x k
p = sym(pi);

% 2. define the equation
eqn = cos(3*x+p/4)==0.7;

% 3. solve for x and return the general solution structure
sol = solve(eqn, x, 'returnconditions', true);

% 4. display the general solution (sol.x) using disp
disp('the general solution for x is:');
pretty(sol.x)

% 5. vpa finds the numerical value of the expression
disp(' ')
disp('the principal numerical solution is:');
disp(vpa(sol.x, 5))
```

Fig. 4.28 Code for solving $\cos\left(3x+\frac{\pi}{4}\right)=0.7$

Figure 4.29 shows the output generated by the code in Fig. 4.28.

Fig. 4.29 Output of the code in Fig. 4.28

```
Command Window
>> SolvingTrigEq
the general solution for x is:
/             / 7 \            \
|         acos| -- |           |
|   pi        \ 10 /   pi k 2  |
| - -- - ---------- + ------   |
|   12        3           3    |
|                              |
|             / 7 \            |
|         acos| -- |           |
|   pi        \ 10 /   pi k 2  |
| - -- + ---------- + ------   |
\   12        3           3    /


the principal numerical solution is:
- 0.52693 + 2.0944*k
0.0033336 + 2.0944*k

>>
```

From Fig. 4.29, we get:

$$\cos\left(3x+\frac{\pi}{4}\right)=0.7 \Rightarrow x=\begin{cases} -\frac{\pi}{12}-\frac{\cos^{-1}\left(\frac{7}{10}\right)}{3}+\frac{2k\pi}{3}=-0.52693+2.0944k \\ -\frac{\pi}{12}+\frac{\cos^{-1}\left(\frac{7}{10}\right)}{3}+\frac{2k\pi}{3}=-0.0033336+2.0944k \end{cases}$$

Example 4.38 Use MATLAB to solve $\sin(x)+\sqrt{2}\cos(x)=1$.

Solution This problem can be solved by modifying the code from Example 4.36 (Fig. 4.30).

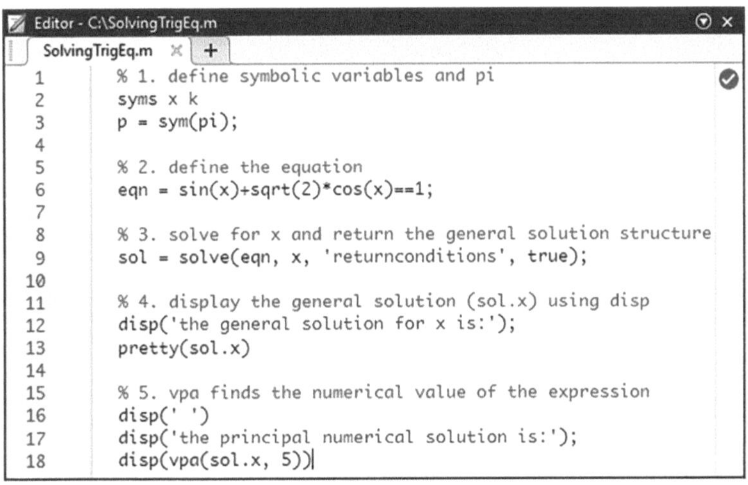

Fig. 4.30 Code for solving $\sin(x)+\sqrt{2}\cos(x)=1$

Figure 4.31 shows the output generated by the code in Fig. 4.30.

Fig. 4.31 Output of the code in Fig. 4.30

```
Command Window
the general solution for x is:
/                 pi                            \
|                 -- + pi k 2                   |
|                  2                            |
|                                               |
|        / 2 sqrt(2)    1  \                    |
| - log| --------- - -i | 1i + pi k 2          |
\        \     3        3  /                    /


the principal numerical solution is:
                          1.5708 + 6.2832*k
- (0.33984 - 5.6843e-14i) + 6.2832*k

>>
```

From Fig. 4.31, we get:

$$\sin(x) + \sqrt{2}\cos(x) = 1 \Rightarrow x = \begin{cases} 1.5708 + 6.2832k \\ -0.33984 + 5.6843 \times 10^{-14}i + 6.2832k \end{cases}$$

$5.6843 \times 10^{-14}i$ is too small, so it can be ignored. Therefore,

$$\sin(x) + \sqrt{2}\cos(x) = 1 \Rightarrow x = \begin{cases} 1.5708 + 6.2832k \\ -0.33984 + 5.6843 \times 10^{-14}i + 6.2832k \end{cases}$$

$$\Rightarrow x = \begin{cases} 1.5708 + 6.2832k \\ -0.33984 + 6.2832k \end{cases}$$

This result matches the one found in Example 4.25.

Example 4.39 Use MATLAB to solve $\sqrt{2x + 1} = x - 1$.

Solution The required code is shown in Fig. 4.32. The resulting output matches the solution found in Example 4.15.

Fig. 4.32 Code for solving $\sqrt{2x + 1} = x - 1$

```
Command Window
>> syms x
>> eqn = sqrt(2*x + 1) == x - 1;
>> solution = solve(eqn, x)

solution =

4

>>
```

Example 4.40 Use MATLAB to solve $|x^2 + 7x - 3| = |x^2 + 3x + 1|$.

Solution Let's start with a manual analysis:

$$|x^2+7x-3|=|x^2+3x+1| \Rightarrow x^2+7x-3=\pm(x^2+3x+1)$$

$$\Rightarrow \begin{cases} x^2+7x-3=+(x^2+3x+1) \Rightarrow x=1 \\ x^2+7x-3=-(x^2+3x+1) \Rightarrow x^2+5x-1=0 \Rightarrow x=\dfrac{-5\pm\sqrt{29}}{2} \Rightarrow x=\begin{cases} 0.1926 \\ -5.1926 \end{cases} \end{cases}$$

Therefore, we expect three solutions: $\{1, 0.1926, -5.1926\}$. Now let's test the code shown in Fig. 4.33. This code, however, predicts only one of the expected solutions. MATLAB's output resulted from analyzing only the positive case: $x^2 + 7x - 3 = +(x^2 + 3x + 1)$.

Fig. 4.33 First attempt to solve $|x^2 + 7x - 3| = |x^2 + 3x + 1|$

```
Command Window
>> syms x
>> eqn=abs(x^2+7*x-3)==abs(x^2+3*x+1);
>> solution = solve(eqn, x)

solution =

1

fx >>
```

Figure 4.34 presents the revised code that successfully predicts all three expected solutions.

Fig. 4.34 Code for solving $|x^2 + 7x - 3| = |x^2 + 3x + 1|$

```
Command Window
>> syms x
>> eqn1=x^2+7*x-3==x^2+3*x+1;
>> eqn2=x^2+7*x-3==-(x^2+3*x+1);
>> sol1=solve(eqn1, x);
>> sol2=solve(eqn2, x);
>> solutions=[sol1; sol2];
>> eval(solutions)

ans =

    1.0000
   -5.1926
    0.1926

fx >>
```

Example 4.40 Use MATLAB to solve $x^2 = 3x - 1$.

Solution The required code is shown in Fig. 4.35.

Fig. 4.35 Code for solving $x^2 = 3x - 1$

```
Command Window
>> syms x
>> eqn=x^2==3*x-1;
>> solution=solve(eqn, x)

solution =

3/2 - 5^(1/2)/2
5^(1/2)/2 + 3/2

>> pretty(solution)
/ 3   sqrt(5) \
| - - ------- |
| 2      2    |
|             |
| sqrt(5)   3 |
| ------- + - |
\    2      2 /

>> eval(solution)

ans =

    0.3820
    2.6180

>>
```

References for Further Study

1. Engineering Mathematics with MATLAB® and Simulink®, Farzin Asadi, Springer, 2025.
2. Applied Numerical Analysis with MATLAB®/Simulink®: For Engineers and Scientists, Farzin Asadi, Springer, 2023.
3. Advanced Engineering Mathematics, Erwin Kreyszig, Wiley, 2020.
4. A Problem-Solving Approach to Single-Variable Calculus (Volume II), Farzin Asadi, Springer, 2026.
5. A Problem-Solving Approach to Multi-Variable Calculus (Volume I), Farzin Asadi, Springer, 2026.
6. A Problem-Solving Approach to Multi-Variable Calculus (Volume II), Farzin Asadi, Springer, 2026.
7. A Problem-Solving Approach to Multi-Variable Calculus (Volume III), Farzin Asadi, Springer, 2026.
8. A Problem-Solving Approach to Ordinary Differential Equations (Volume I), Farzin Asadi, Springer, 2026.

9. A Problem-Solving Approach to Ordinary Differential Equations (Volume II), Farzin Asadi, Springer, 2026.
10. A Problem-Solving Approach to Ordinary Differential Equations (Volume III), Farzin Asadi, Springer, 2026.
11. A Problem-Solving Approach to Ordinary Differential Equations (Volume IV), Farzin Asadi, Springer, 2026.

Chapter 5
Complex Numbers

5.1 Introduction

Complex numbers extend real numbers by including the imaginary unit i ($i^2 = -1 \Rightarrow i = \sqrt{-1}$). Written as $a + bi$, they have a real part (a) and an imaginary part (b). They are essential in mathematics and engineering for solving equations (like $x^2 + 1 = 0$) and analyzing systems such as AC circuits.

The scope of this chapter encompasses the definition, algebra, and geometrical representation of complex numbers.

5.2 Complex Numbers

In mathematics, a complex number is an element of a number system that extends the real numbers with a specific element denoted i, called the imaginary unit and satisfying the equation $i^2 = -1$. Every complex numbers can be expressed in the form $a + bi$, where a and b are real numbers.

For the complex number $a + bi$, a is called the real part, and b is called the imaginary part. The set of complex numbers is denoted by the symbol $\mathbb{C}$.

Two complex numbers $z_1 = a + bi$ and $z_2 = c + di$ are equal when $a = c$ and $b = d$.

5.3 Complex Numbers as Roots of Second-Order Equations

Roots of $ax^2 + bx + c = 0$ equation are $x_{1,2} = \frac{-b \pm \sqrt{\Delta}}{2a}$, where $\Delta = b^2 - 4ac$. When $\Delta < 0$ roots are complex. For instance, for $x^2 + 2x + 10 = 0$ the $\Delta = 2^2 - 4(1)(10) = -36$ and the roots are:

F. Asadi, *A Problem-Solving Approach to Single-Variable Calculus*,
https://doi.org/10.1007/978-3-032-13505-6_5

$$x_{1,2}=\frac{-2\pm\sqrt{2^2-4(1)(10)}}{2}=\frac{-2\pm\sqrt{-36}}{2}=\frac{-2\pm\sqrt{36}\sqrt{-1}}{2}=\frac{-2\pm 6j}{2}=-1\pm 3j$$

Note that $\sqrt{-1}=j$. Mathematicians generally use i for $\sqrt{-1}$. Engineers prefer the j notation to avoid conflict with the notation for electrical current. $a + jb$ and $a + bj$ are the same.

5.4 The Geometric View of Complex Numbers

Figure 5.1 shows how points $A = 1 + 2j$ and $B = -1 - 3j$ can be shown on the complex plane.

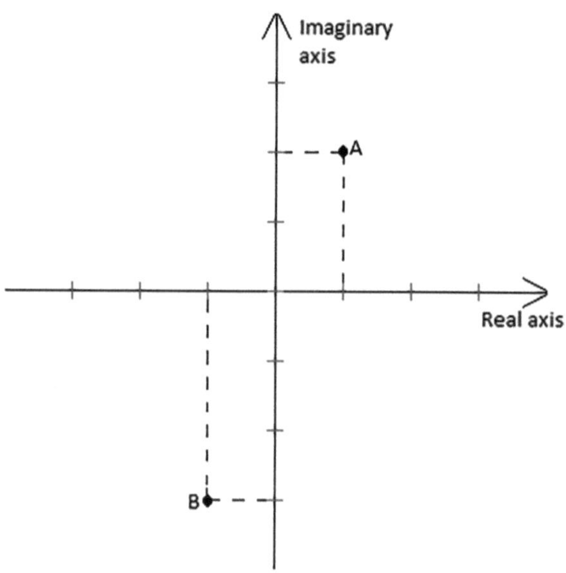

Fig. 5.1 Points $A = 1 + 2j$ and $B = -1 - 3j$

5.5 Euler's Formula

Euler's formula states that, for any real number x, one has:

$$e^{jx} = \cos(x) + j\sin(x)$$

For instance,

(1) $e^{j\frac{\pi}{6}} = \cos\left(\frac{\pi}{6}\right) + j\sin\left(\frac{\pi}{6}\right) = \frac{\sqrt{3}}{2} + j\frac{1}{2}$
(2) $e^{-j\frac{\pi}{4}} = \cos\left(-\frac{\pi}{4}\right) + j\sin\left(-\frac{\pi}{4}\right) = \frac{1}{\sqrt{2}} - j\frac{1}{\sqrt{2}}$
(3) $\sqrt{5}e^{j7.3903} = \sqrt{5}\cos(7.3903) + \sin(7.3903) = 1 + j2$
(4) $e^{2 + 3j} = e^2e^{3j} = e^2(\cos(3) + j\sin(3)) = e^2(-0.9900 + j0.1411) = -7.3151 + j1.0427$

Sometimes the complex number $re^{jx} = r(\cos(x) + j\sin(x)) = r\cos(x) + jr\sin(x)$ is shown as $r \measuredangle x$. For instance, $\sqrt{5}e^{j7.3903}$ can be shown as $\sqrt{5}\measuredangle 7.3903$.

Exercise 5.1 Show that $\cos(x) = \frac{e^{jx}+e^{-jx}}{2}$ and $\sin(x) = \frac{e^{jx}-e^{-jx}}{2j}$.

5.6 Magnitude of a Complex Number

Magnitude of a complex number $a + jb$ is shown as $|a + jb|$ and equals to $\sqrt{a^2 + b^2}$. For instance, $|1 + j2| = \sqrt{1^2 + 2^2} = \sqrt{5} = 2.2361$. Magnitude of $z = re^{j\theta}$ equals to r. For instance, $\left|7e^{j\frac{\pi}{4}}\right| = 7$. Magnitude of division of two complex numbers can be calculated with the aid of $\left|\frac{z_1}{z_2}\right| = \frac{|z_1|}{|z_2|}$. For instance, $\left|\frac{1+2j}{3+4j}\right| = \frac{|1+2j|}{|3+4j|} = \frac{\sqrt{1^2+2^2}}{\sqrt{3^2+4^2}} = \frac{\sqrt{5}}{\sqrt{25}} = 0.447$.

5.7 The Principle Argument of a Complex Number

The principal argument of a complex number z is the angle φ formed between the positive real axis and the line representing the complex number in the complex plane. It is typically measured in radians and is defined within the range $(-\pi, \pi]$. The principal argument is often denoted as $Arg(z)$.

Argument of a complex number z is shown with $arg(z)$ and $arg(z) = \{Arg(z) + 2\pi n | n \in \mathbb{Z}\}$.

For a complex number $z = x + yi$, the principal argument can be calculated using the arctangent function. However, special care must be taken depending on the quadrant in which the complex number lies.

Let's study some numeric examples. The principle argument of $A = 1 + 2j$ is $\varphi = \tan^{-1}\left(\frac{2}{1}\right) = 1.1071\ rad$ (Fig. 5.2). Argument of A is $1.1071 + 2\pi n$, $n \in \mathbb{Z}$.

Fig. 5.2 Point $A = 1 + 2j$

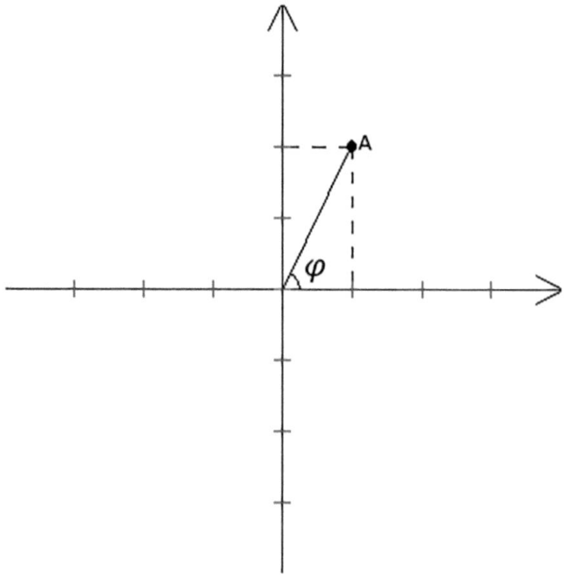

The principle argument of $B = -1 + 2j$ is $\varphi = \frac{\pi}{2} + \tan^{-1}\left(\frac{1}{2}\right) = \frac{\pi}{2} + 0.463 = 2.034$ *rad* (Fig. 5.3). Argument of B is $2.034 + 2\pi n$, $n \in \mathbb{Z}$.

Fig. 5.3 Point $B = -1 + 2j$

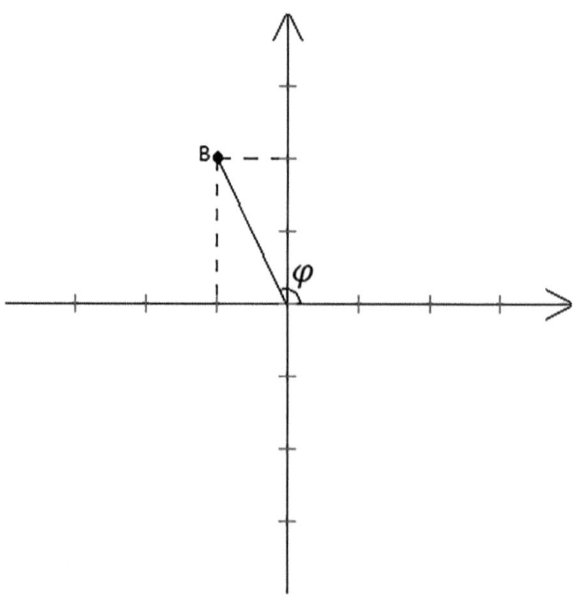

The principle argument of $C = 1 - 2j$ is $\varphi = -\tan^{-1}\left(\frac{2}{1}\right) = -1.1071\ rad$ (Fig. 5.4). Note that the principle argument of C is NOT $5.1761\ rad$ (Fig. 5.5) since $5.1761 \notin (-\pi, \pi]$. Argument of C is $-1.1071 + 2\pi n$, $n \in \mathbb{Z}$.

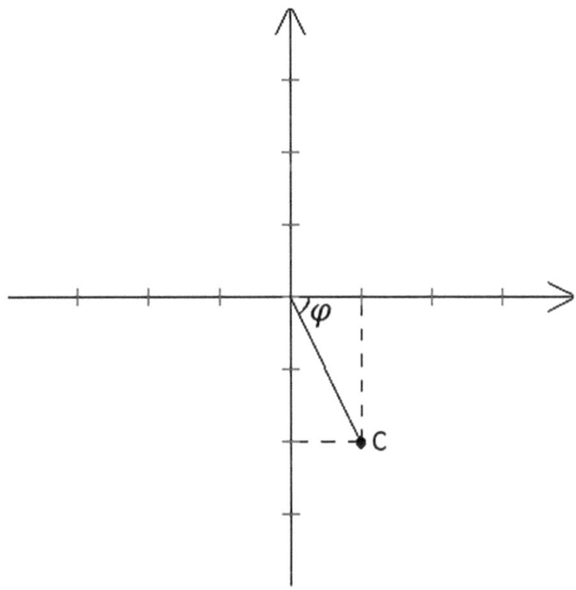

Fig. 5.4 Point $C = 1 - 2j$

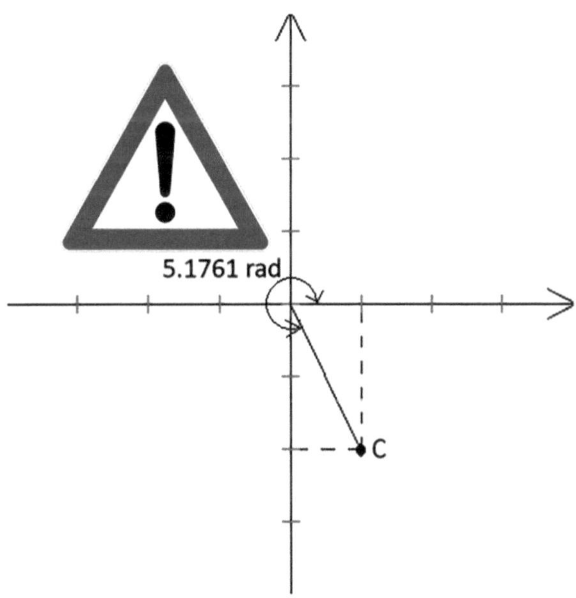

Fig. 5.5 Wrong measurement of principle argument

The principle argument of $D = -1 - 2j$ is $\varphi = -\frac{\pi}{2} - \tan^{-1}\left(\frac{1}{2}\right) = -\frac{\pi}{2} - 0.463 = -2.034\ rad$ (Fig. 5.6). Note that the principle argument of D is NOT $4.2492\ rad$ (Fig. 5.7) since $4.2492 \notin (-\pi, \pi]$. Argument of D is $-2.034 + 2\pi n, n \in \mathbb{Z}$.

Fig. 5.6 Point $D = -1 - 2j$

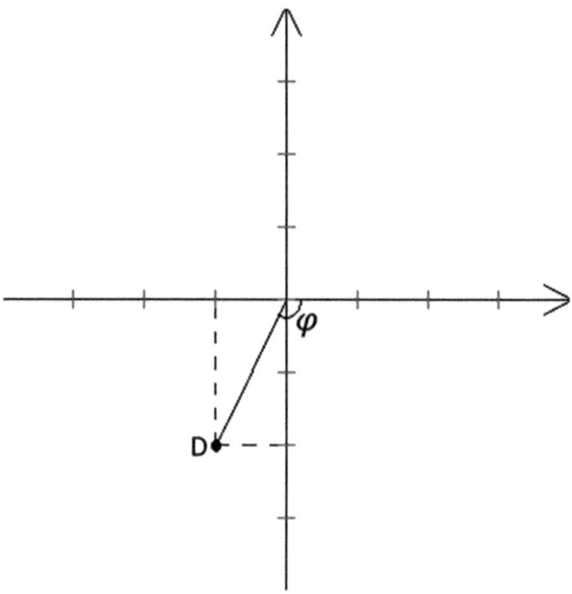

Fig. 5.7 Wrong measurement of principle argument

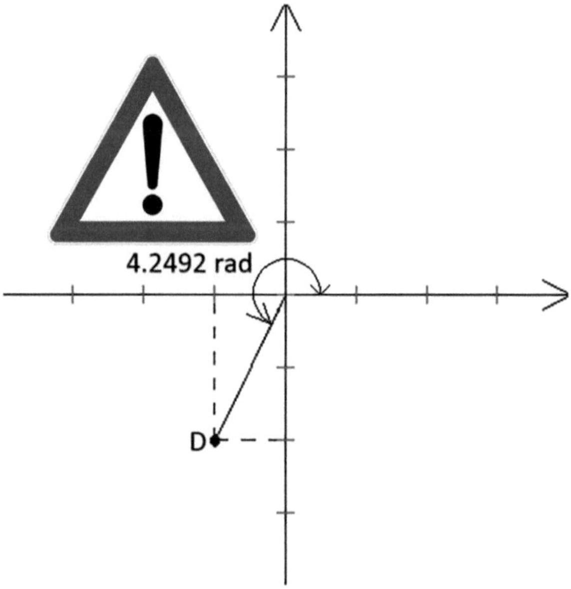

Additionally, if z lies on the positive real axis, the principle argument is 0. If z lies on the negative real axis, the principle argument is π. The principle argument is not defined for $z = 0$. Note that:

$$Arg\left(\frac{z_1}{z_2}\right) = Arg(z_1) - Arg(z_2)$$

$$Arg(z_1.z_2) = Arg(z_1) + Arg(z_2)$$

If the resulting sum $Arg(z_1) + Arg(z_2)$ or subtraction $Arg(z_1) - Arg(z_2)$ exceeds the $(-\pi, \pi]$ range, you may need to adjust the result by adding or subtracting 2π to bring it back into the principal range. For instance,

$$z_1 = -1 + 2j \Rightarrow Arg(z_1) = 2.0344\ rad$$

$$z_2 = -2 + j \Rightarrow Arg(z_2) = 2.6779\ rad$$

$$Arg(z_1.z_2) = 2.0344 + 2.6779 = 4.7123 \notin (-\pi, \pi]$$

$$4.7123 - 2\pi = -1.5709 \in (-\pi, \pi]$$

$$Arg(z_1.z_2) = Arg((-1 + 2j)(-2 + j)) = Arg(-5j) = -\frac{\pi}{2} \approx -1.5709$$

5.8 Polar Form of Complex Numbers

A complex number $z = a + bj$ can be shown in the form of $z = re^{j(\varphi + 2k\pi)}$ $k \in \mathbb{Z}$. r and φ show the modulus (magnitude) and the principle argument, respectively. Note that the polar form of a complex number is not unique since $re^{j(\varphi + 2k\pi)} = re^{j\varphi}e^{j2k\pi} = re^{j\varphi}(\cos(2k\pi) + j\sin(2k\pi)) = re^{j\varphi}(1 + j0) = re^{j\varphi}$, $k \in \mathbb{Z}$. For instance, $\sqrt{5}e^{j1.1071}$, $\sqrt{5}e^{j(1.1071+2\pi)} = \sqrt{5}e^{j7.3903}$, $\sqrt{5}e^{j(1.1071-2\pi)} = \sqrt{5}e^{-j5.1761}$, and $\sqrt{5}e^{j(1.1071+20\pi)} = \sqrt{5}e^{j63.9390}$ all represent $1 + 2j$.

Let's study a numeric example. Polar form of $z = -1 - 2j$ is $\sqrt{(-1)^2 + (-2)^2}\, e^{j(-2.034+2k\pi)} = \sqrt{5}e^{j(-2.034+2k\pi)}, k \in \mathbb{Z}$ (Fig. 5.6).

Example 5.1 Convert $1 + 2j$ to polar form.

Solution

$$|1 + 2j| = \sqrt{1 + 4} = \sqrt{5}$$

$$Arg(1 + 2j) = 1.1071$$

$$1 + 2j = \sqrt{5}e^{j1.1071+2k\pi} \quad k = 0, \pm 1, \pm 2, \pm 3, \ldots$$

Example 5.2 Convert $-1 - 2j$ to polar form.

Solution

$$|-1-2j| = \sqrt{1+4} = \sqrt{5}$$

$$Arg(-1-2j) = -2.034$$

$$-1-2j = \sqrt{5}e^{-j2.034+2k\pi} \quad k = 0, \pm 1, \pm 2, \pm 3, \ldots$$

5.9 Representation of Polar Form on the Complex Plane

Figure 5.8 shows how points $A = 2.83e^{j\frac{3\pi}{4}} = 2.83e^{j135^\circ}$ and $B = 2e^{-j\frac{\pi}{2}} = 2e^{-j90^\circ}$ can be shown on the complex plane.

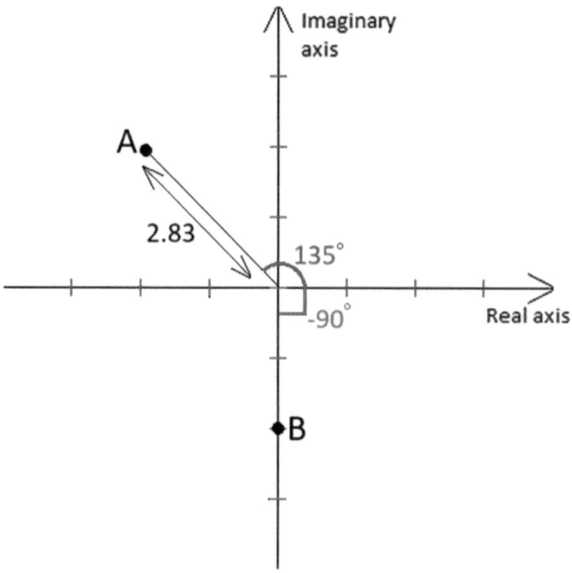

Fig. 5.8 Points $A = 2.83e^{j\frac{3\pi}{4}}$ and $B = 2e^{-j\frac{\pi}{2}}$

5.10 Basic Operations on Complex Numbers

Following examples reviews the basic operations on complex numbers.

(1) $(1 + 2j)(3 + 4j) = 1 \times 3 + 1 \times 4j + 2j \times 3 + 2j \times 4j = 3 + 4j + 6j - 8 = -5 + 10j$

(2) $\frac{(1+2j)}{(3+4j)} = \frac{(1+2j)(3-4j)}{(3+4j)(3-4j)} = \frac{3-4j+6j+8}{3^2+4^2} = \frac{11+2j}{25}$

(3) $\frac{1}{6j} = \frac{1 \times j}{6j \times j} = \frac{1 \times j}{-6} = -\frac{1}{6}j$

(4) $(1 + 2j) + (3 + 4j) = (1 + 3) + (2 + 4)j = 4 + 6j$

(5) $(1 + 2j) - (3 + 4j) = (1 - 3) + (2 - 4)j = -2 - 2j$

(6) $\frac{7e^{j\frac{\pi}{4}}}{12e^{j\frac{\pi}{3}}} = \frac{7}{12}e^{j\left(\frac{\pi}{4}-\frac{\pi}{3}\right)} = \frac{7}{12}e^{j\left(\frac{-\pi}{12}\right)} = 0.5635 - 0.1510j$

(7) $7e^{j\frac{\pi}{4}} \times 12e^{j\frac{\pi}{3}} = 7 \times 12e^{j\left(\frac{\pi}{4}+\frac{\pi}{3}\right)} = 84e^{j\frac{7\pi}{12}} = -21.7408 + 81.1378j$

(8) $7e^{j\frac{\pi}{4}} + 12e^{j\frac{\pi}{3}} = 7\left(\cos\left(\frac{\pi}{4}\right) + j\sin\left(\frac{\pi}{4}\right)\right) + 12\left(\cos\left(\frac{\pi}{3}\right) + j\sin\left(\frac{\pi}{3}\right)\right) = 4.9497 + 4.9497i + 6.0000 + 10.3923i = 10.9497 + 15.3420i$

5.11 Complex Conjugate of a Complex Number

A complex conjugate of a complex number is another complex number that has the same real part as the original complex number and the imaginary part has the same magnitude but opposite sign. The product of a complex number and its complex conjugate is a real number. Complex conjugate of $a + jb$ is $a - jb$. The notation for the complex conjugate of z is either $\overline{z}$ or z^*. In this book, we use the $\overline{z}$. Following properties are important:

$\overline{(a + jb)} = a - jb$. For instance, $\overline{(1 + 2j)} = 1 - 2j$.

$\overline{(a + jb)}(a + jb) = (a + jb)(a - jb) = a^2 + b^2$. For instance, $(1 + 2j)(1 - 2j) = 1^2 + 2^2 = 5$.

$z = re^{j\theta} \Rightarrow \overline{z} = re^{-j\theta}$. For instance, $z = 2e^{j\frac{\pi}{4}} \Rightarrow \overline{z} = 2e^{-j\frac{\pi}{4}}$.

$|z| = |\overline{z}|$. For instance, $|1 - j2| = |1 + j2| = \sqrt{1^2 + 2^2} = \sqrt{5}$.

$\overline{\left(\frac{z_1}{z_2}\right)} = \frac{\overline{z_1}}{\overline{z_2}}$. For instance, $\overline{\left(\frac{1+2j}{3+4j}\right)} = \frac{1-2j}{3-4j}$.

5.12 Geometric Interpretation of Complex Conjugation

Complex conjugation represents a reflection about the real axis. For instance, in Fig. 5.9 point $A = 1 + 2j$ and its complex conjugate $\overline{A} = 1 - 2j$ are shown.

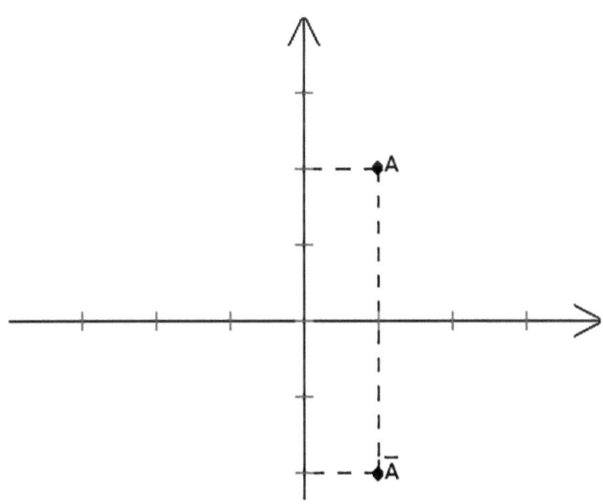

Fig. 5.9 Point A and its complex conjugate

5.13 The Complex Conjugate Root Theorem

In mathematics, the complex conjugate root theorem states that if $P(x)$ is a polynomial in one variable with real coefficients, and $a + jb$ is a root of $P(x)$, then its complex conjugate $a - jb$ is also a root of $P(x)$.

For instance, $P(x) = x^3 + 3x^2 + 7x + 5 = 0$ is polynomial in one variable with real coefficients. Its roots are, -1, $-1 + j2$, and $-1 - j2$.

5.14 Complex Exponential Function

Value of complex exponential functions can be calculated with the aid of the following formula:

$$e^{a+bj} = e^a \times e^{bj} = e^a \times (\cos(b) + j\sin(b))$$

Here are two examples:

(1) $e^{1+2j} = e^1(\cos(2) + j\sin(2)) = e^1(-0.4161 + j0.9093) = -1.1312 + 2.4717j$

(2) $e^{j\frac{\pi}{6}} = \cos\left(\frac{\pi}{6}\right) + j\sin\left(\frac{\pi}{6}\right) = \frac{\sqrt{3}}{2} + j\frac{1}{2}$

If $a < 0$, then $\lim\limits_{x\to} e^{(a+jb)x} = 0$ (note that j shows the imaginary unit). For instance, $\lim\limits_{x\to} e^{(-3+j4)x}$, $\lim\limits_{x\to} e^{(-3-j4)x}$, or $\lim\limits_{x\to} e^{-6x}$ equals to zero.

If $a > 0$, then $\lim\limits_{x\to -} e^{(a+jb)x} = 0$ (note that j shows the imaginary unit). For instance, $\lim\limits_{x\to -} e^{(7+j2)x}$, $\lim\limits_{x\to -} e^{(7-j2)x}$, or $\lim\limits_{x\to -} e^{12x}$equals to zero.

5.15 Polar Form of Complex Numbers

A complex number $z = a + bj$ can be shown in the form of $z = re^{j\theta}$. Note that the polar form of a complex number is not unique since:

$$re^{j(\theta+2k\pi)} = re^{j\theta}e^{j2k\pi} = re^{j\theta}(\cos(2k\pi) + j\sin(2k\pi)) = re^{j\theta}(1 + j0) = re^{j\theta}, k \in \mathbb{Z}$$

For instance, $\sqrt{5}e^{j1.1071}$, $\sqrt{5}e^{j(1.1071+2\pi)} = \sqrt{5}e^{j7.3903}$, $\sqrt{5}e^{j(1.1071-2\pi)} = \sqrt{5}e^{-j5.1761}$, and $\sqrt{5}e^{j(1.1071+20\pi)} = \sqrt{5}e^{j63.9390}$ all represent $1 + 2j$.

5.16 Some of Important Equations in Complex Plane

$|z - z_0| = R$ shows a circle with center of z_0 and radius of R. For instance, equation of a circle with center of $A = 1 + 2j$ and radius of 1 (Fig. 5.10) is $|z - (1 + 2j)| = 1$ or $|z - 1 - 2j| = 1$.

Fig. 5.10 Plot of $|z - (1 + 2j)| = 1$

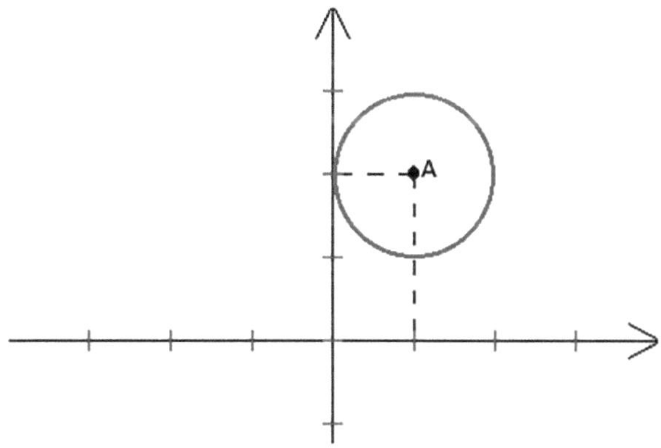

Equation of a circle with center of origin and radius of 1.2 (Fig. 5.11) is $|z - (0 + 0j)| = 1.2$ or $|z| = 1.2$.

Fig. 5.11 Plot of $|z| = 1.2$

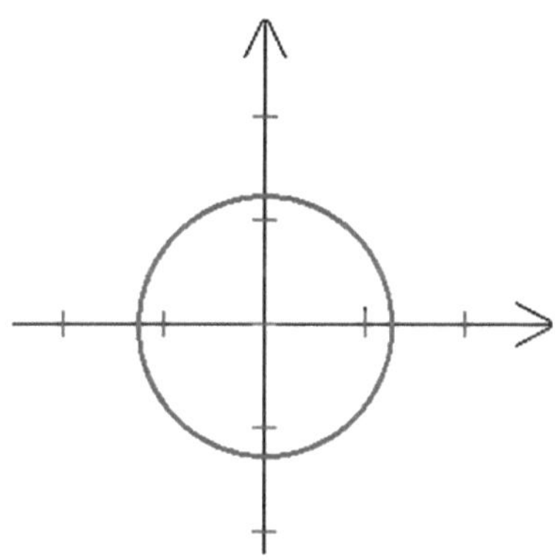

$|z - z_0| < R$ shows points fall inside a circle with center of z_0 and radius of R. $|z - z_0| \leq R$ shows points fall inside and on a circle with center of z_0 and radius of R. For instance, $|z| < 1.2$ includes points inside a circle with center of origin and radius of 1.2 (Fig. 5.12). $|z| \leq 1.2$ includes points inside and on a circle with center of origin and radius of 1.2 (Fig. 5.13).

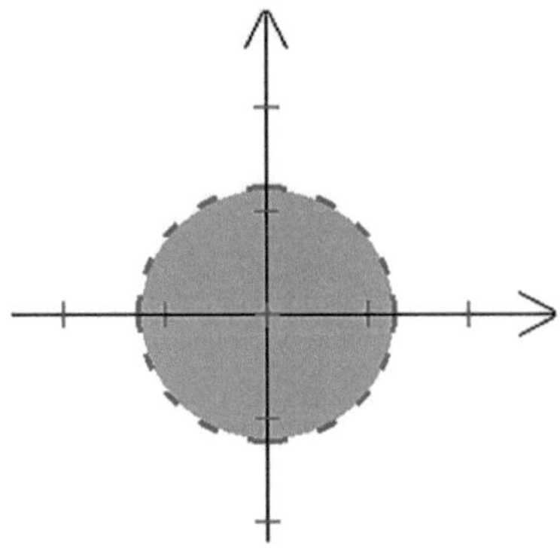

Fig. 5.12 Plot of $|z| < 1.2$

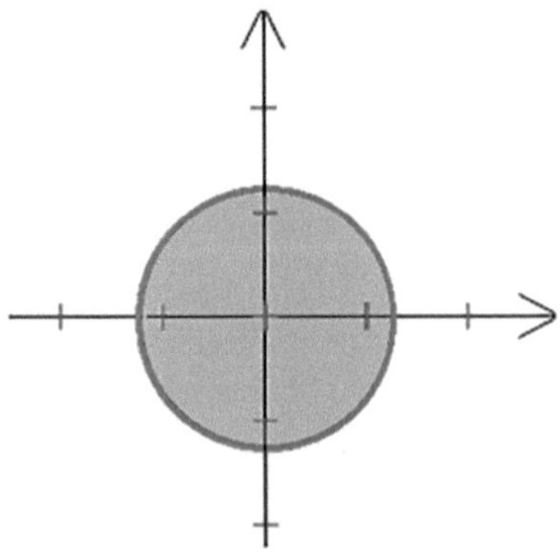

Fig. 5.13 Plot of $|z| \leq 1.2$

$|z - z_0| > R$ shows points fall outside a circle with center of z_0 and radius of R. $|z - z_0| \geq R$ shows points fall outside and on a circle with center of z_0 and radius of R. For instance, $|z - (1 + 2j)| \geq 1$ includes points outside and on a circle with center of $1 + 2j$ and radius of 1 (Fig. 5.14).

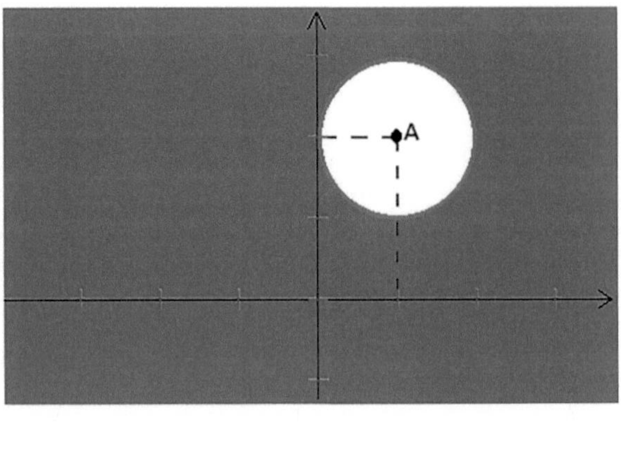

Fig. 5.14 Plot of $|z - (1 + 2j)| \geq 1$

In mathematics, an annulus (pl.: annuli or annuluses) is the region between two concentric circles. $r_1 < |z - z_0| < r_2$ shows the region between two concentric circles with center of z_0 and radiuses r_1 and r_2. For instance, points satisfy $1 \le |z - 1 - 2j| \le 1.5$ are shown in Fig. 5.15.

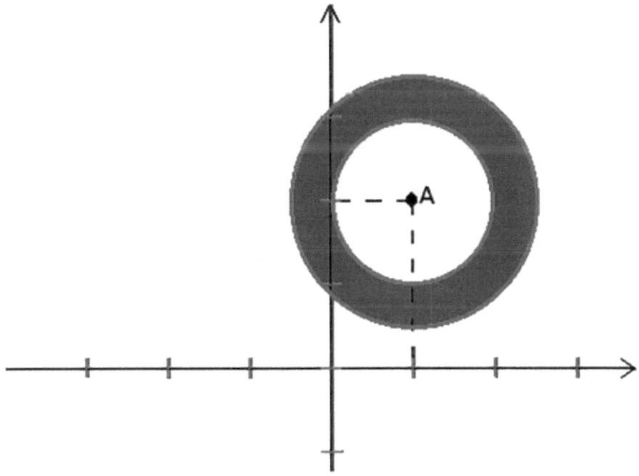

Fig. 5.15 Plot of $1 \le |z - 1 - 2j| \le 1.5$

5.17 Complex Trigonometric Functions

The sine and cosine of a complex variable z are defined as follows:

$$\sin(z) = \frac{e^{jz} - e^{-jz}}{2j}$$

$$\cos(z) = \frac{e^{jz} + e^{-jz}}{2}$$

For example,

$$\sin(1 + 2j) = \frac{e^{j(1+2j)} - e^{-j(1+2j)}}{2j} \Rightarrow \sin(1 + 2j) = \frac{e^{-2+j} - e^{2-j}}{2j}$$

$$e^{-2+j} = e^{-2}(\cos(1) + j\sin(1)) \Rightarrow e^{-2+j} = 0.0731 + 0.1139j$$

$$e^{2-j} = e^{2}(\cos(-1) + j\sin(-1)) \Rightarrow e^{2-j} = 3.9923 - 6.2177j$$

$$\sin(1+2j)=\frac{e^{-2+j}-e^{2-j}}{2j}\Rightarrow\sin(1+2j)=\frac{0.0731+0.1139j-(3.9923-6.2177j)}{2j}$$
$$\Rightarrow\sin(1+2j)=\frac{0.0731+0.1139j-3.9923+6.2177j}{2j}\Rightarrow\sin(1+2j)$$
$$=\frac{-3.9192+6.3316j}{2j}\Rightarrow\sin(1+2j)=\frac{(-3.9192+6.3316j)\times j}{2j\times j}$$
$$\Rightarrow\sin(1+2j)=\frac{-3.9192j+6.3316j^2}{2j^2}\Rightarrow\sin(1+2j)$$
$$=\frac{-3.9192j-6.3316}{-2}\Rightarrow\sin(1+2j)=1.9596j+3.1658$$
$$\Rightarrow\sin(1+2j)=3.1658+1.9596j$$

Following formulas help to calculate the value of complex trigonometric functions easily. In the following formulas sin*h*, cos *h*, tan *h*, cot *h*, sec *h*, and csc*h* denotes the hyperbolic sine, cosine, tangent, cotangent, secant, and cosecant, respectively.

(1) $\sin(a+bj)=\sin(a)\cos h(b)+j\cos(a)\sin h(b)$
(2) $\cos(a+bj)=\cos(a)\cos h(b)-j\sin(a)\sin h(b)$
(3) $\tan(a+bj)=\dfrac{\sin(2a)}{\cos(2a)+\cos h(2b)}+j\dfrac{\sin h(2b)}{\cos(2a)+\cos h(2b)}$
(4) $\cot(a+bj)=\dfrac{\cot(a).\cot h^2(b)-\cot(a)}{\cot^2(a)+\cot h^2(b)}+j\dfrac{-\cot^2(a).\cot h(b)-\cot h(b)}{\cot^2(a)+\cot h^2(b)}$
(5) $\sec(a+bj)=\dfrac{1}{\cos(a+bj)}=\dfrac{\cos(a)\cosh(b)+j\sin(a)\sinh(b)}{\cos^2(a)\cosh^2(b)+\sin^2(a)\sinh^2(b)}$
(6) $\csc(a+bj)=\dfrac{1}{\sin(a+bj)}=\dfrac{\sin(a)\cos h(b)-j\cos(a)\sin h(b)}{\sin^2(a)\cos h^2(b)+\cos^2(a)\sin h^2(b)}$

For example, $\sin(1+2j)=\sin(1)\cos h(2)+j\cos(1)\sin h(2)=3.1658+1.9596j$.

For purely imaginary arguments, i.e., arguments with zero real part,

(1) $\sin(jb)=j\sin h(b)$
(2) $\cos(jb)=\cos h(b)$
(3) $\tan(jb)=j\tan h(b)$
(4) $\cot(jb)=-j\cot h(b)$
(5) $\sec(jb)=\sec h(b)$
(6) $\csc(jb)=-j\csc h(b)$

5.18 Complex Root Function

For any non-zero complex number z and any positive integer n, the equation $w^n=z$ has exactly n distinct complex solutions $w_0, w_1, \ldots, w_{n-1}$.

Procedure for calculation of complex root is shown with an example. Assume that we want to calculate the third root of $-1-3j$, i.e., $\sqrt[3]{-1-3j}$. Given Cartesian form needs to be converted into the polar form. $-1-3j$ lies in second quadrant (Fig. 5.16).

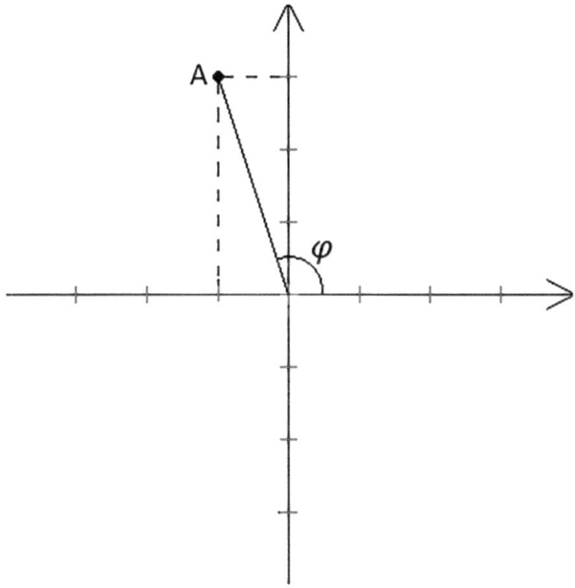

Fig. 5.16 Point $A = -1 + 3j$

$$\varphi = Arg(-1+3j) = 90° + \tan^{-1}\left(\frac{1}{3}\right) = 90° + 18.4349 = 108.4349° = 1.8925\ Rad$$

$$-1+3j = \sqrt{(-1)^2 + (-3)^2} = 3.1623e^{j(1.8925+2k\pi)} \quad k = 0, \pm 1, \pm 2, \pm 3, \ldots$$

$$\sqrt[3]{-1+3j} = \sqrt[3]{3.1623}e^{j\left(\frac{1.8925+2k\pi}{3}\right)} = 1.4678e^{j\left(\frac{1.8925+2k\pi}{3}\right)}$$

$$k=0 \Longrightarrow z_1 = 1.4678e^{j\left(\frac{1.8925+0}{3}\right)} = 1.4678e^{j0.6308} = 1.1853 + 0.8657j$$

$$k=1 \Longrightarrow z_2 = 1.4678e^{j\left(\frac{1.8925+2\pi}{3}\right)} = 1.4678e^{j2.7252} = -1.3424 + 0.5937j$$

$$k=2 \Longrightarrow z_3 = 1.4678e^{j\left(\frac{1.8925+4\pi}{3}\right)} = 1.4678e^{j4.8196} = 0.1571 - 1.4594j$$

Therefore, $\sqrt[3]{-1+3j} = \begin{cases} 1.1853 + 0.8657j \\ -1.3424 + 0.5937j \\ 0.1571 - 1.4594j \end{cases}.$

5.19 Problem Solving with MATLAB®

This section offers MATLAB examples relevant to the problems in this chapter. If you are new to MATLAB, begin with Chap. 11.

Example 5.3 Define the complex number $z_1 = 1 + 2j$ in MATLAB.

Solution You can use the methods shown in Figs. 5.17, 5.18, and 5.19.

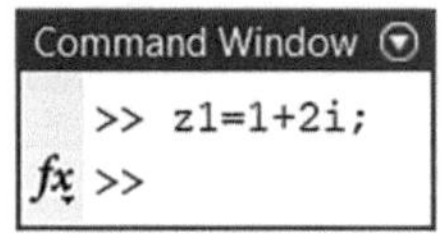

Fig. 5.17 Defining z1 as $1 + 2j$

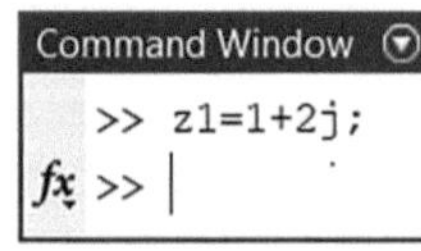

Fig. 5.18 Defining z1 as $1 + 2j$

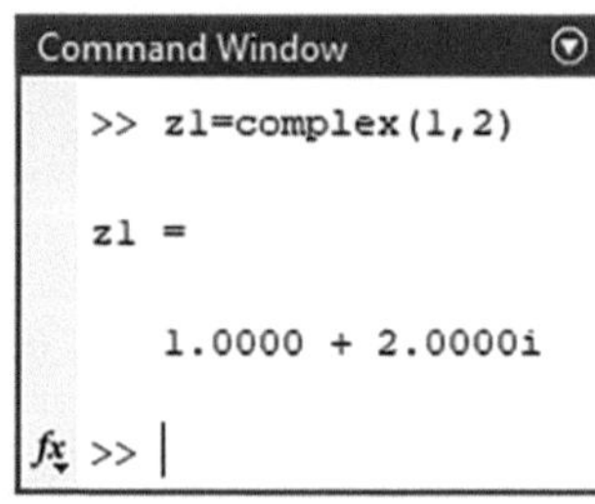

Fig. 5.19 Defining z1 as $1 + 2j$

Example 5.4 Define the complex number $z_2 = 3e^{\frac{j\pi}{3}}$ in MATLAB.

Solution You can use the command shown in Fig. 5.20. According to Fig. 5.20, $z_2 = 3e^{\frac{j\pi}{3}} = 1.5 + 2.5981j$.

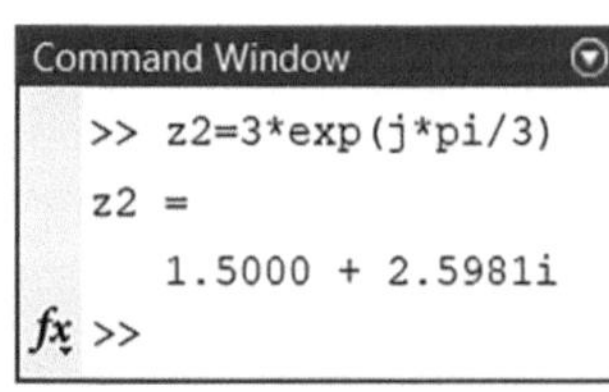

Fig. 5.20 Defining z2 as $3e^{\frac{j\pi}{3}}$

Example 5.5 Use MATLAB to calculate the real part and imaginary part of $z_2 = 3e^{\frac{j\pi}{3}}$.

Solution The commands shown in Fig. 5.21 calculate the real part and the imaginary part of $z_2 = 3e^{\frac{j\pi}{3}}$.

Fig. 5.21 Real and imaginary part of $z_2 = 3e^{j\frac{\pi}{3}}$

```
Command Window
>> z2=3*exp(j*pi/3)
z2 =
   1.5000 + 2.5981i
>> real(z2)
ans =
    1.5000
>> imag(z2)
ans =
    2.5981
fx >>
```

Example 5.6 Given $z_1 = 1 + 2i$ and $z_2 = 3 + 4i$, use MATLAB to calculate $z_1.z_2$, $\frac{z_1}{z_2}$, $z_1 + z_2$, and $z_1 - z_2$.

Solution Figure 5.22 shows the MATLAB code for calculating $z_1.z_2$, $\frac{z_1}{z_2}$, $z_1 + z_2$, and $z_1 - z_2$.

Fig. 5.22 Calculation of $z_1.z_2$, $\frac{z_1}{z_2}$, $z_1 + z_2$, and $z_1 - z_2$ for $z_1 = 1 + 2i$ and $z_2 = 3 + 4i$

```
Command Window
>> z1=1+2j;
>> z2=3+4j;
>> z1*z2
ans =
 -5.0000 +10.0000i
>> z1/z2
ans =
   0.4400 + 0.0800i
>> z1+z2
ans =
   4.0000 + 6.0000i
>> z1-z2
ans =
 -2.0000 - 2.0000i
fx >>
```

Example 5.7 Determine the modulus (magnitude) and principal argument of the complex number $z_1 = 1 - 2i$ using MATLAB.

Solution Figures 5.23 and 5.24 present two approaches to determine the modulus and principal argument of $z_1 = 1 - 2i$. According to the Figs. 5.23 or 5.24, $|z_1| = 2.2361$ and $Arg(z_1) = -1.1071$. Therefore, $z_1 = 1 - 2i = 2.2361e^{i(-1.1071 + 2k\pi)}$, where k is any integer ($k = 0, \pm 1, \pm 2, \pm 3, \ldots$).

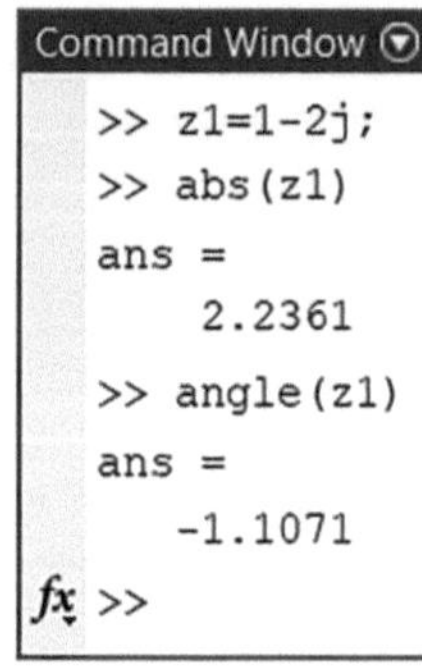

Fig. 5.23 Modulus and principle argument of $z_1 = 1 - 2i$

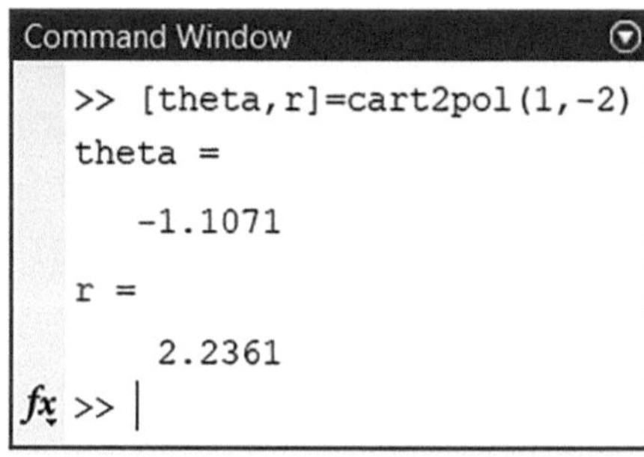

Fig. 5.24 Modulus and principle argument of $z_1 = 1 - 2i$

Example 5.8 Use MATLAB to determine the Cartesian form of $z = 2.2361e^{-j1.1071}$.

Solution You can use either of the methods shown in Figs. 5.25 or 5.26. According to the Figs. 5.25 and 5.26, $z = 2.2361e^{-j1.1071} = 1.0001 - 2.0000j$.

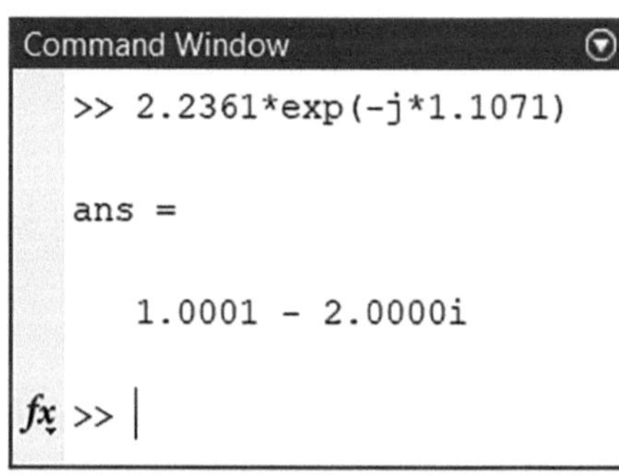

Fig. 5.25 Conversion of $z = 2.2361e^{-j1.1071}$ to Cartesian coordinates

```
Command Window
>> [x,y]=pol2cart(-1.1071,2.2361)
x =
     1.0001
y =
    -2.0000
fx >>
```

Fig. 5.26 Conversion of $z = 2.2361e^{-j1.1071}$ to Cartesian coordinates

Example 5.9 Use MATLAB to calculate the complex conjugate of $z_1 = 1 + 2j$ and $z_2 = 3e^{j\frac{\pi}{3}}$.

Solution Figure 5.27 shows the MATLAB code for calculating the complex conjugates of $z_1 = 1 + 2j$ and $z_2 = 3e^{j\frac{\pi}{3}}$.

```
Command Window
>> z1=1+2j;
>> conj(z1)
ans =
   1.0000 - 2.0000i
>> z2=3*exp(j*pi/3);
>> conj(z2)
ans =
   1.5000 - 2.5981i
fx >>
```

Fig. 5.27 Calculation of complex conjugate of $z_1 = 1 + 2j$ and $z_2 = 3e^{j\frac{\pi}{3}}$

Example 5.10 Given $z_1 = 1 + 2i$, use MATLAB to calculate sin (z_1), cos (z_1), tan (z_1), cot (z_1), sec (z_1), and cosec(z_1).

Solution Figure 5.28 shows the MATLAB code for calculating sin (z_1), cos (z_1), tan (z_1), cot (z_1), sec (z_1), and cosec(z_1).

Fig. 5.28 Calculation of trigonometric ratios for $z_1 = 1 + 2i$

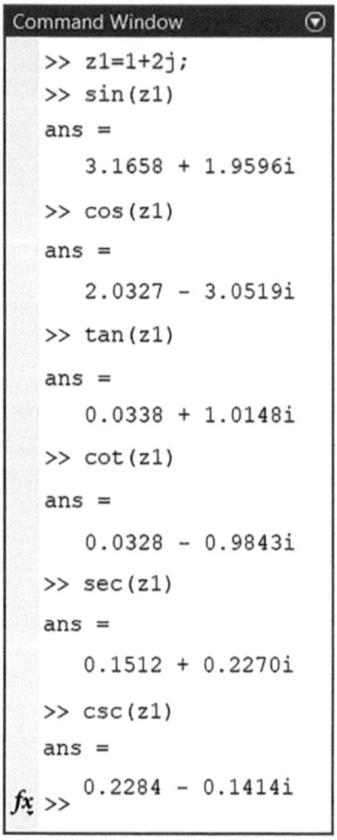

Example 5.11 Given $z_1 = 1 + 2i$ and $z_2 = 3 + 4i$, use MATLAB to calculate ${z_1}^{z_2}$ and e^{z_1}.

Solution Figure 5.29 shows the MATLAB code for calculating ${z_1}^{z_2}$ and e^{z_1}.

Fig. 5.29 Calculation of ${z_1}^{z_2}$ and e^{z_1} for $z_1 = 1 + 2i$ and $z_2 = 3 + 4i$

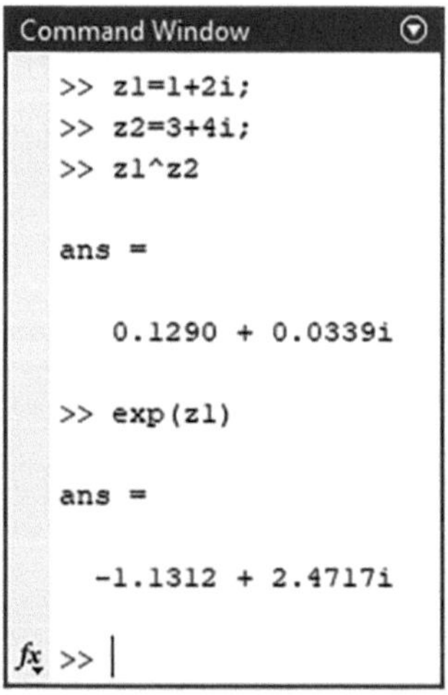

Example 5.12 Given $z = -1 + 3j$, use MATLAB to calculate $\sqrt[3]{z}$.

Solution As shown in Sect. 5.18, $\sqrt[3]{-1+3j} = \begin{cases} 1.1853 + 0.8657j \\ -1.3424 + 0.5937j \\ 0.1571 - 1.4594j \end{cases}$.

The MATLAB command shown in Fig. 5.30 calculates one of the roots only. The remaining two roots can be obtained by multiplying the initial root by $e^{j\frac{2\pi}{3}}$ and $e^{j\frac{4\pi}{3}}$ (Fig. 5.31).

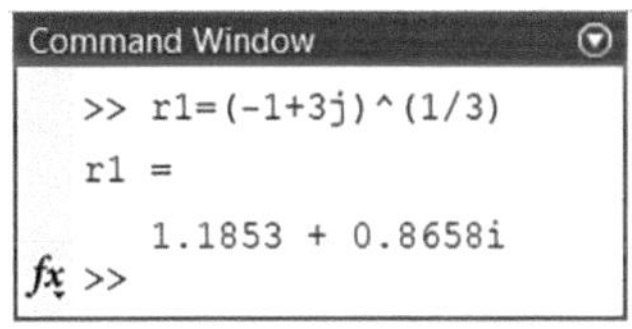

Fig. 5.30 Calculating one of the three cube roots of $-1 + 3j$

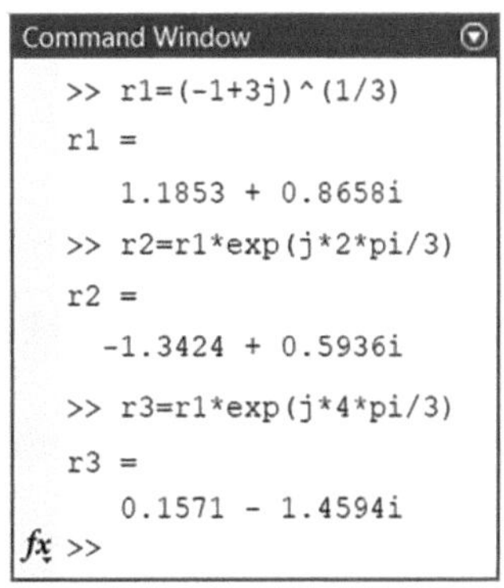

Fig. 5.31 Determining the three complex third roots of $z = -1 + 3j$

Example 5.13 Given $z = -1 + 3j$, use MATLAB to calculate $\sqrt[5]{z}$.

Solution Figure 5.32 shows the MATLAB code for calculating the five values of $\sqrt[5]{-1+3j}$.

Fig. 5.32 Determining the five complex fifth roots of $z = 1 + 3j$

```
Command Window
>> r1=(-1+3j)^(1/5)
r1 =
   1.1698 + 0.4652i
>> r2=r1*exp(j*2*pi/5)
r2 =
  -0.0810 + 1.2563i
>> r3=r1*exp(j*2*2*pi/5)
r3 =
  -1.2198 + 0.3112i
>> r4=r1*exp(j*3*2*pi/5)
r4 =
  -0.6730 - 1.0640i
>> r5=r1*exp(j*4*2*pi/5)
r5 =
   0.8039 - 0.9688i
>>
```

References for Further Study

1. Engineering Mathematics with MATLAB® and Simulink®, Farzin Asadi, Springer, 2025.
2. Applied Numerical Analysis with MATLAB®/Simulink®: For Engineers and Scientists, Farzin Asadi, Springer, 2023.
3. Advanced Engineering Mathematics, Erwin Kreyszig, Wiley, 2020.
4. A Problem-Solving Approach to Single-Variable Calculus (Volume II), Farzin Asadi, Springer, 2026.
5. A Problem-Solving Approach to Multi-Variable Calculus (Volume I), Farzin Asadi, Springer, 2026.
6. A Problem-Solving Approach to Multi-Variable Calculus (Volume II), Farzin Asadi, Springer, 2026.
7. A Problem-Solving Approach to Multi-Variable Calculus (Volume III), Farzin Asadi, Springer, 2026.
8. A Problem-Solving Approach to Ordinary Differential Equations (Volume I), Farzin Asadi, Springer, 2026.
9. A Problem-Solving Approach to Ordinary Differential Equations (Volume II), Farzin Asadi, Springer, 2026.
10. A Problem-Solving Approach to Ordinary Differential Equations (Volume III), Farzin Asadi, Springer, 2026.
11. A Problem-Solving Approach to Ordinary Differential Equations (Volume IV), Farzin Asadi, Springer, 2026.

Chapter 6
Limits

6.1 Introduction

A limit asks the question: What value does the output of a function ($f(x)$) get arbitrarily close to as the input (x) gets arbitrarily close to a certain number (c), but is not equal to c?

In mathematical notation, this is written as:

$$\lim_{x \to c} f(x) = L$$

This means that as x approaches c (from the left side and the right side), the value of $f(x)$ approaches the value L.

The notation $\lim_{x \to 1^-} f(x)$ means the limit of a function as x approaches 1 from the left-hand side. This is also known as a left-hand limit.

In simple terms, you're examining the behavior of a function as the x-values get closer and closer to 1, but always stay slightly less than 1 (e.g., 0.9, 0.99, 0.999, …).

This is in contrast to a right-hand limit, denoted as $\lim_{x \to 1^+} f(x)$, where x approaches 1 from values greater than 1 (e.g., 1.1, 1.01, 1.001, …).

The overall limit of a function at a point exists if and only if both the left-hand and right-hand limits exist and are equal.

If the limits of $f(x)$ or $g(x)$ (or both) as x approaches c do not exist, then we cannot make a general statement about the existence of the limits of $f(x) \pm g(x)$, $f(x) \times g(x)$, or $\frac{f(x)}{g(x)}$. The limit of the combined function might or might not exist; it depends entirely on the specific functions.

Let's consider two functions where the individual limits do not exist: $f(x) = \frac{1}{x-1}$ and $g(x) = \frac{-1}{x-1}$. Neither $\lim_{x \to 1} f(x)$ nor $\lim_{x \to 1} g(x)$ exists. However, the limit of their sum does exist: $\lim_{x \to 1} f(x) + g(x) = \lim_{x \to 1} \frac{1}{x-1} + \frac{-1}{x-1} = \lim_{x \to 1} 0 = 0$.

F. Asadi, *A Problem-Solving Approach to Single-Variable Calculus*,
https://doi.org/10.1007/978-3-032-13505-6_6

Consider these two functions, $f(x) = \frac{1}{x}$ and $g(x) = x$. The limit $\lim_{x \to 0} f(x)$ does not exist. The limit $\lim_{x \to 0} g(x)$ exists and is 0. However, the limit of their product does exist: $\lim_{x \to 0} f(x) \times g(x) = \lim_{x \to 0} \frac{1}{x} \times x = \lim_{x \to 0} 1 = 1$.

6.1.1 *Basic Limit Rules*

Following rules are the foundational tools for evaluating limits analytically.

(1) $\lim_{x \to a} c = c$

(2) $\lim_{x \to a} cf(x) = c \lim_{x \to a} f(x)$

(3) $\lim_{x \to a} f(x) \pm g(x) = \lim_{x \to a} f(x) \pm \lim_{x \to a} g(x)$

(4) $\lim_{x \to a} f(x) g(x) = \lim_{x \to a} f(x) \times \lim_{x \to a} g(x)$

(5) $\lim_{x \to a} \frac{f(x)}{g(x)} = \frac{\lim_{x \to a} f(x)}{\lim_{x \to a} g(x)}, if \lim_{x \to a} g(x) \neq 0$

(6) $\lim_{x \to a} f(x)^n = (\lim_{x \to a} f(x))^n for\ any\ real\ number\ n$

(7) The limit of the n-th root of a function is the n-th root of the limit of the function, provided $\lim_{x \to a} f(x) \geq 0$ when n is an even integer: $\lim_{x \to a} \sqrt[n]{f(x)} = \sqrt[n]{\lim_{x \to a} f(x)}$.

6.1.2 *The Squeeze Theorem*

The Squeeze Theorem, also known as the Sandwich Theorem, is a mathematical principle used to find the limit of a function that's difficult to evaluate directly. It states that if a function, say $f(x)$, is "squeezed" between two other functions, $g(x)$ and $h(x)$, that both approach the same limit at a given point, then $f(x)$ must also approach that same limit at the same point.

In formal terms, if we have three functions $g(x)$, $f(x)$, and $h(x)$ such that:

1. $g(x) \leq f(x) \leq h(x)$ for all x in an open interval containing some point c (except possibly at c itself).
2. $\lim_{x \to c} g(x) = L$
3. $\lim_{x \to c} h(x) = L$

Then, it must be true that: $\lim_{x \to c} f(x) = L$.

6.1.3 The L'Hôpital's Rule

L'Hôpital's Rule allows you to evaluate limits of indeterminate forms, such as $\frac{0}{0}$ or $\frac{\infty}{\infty}$. The rule states: If the limit of the ratio of two functions $f(x)$ and $g(x)$ as x approaches a, i.e., $\lim\limits_{x \to a} \frac{f(x)}{g(x)}$, is an indeterminate form $\frac{0}{0}$ or $\frac{\infty}{\infty}$, then $\lim\limits_{x \to a} \frac{f(x)}{g(x)} = \lim\limits_{x \to a} \frac{f'(x)}{g'(x)}$, where $f'(x)$ and $g'(x)$ are the derivatives of $f(x)$ and $g(x)$, respectively. For example, $\lim\limits_{x \to 2} \left(\frac{2x^3 - 5x^2+3x-2}{2x-4}\right) = \lim\limits_{x \to 2} \left(\frac{6x^2 - 10x+3}{2}\right) = \frac{7}{2}$.

If after applying L'Hôpital's Rule once, the new limit **still** results in an indeterminate form $\frac{0}{0}$ or $\frac{\infty}{\infty}$, you can **differentiate numerator and denominator again** and apply the rule again.

For example, $\lim\limits_{x \to 0} \frac{1-\cos(x)}{x^2} = \frac{0}{0}$. Therefore, we can apply the L'Hôpital's rule:

First application: $\lim\limits_{x \to 0} \frac{1-\cos(x)}{x^2} = \lim\limits_{x \to 0} \frac{\sin(x)}{2x} = \frac{0}{0}$

Second application: $\lim\limits_{x \to 0} \frac{1-\cos(x)}{x^2} = \lim\limits_{x \to 0} \frac{\sin(x)}{2x} = \lim\limits_{x \to 0} \frac{\cos(x)}{2} = \frac{1}{2}$

Let's study another example: $\lim\limits_{x \to 0} \frac{\sin(x) - x + \frac{x^3}{6}}{x^4}$ requires L'Hôpital's Rule four times:

$$\lim_{x \to 0} \frac{\sin(x) - x + \frac{x^3}{6}}{x^4} = \frac{0}{0}$$

First application: $\lim\limits_{x \to 0} \frac{\sin(x) - x + \frac{x^3}{6}}{x^4} = \lim\limits_{x \to 0} \frac{\cos(x) - 1 + \frac{x^2}{2}}{4x^3} = \frac{0}{0}$

Second application: $\lim\limits_{x \to 0} \frac{\sin(x) - x + \frac{x^3}{6}}{x^4} = \lim\limits_{x \to 0} \frac{\cos(x) - 1 + \frac{x^2}{2}}{4x^3} = \lim\limits_{x \to 0} \frac{-\sin(x)+x}{12x^2} = \frac{0}{0}$

Third application: $\lim\limits_{x \to 0} \frac{\sin(x) - x + \frac{x^3}{6}}{x^4} = \lim\limits_{x \to 0} \frac{\cos(x) - 1 + \frac{x^2}{2}}{4x^3} = \lim\limits_{x \to 0} \frac{-\sin(x)+x}{12x^2} = \lim\limits_{x \to 0} \frac{-\cos(x)+1}{24x} = \frac{0}{0}$

Fourth application: $\lim\limits_{x \to 0} \frac{\sin(x) - x + \frac{x^3}{6}}{x^4} = \lim\limits_{x \to 0} \frac{\cos(x) - 1 + \frac{x^2}{2}}{4x^3} = \lim\limits_{x \to 0} \frac{-\sin(x)+x}{12x^2} = \lim\limits_{x \to 0} \frac{-\cos(x)+1}{24x} = \lim\limits_{x \to 0} \frac{\sin(x)}{24} = \frac{0}{24} = 0$

6.1.4 Asymptotically Equivalence

Two functions, $f(x)$ and $g(x)$, are said to be asymptotically equivalent as $x \to a$ if their ratio approaches 1, i.e., $\lim\limits_{x \to a} \frac{f(x)}{g(x)} = 1$. This is denoted as $f(x) \sim g(x)$ as $x \to a$.

Some common equivalent infinitesimals as $u \to 0$ are:

(1) $\sin(u) \sim \begin{cases} u \\ u - \frac{u^3}{6} \end{cases}$

(2) $\sin^n(u) = (sin(u))^n \sim u^n$

(3) $\sin^{-1}(u) \sim \begin{cases} u \\ u + \frac{u^3}{6} \end{cases}$

(4) $\cos(u) \sim 1 - \frac{u^2}{2}$

(5) $\cos^n(u) \sim 1 - \frac{nu^2}{2}$

(6) $\tan(u) \sim \begin{cases} u \\ u + \frac{u^3}{3} \end{cases}$

(7) $\tan^{-1}(u) \sim \begin{cases} u \\ u - \frac{u^3}{3} \end{cases}$

(8) $e^u \sim 1 + u$

(9) $\ln(1 + u) \sim u$

(10) $(1 + u)^n \sim 1 + nu$

(11) $\sqrt[n]{1+u} \sim 1 + \frac{u}{n}$

(12) For a polynomial $P(x) = a_nx^n + a_{n-1}x^{n-1} + a_{n-2}x^{n-2} + \ldots + a_0$, its behavior as $x \to 0$ is dominated by the term with the lowest degree. For instance, as $x \to 0$, $4x^3 - 2x^2 - 7x \sim -7x$, or $4x^3 - 2x^2 - 7x + 12 \sim 12$.

(13) For a polynomial $P(x) = a_nx^n + a_{n-1}x^{n-1} + a_{n-2}x^{n-2} + \ldots + a_0$, its behavior as $x \to \infty$ is dominated by the term with the highest degree. For instance, as $x \to +\infty$ or $x \to -\infty$, $4x^3 - 2x^2 - 7x \sim 4x^3$.

(14) When $x \to \infty$, $\sqrt[n]{a_nx^n + a_{n-1}x^{n-1} + a_{n-2}x^{n-2} + \ldots + a_0} \sim \begin{cases} \sqrt[n]{a_n}\left|x + \dfrac{a_{n-1}}{na_n}\right| & a_n > 0, n \text{ even} \\ \sqrt[n]{a_n}\left(x + \dfrac{a_{n-1}}{na_n}\right) & n \text{ odd} \end{cases}$.

This formula essentially gives you a more precise approximation than just using the leading term, $\sqrt[n]{a_nx^n}$, which simplifies to $\sqrt[n]{a_n}x$. Note that the radical's order, n, equals the highest degree of the polynomial's leading term inside the radical. For instance, when $x \to \infty$, $\sqrt{x^2 + 3x + 7} \sim \sqrt{1}\left|x + \frac{-3}{2}\right| = |x - 1.5|$, $\sqrt[3]{-x^3 - 3x^2 + x + 3} \sim \sqrt[3]{-1}\left(x + \frac{-3}{-3}\right) = -(x + 1)$, and $\sqrt[4]{3x^4 - 3x^2 + x + 3} \sim \sqrt[4]{3}\left|x + \frac{0}{12}\right| = \sqrt[4]{3}|x|$.

When using equivalent infinitesimals, always check the **lowest-order (i.e., first-order) terms** first. If those terms do **not cancel**, the first-order approximation is sufficient. If they do **cancel and give 0**, then you must expand further until you reach the first **non-zero term**. For instance,

(1) $\lim\limits_{x\to0} \frac{\sin(x)}{2x} = \lim\limits_{x\to0} \frac{x}{2x} = \frac{1}{2}$

(2) $\lim\limits_{x\to0} \dfrac{\sin(x) - \tan(x)}{2x^3} = \begin{cases} \lim\limits_{x\to0} \dfrac{x - x}{2x^3} = \lim\limits_{x\to0} \dfrac{0}{2x^3} (!) \\ \lim\limits_{x\to0} \dfrac{x - \frac{x^3}{6} - \left(x + \frac{x^3}{3}\right)}{2x^3} = \lim\limits_{x\to0} \dfrac{-\frac{x^3}{6} - \frac{x^3}{3}}{2x^3} = \lim\limits_{x\to0} \dfrac{-\frac{3x^3}{6}}{2x^3} = -\dfrac{1}{4} \end{cases}$

6.1.5 The Indeterminate Form 0 × ∞

To resolve this indeterminate form, the primary strategy is to algebraically rewrite the product into a quotient ($\frac{0}{0}$ or $\frac{\infty}{\infty}$) so that L'Hôpital's Rule can be applied.

If you have a limit of the form $\lim_{x\to a} f(x)g(x)$ where $\lim_{x\to a} f(x) = 0$ and $\lim_{x\to a} f(x) = \pm\infty$, you have two options for conversion:

1. **Convert to $\frac{\infty}{\infty}$ Form:** Leave the function that goes to infinity ($g(x)$) in the numerator and move the function that goes to zero ($f(x)$) to the denominator by taking its reciprocal:

$$\lim_{x\to a} f(x)g(x) = \lim_{x\to a} \frac{1}{\frac{1}{f(x)}} g(x) = \lim_{x\to a} \frac{g(x)}{\frac{1}{f(x)}} = \frac{\pm\infty}{\pm\infty}$$

   Since $f(x) \to 0$, its reciprocal, $\frac{1}{f(x)}$, must approach $\pm\infty$. This creates the $\frac{\infty}{\infty}$ form.

2. **Convert to $\frac{0}{0}$ Form:** Leave the function that goes to zero ($f(x)$) in the numerator and move the function that goes to infinity ($g(x)$) to the denominator by taking its reciprocal:

$$\lim_{x\to a} f(x)g(x) = \lim_{x\to a} f(x)\frac{1}{\frac{1}{g(x)}} = \lim_{x\to a} \frac{f(x)}{\frac{1}{g(x)}} = \frac{0}{0}$$

   Since $g(x) \to \pm\infty$, its reciprocal, $\frac{1}{g(x)}$, must approach 0. This creates the $\frac{0}{0}$ form.

Once the limit is converted to one of these two quotient forms, you can repeatedly apply L'Hôpital's Rule until the limit can be evaluated.

For example, $\lim_{x\to 0^+} x^2 \ln(x) = 0 \times -\infty$. Therefore,

$$\lim_{x\to 0^+} x^2 \ln(x) = \lim_{x\to 0^+} \frac{\ln(x)}{\frac{1}{x^2}} = \lim_{x\to 0^+} \frac{\ln(x)}{x^{-2}} = \frac{-\infty}{+\infty}$$

$$\lim_{x\to 0^+} x^2 \ln(x) = \lim_{x\to 0^+} \frac{\ln(x)}{\frac{1}{x^2}} = \lim_{x\to 0^+} \frac{\ln(x)}{x^{-2}} = \lim_{x\to 0^+} \frac{\frac{1}{x}}{-2x^{-3}} = \lim_{x\to 0^+} \frac{1}{-2x^{-2}} = \lim_{x\to 0^+} \frac{x^2}{-2} = 0$$

6.1.6 The Indeterminate Form $1^{\pm\infty}$

The following formulas are useful for solving limits of the form $1^{\pm\infty}$:

1. $\lim_{x\to a} f(x)^{g(x)} = 1^{\pm\infty} \Rightarrow \lim_{x\to a} f(x)^{g(x)} = e^{\lim_{x\to a} g(x)(f(x)-1)}$
2. $\lim_{x\to\infty} \left(1 + \frac{1}{x}\right)^x = e$

3. $\lim_{x \to 0} (1 + x)^{\frac{1}{x}} = e$
4. $\lim_{x \to \infty} \left(1 + \frac{a}{x}\right)^{bx} = e^{ab}$
5. $\lim_{x \to 0} (1 + ax)^{\frac{b}{x}} = e^{ab}$

6.1.7 The Indeterminate Form 0^0

Limits of the indeterminate form 0^0 are solved by using logarithmic differentiation to convert the expression into a product or quotient form, which can then typically be handled using L'Hôpital's Rule. The 0^0 form arises in limits of the type:

$$\lim_{x \to a} f(x)^{g(x)}, \quad \text{where } \lim_{x \to a} f(x) = 0 \quad \text{and} \quad \lim_{x \to a} g(x) = 0.$$

Here is the step-by-step method for solving these limits.

Let L be the value of the limit:

$$L = \lim_{x \to a} f(x)^{g(x)}$$

Take the natural logarithm (ln) of the expression:

$$\ln(L) = \ln\left(\lim_{x \to a} f(x)^{g(x)} \right) = \lim_{x \to a} \ln\left(f(x)^{g(x)}\right) = \lim_{x \to a} g(x) \ln(f(x))$$

When $x \to a$, the new expression $g(x)$ ln $(f(x))$ approaches: (0). $(\ln(0^+)) = 0.\ (-\infty)$.

This is a new indeterminate form, $0.\ (-\infty)$, which can be easily converted into a quotient suitable for L'Hôpital's Rule. Rewrite the product as a quotient by moving one of the terms to the denominator with a negative exponent (its reciprocal):

$$\ln(L) = \lim_{x \to a} g(x) \ln(f(x)) = \lim_{x \to a} \frac{1}{\frac{1}{g(x)}} \ln(f(x)) = \lim_{x \to a} \frac{\ln(f(x))}{\frac{1}{g(x)}}$$

This is the indeterminate form $\frac{\infty}{\infty}$, which is ready for L'Hôpital's Rule. Apply L'Hôpital's Rule to the quotient to find the value, say M:

$$M = \lim_{x \to a} \frac{\frac{d}{dx}(\ln(f(x)))}{\frac{d}{dx}\left(\frac{1}{g(x)}\right)}$$

Remember that the value M you found is the limit of $\ln(L)$, not L itself:

$$\ln(L) = M$$

Therefore, the original limit L is found by exponentiating the result:

$$L = e^M$$

While the indeterminate form $0. (-\infty)$ can theoretically be converted to either $\frac{0}{0}$ or $\frac{-\infty}{\infty}$, we strongly recommend converting to the $\frac{-\infty}{\infty}$ form (by rewriting the product as $\frac{\ln(f(x))}{\frac{1}{g(x)}}$). Although $\frac{0}{0}$ is mathematically valid (i.e., $\frac{g(x)}{\frac{1}{\ln(f(x))}}$), its resulting denominator typically involves complex derivatives like $-\frac{1}{(\ln(f(x)))^2}\frac{f'(x)}{f(x)}$, which often leads to significantly more complicated algebraic manipulation when applying L'Hôpital's Rule. Converting to $\frac{-\infty}{\infty}$ usually keeps the derivatives simpler and minimizes the risk of algebraic errors.

For example, for $\lim_{x\to 0^+} x^x$:

$$L = \lim_{x\to 0^+} x^x \Rightarrow \ln(L) = \lim_{x\to 0^+} x\ln(x) = 0 \times -\infty$$

$$\ln(L) = \lim_{x\to 0^+} x\ln(x) = \lim_{x\to 0^+} \frac{\ln(x)}{\frac{1}{x}} = \frac{-\infty}{\infty} \Rightarrow \ln(L) = \lim_{x\to 0^+} \frac{\ln(x)}{\frac{1}{x}} = \lim_{x\to 0^+} \frac{\frac{1}{x}}{\frac{-1}{x^2}} = \lim_{x\to 0^+} -x = 0$$

$$\ln(L) = 0 \Rightarrow L = 1 \Rightarrow \lim_{x\to 0^+} x^x = 1$$

As an exercise show that $\lim_{x\to 0^+} x^{\tan(x)} = 1$.

6.1.8 The Indeterminate Form $\infty - \infty$

The indeterminate form $\infty - \infty$ is primarily treated by algebraically manipulating the expression to convert it into the indeterminate form $\frac{0}{0}$ or $\frac{\infty}{\infty}$. Once in a quotient form, L'Hôpital's Rule can be applied.

Here are the three most common strategies for the $\infty - \infty$ form, depending on the structure of the limit:

6.1.8.1 Finding a Common Denominator (for Rational Expressions)

This method is used when the expression is the difference between two rational functions (fractions). Assume $\lim_{x\to a} \frac{f(x)}{g(x)} - \frac{h(x)}{k(x)} = \infty - \infty$ Combine the terms into a single fraction:

$$\lim_{x\to a}\frac{f(x)}{g(x)}-\frac{h(x)}{k(x)}=\lim_{x\to a}\frac{f(x)k(x)-h(x)g(x)}{g(x)k(x)}$$

This combined fraction will often result in the $\frac{0}{0}$ or $\frac{\infty}{\infty}$ form, allowing you to use L'Hôpital's Rule. For example, $\lim_{x\to 0}\left(\frac{1}{x}-\frac{1}{\sin(x)}\right)=\infty-\infty$ Let's combine using a common denominator (x. sin (x)):

$$\lim_{x\to 0}\left(\frac{1}{x}-\frac{1}{\sin(x)}\right)=\lim_{x\to 0}\left(\frac{\sin(x)-x}{x\sin(x)}\right)=\frac{0}{0}$$

Apply L'Hôpital's Rule (First Time): $\lim_{x\to 0}\left(\frac{1}{x}-\frac{1}{\sin(x)}\right)=\lim_{x\to 0}\left(\frac{\sin(x)-x}{x\sin(x)}\right)=\lim_{x\to 0}\left(\frac{\cos(x)-1}{\sin(x)+x\cos(x)}\right)=\frac{0}{0}$

Apply L'Hôpital's Rule (Second Time): $\lim_{x\to 0}\left(\frac{1}{x}-\frac{1}{\sin(x)}\right)=\lim_{x\to 0}\left(\frac{\sin(x)-x}{x\sin(x)}\right)=\lim_{x\to 0}\left(\frac{\cos(x)-1}{\sin(x)+x\cos(x)}\right)=\lim_{x\to 0}\left(\frac{-\sin(x)}{\cos(x)+\cos(x)-x\sin(x)}\right)=\frac{0}{2}=0$

6.1.8.2 Rationalizing (for Radical Expressions)

This technique is used when the expression involves a difference of square roots:

$$\lim_{x\to\infty}\left(\sqrt{f(x)}-\sqrt{g(x)}\right)$$

Multiply the expression by its conjugate:

$$\lim_{x\to\infty}\left(\sqrt{f(x)}-\sqrt{g(x)}\right)=\lim_{x\to\infty}\left(\left(\sqrt{f(x)}-\sqrt{g(x)}\right)\left(\frac{\sqrt{f(x)}+\sqrt{g(x)}}{\sqrt{f(x)}+\sqrt{g(x)}}\right)\right)$$
$$=\lim_{x\to\infty}\left(\frac{f(x)-g(x)}{\sqrt{f(x)}+\sqrt{g(x)}}\right)$$

This will yield the form $\frac{\infty}{\infty}$ or $\frac{0}{0}$ and is often solvable by dividing the numerator and denominator by the highest power of x.

For example, $\lim_{x\to+\infty}\left(\sqrt{x^2+3x}-x\right)=\infty-\infty$ Therefore,

$$\lim_{x\to+\infty}\left(\sqrt{x^2+3x}-x\right)=\lim_{x\to+\infty}\left(\sqrt{x^2+3x}-x\right)\times\frac{\sqrt{x^2+3x}+x}{\sqrt{x^2+3x}+x}$$

$$=\lim_{x\to+\infty}\left(\frac{x^2+3x-x^2}{\sqrt{x^2+3x}+x}\right)=\lim_{x\to+\infty}\left(\frac{3x}{\sqrt{x^2\left(1+\frac{3x}{x^2}\right)}+x}\right)$$

$$=\lim_{x\to+\infty}\left(\frac{3x}{\sqrt{x^2\left(1+\frac{3}{x}\right)}+x}\right)=\lim_{x\to+\infty}\left(\frac{3x}{|x|+x}\right)$$

$$=\lim_{x\to+\infty}\left(\frac{3x}{x+x}\right)=\lim_{x\to+\infty}\left(\frac{3x}{2x}\right)=\lim_{x\to+\infty}\left(\frac{3}{2}\right)=\frac{3}{2}$$

Therefore, $\lim_{x\to+\infty}\left(\sqrt{x^2+3x}-x\right)=\frac{3}{2}$.

6.1.8.3 Factoring or Converting to a Quotient (for Logarithms and Exponentials)

This applies to expressions involving logarithms or other functions where a common factor can be isolated.

For example, $\lim_{x\to 0^+}(\ln(x^2+x)-\ln(x))=-\infty-(-\infty)=\infty-\infty$ Therefore,

$$\lim_{x\to 0^+}\left(\ln\left(x^2+x\right)-\ln(x)\right)=\lim_{x\to 0^+}\left(\ln\left(\frac{x^2+x}{x}\right)\right)=\lim_{x\to 0^+}(\ln(x+1))=\ln(1)=0$$

As another example, consider $\lim_{x\to\infty}(x-e^x)=\infty-\infty$ Factor out the dominant term, e^x (since exponential growth is faster than polynomial growth):

$$\lim_{x\to\infty}(x-e^x)=\lim_{x\to\infty}\left(e^x\left(\frac{x}{e^x}-1\right)\right)$$

The limit $\lim_{x\to\infty}\frac{x}{e^x}=0$ (a classic result from L'Hôpital's Rule proving exponential growth dominance). Therefore,

$$\lim_{x\to\infty}(x-e^x)=\lim_{x\to\infty}\left(e^x\left(\frac{x}{e^x}-1\right)\right)=\lim_{x\to\infty}(e^x(0-1))=\lim_{x\to\infty}-e^x=-\infty$$

6.1.9 Function Growth Rate

ln(x) grows slower than any function of the form x^p(e.g., $\sqrt{x}, x^{0.001}$, or x).

$\ln(x)$ grows much more slowly than any exponential function a^x, where $a > 1$ (e.g., e^x or 2^x).

In order of increasing growth rate, we have:

$$\ln(x) \ll \text{Polynomials } (x^p) \ll \text{Exponentials } (a^x) \ll x!$$

For example, $\lim_{x \to +\infty} e^x - \ln(x) = \lim_{x \to +\infty} e^x \left(1 - \frac{\ln(x)}{e^x}\right) = \lim_{x \to +\infty} e^x (1 - 0) = \lim_{x \to +\infty} e^x = +\infty$ and $\lim_{x \to +\infty} \frac{x^3 + 3x^2}{e^x} = 0$.

6.2 Solved Problems

Example 6.1 Find the value of $\lim_{x \to 1^-} f(x)$, $\lim_{x \to 1^+} f(x)$, $\lim_{x \to 2^-} f(x)$, $\lim_{x \to 2^+} f(x)$, $\lim_{x \to 3^-} f(x)$, and $\lim_{x \to 3^+} f(x)$ in Fig. 6.1.

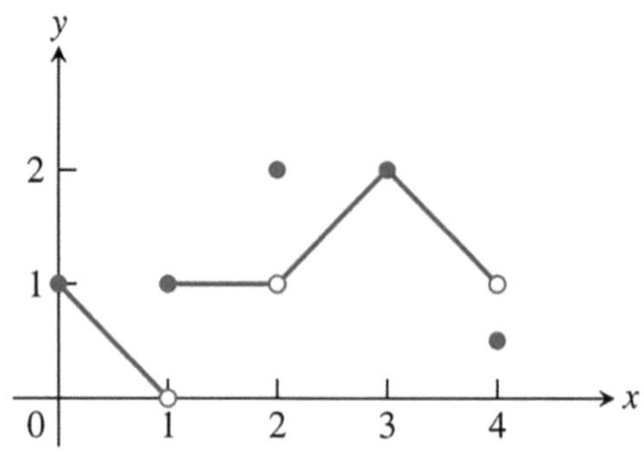

Fig. 6.1 Function for Example 6.1

Solution $\lim_{x \to 1^-} f(x) = 0$, $\lim_{x \to 1^+} f(x) = 1$, $\lim_{x \to 2^-} f(x) = 1$, $\lim_{x \to 2^+} f(x) = 1$, $\lim_{x \to 3^-} f(x) = 2$ and $\lim_{x \to 3^+} f(x) = 2$.

Example 6.2 Given the graph of the function $f(x) = \tan(x)$ in Fig. 6.2, determine the value of $\lim_{x \to \frac{\pi}{2}^-} f(x)$, $\lim_{x \to \frac{\pi}{2}^+} f(x)$, $\lim_{x \to \pi^-} f(x)$, and $\lim_{x \to \pi^+} f(x)$.

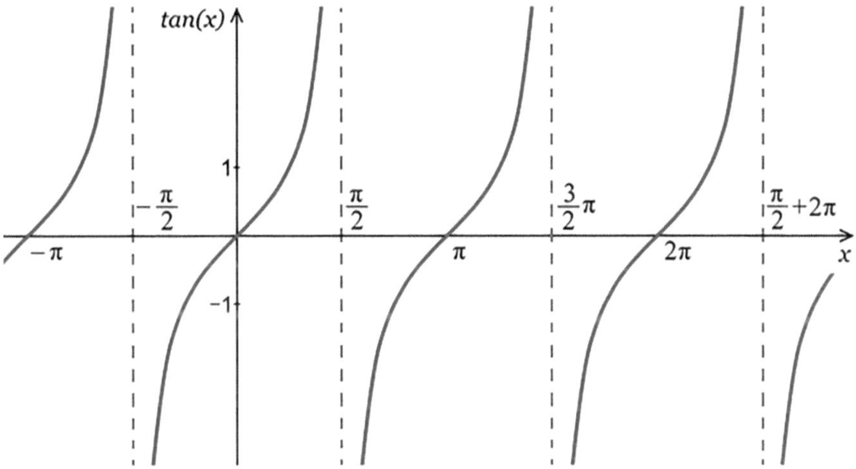

Fig. 6.2 Function for Example 6.2

Solution $\lim_{x\to\frac{\pi}{2}^-} f(x) = +\infty$, $\lim_{x\to\frac{\pi}{2}^+} f(x) = -\infty$, $\lim_{x\to\pi^-} f(x) = 0$, and $\lim_{x\to\pi^+} f(x) = 0$.

Example 6.3 Given the graph of the function $f(x) = |\tan(x)|$ in Fig. 6.3, determine the value of $\lim_{x\to\frac{\pi}{2}^-} f(x)$, $\lim_{x\to\frac{\pi}{2}^+} f(x)$, $\lim_{x\to\pi^-} f(x)$, and $\lim_{x\to\pi^+} f(x)$.

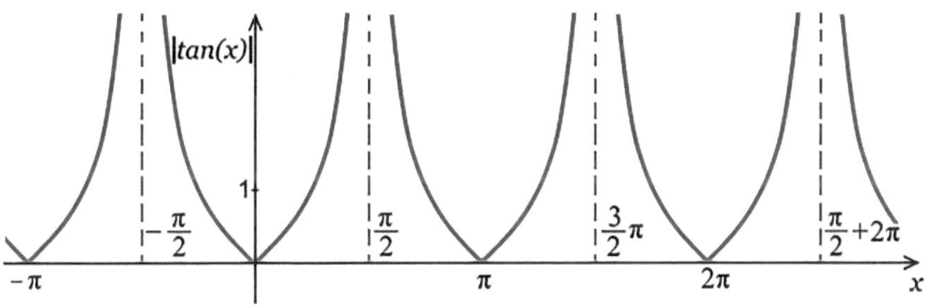

Fig. 6.3 Plot of $f(x) = |\tan(x)|$

Solution $\lim_{x\to\frac{\pi}{2}^-} f(x) = +\infty$, $\lim_{x\to\frac{\pi}{2}^+} f(x) = +\infty$, $\lim_{x\to\pi^-} f(x) = 0$, and $\lim_{x\to\pi^+} f(x) = 0$.

Example 6.4 Given the graph of the function $f(x)$ in Fig. 6.4, determine the value of $\lim_{x\to 1^+} f(1-x^2) + f(x^2) + f(x^2-1)$.

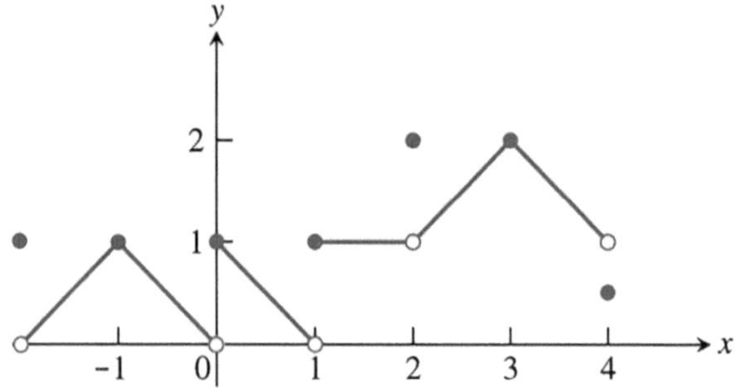

Fig. 6.4 Function for Example 6.4

Solution

$$\lim_{x\to 1^+} f\left(1-x^2\right) + f\left(x^2\right) + f\left(x^2-1\right) = \lim_{x\to 1^+} f\left(1-x^2\right) + \lim_{x\to 1^+} f\left(x^2\right)$$
$$+ \lim_{x\to 1^+} f\left(x^2-1\right) = \lim_{t\to 0^-} f(t) + \lim_{t\to 1^+} f(t) + \lim_{t\to 0^+} f(t) = 0+1+1=2.$$

Example 6.5 Given the graph of the function $f(x)$ in Fig. 6.5, determine the value of **(A)** $\lim_{x\to 1^+} f(f(f(x)))$, **(B)** $\lim_{x\to 1^-} f(f(f(x)))$, **(C)** $\lim_{x\to 1^-} f\left(\frac{1}{2x-1}\right)$, **(C)** $\lim_{x\to 1^+} f\left(\frac{1}{2x-1}\right)$.

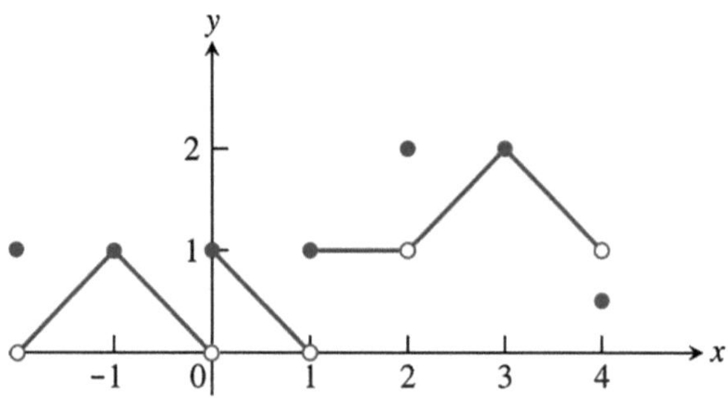

Fig. 6.5 Function for Example 6.5

Solution We need to evaluate the nested limit $\lim_{x\to 1^+} f(f(f(x),))$. We do this by working from the innermost function outward.

(A) $\lim_{x\to 1^+} f(x) = 1$

$$\lim_{x\to 1^+} f(f(f(x))) = \lim_{x\to 1^+} f(f(1)) = f(f(1)) = f(1) = 1$$

(B) $\lim_{x\to 1^-} f(x) = 0^+$

$$\lim_{x\to 1^-} f(f(f(x))) = \lim_{t\to 0^+} f(f(t)) = \lim_{u\to 1^-} f(u) = 0$$

(C) $\lim\limits_{x \to 1^-} f\left(\frac{1}{2x-1}\right) = \lim\limits_{t \to 1^+} f(t) = 1$

(D) $\lim\limits_{x \to 1^+} f\left(\frac{1}{2x-1}\right) = \lim\limits_{t \to 1^-} f(t) = 0$

Example 6.6 Calculate $\lim\limits_{x \to 2} x^2 + 3x$.

Solution

$$\lim_{x \to 2} x^2 + 3x = 2^2 + 3 \times 2 = 10$$

Example 6.7 Calculate $\lim\limits_{x \to 0} \left(\frac{x+1}{x+3}\right)$.

$$\lim_{x \to 4} \left(\frac{x+1}{x+3}\right) = \frac{5}{7}$$

Example 6.8 Calculate $\lim\limits_{x \to 3^+} \left|x^2 - 3x\right|$.

Solution To calculate the limit, we need to determine the sign of the expression inside the absolute value, $x^2 - 3x$, as x approaches 3 from the right ($x \to 3^+$). The roots of $x^2 - 3x$ are $x = 0$ and $x = 3$. The sign table is shown in Fig. 6.6.

Fig. 6.6 Sign table for $x^2 - 3x$

x	$-\infty$		0		3		$+\infty$
$x^2 - 3x$		+	0	−	0	+	

The function can be written as:

$$y = \begin{cases} x^2 - 3x & x \leq 0 \; or \; x \geq 3 \\ -(x^2 - 3x) & 0 < x < 3 \end{cases}$$

Therefore, $\lim\limits_{x \to 3^+} \left|x^2 - 3x\right| = \lim\limits_{x \to 3^+} x^2 - 3x = 3^2 - 3 \times 3 = 0$.

Example 6.9 Calculate $\lim\limits_{x \to 3^-} \left|x^2 - 3x\right|$.

Solution To calculate the limit, we need to determine the sign of the expression inside the absolute value, $x^2 - 3x$, as x approaches 3 from the left ($x \to 3^-$). We can do this by picking a number slightly less than 3, such as 2.9, and substituting it into the expression (The technique from the previous problem can also be applied).

$$x^2 - 3x \overset{x=2.9}{\Rightarrow} 2.9^2 - 3 \times 2.9 = -0.29 < 0$$

$$\lim_{x \to 3^-} \left|x^2 - 3x\right| = \lim_{x \to 3^-} -\left(x^2 - 3x\right) = 0$$

Example 6.10 Calculate (**A**) $\lim\limits_{x \to 3^-} \frac{|x^2 - 3x|}{x - 3}$ (**B**) $\lim\limits_{x \to 3^+} \frac{|x^2 - 3x|}{x - 3}$.

Solution The sign table for $|x^2 - 3x|$ is shown in Fig. 6.7.

Fig. 6.7 Sign table for $x^2 - 3x$

x	$-\infty$		0		3		$+\infty$
$x^2 - 3x$		+	0	−	0	+	

According to the sign table in Fig. 6.7,

$$\lim_{x \to 3^-} \frac{|x^2 - 3x|}{x - 3} = \lim_{x \to 3^-} \frac{-(x^2 - 3x)}{x - 3} = \lim_{x \to 3^+} \frac{-x(x-3)}{x - 3} = \lim_{x \to 3^-} -x = -3$$

$$\lim_{x \to 3^+} \frac{|x^2 - 3x|}{x - 3} = \lim_{x \to 3^+} \frac{x^2 - 3x}{x - 3} = \lim_{x \to 3^+} \frac{x(x-3)}{x - 3} = \lim_{x \to 3^-} x = 3$$

Since the left-hand limit and the right-hand limit are not equal, the two-sided limit $\lim\limits_{x \to 3} \frac{|x^2 - 3x|}{x - 3}$ does not exist.

Example 6.11 Calculate (**A**) [1.5] (**B**) [−1.5]

Solution The graph of the function $y = [x]$ is shown in Fig. 6.8. (**A**) $[1.5] = 1$ (**B**) $[-1.5] = -2$.

Fig. 6.8 Plot of $y = [x]$

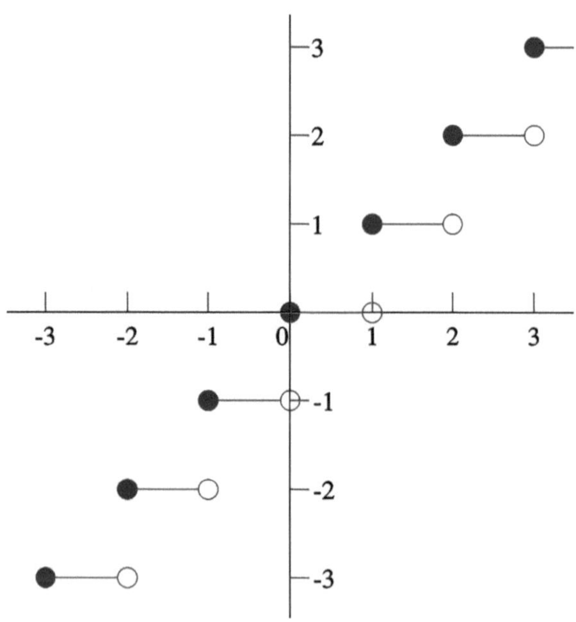

Example 6.12 Calculate (**A**) $\lim\limits_{x \to \frac{\pi}{2}^-} [\cos(x)]$ (**B**) $\lim\limits_{x \to \frac{\pi}{2}^+} [\cos(x)]$.

Solution

(A) As x approaches $\frac{\pi}{2}$ from the left side, the value of cos(x) is positive and approaches 0. For example, if x is slightly less than $\frac{\pi}{2}$, cos(x) could be 0.1, 0.01, 0.001, and so on (Fig. 6.9). Therefore, $[\cos(x)] = 0$ and $\lim\limits_{x \to \frac{\pi}{2}^-} [cos(x)] = \lim\limits_{x \to \frac{\pi}{2}^-} 0 = 0$.

(B) As x approaches $\frac{\pi}{2}$ from the right side, the value of cos(x) is negative and approaches 0. For example, if x is slightly greater than $\frac{\pi}{2}$, cos(x) could be $-$ 0.1, $-$ 0.01, $-$ 0.001, and so on (Fig. 6.9). Therefore, $[\cos(x)] = -1$ and $\lim\limits_{x \to \frac{\pi}{2}^+} [\cos(x)] = \lim\limits_{x \to \frac{\pi}{2}^+} -1 = -1$.

Since the left-hand limit and the right-hand limit are not equal, the two-sided limit $\lim\limits_{x \to \frac{\pi}{2}^+} [\cos(x)]$ does not exist.

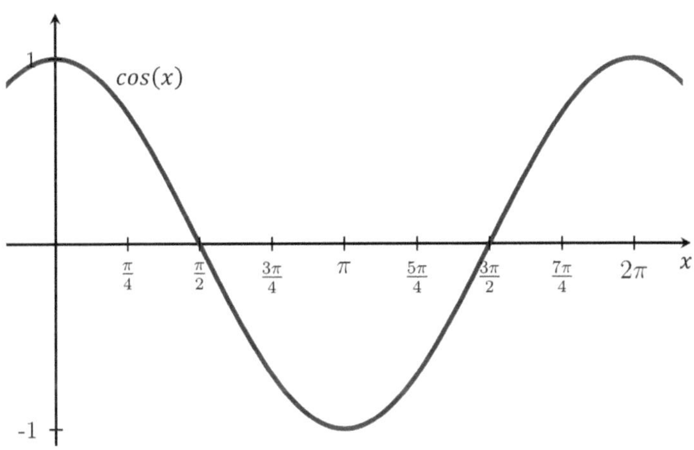

Fig. 6.9 Plot of $y = \cos(x)$

Example 6.13 Calculate (**A**) $\lim\limits_{x \to \frac{\pi}{2}^-} [\sin(2x)] + [\sin(3x)]$ (**B**) $\lim\limits_{x \to \frac{\pi}{2}^+} [\sin(2x)] + [\sin(3x)]$.

Solution

$$\lim_{x \to \frac{\pi}{2}^-} [\sin(2x)] + [\sin(3x)] = \lim_{x \to \frac{\pi}{2}^-} [\sin(2x)] + \lim_{x \to \frac{\pi}{2}^-} [\sin(3x)] = \lim_{u \to 0^+} [u] + \lim_{u \to -1^+} [u] = 0 - 1 = -1 \quad \lim_{x \to \frac{\pi}{2}^+} [\sin(2x)] + [\sin(3x)] = \lim_{x \to \frac{\pi}{2}^+} [\sin(2x)] + \lim_{x \to \frac{\pi}{2}^+} [\sin(3x)] = \lim_{u \to 0^-} [u] + \lim_{u \to -1^+} [u] = -1 - 1 = -2$$

Example 6.14 Calculate $\lim_{x\to 0}\frac{\cos(2x)+[-\sin^4(x)]}{\sin^2(x)+[\sin^6(x)]}$.

Solution

$$\lim_{x\to 0}\frac{\cos(2x)+[-\sin^4(x)]}{\sin^2(x)+[\sin^6(x)]}=\lim_{x\to 0}\frac{\cos(2x)-1}{\sin^2(x)+0}=\lim_{x\to 0}\frac{\cos(2x)-1}{\sin^2(x)}$$

$$=\lim_{x\to 0}\frac{\frac{\cos(2x)-1}{2}}{\frac{\sin^2(x)}{2}}=\lim_{x\to 0}\frac{-\frac{1-\cos(2x)}{2}}{\frac{\sin^2(x)}{2}}=\lim_{x\to 0}\frac{-\sin^2(x)}{\frac{\sin^2(x)}{2}}=\lim_{x\to 0}-2=-2$$

Example 6.15 Calculate $\lim_{x\to 0}\frac{\cos(2x)+[\sin^4(x)]}{\sin^2(x)+[\sin^6(x)]}$.

Solution

$$\lim_{x\to 0}\frac{\cos(2x)+[\sin^4(x)]}{\sin^2(x)+[\sin^6(x)]}=\lim_{x\to 0}\frac{\cos(2x)+0}{\sin^2(x)+0}=\lim_{x\to 0}\frac{\cos(2x)}{\sin^2(x)}=\frac{1}{0^+}=+\infty$$

Example 6.16 Calculate **(A)** $\lim_{x\to 2^-}[x-2]$ **(B)** $\lim_{x\to 2^+}[x-2]$.

Solution

(A) $\lim_{x\to 2^-}[x-2]=\lim_{x\to 2^-}-1=-1$

(B) $\lim_{x\to 2^+}[x-2]=\lim_{x\to 2^+}0=0$

Example 6.17 Given $f(x)=(x+a)[x+3]$ and the condition $\lim_{x\to 3^+}f(x)-\lim_{x\to 3^-}f(x)=12$, determine the value of the constant a.

Solution

$$\lim_{x\to 3^+}f(x)=\lim_{x\to 3^+}(x+a)[x+3]=\lim_{x\to 3^+}(x+a)\times 6=6(3+a)=18+6a$$

$$\lim_{x\to 3^-}f(x)=\lim_{x\to 3^-}(x+a)[x+3]=\lim_{x\to 3^-}(x+a)\times 5=5(3+a)=15+5a$$

$$\lim_{x\to 3^+}f(x)-\lim_{x\to 3^-}f(x)=12\Rightarrow 18+6a-(15+5a)=12\Rightarrow 3+a=12\Rightarrow a=9$$

Example 6.18 Calculate **(A)** $\lim_{x\to 0^-}\frac{|3x|+[x]}{|4x|-[x]}$ **(B)** $\lim_{x\to 0^+}\frac{|3x|+[x]}{|4x|-[x]}$.

Solution

$$\lim_{x\to 0^-}\frac{|3x|+[x]}{|4x|-[x]}=\lim_{x\to 0^-}\frac{-3x+(-1)}{-4x-(-1)}=\frac{-3\times 0+(-1)}{-4\times 0-(-1)}=-1$$

$$\lim_{x\to 0^+}\frac{|3x|+[x]}{|4x|-[x]}=\lim_{x\to 0^+}\frac{3x+0}{4x-0}=\lim_{x\to 0^+}\frac{3}{4}=\frac{3}{4}$$

Example 6.19 Calculate $\lim_{x\to -7^-}(x+7)\left[\frac{1}{x+7}\right]$.

Solution

$$\lim_{x\to -7^-}(x+7)\left[\frac{1}{x+7}\right]=\lim_{y\to 0^-}y\left[\frac{1}{y}\right]$$

We know that for any number u, the floor function satisfies the inequality $u-1<[u]\le u$. Applying this to our expression with $=\frac{1}{y}$, we get $\frac{1}{y}-1<\left[\frac{1}{y}\right]\le\frac{1}{y}$. Multiplying the entire inequality by y (which is a negative number since $y\to 0^-$) reverses the inequality signs:

$$\frac{1}{y}-1<\left[\frac{1}{y}\right]\le\frac{1}{y}\overset{y<0}{\Rightarrow}y\times\left(\frac{1}{y}-1\right)>y\times\left[\frac{1}{y}\right]\ge y\times\frac{1}{y}\Rightarrow 1<y\left[\frac{1}{y}\right]\le 1-y$$

As $y\to 0^-$, the term $1-y$ approaches 1. By the Squeeze Theorem, since the expression is squeezed between 1 and a value approaching 1, the limit is 1.

Exercise 6.1 Calculate $\lim_{x\to -7^+}(x+7)\left[\frac{1}{x+7}\right]$. (Ans. 1)

Example 6.20 Given $f(x)=\begin{cases}ax+1 & x<3\\ x^2+4a & x\ge 3\end{cases}$ and the condition $\lim_{x\to 3^+}f(x)-\lim_{x\to 3^-}f(x)=-2$, determine the value of the constant a.

Solution

$$\lim_{x\to 3^+}f(x)-\lim_{x\to 3^-}f(x)=-2\Rightarrow 9+4a-(3a+1)=-2\Rightarrow 8+a=-2\Rightarrow a=-10$$

Example 6.21 Given $f(x)=\begin{cases}3x+1 & x>0\\ x^2+4 & x\le 0\end{cases}$, calculate the value of $\lim_{x\to 0^-}f(x^2+2x-[x])$.

Solution

$$\lim_{x\to 0^-}f\left(x^2+2x-[x]\right)=\lim_{x\to 0^-}f\left(x^2+2x-(-1)\right)=\lim_{x\to 0^-}f\left(x^2+2x+1\right)$$

$$=\lim_{x\to 0^-}f\left((x+1)^2\right)=\lim_{x\to 0^-}\left(3(x+1)^2+1\right)=4$$

Example 6.22 Given $f(x)=\begin{cases}3x+4 & x>0\\ 5x^2-1 & x\le 0\end{cases}$, calculate the value of $\lim_{x\to 0^+}f(x^2-2x-[x])$.

Solution
$\lim_{x\to 0^+} f(x^2-2x-[x]) = \lim_{x\to 0^+} f(x^2-2x-0) = \lim_{x\to 0^+} f(x^2-2x) = \lim_{u\to 0^-} f(u) = \lim_{u\to 0^-} (5u^2-1) = -1$

Example 6.23 Given $f(x) = \begin{cases} x+1 & x\notin\mathbb{Z} \\ 2 & x\in\mathbb{Z} \end{cases}$, calculate the value of $\lim_{x\to 0^+} f(x) + \lim_{x\to 1} f(x)$.

Solution Recall that the **limit** $\lim_{x\to a} f(x)$ depends on the values of *f*(*x*) **near** *a*, but **not at** *a* **itself**. Therefore, $\lim_{x\to 0^+} f(x) + \lim_{x\to 1} f(x) = \lim_{x\to 0^+} (x+1) + \lim_{x\to 1} (x+1) = 1+2=3$.

Example 6.24 Given *f*(*x*) = *a*[*x*] + (3*a* − 1)[−*x*], $g(x) = \begin{cases} a & x\notin\mathbb{Z} \\ -a & x\in\mathbb{Z} \end{cases}$ and the condition that $\lim_{x\to 1} f(x)$ exists, determine the value of $\lim_{x\to 4a} g(x)$.

Solution

$$\lim_{x\to 1} f(x) = \lim_{x\to 1} a[x] + (3a-1)[-x] = \begin{cases} \lim_{x\to 1^-} a[x]+(3a-1)[-x] \\ \lim_{x\to 1^+} a[x]+(3a-1)[-x] \end{cases}$$

$$\lim_{x\to 1^-} a[x] + (3a-1)[-x] = \lim_{x\to 1^-} a\times 0 + (3a-1)\times -1 = \lim_{x\to 1^-} (1-3a) = 1-3a$$

$$\lim_{x\to 1^+} a[x] + (3a-1)[-x] = \lim_{x\to 1^+} a\times 1 + (3a-1)\times -2 = \lim_{x\to 1^-} (2-5a) = 2-5a$$

$$\lim_{x\to 1} f(x) \text{ exists} \Rightarrow \lim_{x\to 1^-} f(x) = \lim_{x\to 1^+} f(x) \Rightarrow 1-3a = 2-5a \Rightarrow 2a=1 \Rightarrow a=\frac{1}{2}$$

$$g(x) = \begin{cases} a & x\notin\mathbb{Z} \\ -a & x\in\mathbb{Z} \end{cases} a=12_{a=\frac{1}{2}} g(x) = \begin{cases} \frac{1}{2} & x\notin\mathbb{Z} \\ -\frac{1}{2} & x\in\mathbb{Z} \end{cases}$$

$$\lim_{x\to 4a} g(x) = \lim_{x\to 4\times\frac{1}{2}} g(x) = \lim_{x\to 2} g(x) = \frac{1}{2}$$

Example 6.25 Given $f(x) = \begin{cases} [x] & x\notin\mathbb{Q} \\ 1 & x\in\mathbb{Q} \end{cases}$, calculate $\lim_{x\to\sqrt{3}} f(x)$. $\mathbb{Q}$ represents the set of rational numbers.

Solution As *x* approaches $\sqrt{3}$, there are infinitely many rational numbers arbitrarily close to $\sqrt{3}$. For all these numbers, *f*(*x*) is always 1. Therefore, $\lim_{x\to\sqrt{3}} f(x) = \lim_{x\to\sqrt{3}} 1 = 1$.

As *x* approaches $\sqrt{3}$, there are also infinitely many irrational numbers arbitrarily close to $\sqrt{3}$. For these numbers, *f*(*x*) = [*x*]. Therefore, $\lim_{x\to\sqrt{3}} f(x) = \lim_{x\to\sqrt{3}} [x] = 1$.

The function approaches the same value for both rational and irrational inputs. Therefore, $\lim_{x\to\sqrt{3}} f(x) = 1$.

Example 6.26 Given $f(x) = \begin{cases} [x] & x \notin \mathbb{Q} \\ 1 & x \in \mathbb{Q} \end{cases}$, calculate $\lim_{x\to 3} f(x)$. $\mathbb{Q}$ represents the set of rational numbers.

Solution As x approaches 3, there are infinitely many rational numbers arbitrarily close to 3. For all these numbers, $f(x)$ is always 1. Therefore, $\lim_{x\to 3} f(x) = \lim_{x\to 3} 1 = 1$.

As x approaches 3, there are also infinitely many irrational numbers arbitrarily close to 3. For these numbers, $f(x) = [x]$. Therefore, $\lim_{x\to 3} f(x) = \lim_{x\to 3} [x] = \begin{cases} \lim_{x\to 3^-} [x] = 2 \\ \lim_{x\to 3^+} [x] = 3 \end{cases}$.

The limit of the function does not exist because its value approaches different numbers based on whether the input is a rational or irrational number, as well as whether it is approached from the left or the right.

Example 6.27 Given $1 - x^2 \le 5 - 2f(x) \le \cos(x)$ for all x, calculate $\lim_{x\to 0} f(x)$.

Solution $\lim_{x\to 0} 1 - x^2 = \lim_{x\to 0} \cos(x) = 1$. Now we can apply the Squeeze Theorem to the original inequality: $\lim_{x\to 0} 1 - x^2 = \lim_{x\to 0} 5 - 2f(x) = \lim_{x\to 0} \cos(x) \Rightarrow \lim_{x\to 0} 5 - 2f(x) = 1 \Rightarrow \lim_{x\to 0} 5 - 2\lim_{x\to 0} f(x) = 1 \Rightarrow 5 - 2\lim_{x\to 0} f(x) = 1 \Rightarrow \lim_{x\to 0} f(x) = 2$

Example 6.28 Given $\frac{\sin(\pi x)}{1-x} \le f(x) \le g(x)$ for all x and the condition $\lim_{x\to 1}\left(\frac{\sin(\pi x)}{1-x} - g(x)\right) = 0$, calculate $\lim_{x\to 1} f(x)$.

Solution

$$\lim_{x\to 1}\left(\frac{\sin(\pi x)}{1-x} - g(x)\right) = 0 \Rightarrow \lim_{x\to 1}\left(\frac{\sin(\pi x)}{1-x}\right) - \lim_{x\to 1}(g(x)) = 0 \Rightarrow \lim_{x\to 1}(g(x)) = \lim_{x\to 1}\left(\frac{\sin(\pi x)}{1-x}\right)$$

Now we can apply the Squeeze Theorem to the original inequality:

$$\lim_{x\to 1} f(x) = \lim_{x\to 1}\left(\frac{\sin(\pi x)}{1-x}\right) = \lim_{x\to 1}(g(x)) \xrightarrow{\text{L'Hôpital's}} \lim_{x\to 1} f(x) = \lim_{x\to 1}\left(\frac{\pi\cos(\pi x)}{-1}\right)$$
$$= \lim_{x\to 1}(g(x)) \Rightarrow \lim_{x\to 1} f(x) = \frac{\pi \times -1}{-1} = \lim_{x\to 1}(g(x)) \Rightarrow \lim_{x\to 1} f(x) = \pi = \lim_{x\to 1}(g(x))$$

Therefore, $\lim_{x\to 1} f(x) = \pi$.

Example 6.29

$$\text{Calculate } \lim_{x\to 0} \frac{\sin(x)}{x}.$$

Solution When you try to directly substitute $x = 0$ into the expression, you get the indeterminate form $\frac{0}{0}$. Using L'Hôpital's rule, we have: $\lim_{x\to 0} \frac{\sin(x)}{x} = \lim_{x\to 0} \frac{\cos(x)}{1} = 1$.

Example 6.30

$$\text{Calculate } \lim_{x\to +\infty} \frac{2x+3}{3x+4}.$$

Solution When you try to directly substitute $x = +\infty$ into the expression, you get the indeterminate form $\frac{\infty}{\infty}$. Using L'Hôpital's rule, we have: $\lim_{x\to\infty} \frac{2x+3}{3x+4} = \lim_{x\to\infty} \frac{2}{3} = \frac{2}{3}$.

Example 6.31

$$\text{Calculate } \lim_{x\to 4} \frac{x^2-16}{x-4}.$$

Solution When you try to directly substitute $x = 4$ into the expression, you get the indeterminate form $\frac{0}{0}$. Using L'Hôpital's rule, we have: $\lim_{x\to 4} \frac{x^2-16}{x-4} = \lim_{x\to 4} \frac{(x+4)(x-4)}{x-4} = \lim_{x\to 4} (x+4) = 8$

Example 6.32 Calculate $\lim_{x\to 1} \frac{(x-1)^2}{(\sqrt{x}-1)(\sqrt[3]{x}-1)}$.

Solution When you try to directly substitute $x = 1$ into the expression, you get the indeterminate form $\frac{0}{0}$.

$$\begin{aligned}
\lim_{x\to 1} \frac{(x-1)^2}{(\sqrt{x}-1)(\sqrt[3]{x}-1)} &= \lim_{x\to 1} \frac{((\sqrt{x}+1)(\sqrt{x}-1))^2}{(\sqrt{x}-1)(\sqrt[3]{x}-1)} = \lim_{x\to 1} \frac{(\sqrt{x}+1)^2(\sqrt{x}-1)}{\sqrt[3]{x}-1} \\
&= \lim_{x\to 1} \frac{(\sqrt{x}+1)^2(\sqrt[3]{x}-1)\left(\sqrt[3]{x^2}+\sqrt[3]{x}+1\right)}{\sqrt[3]{x}-1} \\
&= \lim_{x\to 1} (\sqrt{x}+1)^2\left(\sqrt[3]{x^2}+\sqrt[3]{x}+1\right) = 6
\end{aligned}$$

$$\text{Recall that} \begin{cases} a^3-b^3 = (a-b)\left(a^2+ab+b^2\right) \\ a^3+b^3 = (a+b)\left(a^2-ab+b^2\right) \end{cases}.$$

Example 6.33 Calculate $\lim_{x\to 1} \frac{(x-1)^2}{(\sqrt{x}-1)(\sqrt[3]{x}-1)}$.

Solution When you try to directly substitute $x = 1$ into the expression, you get the indeterminate form $\frac{0}{0}$. Using L'Hôpital's rule, we have:

$$\lim_{x\to 1} \frac{(x-1)^2}{(\sqrt{x}-1)(\sqrt[3]{x}-1)} = \lim_{x\to 1} \frac{2(x-1)}{\frac{1}{2\sqrt{x}}(\sqrt[3]{x}-1)+(\sqrt{x}-1)\left(\frac{1}{3\sqrt[3]{x^2}}\right)} = \frac{0}{0}$$

Applying L'Hôpital's rule a second time:

$$\lim_{x\to 1} \frac{(x-1)^2}{(\sqrt{x}-1)(\sqrt[3]{x}-1)} = \lim_{x\to 1} \frac{2(x-1)}{\frac{1}{2\sqrt{x}}(\sqrt[3]{x}-1)+(\sqrt{x}-1)\left(\frac{1}{3\sqrt[3]{x^2}}\right)} =$$

$$\lim_{x\to 1} \frac{2}{-\frac{1}{4\sqrt{x^3}}(\sqrt[3]{x}-1)+\frac{1}{2\sqrt{x}}\times\frac{1}{3\sqrt[3]{x^2}}+\frac{1}{2\sqrt{x}}\left(\frac{1}{3\sqrt[3]{x^2}}\right)+(\sqrt{x}-1)\left(\frac{-2}{9\sqrt[3]{x^5}}\right)} = \frac{2}{0+\frac{1}{6}+\frac{1}{6}+0} = \frac{2}{\frac{2}{6}} = \frac{12}{2} = 6$$

Example 6.34

$$\text{Caculate } \lim_{x\to 0} \frac{(1+x)(1+x^2)(1+x^4)(1+x^8)(1+x^{16})-1}{x}.$$

Solution When you try to directly substitute $x = 0$ into the expression, you get the indeterminate form $\frac{0}{0}$.

$$\lim_{x\to 0} \frac{(1+x)(1+x^2)(1+x^4)(1+x^8)(1+x^{16})-1}{x}$$

$$= \lim_{x\to 0} \frac{(1-x)(1+x)(1+x^2)(1+x^4)(1+x^8)(1+x^{16})-(1-x)\times 1}{(1-x)x}$$

$$= \lim_{x\to 0} \frac{(1-x^2)(1+x^2)(1+x^4)(1+x^8)(1+x^{16})-(1-x)}{(1-x)x}$$

$$= \lim_{x\to 0} \frac{(1-x^4)(1+x^4)(1+x^8)(1+x^{16})-(1-x)}{(1-x)x}$$

$$= \lim_{x\to 0} \frac{(1-x^8)(1+x^8)(1+x^{16})-(1-x)}{(1-x)x}$$

$$= \lim_{x\to 0} \frac{(1-x^{16})(1+x^{16})-(1-x)}{(1-x)x} = \lim_{x\to 0} \frac{(1-x^{32})-(1-x)}{(1-x)x} = \lim_{x\to 0} \frac{x-x^{32}}{(1-x)x}$$

$$= \lim_{x\to 0} \frac{1-x^{31}}{1-x} = 1$$

Example 6.35 Calculate $\lim_{x\to 0} \frac{(1+x)(1+x^2)(1+x^4)(1+x^8)(1+x^{16})-1}{x}$.

Solution When you try to directly substitute $x = 0$ into the expression, you get the indeterminate form $\frac{0}{0}$. Using L'Hôpital's rule, we have:

$$\lim_{x\to 0} \frac{(1+x)(1+x^2)(1+x^4)(1+x^8)(1+x^{16})-1}{x} = \lim_{x\to 0} \frac{(1+x^2)(1+x^4)(1+x^8)(1+x^{16})+(1+x)(2x)(1+x^4)(1+x^8)(1+x^{16})+\ldots}{1} =$$

$$\lim_{x\to 0} (1+x^2)(1+x^4)(1+x^8)(1+x^{16})+(1+x)(2x)(1+x^4)(1+x^8)(1+x^{16})+\ldots = 1+0+0+0+0 = 1$$

Example 6.36 Calculate $\lim\limits_{x\to\infty}\frac{x^2}{e^x}$.

Solution When you try to directly substitute $x=\infty$ into the expression, you get the indeterminate form $\frac{\infty}{\infty}$. Using L'Hôpital's rule, we have:

$\lim\limits_{x\to\infty}\frac{x^2}{e^x}=\lim\limits_{x\to\infty}\frac{2x}{e^x}=\lim\limits_{x\to\infty}\frac{2}{e^x}=0$

Example 6.37 Calculate $\lim\limits_{x\to\frac{\pi}{4}}\frac{(tan(x)-1)^2}{1+cos(4x)}$.

Solution When you try to directly substitute $x=\frac{\pi}{4}$ into the expression, you get the indeterminate form $\frac{0}{0}$. Using L'Hôpital's rule, we have:

$$\lim_{x\to\frac{\pi}{4}}\frac{(\tan(x)-1)^2}{1+\cos(4x)}=\lim_{x\to\frac{\pi}{4}}\frac{2(1+\tan^2(x))(\tan(x)-1)}{-4\sin(4x)}=\lim_{x\to\frac{\pi}{4}}\frac{(1+\tan^2(x))(\tan(x)-1)}{-2\sin(4x)}$$
$$=\lim_{x\to\frac{\pi}{4}}\frac{(1+\tan^2(x))}{-2}\lim_{x\to\frac{\pi}{4}}\frac{\tan(x)-1}{\sin(4x)}=-\lim_{x\to\frac{\pi}{4}}\frac{\tan(x)-1}{\sin(4x)}=\frac{0}{0}$$

Applying L'Hôpital's rule a second time:

$$\lim_{x\to\frac{\pi}{4}}\frac{(\tan(x)-1)^2}{1+\cos(4x)}=-\lim_{x\to\frac{\pi}{4}}\frac{\tan(x)-1}{\sin(4x)}=-\lim_{x\to\frac{\pi}{4}}\frac{1+\tan^2(x)}{4\cos(4x)}=\frac{1}{2}$$

Example 6.38 Calculate $\lim\limits_{x\to 1}\frac{\cos\left(\frac{\pi}{2}x\right)}{1-x}$.

Solution When you try to directly substitute $x=1$ into the expression, you get the indeterminate form $\frac{0}{0}$. Using L'Hôpital's rule, we have:

$$\lim_{x\to 1}\frac{\cos\left(\frac{\pi}{2}x\right)}{1-x}=\lim_{x\to 1}\frac{-\frac{\pi}{2}\sin\left(\frac{\pi}{2}x\right)}{-1}=\lim_{x\to 1}\frac{\pi}{2}=\frac{\pi}{2}$$

Exercise 6.2 Calculate $\lim\limits_{x\to 1}\frac{\sin(\pi x)}{x-1}$. (Ans. $-\pi$)

Example 6.39 Given that $\lim\limits_{x\to\infty}\frac{ax+9}{1-x+\sqrt{x+1}}=3$, determine the value of $\lim\limits_{x\to 3}\frac{ax+9}{1-x+\sqrt{x+1}}$.

Solution Recall that as x approaches infinity, a polynomial's limit is determined by its highest-degree term. Thus,

$$\lim_{x\to\infty}\frac{ax+9}{1-x+\sqrt{x+1}}=3\Longrightarrow\lim_{x\to\infty}\frac{ax}{-x}=3\Longrightarrow\lim_{x\to\infty}-a=3\Longrightarrow a=-3$$

$$\lim_{x\to 3}\frac{-3x+9}{1-x+\sqrt{x+1}}=\frac{0}{0}\Longrightarrow\lim_{x\to 3}\frac{-3x+9}{1-x+\sqrt{x+1}}=\lim_{x\to 3}\frac{-3}{-1+\frac{1}{2\sqrt{x+1}}}=\frac{-3}{-1+\frac{1}{2\sqrt{3+1}}}=4$$

Example 6.40 Given that $\lim\limits_{x\to-\infty}\frac{2x-\sqrt{x^2+6x}}{ax-2}=3$, determine the value of $\lim\limits_{x\to 2}\frac{2x-\sqrt{x^2+6x}}{ax-2}$.

Solution

$$\lim_{x\to-\infty}\frac{2x-\sqrt{x^2+6x}}{ax-2}=3\Rightarrow\lim_{x\to-\infty}\frac{2x-\sqrt{x^2}}{ax}=3\Rightarrow\lim_{x\to-\infty}\frac{2x-|x|}{ax}$$
$$=3\Rightarrow\lim_{x\to-\infty}\frac{2x-(-x)}{ax}=3\Rightarrow\lim_{x\to-\infty}\frac{3x}{ax}=3\Rightarrow\lim_{x\to-\infty}\frac{3}{a}=3\Rightarrow a=1$$

$\lim\limits_{x\to 2}\frac{2x-\sqrt{x^2+6x}}{ax-2}=\lim\limits_{x\to 2}\frac{2x-\sqrt{x^2+6x}}{x-2}=\lim\limits_{x\to 2}\frac{2-\frac{2x+6}{2\sqrt{x^2+6x}}}{1}=2-\frac{2\times 2+6}{2\sqrt{2^2+6\times 2}}=2-\frac{10}{2\sqrt{16}}=2-\frac{10}{8}=\frac{6}{8}=\frac{3}{4}=0.75$

Example 6.41 Given that $f(x+5)=\frac{\sqrt{1-x}-3}{\sqrt[3]{x}+2}$, determine the value of $\lim\limits_{x\to-3}f(x)$.

Solution $\lim\limits_{x\to-3}f(x)=\lim\limits_{x\to-8}\frac{\sqrt{1-x}-3}{\sqrt[3]{x}+2}=\frac{0}{0}$. Let's use L'Hôpital's rule:

$$\lim_{x\to-3}f(x)=\lim_{x\to-8}\frac{\sqrt{1-x}-3}{\sqrt[3]{x}+2}=\lim_{x\to-8}\frac{-\frac{1}{2\sqrt{1-x}}}{\frac{1}{3\sqrt[3]{x^2}}}=\frac{-\frac{1}{2\sqrt{1+8}}}{\frac{1}{3\sqrt[3]{64}}}=\frac{-\frac{1}{6}}{\frac{1}{12}}=-2$$

Example 6.42 Given that $f(x)=\frac{x^2-2}{x-1}+\frac{x^2+x}{x^2-1}$ and $\lim\limits_{x\to a}f^{-1}(x)=1$, determine the value of a.

Solution

$$\lim_{x\to a}f^{-1}(x)=1\Rightarrow\lim_{x\to 1}f(x)=a\Rightarrow\lim_{x\to 1}\left(\frac{x^2-2}{x-1}+\frac{x^2+x}{x^2-1}\right)$$
$$=a\Rightarrow\lim_{x\to 1}\left(\frac{x^2-2}{x-1}+\frac{x(x+1)}{(x+1)(x-1)}\right)=a\Rightarrow\lim_{x\to 1}\left(\frac{x^2-2}{x-1}+\frac{x}{x-1}\right)$$
$$=a\Rightarrow\lim_{x\to 1}\left(\frac{x^2-2+x}{x-1}\right)=a\Rightarrow\frac{0}{0}=a$$

Using L'Hôpital's rule, we have:
$\lim\limits_{x\to 1}\left(\frac{x^2-2+x}{x-1}\right)=a\Rightarrow\lim\limits_{x\to 1}\left(\frac{2x+1}{1}\right)=a\Rightarrow 3=a\Rightarrow a=3$

Example 6.43 Given that $f\left(x+\frac{\pi}{4}\right)=\frac{[x]-1}{1-\tan(x)}$, determine the value of $\lim\limits_{x\to\frac{\pi}{2}^-}f(x)$.

Solution
$\lim\limits_{x\to\frac{\pi}{2}^-}f(x)=\lim\limits_{x\to\frac{\pi}{4}^-}\frac{[x]-1}{1-\tan(x)}=\lim\limits_{x\to\frac{\pi}{4}^-}\frac{0-1}{1-\tan(x)}=\lim\limits_{x\to\frac{\pi}{4}^-}\frac{-1}{1-\tan(x)}=-\frac{1}{1-1^-}=\frac{-1}{0^+}=-\infty$

Example 6.44 Given that $\lim\limits_{x\to+\infty}\frac{|x^2-4|}{ax^2-x+2}=-1$, determine the value of (**A**) $\lim\limits_{x\to-2^-}\frac{|x^2-4|}{ax^2-x+2}$ (**B**) $\lim\limits_{x\to-2^+}\frac{|x^2-4|}{ax^2-x+2}$.

Solution

$$\lim_{x\to+\infty}\frac{|x^2-4|}{ax^2-x+2}=-1\Rightarrow\lim_{x\to+\infty}\frac{x^2-4}{ax^2-x+2}=-1\Rightarrow\lim_{x\to+\infty}\frac{x^2}{ax^2}$$
$$=-1\Rightarrow\lim_{x\to+\infty}\frac{1}{a}=-1\Rightarrow a=-1$$

$$\lim_{x\to-2^-}\frac{|x^2-4|}{ax^2-x+2}=\lim_{x\to-2^-}\frac{|x^2-4|}{-x^2-x+2}=\lim_{x\to-2^-}\frac{-(x^2-4)}{-x^2-x+2}=\frac{0}{0}$$

Let's use L'Hôpital's rule:

$$\lim_{x\to-2^-}\frac{|x^2-4|}{ax^2-x+2}=\lim_{x\to-2^-}\frac{-(x^2-4)}{-x^2-x+2}=\lim_{x\to-2^-}\frac{-2x}{-2x-1}=\frac{4}{3}$$

By applying the same method, you can show that $\lim\limits_{x\to-2^+}\frac{|x^2-4|}{ax^2-x+2}=-\frac{4}{3}$.

Example 6.45 Calculate $\lim\limits_{x\to0}\frac{\sqrt{x+9}-3}{x}$.

Solution When you try to directly substitute $x=0$ into the expression, you get the indeterminate form $\frac{0}{0}$.

$$\lim_{x\to0}\frac{\sqrt{x+9}-3}{x}=\lim_{x\to0}\frac{\sqrt{x+9}-3}{x}\times\frac{\sqrt{x+9}+3}{\sqrt{x+9}+3}=\lim_{x\to0}\frac{x+9-9}{x\left(\sqrt{x+9}+3\right)}$$
$$=\lim_{x\to0}\frac{x}{x\left(\sqrt{x+9}+3\right)}=\lim_{x\to0}\frac{1}{\sqrt{x+9}+3}=\lim_{x\to0}\frac{1}{6}=\frac{1}{6}$$

Example 6.46 Calculate $\lim\limits_{x\to0}\frac{\sqrt{x+9}-3}{x}$.

Solution When you try to directly substitute $x=0$ into the expression, you get the indeterminate form $\frac{0}{0}$. Using L'Hôpital's rule, we have:

$$\lim_{x\to0}\frac{\sqrt{x+9}-3}{x}=\lim_{x\to0}\frac{\frac{1}{2\sqrt{x+9}}}{1}=\lim_{x\to0}\frac{1}{2\sqrt{x+9}}=\frac{1}{2\times3}=\frac{1}{6}$$

Example 6.47 Calculate $\lim\limits_{x\to0}\frac{\sin(6x)}{\sin(2x)}$.

Solution When you try to directly substitute $x = 0$ into the expression, you get the indeterminate form $\frac{0}{0}$. Recall that $sin(x) \sim x$ as $x \to 0$. Therefore,

$$\lim_{x\to 0} \frac{\sin(6x)}{\sin(2x)} = \lim_{x\to 0} \frac{6x}{2x} = \lim_{x\to 0} 3 = 3$$

Example 6.48 Calculate $\lim_{x\to 0} \frac{\sin(\sqrt[8]{x})}{x+\sqrt{3x}+\sqrt{x+2\sqrt[4]{x}}}$.

Solution When you try to directly substitute $x = 0$ into the expression, you get the indeterminate form $\frac{0}{0}$. As $x \to 0$,

$$\sin\left(\sqrt[8]{x}\right) \sim \sqrt[8]{x}$$

$$x + \sqrt{3x} + \sqrt{x + 2\sqrt[4]{x}} \sim x + \sqrt{3x} + \sqrt{2\sqrt[4]{x}} = x + \sqrt{3x} + \sqrt{2}\sqrt{\sqrt[4]{x}} = x + \sqrt{3x} + \sqrt{2}\sqrt[8]{x} \sim \sqrt{2}\sqrt[8]{x}$$

Therefore, $\lim_{x\to 0} \frac{\sin(\sqrt[8]{x})}{x+\sqrt{3x}+\sqrt{x+2\sqrt[4]{x}}} = \lim_{x\to 0} \frac{\sqrt[8]{x}}{\sqrt{2}\sqrt[8]{x}} = \frac{1}{\sqrt{2}}$.

Example 6.49 Calculate $\lim_{x\to 0} \frac{\sin(2x)}{x^5 - 2x}$.

Solution When you try to directly substitute $x = 0$ into the expression, you get the indeterminate form $\frac{0}{0}$. As $x \to 0$, $\sin(2x) \sim 2x$ and $x^5 - 2x \sim -2x$. Therefore, $\lim_{x\to 0} \frac{\sin(2x)}{x^5 - 2x} = \lim_{x\to 0} \frac{2x}{-2x} = \lim_{x\to 0} -1 = -1$.

Example 6.50 Calculate $\lim_{x\to 0} \frac{(1+4x)^5 - \sqrt[5]{1+3x}}{x}$.

Solution When you try to directly substitute $x = 0$ into the expression, you get the indeterminate form $\frac{0}{0}$. As $x \to 0$, $(1 + 4x)^5 \sim 1 + 5 \times 4x$ and $\sqrt[5]{1 + 3x} \sim 1 + \frac{1}{5} \times 3x$. Therefore, $\lim_{x\to 0} \frac{(1+4x)^5 - \sqrt[5]{1+3x}}{x} = \lim_{x\to 0} \frac{1+20x - \left(1+\frac{3x}{5}\right)}{x} = \lim_{x\to 0} \frac{\frac{97x}{5}}{x} = \lim_{x\to 0} \frac{97}{5} = \frac{97}{5}$.

Example 6.51 Calculate $\lim_{x\to 0} \frac{\sqrt{12(\tan(x) - x)}}{\sqrt{x}\tan(x)}$.

Solution When you try to directly substitute $x = 0$ into the expression, you get the indeterminate form $\frac{0}{0}$. We use equivalent infinitesimals:

$$\lim_{x\to 0} \frac{\sqrt{12(\tan(x) - x)}}{\sqrt{x}\tan(x)} = \lim_{x\to 0} \frac{\sqrt{12\left(x+\frac{x^3}{3} - x\right)}}{\sqrt{x} \times x} = \lim_{x\to 0} \frac{\sqrt{4x^3}}{x^{\frac{3}{2}}} = \lim_{x\to 0} \frac{2x^{\frac{3}{2}}}{x^{\frac{3}{2}}} = 2$$

Example 6.52 Calculate $\lim_{x\to 0} \frac{\sqrt{1+\sin(x)} - \sqrt{1+\tan(x)}}{x^3}$.

Solution When you try to directly substitute $x = 0$ into the expression, you get the indeterminate form $\frac{0}{0}$. We use equivalent infinitesimals:

$$\lim_{x\to 0}\frac{\sqrt{1+\sin(x)}-\sqrt{1+\tan(x)}}{x^3}=\lim_{x\to 0}\frac{1+\frac{\sin(x)}{2}-\left(1+\frac{\tan(x)}{2}\right)}{x^3}=\lim_{x\to 0}\frac{\sin(x)-\tan(x)}{2x^3}$$

$$=\lim_{x\to 0}\frac{x-\frac{x^3}{6}-\left(x+\frac{x^3}{3}\right)}{2x^3}=\lim_{x\to 0}\frac{-\frac{x^3}{6}-\frac{x^3}{3}}{2x^3}=\lim_{x\to 0}\frac{-\frac{3x^3}{6}}{2x^3}=-\frac{1}{4}$$

Example 6.53 Calculate $\lim\limits_{x\to 0^-}\frac{\tan(2x)-\sqrt{1-\cos(x)}}{-3x+x^2+x^6+x^{10}+x^{14}}$.

Solution When you try to directly substitute $x = 0$ into the expression, you get the indeterminate form $\frac{0}{0}$. We use equivalent infinitesimals:

$$\lim_{x\to 0^-}\frac{\tan(2x)-\sqrt{1-\cos(x)}}{-3x+x^2+x^6+x^{10}+x^{14}}=\lim_{x\to 0^-}\frac{2x-\sqrt{\frac{x^2}{2}}}{-3x}=\lim_{x\to 0^-}\frac{2x-\frac{|x|}{\sqrt{2}}}{-3x}=\lim_{x\to 0^-}\frac{2x-\frac{-x}{\sqrt{2}}}{-3x}=-\frac{2+\frac{1}{\sqrt{2}}}{3}=-0.9024$$

Exercise 6.3 Calculate (**A**) $\lim\limits_{x\to 0^-}\frac{\tan(2x)\times\sqrt{1-\cos(x)}}{x^2+x^6+x^{10}+x^{14}}$ (**B**) $\lim\limits_{x\to 0^+}\frac{\tan(2x)\times\sqrt{1-\cos(x)}}{x^2+x^6+x^{10}+x^{14}}$ (Ans. **A**) $-\sqrt{2}$ (**B**) $\sqrt{2}$)

Example 6.54 Calculate $\lim\limits_{x\to 3^-}\frac{|x^2-5x+6|}{2x-\sqrt{x^3+9}}$.

Solution $\lim\limits_{x\to 3^-}\frac{|x^2-5x+6|}{2x-\sqrt{x^3+9}}=\frac{0}{0}$. Therefore, we can use L'Hôpital's rule.

$$x^2-5x+6\overset{x=2.9}{\Rightarrow}2.9^2-5\times 2.9+6=-0.09$$

$$\lim_{x\to 3^-}\frac{|x^2-5x+6|}{2x-\sqrt{x^3+9}}=\lim_{x\to 3^-}\frac{-(x^2-5x+6)}{2x-\sqrt{x^3+9}}=\lim_{x\to 3^-}\frac{-x^2+5x-6}{2x-\sqrt{x^3+9}}=\lim_{x\to 3^-}\frac{-2x+5}{2-\frac{3x^2}{2\sqrt{x^3+9}}}=\frac{-6+5}{2-\frac{3\times 9}{2\sqrt{27+9}}}=\frac{-1}{2-\frac{27}{2\times 6}}=\frac{-12}{-3}=4$$

Example 6.55 Calculate $\lim\limits_{x\to\frac{3\pi}{4}}\frac{\tan^2(x)-1}{\cos(2x)}$.

Solution $\lim\limits_{x\to\frac{3\pi}{4}}\frac{\tan^2(x)-1}{\cos(2x)}=\frac{0}{0}$. Therefore, we can use L'Hôpital's rule.

$$\lim_{x\to\frac{3\pi}{4}}\frac{\tan^2(x)-1}{\cos(2x)}=\lim_{x\to\frac{3\pi}{4}}\frac{2\tan(x)\left(1+\tan^2(x)\right)}{-2\sin(2x)}=\frac{2\times -1\times\left(1+(-1)^2\right)}{-2\times -1}=-2$$

Example 6.56 Calculate $\lim\limits_{x\to\frac{3\pi}{4}}\frac{\tan^2(x)-1}{\cos(2x)}$.

Solution This problem was previously solved using L'Hôpital's rule. Here, we will use the method of equivalent infinitesimals.

$$\lim_{x\to\frac{3\pi}{4}} \frac{\tan^2(x)-1}{\cos(2x)} = \frac{0}{0}$$

$$x \to \frac{3\pi}{4} \Rightarrow x - \frac{3\pi}{4} \to 0 \overset{u=x-\frac{3\pi}{4}}{\Rightarrow} u \to 0$$

$$\lim_{x\to\frac{3\pi}{4}} \frac{\tan^2(x)-1}{\cos(2x)} = \lim_{u\to 0} \frac{\tan^2\left(u+\frac{3\pi}{4}\right)-1}{\cos\left(2\left(u+\frac{3\pi}{4}\right)\right)} = \lim_{u\to 0} \frac{\tan^2\left(u+\frac{3\pi}{4}\right)-1}{\cos\left(2u+\frac{3\pi}{2}\right)}$$

Let's simplify the numerator and denominator. Recall that $\tan(A+B) = \frac{\tan(A)+\tan(B)}{1-\tan(A)\tan(B)}$. Therefore,

$$\tan\left(u+\frac{3\pi}{4}\right) = \frac{\tan(u)+\tan\left(\frac{3\pi}{4}\right)}{1-\tan(u)\tan\left(\frac{3\pi}{4}\right)} = \frac{\tan(u)-1}{1-\tan(u)\times -1} = \frac{\tan(u)-1}{1+\tan(u)}$$

$$\tan^2\left(u+\frac{3\pi}{4}\right) - 1 = \left(\frac{\tan(u)-1}{1+\tan(u)}\right)^2 - 1 = \frac{(\tan(u)-1)^2}{(1+\tan(u))^2} - 1$$

$$= \frac{(\tan(u)-1)^2-(1+\tan(u))^2}{(1+\tan(u))^2} = \frac{-4\tan(u)}{(1+\tan(u))^2}$$

$$\cos\left(2u+\frac{3\pi}{2}\right) = \cos\left(2u+\pi+\frac{\pi}{2}\right) = -\cos\left(2u+\frac{\pi}{2}\right) = -\sin(-2u) = \sin(2u)$$

We can now substitute the results back into the original limit expression:

$$\lim_{x\to\frac{3\pi}{4}} \frac{\tan^2(x)-1}{\cos(2x)} = \lim_{u\to 0} \frac{\tan^2\left(u+\frac{3\pi}{4}\right)-1}{\cos\left(2\left(u+\frac{3\pi}{4}\right)\right)} = \lim_{u\to 0} \frac{\tan^2\left(u+\frac{3\pi}{4}\right)-1}{\cos\left(2u+\frac{3\pi}{2}\right)}$$

$$= \lim_{u\to 0} \frac{\frac{-4\tan(u)}{(1+\tan(u))^2}}{\sin(2u)} = \lim_{u\to 0} \frac{\frac{-4u}{(1+u)^2}}{2u} = \lim_{u\to 0} \frac{-4u}{2u(1+u)^2} = \lim_{u\to 0} \frac{-2}{(1+u)^2} = -2$$

The obtained result is consistent with the one from the previous example.

Example 6.57 Calculate $\lim\limits_{x\to 1} \frac{\sin(\pi x)}{x^2-\sqrt{x}}$.

Solution $\lim\limits_{x\to 1} \frac{\sin(\pi x)}{x^2-\sqrt{x}} = \frac{0}{0}$. Therefore, we can use L'Hôpital's rule.

$$\lim_{x\to 1} \frac{\sin(\pi x)}{x^2-\sqrt{x}} = \lim_{x\to 1} \frac{\pi\cos(\pi x)}{2x-\frac{1}{2\sqrt{x}}} = \frac{-\pi}{2-\frac{1}{2}} = -\frac{2\pi}{3}$$

Example 6.58 Calculate $\lim\limits_{x\to 1} \frac{\sin(\pi x)}{x^2-\sqrt{x}}$.

Solution This problem was previously solved using L'Hôpital's rule. Here, we will use the method of equivalent infinitesimals.

$$\lim_{x\to 1}\frac{\sin(\pi x)}{x^2-\sqrt{x}}=\frac{0}{0}$$

$$x\to 1 \Rightarrow x-1\to 0 \overset{u=x-1}{\Rightarrow} u\to 0$$

$$\lim_{x\to 1}\frac{\sin(\pi x)}{x^2-\sqrt{x}}=\lim_{u\to 0}\frac{\sin(\pi(u+1))}{(u+1)^2-\sqrt{u+1}}=\lim_{u\to 0}\frac{\sin(\pi u+\pi)}{(u+1)^2-\sqrt{u+1}}$$
$$=\lim_{u\to 0}\frac{-\sin(\pi u)}{(u+1)^2-\sqrt{u+1}}=\lim_{u\to 0}\frac{-\pi u}{1+2u-\left(1+\frac{u}{2}\right)}=\lim_{u\to 0}\frac{-\pi u}{\frac{3u}{2}}=-\frac{2\pi}{3}$$

The obtained result is consistent with the one from the previous example.

Example 6.59 Calculate $\lim\limits_{x\to 0}\frac{\cos(x)-\cos(2x)}{x^2}$.

Solution $\lim\limits_{x\to 0}\frac{\cos(x)-\cos(2x)}{x^2}=\frac{0}{0}$. Therefore, we can use L'Hôpital's rule.

$$\lim_{x\to 0}\frac{\cos(x)-\cos(2x)}{x^2}=\lim_{x\to 0}\frac{-\sin(x)+2\sin(2x)}{2x}=\lim_{x\to 0}\frac{-\cos(x)+4\cos(2x)}{2}=\frac{3}{2}$$

Example 6.60 Calculate $\lim\limits_{x\to 0}\frac{\cos(x)-\cos(2x)}{x^2}$.

Solution This problem was previously solved using L'Hôpital's rule. Here, we will use the method of equivalent infinitesimals.

$$\lim_{x\to 0}\frac{\cos(x)-\cos(2x)}{x^2}=\frac{0}{0}$$

$$\lim_{x\to 0}\frac{\cos(x)-\cos(2x)}{x^2}=\lim_{x\to 0}\frac{1-\frac{x^2}{2}-\left(1-\frac{(2x)^2}{2}\right)}{x^2}=\lim_{x\to 0}\frac{-\frac{x^2}{2}+\frac{4x^2}{2}}{x^2}=\lim_{x\to 0}\frac{\frac{3x^2}{2}}{x^2}=\frac{3}{2}$$

Example 6.61 Calculate $\lim\limits_{x\to 9}\frac{3-\sqrt{x}}{5-\sqrt{2x+7}}$.

Solution $\lim\limits_{x\to 9}\frac{3-\sqrt{x}}{5-\sqrt{2x+7}}=\frac{0}{0}$. We will use the method of equivalent infinitesimals; solving this using L'Hôpital's rule is left as an exercise for the reader.

$$\lim_{x\to 9}\frac{3-\sqrt{x}}{5-\sqrt{2x+7}}=\lim_{u\to 0}\frac{3-\sqrt{u+9}}{5-\sqrt{2(u+9)+7}}=\lim_{u\to 0}\frac{3-\sqrt{u+9}}{5-\sqrt{2u+25}}$$
$$=\lim_{u\to 0}\frac{3-\sqrt{9\left(\frac{u}{9}+1\right)}}{5-\sqrt{25\left(\frac{2}{25}u+1\right)}}=\lim_{u\to 0}\frac{3-3\sqrt{\frac{u}{9}+1}}{5-5\sqrt{\frac{2}{25}u+1}}$$
$$=\lim_{u\to 0}\frac{3-3\left(1+\frac{1}{2}\times\frac{u}{9}\right)}{5-5\left(1+\frac{1}{2}\times\frac{2}{25}u\right)}=\lim_{u\to 0}\frac{\frac{u}{6}}{-\frac{u}{5}}=\lim_{u\to 0}-\frac{5}{6}=-\frac{5}{6}$$

Example 6.62 Calculate $\lim_{x\to 2^-}\frac{2-\sqrt[3]{x+6}}{\sqrt{x^2-4x+4}}$.

Solution $\lim_{x\to 2^-}\frac{2-\sqrt[3]{x+6}}{\sqrt{x^2-4x+4}}=\frac{0}{0}$. We will use the method of equivalent infinitesimals; solving this using L'Hôpital's rule is left as an exercise for the reader.

$$\lim_{x\to 2^-}\frac{2-\sqrt[3]{x+6}}{\sqrt{x^2-4x+4}}=\lim_{x\to 2^-}\frac{2-\sqrt[3]{x+6}}{\sqrt{(x-2)^2}}=\lim_{x\to 2^-}\frac{2-\sqrt[3]{x+6}}{|x-2|}=\lim_{x\to 2^-}\frac{2-\sqrt[3]{x+6}}{-(x-2)}$$

$$=\lim_{u\to 0^-}\frac{2-\sqrt[3]{u+2+6}}{-(u+2-2)}=\lim_{u\to 0^-}\frac{2-\sqrt[3]{u+8}}{-u}=\lim_{u\to 0^-}\frac{2-\sqrt[3]{8\left(\frac{u}{8}+1\right)}}{-u}$$
$$=\lim_{u\to 0^-}\frac{2-2\sqrt[3]{1+\frac{u}{8}}}{-u}=\lim_{u\to 0^-}\frac{2-2\left(1+\frac{u}{24}\right)}{-u}=\lim_{u\to 0^-}\frac{-\frac{u}{12}}{-u}=\lim_{u\to 0^-}\frac{1}{12}=\frac{1}{12}$$

Exercise 6.4 Calculate $\lim_{x\to 2^+}\frac{2-\sqrt[3]{x+6}}{\sqrt{x^2-4x+4}}$. (Ans. $-\frac{1}{12}$)

Example 6.63 Calculate $\lim_{x\to\frac{\pi}{8}}\frac{\sqrt{\cos(2x)}-\sqrt{\sin(2x)}}{\cos\left(2x+\frac{\pi}{4}\right)}$.

Solution $\lim_{x\to\frac{\pi}{8}}\frac{\sqrt{\cos(2x)}-\sqrt{\sin(2x)}}{\cos\left(2x+\frac{\pi}{4}\right)}=\frac{0}{0}$. Therefore, using L'Hôpital's rule is appropriate.

$$\lim_{x\to\frac{\pi}{8}}\frac{\sqrt{\cos(2x)}-\sqrt{\sin(2x)}}{\cos\left(2x+\frac{\pi}{4}\right)}=\lim_{x\to\frac{\pi}{8}}\frac{\frac{-\sin(2x)}{\sqrt{\cos(2x)}}-\frac{\cos(2x)}{\sqrt{\sin(2x)}}}{-2\sin\left(2x+\frac{\pi}{4}\right)}=\frac{\frac{-\sin\left(\frac{\pi}{4}\right)}{\sqrt{\cos\left(\frac{\pi}{4}\right)}}-\frac{\cos\left(\frac{\pi}{4}\right)}{\sqrt{\sin\left(\frac{\pi}{4}\right)}}}{-2\sin\left(\frac{\pi}{4}+\frac{\pi}{4}\right)}=\frac{\frac{-\frac{\sqrt{2}}{2}}{\sqrt{\frac{\sqrt{2}}{2}}}-\frac{\frac{\sqrt{2}}{2}}{\sqrt{\frac{\sqrt{2}}{2}}}}{-2}=\frac{\frac{\sqrt{2}}{\sqrt{\frac{\sqrt{2}}{2}}}}{2}=\frac{\frac{\sqrt{2}}{\frac{\sqrt[4]{2}}{\sqrt{2}}}}{2}=\frac{\sqrt{2}}{2\times\frac{\sqrt[4]{2}}{\sqrt{2}}}=\frac{\sqrt{2}}{\sqrt{2}\times\sqrt[4]{2}}=\frac{1}{\sqrt[4]{2}}\approx 0.8409$$

Example 6.64 Calculate $\lim_{x\to\frac{\pi}{8}}\frac{\sqrt{\cos(2x)}-\sqrt{\sin(2x)}}{\cos\left(2x+\frac{\pi}{4}\right)}$.

Solution This problem was previously solved using L'Hôpital's rule. Here, we will use the method of equivalent infinitesimals.

$$\lim_{x\to\frac{\pi}{8}} \frac{\sqrt{\cos(2x)} - \sqrt{\sin(2x)}}{\cos\left(2x + \frac{\pi}{4}\right)} = \frac{0}{0}$$

$$\lim_{x\to\frac{\pi}{8}} \frac{\sqrt{\cos(2x)} - \sqrt{\sin(2x)}}{\cos\left(2x + \frac{\pi}{4}\right)} = \lim_{u\to 0} \frac{\sqrt{\cos\left(2\left(u + \frac{\pi}{8}\right)\right)} - \sqrt{\sin\left(2\left(u + \frac{\pi}{8}\right)\right)}}{\cos\left(2\left(u + \frac{\pi}{8}\right) + \frac{\pi}{4}\right)}$$

$$= \lim_{u\to 0} \frac{\sqrt{\cos\left(2u + \frac{\pi}{4}\right)} - \sqrt{\sin\left(2u + \frac{\pi}{4}\right)}}{cos\left(2u + \frac{\pi}{2}\right)}$$

$$= \lim_{u\to 0} \frac{\sqrt{cos(2u)\,cos\left(\frac{\pi}{4}\right) - sin(2u)\,sin\left(\frac{\pi}{4}\right)} - \sqrt{sin(2u)\,sin\left(\frac{\pi}{4}\right) + cos(2u)\,cos\left(\frac{\pi}{4}\right)}}{sin(-2u)}$$

$$= \lim_{u\to 0} \frac{\sqrt{\cos(2u)\frac{\sqrt{2}}{2} - \sin(2u)\frac{\sqrt{2}}{2}} - \sqrt{\sin(2u)\frac{\sqrt{2}}{2} + \cos(2u)\frac{\sqrt{2}}{2}}}{-\sin(2u)}$$

$$= \sqrt{\frac{\sqrt{2}}{2}} \lim_{u\to 0} \frac{\sqrt{\cos(2u) - \sin(2u)} - \sqrt{\sin(2u) + \cos(2u)}}{-\sin(2u)}$$

$$= \frac{\sqrt[4]{2}}{\sqrt{2}} \lim_{u\to 0} \frac{\sqrt{1 - \frac{(2u)^2}{2} - 2u} - \sqrt{2u + 1 - \frac{(2u)^2}{2}}}{-2u}$$

$$= \frac{\sqrt[4]{2}}{\sqrt{2}} \lim_{u\to 0} \frac{\sqrt{1 - 2u^2 - 2u} - \sqrt{2u + 1 - 2u^2}}{-2u}$$

$$= \frac{\sqrt[4]{2}}{\sqrt{2}} \lim_{u\to 0} \frac{1 - u^2 - u - (1 + u - u^2)}{-2u} = \frac{\sqrt[4]{2}}{\sqrt{2}} \lim_{u\to 0} \frac{-2u}{-2u} = \frac{\sqrt[4]{2}}{\sqrt{2}} \lim_{u\to 0} 1 = \frac{\sqrt[4]{2}}{\sqrt{2}}$$

$$\approx 0.8409$$

The obtained result is consistent with the one from the previous example.

Example 6.65 Calculate $\lim\limits_{x\to 0} \frac{\cos^2(x) - \sqrt{\cos(x)}}{x^2}$.

Solution $\lim\limits_{x\to 0} \frac{\cos^2(x) - \sqrt{\cos(x)}}{x^2} = \frac{0}{0}$. We will use the method of equivalent infinitesimals; solving this using L'Hôpital's rule is left as an exercise for the reader.

$$\lim_{x\to 0} \frac{\cos^2(x) - \sqrt{\cos(x)}}{x^2} = \lim_{x\to 0} \frac{\left(1 - \frac{x^2}{2}\right)^2 - \sqrt{1 - \frac{x^2}{2}}}{x^2}$$

$$= \lim_{x\to 0} \frac{\left(1 - x^2 + \frac{x^4}{4}\right) - \left(1 - \frac{x^2}{4}\right)}{x^2} = \lim_{x\to 0} \frac{-\frac{3}{4}x^2 + \frac{x^4}{4}}{x^2} = \lim_{x\to 0}\left(-\frac{3}{4} + \frac{x^2}{4}\right) = -\frac{3}{4}$$

Example 6.66 Calculate $\lim\limits_{x\to 0}([4x]+[-4x])\frac{1-\cos^3(x)}{1-\sqrt{1+x^2}}$.

Solution Recall that $[u]+[-u]=\begin{cases}0 & u\in\mathbb{Z}\\ -1 & u\notin\mathbb{Z}\end{cases}$. As x approaches 0, but is not equal to 0, $4x$ is not an integer. Therefore, $[4x]+[-4x]=-1$ and $\lim\limits_{x\to 0}([4x]+[-4x])\frac{1-\cos^3(x)}{1-\sqrt{1+x^2}}=\lim\limits_{x\to 0}-\frac{1-\cos^3(x)}{1-\sqrt{1+x^2}}=\frac{0}{0}$, which means that L'Hôpital's rule can be appropriately applied.

$\lim\limits_{x\to 0}([4x]+[-4x])\frac{1-\cos^3(x)}{1-\sqrt{1+x^2}}=\lim\limits_{x\to 0}-\frac{1-\cos^3(x)}{1-\sqrt{1+x^2}}=\lim\limits_{x\to 0}-\frac{3\sin(x)\cos^2(x)}{-\frac{1}{2}\times 2x\times\frac{1}{\sqrt{1+x^2}}}=\lim\limits_{x\to 0}\frac{3\sqrt{1+x^2}\cos^2(x)\sin(x)}{x}=$
$\lim\limits_{x\to 0}\left(3\sqrt{1+x^2}\cos^2(x)\right)\lim\limits_{x\to 0}\frac{\sin(x)}{x}=3\lim\limits_{x\to 0}\frac{\sin(x)}{x}=3\lim\limits_{x\to 0}\frac{\cos(x)}{1}=3\times 1=3$

Example 6.67 Calculate $\lim\limits_{x\to 0}([4x]+[-4x])\frac{1-\cos^3(x)}{1-\sqrt{1+x^2}}$.

Solution Recall that $[u]+[-u]=\begin{cases}0 & u\in\mathbb{Z}\\ -1 & u\notin\mathbb{Z}\end{cases}$. As x approaches 0, but is not equal to 0, $4x$ is not an integer. Therefore, $[4x]+[-4x]=-1$ and $\lim\limits_{x\to 0}([4x]+[-4x])\frac{1-\cos^3(x)}{1-\sqrt{1+x^2}}=\lim\limits_{x\to 0}-\frac{1-\cos^3(x)}{1-\sqrt{1+x^2}}=\frac{0}{0}$. We will use the method of equivalent infinitesimals (Recall that $(a+b)^3=a^3+3a^2b+3ab^2+b^3$):

$$\lim_{x\to 0}([4x]+[-4x])\frac{1-\cos^3(x)}{1-\sqrt{1+x^2}}=\lim_{x\to 0}-\frac{1-\cos^3(x)}{1-\sqrt{1+x^2}}=\lim_{x\to 0}-\frac{1-\left(1-\frac{x^2}{2}\right)^3}{1-\left(1+\frac{x^2}{2}\right)}$$

$$=\lim_{x\to 0}\frac{1-\left(1-3\frac{x^2}{2}+3\frac{x^4}{4}-\frac{x^6}{8}\right)}{\frac{x^2}{2}}=\lim_{x\to 0}\frac{3\frac{x^2}{2}-3\frac{x^4}{4}+\frac{x^6}{8}}{\frac{x^2}{2}}=\lim_{x\to 0}\left(3-\frac{3}{2}x^2+\frac{x^2}{4}\right)=3$$

Exercise 6.5 Calculate $\lim\limits_{x\to\frac{\pi}{3}}\left(\left[\sin\left(x-\frac{\pi}{3}\right)\right]\cos(3x)+[\tan^2(x)]\right)$. (Ans. 3)

Example 6.68 Calculate $\lim\limits_{x\to 0^+}\left[\frac{\sin(x)}{x}\right]\cot(x)$.

Solution We know the fundamental limit $\lim\limits_{x\to 0^+}\frac{\sin(x)}{x}=1$. However, for any small positive value of x (i.e., as $x\to 0^+$), the inequality $\sin(x)<x$ holds. This means the value of the ratio $\frac{\sin(x)}{x}$ is always slightly less than 1. For example, if $x=0.1$, $\frac{\sin(0.1)}{0.1}\approx\frac{0.09983}{0.1}\approx 0.9983$. Since the value is less than 1, the floor function of this value will be 0. Therefore, for any x in the interval $(0,\epsilon)$ for a small ϵ, the term $\left[\frac{\sin(x)}{x}\right]$ is exactly 0.

The second terms is $\cot(x)=\frac{\cos(x)}{\sin(x)}$. As $x\to 0^+$, $\lim\limits_{x\to 0^+}\cot(x)=+\infty$

The limit is now in the form of $0 \times \infty$, which is an indeterminate form. However, the "0" from the floor function is an exact value for the entire interval, not a value that approaches zero.

Substituting our findings back into the original limit expression: $\lim_{x \to 0^+} \left[\frac{\sin(x)}{x}\right] \cot(x) = \lim_{x \to 0^+} 0 \times \cot(x) = 0$. Since the first term is exactly zero for all small positive x, the product is also zero.

By applying the same technique, it can be demonstrated that the limit $\lim_{x \to 0^-} \left[\frac{\sin(x)}{x}\right] \cot(x) = 0$.

Exercise 6.6 Calculate $\lim_{x \to 0^-} \left[-\frac{\sin(x)}{x}\right] \cot(x)$. (Ans. $+\infty$)

Example 6.69 Calculate $\lim_{x \to +\infty} x\left[\frac{1}{x}\right]$.

Solution As $x \to +\infty$, the term $\frac{1}{x}$ approaches 0 from the positive side. We can write this as $\frac{1}{x} \to 0^+$. For any number between 0 and 1 (exclusive of 1), the floor is 0. Therefore,

$$\lim_{x \to +\infty} x\left[\frac{1}{x}\right] = \lim_{x \to +\infty} x \times \left[0^+\right] = \lim_{x \to +\infty} x \times 0 = 0$$

Example 6.70 Calculate $\lim_{x \to -\infty} x\left[\frac{1}{x}\right]$.

Solution As $x \to -\infty$, the term $\frac{1}{x}$ approaches 0 from the negative side. We can write this as $\frac{1}{x} \to 0^-$. For any number between -1 and 0 (exclusive of 0), the floor is -1. Therefore,

$$\lim_{x \to -\infty} x\left[\frac{1}{x}\right] = \lim_{x \to -\infty} x \times \left[0^-\right] = \lim_{x \to -\infty} x \times -1 = +\infty$$

Example 6.71 Calculate $\lim_{x \to 0} \frac{2\sin^2(x)}{x^3 \sin\left(\frac{1}{2x}\right)}$.

Solution

$$\lim_{x \to 0} \frac{2\sin^2(x)}{x^3 \sin\left(\frac{1}{2x}\right)} = \lim_{x \to 0} \frac{2x^2}{x^3 \sin\left(\frac{1}{2x}\right)} = \lim_{x \to 0} \frac{2}{x \sin\left(\frac{1}{2x}\right)}$$

As $x \to 0$, $\sin\left(\frac{1}{2x}\right)$ oscillates between -1 and 1. Thus, the denominator $x \sin\left(\frac{1}{2x}\right)$ goes to 0 but changes sign. Therefore, the whole expression ($\frac{2}{x \sin\left(\frac{1}{2x}\right)}$) becomes unbounded and takes values near both $+\infty$ and $-\infty$. This leads to the result that $\lim_{x \to 0} \frac{2x^2}{x^3 \sin\left(\frac{1}{2x}\right)}$ does not exist.

Example 6.72 Calculate $\lim_{x \to \frac{\pi}{2}} \frac{\tan(x)}{\cot(x)}$.

Solution

$$\lim_{x\to\frac{\pi}{2}} \frac{\tan(x)}{\cot(x)} = \begin{cases} \lim_{x\to\frac{\pi}{2}^+} \frac{\tan(x)}{\cot(x)} = \frac{-\infty}{0^-} = +\infty \\ \lim_{x\to\frac{\pi}{2}^-} \frac{\tan(x)}{\cot(x)} = \frac{-\infty}{0^-} = +\infty \end{cases}$$

Example 6.73 Calculate $\lim_{x\to+\infty} \frac{2x^5+3x^4-6x+7}{8x^3+9x^2+10x+11}$.

Solution $\lim_{x\to+\infty} \frac{2x^5+3x^4-6x+7}{8x^3+9x^2+10x+11} = \lim_{x\to+\infty} \frac{2x^5}{8x^3} = \lim_{x\to+\infty} \frac{1}{4}x^2 = +\infty$

Example 6.74 Calculate $\lim_{x\to+\infty} \frac{0.000002x^5+3\times10^{2025}x^4-6x+7}{8x^5+9x^2+10x+11\times10^{7530}}$.

Solution

$$\lim_{x\to+\infty} \frac{0.000002x^5+3\times10^{2025}x^4-6x+7}{8x^5+9x^2+10x+11\times10^{7530}} = \lim_{x\to+\infty} \frac{0.000002x^5}{8x^5}$$
$$= \lim_{x\to+\infty} \frac{0.000002}{8} = \frac{0.000002}{8}$$

Example 6.75 Calculate $\lim_{x\to-\infty} \frac{2x^5+3x^4-6x+7}{8x^3+9x^2+10x+11}$.

Solution $\lim_{x\to-\infty} \frac{2x^5+3x^4-6x+7}{8x^3+9x^2+10x+11} = \lim_{x\to-\infty} \frac{2x^5}{8x^3} = \lim_{x\to-\infty} \frac{1}{4}x^2 = +\infty$

Example 6.76 Calculate $\lim_{x\to-\infty} \frac{8x^3+9x^2+10x+11}{2x^5+3x^4-6x+7}$.

Solution $\lim_{x\to-\infty} \frac{8x^3+9x^2+10x+11}{2x^5+3x^4-6x+7} = \lim_{x\to-\infty} \frac{8x^3}{2x^5} = \lim_{x\to-\infty} \frac{4}{x^2} = 0$

Example 6.77 Calculate $\lim_{x\to+\infty} \frac{8x^5+9x^2+10x+11}{2x^5+3x^4-6x+7}$.

Solution

$$\lim_{x\to+\infty} \frac{8x^5+9x^2+10x+11}{2x^5+3x^4-6x+7} = \lim_{x\to+\infty} \frac{8x^5}{2x^5} = \lim_{x\to+\infty} 4 = 4$$

Example 6.78 Calculate $\lim_{x\to+\infty} \frac{x^6-[x^6]}{3x^6+10^{2025}x^5+7}$.

Solution

$$\lim_{x\to+\infty} \frac{x^6-[x^6]}{3x^6+10^{2025}x^5+7} = \lim_{x\to+\infty} \frac{\frac{x^6-[x^6]}{x^6}}{\frac{3x^6+10^{2025}x^5+7}{x^6}} = \lim_{x\to+\infty} \frac{\frac{x^6}{x^6}-\frac{[x^6]}{x^6}}{3+\frac{10^{2025}}{x}+\frac{7}{x^6}}$$
$$= \lim_{x\to+\infty} \frac{1-\frac{[x^6]}{x^6}}{3} = \frac{1}{3} - \frac{1}{3}\lim_{x\to+\infty} \frac{[x^6]}{x^6}$$

$$\forall u : u-1<[u]\le u \Rightarrow x^6-1<\left[x^6\right]\le x^6 \Rightarrow \frac{x^6-1}{x^6}<\frac{\left[x^6\right]}{x^6}$$

$$\le\frac{x^6}{x^6}\Rightarrow 1-\frac{1}{x^6}<\frac{\left[x^6\right]}{x^6}\le 1\Rightarrow\begin{cases}\lim\limits_{x\to+\infty}\left(1-\frac{1}{x^6}\right)=1\\ \lim\limits_{x\to+\infty}(1)=1\end{cases}\Rightarrow \lim_{x\to+\infty}\frac{\left[x^6\right]}{x^6}=1$$

$$\lim_{x\to+\infty}\frac{x^6-\left[x^6\right]}{3x^6+10^{2025}x^5+7}=\frac{1}{3}-\frac{1}{3}\lim_{x\to\infty}\frac{\left[x^6\right]}{x^6}=\frac{1}{3}-\frac{1}{3}=0$$

Example 6.79 Calculate $\lim\limits_{x\to 2^-}\left(\frac{x+2}{x^2-2x}+\frac{2[x]}{2-x}\right)$.

Solution

$$\lim_{x\to 2^-}\left(\frac{x+2}{x^2-2x}+\frac{2[x]}{2-x}\right)=\lim_{x\to 2^-}\frac{x+2}{x^2-2x}+\lim_{x\to 2^-}\frac{2[x]}{2-x}=\lim_{x\to 2^-}\frac{4}{0^-}+\lim_{x\to 2^-}\frac{2\times 1}{0^+}=-\infty+\infty$$

$$\begin{aligned}\lim_{x\to 2^-}\left(\frac{x+2}{x^2-2x}+\frac{2[x]}{2-x}\right)&=\lim_{x\to 2^-}\left(\frac{x+2}{x(x-2)}+\frac{2[x]}{2-x}\right)=\lim_{x\to 2^-}\left(\frac{x+2-2x[x]}{x(x-2)}\right)\\&=\lim_{x\to 2^-}\left(\frac{x+2-2x\times 1}{x(x-2)}\right)=\lim_{x\to 2^-}\left(\frac{x+2-2x}{x(x-2)}\right)\\&=\lim_{x\to 2^-}\left(\frac{2-x}{x(x-2)}\right)=\lim_{x\to 2^-}\left(\frac{-1}{x}\right)=-\frac{1}{2}\end{aligned}$$

Example 6.80 Calculate $\lim\limits_{x\to 0}\left(\frac{x+2}{x^2+x}-\frac{3x-4}{x^2-2x}\right)$.

Solution

$$\lim_{x\to 0}\left(\frac{x+2}{x^2+x}-\frac{3x-4}{x^2-2x}\right)=\frac{2}{0}-\frac{4}{0}=\infty-\infty$$

$$\lim_{x\to 0}\left(\frac{x+2}{x^2+x}-\frac{3x-4}{x^2-2x}\right)=\lim_{x\to 0}\left(\frac{x+2}{x(x+1)}-\frac{3x-4}{x(x-2)}\right)=\lim_{x\to 0}\left(\frac{(x+2)(x-2)-(3x-4)(x+1)}{x(x+1)(x-2)}\right)=$$
$$\lim_{x\to 0}\left(\frac{x^2-4-\left(3x^2-x-4\right)}{x(x+1)(x-2)}\right)=\lim_{x\to 0}\left(\frac{x^2-4-3x^2+x+4}{x(x+1)(x-2)}\right)=\lim_{x\to 0}\left(\frac{-2x^2+x}{x(x+1)(x-2)}\right)=\lim_{x\to 0}\left(\frac{-2x+1}{(x+1)(x-2)}\right)=$$
$$\frac{1}{1\times -2}=-\frac{1}{2}=-0.5$$

Example 6.81 Calculate $\lim\limits_{x\to -2}\left(\frac{3}{2x^2+5x+2}-\frac{4}{x^2-4}\right)$.

Solution

$$\lim_{x\to -2}\left(\frac{3}{2x^2+5x+2}-\frac{4}{x^2-4}\right)=\frac{3}{0}-\frac{4}{0}=\infty-\infty$$

$$\lim_{x\to-2}\left(\frac{3}{2x^2+5x+2}-\frac{4}{x^2-4}\right)=\lim_{x\to-2}\left(\frac{3}{2(x+2)(x+0.5)}-\frac{4}{(x+2)(x-2)}\right)=\lim_{x\to-2}\left(\frac{3(x-2)-4\times2(x+0.5)}{2(x+2)(x+0.5)(x-2)}\right)=$$
$$\lim_{x\to-2}\left(\frac{3(x-2)-8(x+0.5)}{2(x+2)(x+0.5)(x-2)}\right)=\lim_{x\to-2}\left(\frac{3x-6-8x-4}{2(x+2)(x+0.5)(x-2)}\right)=\lim_{x\to-2}\left(\frac{-5x-10}{2(x+2)(x+0.5)(x-2)}\right)=$$
$$\lim_{x\to-2}\left(\frac{-5(x+2)}{2(x+2)(x+0.5)(x-2)}\right)=\lim_{x\to-2}\left(\frac{-5}{2(x+0.5)(x-2)}\right)=\frac{-5}{2\times-1.5\times-4}=-\frac{5}{12}=-0.4167$$

Example 6.82 Calculate $\lim\limits_{x\to+\infty}\frac{\sqrt{x+6\sqrt{x+1}}-\sqrt{x+7}}{\sqrt{x+2\sqrt{2x-\sqrt{x}}}}$.

Solution Recall that as x approaches infinity, a polynomial's limit is determined by its highest-degree term. Thus,

$$\lim_{x\to+\infty}\frac{\sqrt{x+6\sqrt{x+1}}-\sqrt{x+7}}{\sqrt{x+2\sqrt{2x-\sqrt{x}}}}=\lim_{x\to+\infty}\frac{\sqrt{x}-\sqrt{x}}{\sqrt{x+2\sqrt{2x}}}=\lim_{x\to+\infty}\frac{\sqrt{x}-\sqrt{x}}{\sqrt{x}}=0$$

Example 6.83 Calculate $\lim\limits_{x\to+\infty}\frac{\sqrt{x+6\sqrt{x}}+\sqrt{x}}{\sqrt{2x+\sqrt{x-\sqrt{x}}}}$.

Solution

$$\lim_{x\to+\infty}\frac{\sqrt{x+6\sqrt{x}}+\sqrt{x}}{\sqrt{2x+\sqrt{x-\sqrt{x}}}}=\lim_{x\to+\infty}\frac{\sqrt{x}+\sqrt{x}}{\sqrt{2x+\sqrt{x}}}=\lim_{x\to+\infty}\frac{\sqrt{x}+\sqrt{x}}{\sqrt{2x}}$$
$$=\lim_{x\to+\infty}\frac{2\sqrt{x}}{\sqrt{2x}}=\lim_{x\to+\infty}\frac{2}{\sqrt{2}}=\frac{2}{\sqrt{2}}$$

Example 6.84 Calculate $\lim\limits_{x\to+\infty}\frac{\sqrt{x^3+3x^2+1}-\sqrt{x^3-x^2+1}}{\sqrt{2x+1}}$.

Solution The terms $\sqrt{x^3+3x^2+1}$ and $\sqrt{x^3-x^2+1}$ are both equivalent to $\sqrt{x^3}$ as $x\to+\infty$. Their difference, however, is not equivalent to zero. The "next-order" terms ($3x^2+1$ and $-x^2+1$) are crucial and determine the final value of the limit.

$$\lim_{x\to+\infty}\frac{\sqrt{x^3+3x^2+1}-\sqrt{x^3-x^2+1}}{\sqrt{2x+1}}=\lim_{x\to+\infty}\frac{\sqrt{x^3\left(1+\frac{3x^2+1}{x^3}\right)}-\sqrt{x^3\left(1+\frac{-x^2+1}{x^3}\right)}}{\sqrt{2x+1}}=\lim_{x\to+\infty}\frac{\sqrt{x^3}\left(\sqrt{1+\frac{3x^2+1}{x^3}}-\sqrt{1+\frac{1-x^2}{x^3}}\right)}{\sqrt{2x+1}}=$$
$$\lim_{x\to+\infty}\frac{\sqrt{x^3}\left(1+\frac{3x^2+1}{2x^3}-\left(1+\frac{1-x^2}{2x^3}\right)\right)}{\sqrt{2x}}=\lim_{x\to+\infty}\frac{|x|\sqrt{x}\left(\frac{3x^2+1}{2x^3}-\frac{1-x^2}{2x^3}\right)}{\sqrt{2x}}=\lim_{x\to+\infty}\frac{x\left(\frac{4x^2}{2x^3}\right)}{\sqrt{2}}=\lim_{x\to+\infty}\frac{2}{\sqrt{2}}=\frac{2}{\sqrt{2}}=\sqrt{2}\approx1.4142$$

Exercise 6.7 Calculate $\lim\limits_{x\to+\infty}\frac{\sqrt{x^3+3x^2+1}-\sqrt{x^3-x^2+1}}{\sqrt{2x+1}}$ by multiplying the expression by its conjugate, $\frac{\sqrt{x^3+3x^2+1}+\sqrt{x^3-x^2+1}}{\sqrt{x^3+3x^2+1}+\sqrt{x^3-x^2+1}}$.

Example 6.85 Calculate $\lim\limits_{x\to+\infty}\frac{x^6-[x^6]}{3x^6+10^{2025}x^5+7}$.

Solution

$$\forall u:0\le u-[u]<1\Longrightarrow0\le x^6-\left[x^6\right]<1$$

$$\lim_{x\to+\infty}\frac{x^6-\left[x^6\right]}{3x^6+10^{2025}x^5+7}=\lim_{x\to+\infty}\frac{\text{a number between 0 and 1}}{3x^6}=\lim_{x\to+\infty}\frac{\text{a number between 0 and 1}}{\infty}=0$$

Example 6.86 Calculate $\lim\limits_{x\to+\infty}\frac{x^6-[x^5]}{3x^6+10^{2025}x^5+7}$.

Solution

$$x\to+\infty \Longrightarrow x^6 \gg [x^5] \Longrightarrow \lim_{x\to+\infty}\frac{x^6-[x^5]}{3x^6+10^{2025}x^5+7}=\lim_{x\to+\infty}\frac{x^6}{3x^6}=\lim_{x\to+\infty}\frac{1}{3}=\frac{1}{3}$$

Example 6.87 Calculate $\lim\limits_{x\to\infty}\frac{x^5-[x^6]}{3x^6+10^{2025}x^5+7}$.

Solution

$$x\to+\infty \Longrightarrow [x^6] \gg x^5 \Longrightarrow \lim_{x\to+\infty}\frac{x^5-[x^6]}{3x^6+10^{2025}x^5+7}=\lim_{x\to+\infty}\frac{[x^6]}{3x^6}=\lim_{x\to+\infty}\frac{x^6}{3x^6}$$
$$=\lim_{x\to+\infty}\frac{1}{3}=\frac{1}{3}$$

Example 6.88 Calculate $\lim\limits_{x\to\infty}\frac{x^5-[x^6]}{3x^7+10^{2025}x^5+7}$.

Solution $x\to+\infty \Longrightarrow [x^6] \gg x^5 \Longrightarrow \lim\limits_{x\to+\infty}\frac{x^5-[x^6]}{3x^7+10^{2025}x^5+7}=\lim\limits_{x\to+\infty}\frac{[x^6]}{3x^7}=\lim\limits_{x\to\infty}\frac{x^6}{3x^7}=\lim\limits_{x\to+\infty}\frac{1}{3x}=0$

Example 6.89 Calculate $\lim\limits_{x\to+\infty}\frac{x^5-[x^6]}{3x^4+10^{2025}x^2+7}$.

Solution $x\to+\infty \Longrightarrow [x^6] \gg x^5 \Longrightarrow \lim\limits_{x\to+\infty}\frac{x^5-[x^6]}{3x^4+10^{2025}x^2+7}=\lim\limits_{x\to+\infty}\frac{[x^6]}{3x^4}=\lim\limits_{x\to+\infty}\frac{x^6}{3x^4}=\lim\limits_{x\to+\infty}\frac{x^2}{3}=\infty$

Example 6.90 Calculate $\lim\limits_{x\to-\infty}\sqrt[3]{27x^3+8x^2}+\sqrt{9x^2+2x}$.

Solution

$$\lim_{x\to-\infty}\sqrt[3]{27x^3+8x^2}+\sqrt{9x^2+2x}=\sqrt[3]{27}\left(x+\frac{8}{3\times 27}\right)+\sqrt{9}\left|x+\frac{2}{2\times 9}\right|=3\left(x+\frac{8}{81}\right)$$
$$+3\left|x+\frac{1}{9}\right|$$

$$x\to-\infty \Longrightarrow 3\left(x+\frac{8}{81}\right)+3\left|x+\frac{1}{9}\right|=3\left(x+\frac{8}{81}\right)-3\left(x+\frac{1}{9}\right)=-\frac{3}{81}=-\frac{1}{27}$$

Example 6.91 Calculate $\lim\limits_{x\to-\infty}\frac{1}{x-\sqrt{3x^2+2x}}$.

Solution $\lim\limits_{x\to-\infty}\frac{1}{x-\sqrt{3x^2+2x}}=\lim\limits_{x\to-\infty}\frac{1}{x-\sqrt{3}\left|x+\frac{2}{2\times 3}\right|}=\lim\limits_{x\to-\infty}\frac{1}{x-\sqrt{3}\left|x+\frac{1}{3}\right|}=\lim\limits_{x\to-\infty}\frac{1}{x-\sqrt{3}\times-\left(x+\frac{1}{3}\right)}=$
$\lim\limits_{x\to-\infty}\frac{1}{x+\sqrt{3}\left(x+\frac{1}{3}\right)}=\lim\limits_{x\to-\infty}\frac{1}{\left(1+\sqrt{3}\right)x+\frac{\sqrt{3}}{3}}=\lim\limits_{x\to-\infty}\frac{1}{\left(1+\sqrt{3}\right)x}=0$

Exercise 6.8 Calculate $\lim\limits_{x\to+\infty}\frac{1}{x-\sqrt{x^2+2x}}$. (Ans. -1)

Example 6.92 Calculate $\lim\limits_{x\to-\infty}\frac{\sqrt{x^2+4x}-\sqrt{x^2+2x}}{\sqrt[3]{x^3+3x}-\sqrt[3]{x^3+6x^2}}$.

Solution

$$\lim_{x\to-\infty}\frac{\sqrt{x^2+4x}-\sqrt{x^2+2x}}{\sqrt[3]{x^3+3x}-\sqrt[3]{x^3+6x^2}}=\lim_{x\to-\infty}\frac{\left|x+\frac{4}{2\times 1}\right|-\left|x+\frac{2}{2\times 1}\right|}{\left(x+\frac{0}{3\times 1}\right)-\left(x+\frac{6}{3\times 1}\right)}$$

$$=\lim_{x\to-\infty}\frac{|x+2|-|x+1|}{x-(x+2)}=\lim_{x\to-\infty}\frac{-(x+2)-(-(x+1))}{x-(x+2)}=\lim_{x\to-\infty}\frac{-1}{-2}=\frac{1}{2}$$

Example 6.93 Calculate $\lim\limits_{x\to+\infty}\left(\sqrt{x+3}-\sqrt{x+7}\right)\left(\sqrt{x+1}+\sqrt{x+9}\right)$.

Solution

$$\lim_{x\to+\infty}\left(\sqrt{x+3}-\sqrt{x+7}\right)\left(\sqrt{x+1}+\sqrt{x+9}\right)=(\infty-\infty)(\infty+\infty)$$

When $x\to\infty$, $\sqrt[n]{a_nx^n+a_{n-1}x^{n-1}+a_{n-2}x^{n-2}+\ldots+a_0}\sim\begin{cases}\sqrt[n]{a_n}\left|x+\dfrac{a_{n-1}}{na_n}\right| & a_n>0, n\ even\\ \sqrt[n]{a_n}\left(x+\dfrac{a_{n-1}}{na_n}\right) & n\ odd\end{cases}$.

However, in this example, the radical's order n does not match the degree of the polynomial's leading term inside the radical. Therefore, this formula cannot be applied here.

$\lim\limits_{x\to+\infty}\left(\sqrt{x+3}-\sqrt{x+7}\right)\left(\sqrt{x+1}+\sqrt{x+9}\right)=\lim\limits_{x\to+\infty}\left(\sqrt{x+3}-\sqrt{x+7}\right)\left(\sqrt{x+1}+\sqrt{x+9}\right)\times$
$\frac{\left(\sqrt{x+3}+\sqrt{x+7}\right)}{\left(\sqrt{x+3}+\sqrt{x+7}\right)}=\lim\limits_{x\to+\infty}\frac{\left(\sqrt{x+3}^2-\sqrt{x+7}^2\right)\left(\sqrt{x+1}+\sqrt{x+9}\right)}{\left(\sqrt{x+3}+\sqrt{x+7}\right)}=\lim\limits_{x\to+\infty}\frac{(x+3-(x+7))\left(\sqrt{x+1}+\sqrt{x+9}\right)}{\left(\sqrt{x+3}+\sqrt{x+7}\right)}=\lim\limits_{x\to+\infty}\frac{-4\left(\sqrt{x+1}+\sqrt{x+9}\right)}{\left(\sqrt{x+3}+\sqrt{x+7}\right)}=$
$\lim\limits_{x\to+\infty}\frac{-4(\sqrt{x}+\sqrt{x})}{(\sqrt{x}+\sqrt{x})}=\lim\limits_{x\to+\infty}\frac{-4\times 2\sqrt{x}}{2\sqrt{x}}=-4$

Example 6.94 Calculate $\lim\limits_{x\to 0}\left(\frac{1}{\cos(2x)-1}-\frac{1}{x}\right)$.

Solution

$$\lim_{x\to 0}\left(\frac{1}{\cos(2x)-1}-\frac{1}{x}\right)=\frac{1}{0}-\frac{1}{0}=\infty-\infty$$

$$\lim_{x\to 0}\left(\frac{1}{\cos(2x)-1}-\frac{1}{x}\right)=\lim_{x\to 0}\left(\frac{x-\cos(2x)+1}{(\cos(2x)-1)x}\right)=\frac{0}{0}$$

Let's use L'Hôpital's rule:

$$\lim_{x\to 0}\left(\frac{1}{\cos(2x)-1}-\frac{1}{x}\right)=\lim_{x\to 0}\left(\frac{x-\cos(2x)+1}{(\cos(2x)-1)x}\right)=\lim_{x\to 0}\left(\frac{1+2\sin(2x)}{-2\sin(2x)x+\cos(2x)-1}\right)=\frac{1}{0}=\infty$$

Thus, the given problem does not have a finite limit.

Example 6.95 Calculate $\lim\limits_{x\to+\infty}\left(1-\frac{3}{x}\right)^{4x}$.

Solution The limit is in the indeterminate form 1^{∞}. It can be solved using the standard formula for this type of limit, which states that $\lim\limits_{x\to a} f(x)^{g(x)} = e^{\lim\limits_{x\to a} g(x)(f(x)-1)}$ when $f(x) \to 1$ and $g(x) \to \infty$. Therefore, $\lim\limits_{x\to+\infty} \left(1-\frac{3}{x}\right)^{4x} = e^{\lim\limits_{x\to+\infty} 4x\left(1-\frac{3}{x}-1\right)} = e^{\lim\limits_{x\to+\infty} -12} = e^{-12}$.

Example 6.96 Calculate $\lim\limits_{x\to 0} (1+\sin(x))^{\cot(x)}$.

Solution $\lim\limits_{x\to 0}(1+\sin(x))^{\cot(x)} = e^{\lim\limits_{x\to 0}(1+\sin(x)-1)\cot(x)} = e^{\lim\limits_{x\to 0}\sin(x)\times\cot(x)} = e^{\lim\limits_{x\to 0}\cos(x)} = e^{1} \approx 2.7183$

Example 6.97 Calculate $\lim\limits_{x\to\frac{\pi}{2}^-} \left(\sqrt{1+cot(x)}\right)^{tan(x)}$.

Solution $\lim\limits_{x\to\frac{\pi}{2}^-} \left(\sqrt{1+\cot(x)}\right)^{\tan(x)} = \lim\limits_{x\to\frac{\pi}{2}^-} \left((1+\cot(x))^{\frac{1}{2}}\right)^{\tan(x)} = \lim\limits_{x\to\frac{\pi}{2}^-} (1+\cot(x))^{\frac{\tan(x)}{2}} = e^{\lim\limits_{x\to\frac{\pi}{2}^-} \frac{\tan(x)}{2}\times(1+\cot(x)-1)} = e^{\lim\limits_{x\to\frac{\pi}{2}^-} \frac{\tan(x)}{2}\times\cot(x)} = e^{\lim\limits_{x\to\frac{\pi}{2}^-} \frac{1}{2}} = e^{\frac{1}{2}} \approx 1.6487$

Example 6.98 Calculate $\lim\limits_{x\to 0}\left(\frac{1}{x^2} - \frac{\cos(3x)}{x^2}\right)$.

Solution

$$\lim_{x\to 0}\left(\frac{1}{x^2} - \frac{\cos(3x)}{x^2}\right) = \infty - \infty$$

$$\lim_{x\to 0}\left(\frac{1}{x^2} - \frac{\cos(3x)}{x^2}\right) = \lim_{x\to 0}\left(\frac{1-\cos(3x)}{x^2}\right) = \frac{0}{0}$$

Applying L'Hôpital's Rule yields:

$$\lim_{x\to 0}\left(\frac{1}{x^2} - \frac{\cos(3x)}{x^2}\right) = \lim_{x\to 0}\left(\frac{1-\cos(3x)}{x^2}\right) = \lim_{x\to 0}\left(\frac{3\sin(3x)}{2x}\right) = \frac{0}{0}$$

Applying L'Hôpital's Rule once more gives:

$$\lim_{x\to 0}\left(\frac{1}{x^2} - \frac{\cos(3x)}{x^2}\right) = \lim_{x\to 0}\left(\frac{1-\cos(3x)}{x^2}\right) = \lim_{x\to 0}\left(\frac{3\sin(3x)}{2x}\right) = \lim_{x\to 0}\left(\frac{9\cos(3x)}{2}\right) = \frac{9}{2}$$

Example 6.99 The code in Fig. 6.10 calculates $\lim\limits_{x\to 0}\left(\frac{1-\cos(3x)}{x^2}\right)$. According to Fig. 6.10, the limit $\lim\limits_{x\to 0}\left(\frac{1-\cos(3x)}{x^2}\right)$ is $\frac{9}{2}$.

Fig. 6.10 Code for calculating $\lim_{x\to 0}\left(\frac{1-\cos(3x)}{x^2}\right)$

```
>> syms x
>> f=(1-cos(3*x))/x^2;
>> limit(f,x,0)

ans =

9/2

fx >>
```

Example 6.100 The code in Fig. 6.11 calculates $\lim_{x\to 0^+}\left(\frac{|x|}{x}\right)$. According to Fig. 6.11, the limit $\lim_{x\to 0^+}\left(\frac{|x|}{x}\right)$ is 1.

Fig. 6.11 Code for calculating $\lim_{x\to 0^+}\left(\frac{|x|}{x}\right)$

```
>> syms x
>> f=abs(x)/x;
>> L_right=limit(f,x,0,'right')

L_right =

1

fx >>
```

Example 6.101 The code in Fig. 6.12 calculates $\lim_{x\to 0^-}\left(\frac{|x|}{x}\right)$. According to Fig. 6.12, the limit $\lim_{x\to 0^-}\left(\frac{|x|}{x}\right)$ is -1.

Fig. 6.12 Code for calculating $\lim_{x\to 0^-}\left(\frac{|x|}{x}\right)$

```
>> syms x
>> f=abs(x)/x;
>> L_left=limit(f,x,0,'left')

L_left =

-1

fx >>
```

Example 6.102 The code in Fig. 6.13 shows that $\lim_{x\to 0}\left(\frac{|x|}{x}\right)$ does not exist. This is expected since the left-hand limit and the right-hand limit are not equal.

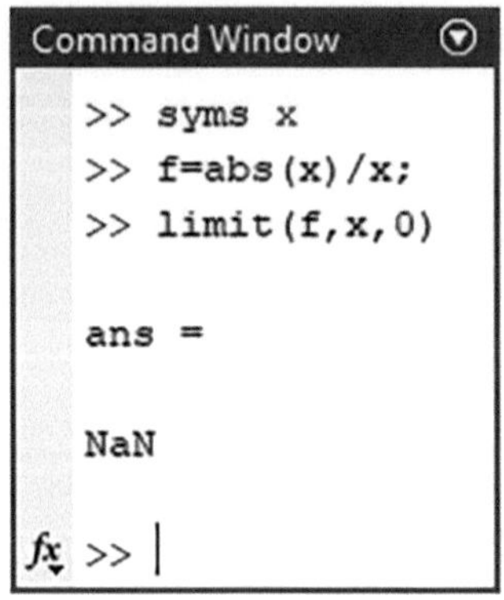

Fig. 6.13 Code for calculating $\lim_{x \to 0} \left(\frac{|x|}{x} \right)$

Example 6.103 The code in Fig. 6.14 calculates $\lim_{x \to -\infty} \frac{x^6 - \lfloor x^6 \rfloor}{3x^6 + 10^{2025}x^5 + 7}$. According to Fig. 6.14, the limit $\lim_{x \to -\infty} \frac{x^6 - \lfloor x^6 \rfloor}{3x^6 + 10^{2025}x^5 + 7}$ is 0.

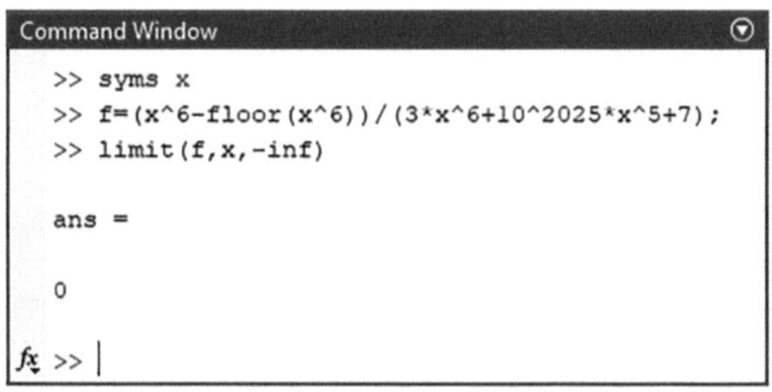

Fig. 6.14 Code for calculating $\lim_{x \to -\infty} \frac{x^6 - \lfloor x^6 \rfloor}{3x^6 + 10^{2025}x^5 + 7}$

Example 6.104 The code in Fig. 6.15 calculates $\lim_{x \to \infty} x^2 e^{-x}$. According to Fig. 6.15, the limit $\lim_{x \to \infty} x^2 e^{-x}$ is 0.

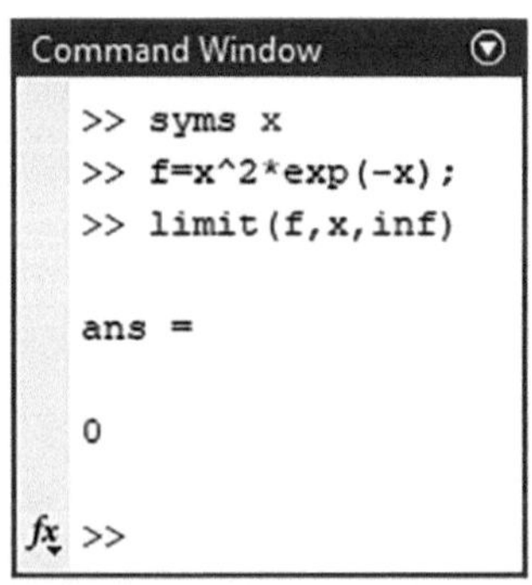

Fig. 6.15 Code for calculating $\lim_{x \to \infty} x^2 e^{-x}$

Example 6.105 The code in Fig. 6.16 calculates $\lim_{x\to -\infty} x^2 e^{-x}$. According to Fig. 6.16, the limit $\lim_{x\to -\infty} x^2 e^{-x}$ is ∞.

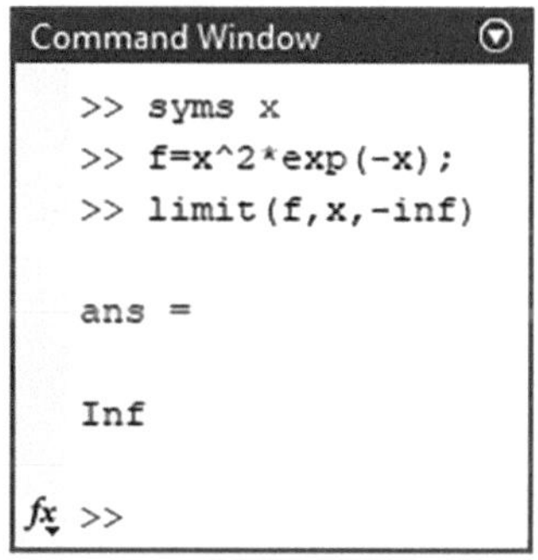

Fig. 6.16 Code for calculating $\lim_{x\to -\infty} x^2 e^{-x}$

References for Further Study

1. Engineering Mathematics with MATLAB® and Simulink®, Farzin Asadi, Springer, 2025.
2. Applied Numerical Analysis with MATLAB®/Simulink®: For Engineers and Scientists, Farzin Asadi, Springer, 2023.
3. Advanced Engineering Mathematics, Erwin Kreyszig, Wiley, 2020.
4. A Problem-Solving Approach to Single-Variable Calculus (Volume II), Farzin Asadi, Springer, 2026.
5. A Problem-Solving Approach to Multi-Variable Calculus (Volume I), Farzin Asadi, Springer, 2026.
6. A Problem-Solving Approach to Multi-Variable Calculus (Volume II), Farzin Asadi, Springer, 2026.
7. A Problem-Solving Approach to Multi-Variable Calculus (Volume III), Farzin Asadi, Springer, 2026.
8. A Problem-Solving Approach to Ordinary Differential Equations (Volume I), Farzin Asadi, Springer, 2026.
9. A Problem-Solving Approach to Ordinary Differential Equations (Volume II), Farzin Asadi, Springer, 2026.
10. A Problem-Solving Approach to Ordinary Differential Equations (Volume III), Farzin Asadi, Springer, 2026.
11. A Problem-Solving Approach to Ordinary Differential Equations (Volume IV), Farzin Asadi, Springer, 2026.

Chapter 7
Derivative

7.1 Introduction

A function $f(x)$ is formally defined as continuous at a point $x = c$ if the following three conditions are met:

1. $f(c)$ is defined. The point must be in the function's domain.
2. $\lim_{x \to c} f(x)$ exists. The limit as x approaches c must be a finite real number. This requires the left-hand limit and the right-hand limit to be equal.
3. $\lim_{x \to c} f(x) = f(c)$ The value of the limit must be equal to the value of the function at that point.

In simpler terms, a function is continuous at a point if there are no jumps, holes, or gaps in its graph at that location. A function is continuous over an interval if it is continuous at every point within that interval.

A function $f(x)$ is considered differentiable at a point $x = a$ if its derivative, denoted as $f'(a)$, exists at that point. The formal definition of the derivative is given by the limit of the difference quotient:

$$f'(a) = \lim_{x \to a} \frac{f(x) - f(a)}{x - a} = \lim_{h \to 0} \frac{f(a+h) - f(a)}{h}$$

For the function to be differentiable at $x = a$, this limit must exist and be a finite number.

While differentiability guarantees continuity, continuity does not guarantee differentiability. A function can be continuous everywhere but fail to be differentiable at certain points. The most common example is the absolute value function, $f(x) = |x|$, at $x = 0$.

If a function is differentiable at a point, it must be continuous at that point.

A function cannot have a derivative at a point where it is not continuous.

F. Asadi, *A Problem-Solving Approach to Single-Variable Calculus*,
https://doi.org/10.1007/978-3-032-13505-6_7

All derivatives in this chapter are with respect to x unless explicitly stated otherwise. Here are the fundamental rules and formulas for differentiation:

1. $y'(x) = \lim\limits_{h \to 0} \frac{y(x+h) - y(x)}{h}$
2. $y'(x^+) = \lim\limits_{h \to 0^+} \frac{y(x+h) - y(x)}{h}$
3. $y'(x^-) = \lim\limits_{h \to 0^-} \frac{y(x+h) - y(x)}{h}$
4. $y''(x) = \lim\limits_{h \to 0} \frac{y(x+h) - 2y(x) + y(x-h)}{h^2}$
5. $(u(x).\, v(x))' = u'(x)v(x) + u(x)v'(x)$
6. $\left(\frac{u(x)}{v(x)}\right)' = \frac{u'(x)v(x) - v'(x)u(x)}{v^2(x)}$
7. $(u(v(x)))' = v'(x)u'(v(x))$
8. $(a^u)' = u'a^u \ln(a)$
9. $(e^u)' = u'e^u$
10. $u(x) > 0 \Rightarrow (\ln(u))' = \frac{u'}{u}$
11. $u(x) > 0 \Rightarrow (\log_a(u))' = \left(\frac{\ln(u)}{\ln(a)}\right)' = \frac{1}{\ln(a)}(\ln(u))' = \frac{1}{\ln(a)} \times \frac{u'}{u}$
12. $u(x) > 0 \Rightarrow \left(u(x)^{v(x)}\right)' = u(x)^{v(x)}\left(v'(x)\ln(u(x)) + v(x)\frac{u'(x)}{u(x)}\right)$
13. $\frac{d}{dx}(|f(x)|) = f'(x)\frac{f(x)}{|f(x)|} = f'(x).\, sgn(f(x))$ This formula holds true for all values of x where $f(x) \neq 0$ and $f(x)$ is differentiable.
14. $y = \sin(x) \Rightarrow y' = \cos(x)$
15. $y = \sin(u) \Rightarrow y' = u'\cos(u)$
16. $y = \cos(x) \Rightarrow y' = -\sin(x)$
17. $y = \cos(u) \Rightarrow y' = -u'\sin(u)$
18. $y = \tan(x) \Rightarrow y' = 1 + \tan^2(x)$
19. $y = \tan(u) \Rightarrow y' = u'(1 + \tan^2(u))$
20. $y = \cot(x) \Rightarrow y' = -(1 + \cot^2(x))$
21. $y = \cot(u) \Rightarrow y' = -u'(1 + \cot^2(u))$
22. $-1 < u < 1 \Rightarrow y = \sin^{-1}(u) \Rightarrow y' = \frac{u'}{\sqrt{1-u^2}}$
23. $-1 < u < 1 \Rightarrow y = \cos^{-1}(u) \Rightarrow y' = -\frac{u'}{\sqrt{1-u^2}}$
24. $y = \sec^{-1}(u) \Rightarrow y' = \frac{u'}{|u|\sqrt{u^2-1}}$
25. $y = \csc^{-1}(u) \Rightarrow y' = -\frac{u'}{|u|\sqrt{u^2-1}}$
26. $y = \sinh(u) \Rightarrow y' = u'\cosh(u)$
27. $y = \cosh(u) \Rightarrow y' = u'\sinh(u)$
28. $y = \tanh(u) \Rightarrow y' = u'\,\mathrm{sec\,h}^2(u)$
29. $y = \coth(u) \Rightarrow y' = -u'\,\mathrm{csc\,h}^2(u)$
30. $y = \mathrm{csch}(u) \Rightarrow y' = -u'\,\mathrm{csch}(u)\coth(u)$
31. $y = \mathrm{sech}(u) \Rightarrow y' = -u'\,\mathrm{sech}(u)\tanh(u)$
32. $u \in \mathbb{R} \Rightarrow y = \sinh^{-1}(u) \Rightarrow y' = \frac{u'}{\sqrt{u^2+1}}$
33. $u > 1 \Rightarrow y = \cosh^{-1}(u) \Rightarrow y' = \frac{u'}{\sqrt{u^2-1}}$

34. $|u| < 1 \Rightarrow y = \tanh^{-1}(u) = \frac{1}{2}\ln\left(\frac{1+u}{1-u}\right) \Rightarrow y' = \frac{u'}{1-u^2}$
35. $|u| > 1 \Rightarrow y = \coth^{-1}(u) = \frac{1}{2}\ln\left(\frac{u+1}{u-1}\right) \Rightarrow y' = \frac{u'}{1-u^2}$
36. $f(x, y) = 0 \Rightarrow y' = -\frac{f'_x}{f'_y}$

$f(x, y)$: The given implicit function
y': The derivative $\frac{dy}{dx}$ you are trying to find.
f'_x: This is the partial derivative of $f(x, y)$ with respect to x. To find it, you treat y as a constant and differentiate the entire function $f(x, y)$ only with respect to x.
f'_y : This is the partial derivative of $f(x, y)$ with respect to y. To find it, you treat x as a constant and differentiate the entire function $f(x, y)$ only with respect to y.
The formulas below summarize the key properties of logarithms:

1. $\log_a xy = \log_a x + \log_a y$
2. $\log_a \frac{x}{y} = \log_a x - \log_a y$
3. $\log_a x^y = y\log_a x$
4. $\log_a x = \frac{\log_b x}{\log_b a}$

If a function f is differentiable and has an inverse function f^{-1}, the derivative of the inverse function at a point x can be found using the following formula:

$$(f^{-1})'(x) = \frac{1}{f'(f^{-1}(x))}$$

Here is the proof:

$$f(f^{-1}(x)) = x \Rightarrow \frac{df(f^{-1}(x))}{dx} = \frac{dx}{dx} \Rightarrow (f^{-1}(x))' \times f(f^{-1}(x)) = 1$$
$$\Rightarrow (f^{-1}(x))' = \frac{1}{f(f^{-1}(x))}$$

$f'(f^{-1}(x))$ is the derivative of the original function f evaluated at the point $y = f^{-1}(x)$.

The derivative of the inverse function at a point is the reciprocal of the derivative of the original function at the corresponding point.

For example, for $f(x) = x^3 + x$, $f(1) = 2$, $f^{-1}(2) = 1$ and $f'(x) = 3x^2 + 1$.

$$(f^{-1})'(x) = \frac{1}{f'(f^{-1}(x))} \Rightarrow (f^{-1})'(2) = \frac{1}{f'(f^{-1}(2))} = \frac{1}{f'(1)} = \frac{1}{3 \times 1^2 + 1} = \frac{1}{4}$$

In a polar coordinate system, a point is represented by (r, θ). To find the derivative of a function $r = f(\theta)$ with respect to θ, we use the standard rules of calculus, but the physical interpretation relates to the slope of the tangent line in the Cartesian plane. If you have a polar curve $r = f(\theta)$, the Cartesian coordinates are:

$$x = r\cos(\theta) = f(\theta)\ \cos(\theta)$$
$$y = r\sin(\theta) = f(\theta)\ \sin(\theta)$$

To find the slope of the tangent line in the (x, y) plane, we use the chain rule to calculate $\frac{dy}{dx} = \frac{\frac{dy}{d\theta}}{\frac{dx}{d\theta}}$:

$$\frac{dy}{dx} = \frac{\frac{dy}{d\theta}}{\frac{dx}{d\theta}} = \frac{f'(\theta)\sin(\theta) + f(\theta)\cos(\theta)}{f'(\theta)\cos(\theta) - f(\theta)\sin(\theta)} = \frac{\frac{df(\theta)}{d\theta}\sin(\theta) + f(\theta)\cos(\theta)}{\frac{df(\theta)}{d\theta}\cos(\theta) - f(\theta)\sin(\theta)}$$

If you have a polar curve $\theta = g(r)$, the Cartesian coordinates are:

$$x = r\cos(\theta) = r\ \cos(g(r))$$
$$y = r\sin(\theta) = r\ \sin(g(r))$$

To find the slope of the tangent line in the (x, y) plane, we use the chain rule to calculate $\frac{dy}{dx} = \frac{\frac{dy}{dr}}{\frac{dx}{dr}}$:

$$\frac{dy}{dx} = \frac{\frac{dy}{dr}}{\frac{dx}{dr}} = \frac{\sin(g(r)) + rg'(r)\cos(g(r))}{\cos(g(r)) - rg'(r)\sin(g(r))} = \frac{\sin(g(r)) + r\frac{dg(r)}{dr}\cos(g(r))}{\cos(g(r)) - r\frac{dg(r)}{dr}\sin(g(r))}$$

Let's study an example. The slope of the tangent line to the Cardioid $r = 1 + \cos(\theta)$ at $\theta = \frac{\pi}{2}$ is:

$$\theta = \frac{\pi}{2} \overset{r=1+\cos(\theta)}{\Rightarrow} r = 1 + \cos\left(\frac{\pi}{2}\right) = 1 + 0 = 1$$

$$\begin{cases} x = r\cos(\theta) \\ y = r\sin(\theta) \end{cases} \overset{\substack{r=1 \\ \theta=\frac{\pi}{2}}}{\Rightarrow} \begin{cases} x = r\cos(\theta) = 1\cos\left(\frac{\pi}{2}\right) = 0 \\ y = r\sin(\theta) = 1\sin\left(\frac{\pi}{2}\right) = 1 \end{cases}$$

$$\frac{dy}{dx} = \frac{\frac{dy}{d\theta}}{\frac{dx}{d\theta}} = \frac{f'(\theta)\sin(\theta) + f(\theta)\cos(\theta)}{f'(\theta)\cos(\theta) - f(\theta)\sin(\theta)} = \frac{\frac{df(\theta)}{d\theta}\sin(\theta) + f(\theta)\cos(\theta)}{\frac{df(\theta)}{d\theta}\cos(\theta) - f(\theta)\sin(\theta)}$$
$$= \frac{(-\sin(\theta))\sin(\theta) + (1 + \cos(\theta))\cos(\theta)}{(-\sin(\theta))\cos(\theta) - (1 + \cos(\theta))\sin(\theta)} = \frac{\cos(\theta) + \cos^2(\theta) - \sin^2(\theta)}{-\sin(\theta) - 2\sin(\theta)\cos(\theta)}$$
$$\overset{\theta=\frac{\pi}{2}}{\Rightarrow} \frac{dy}{dx} = \frac{\frac{dy}{d\theta}}{\frac{dx}{d\theta}} = \frac{\cos\left(\frac{\pi}{2}\right) + \cos^2\left(\frac{\pi}{2}\right) - \sin^2\left(\frac{\pi}{2}\right)}{-\sin\left(\frac{\pi}{2}\right) - 2\sin\left(\frac{\pi}{2}\right)\cos\left(\frac{\pi}{2}\right)} = \frac{0 + 0 - 1}{-1 - 2\times 0} = 1$$

Consequently, the slope of the tangent line at the polar point $\left(r = 1, \theta = \frac{\pi}{2}\right)$, which corresponds to the Cartesian coordinates $(0, 1)$, is $m = 1$.

Let's study another example. We want to find the slope of the tangent line to the curve defined by $\theta(r) = \frac{1}{r}$ at $r = 1$.

$$r = 1 \overset{\theta(r)=\frac{1}{r}}{\Longrightarrow} \theta = \frac{1}{1} = 1$$

$$\begin{cases} x = r\cos(\theta) \\ y = r\sin(\theta) \end{cases} \overset{\substack{r=1 \\ \theta=1}}{\Longrightarrow} \begin{cases} x = r\cos(\theta) = 1\cos(1) = 0.5403 \\ y = r\sin(\theta) = 1\sin(1) = 0.8415 \end{cases}$$

$$\frac{dy}{dx} = \frac{\frac{dy}{dr}}{\frac{dx}{dr}} = \frac{\sin(g(r)) + rg'(r)\cos(g(r))}{\cos(g(r)) - rg'(r)\sin(g(r))} = \frac{\sin(g(r)) + r\frac{dg(r)}{dr}\cos(g(r))}{\cos(g(r)) - r\frac{dg(r)}{dr}\sin(g(r))}$$

$$= \frac{\sin\left(\frac{1}{r}\right) + r \times -\frac{1}{r^2} \times \cos\left(\frac{1}{r}\right)}{\cos\left(\frac{1}{r}\right) - r \times -\frac{1}{r^2} \times \sin\left(\frac{1}{r}\right)} = \frac{\sin\left(\frac{1}{r}\right) - \frac{1}{r} \times \cos\left(\frac{1}{r}\right)}{\cos\left(\frac{1}{r}\right) + \frac{1}{r} \times \sin\left(\frac{1}{r}\right)}$$

$$\overset{r=1}{\Longrightarrow} \frac{dy}{dx} = \frac{\frac{dy}{dr}}{\frac{dx}{dr}} = \frac{\sin\left(\frac{1}{1}\right) - \frac{1}{1} \times \cos\left(\frac{1}{1}\right)}{\cos\left(\frac{1}{1}\right) + \frac{1}{1} \times \sin\left(\frac{1}{1}\right)} = \frac{\sin(1) - \cos(1)}{\cos(1) + \sin(1)} = \frac{0.8415 - 0.5403}{0.5403 + 0.8415}$$
$$= 0.2180$$

Consequently, the slope of the tangent line at the polar point $(r = 1, \theta = 1)$, which corresponds to the Cartesian coordinates $(0.5403, 0.8415)$, is $m = 0.2180$.

7.2 Solved Problems

Example 7.1 Find the points at which the derivative of the function shown in Fig. 7.1 fails to exist.

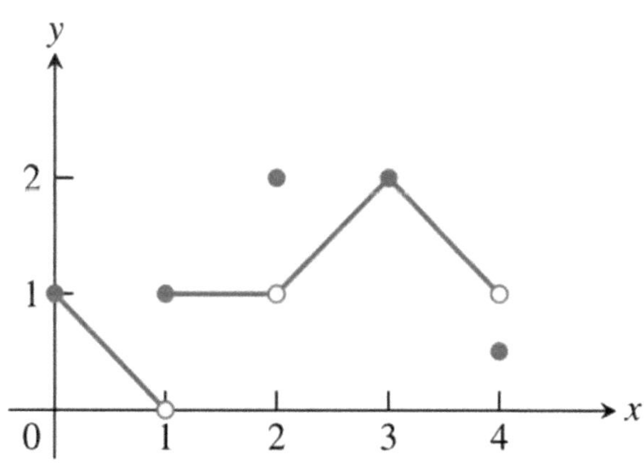

Fig. 7.1 Function for Example 7.1

Solution A derivative does not exist at a point for several key reasons, including:

Discontinuities: If a function has a jump, a hole (a removable discontinuity), or an infinite discontinuity (a vertical asymptote) at a point, it cannot be differentiable there. For a function to be differentiable at a point, it must first be continuous at that point.

Sharp Corners or Cusps: At points where a graph has a sharp turn, like at the vertex of $y = |x|$, the left-hand and right-hand derivatives are not equal. This is because the slope of the tangent line changes abruptly.

Vertical Tangents: The derivative represents the slope of the tangent line. If the tangent line at a point is vertical, its slope is undefined (it is infinite). Therefore, the derivative does not exist. A classic example is the function $f(x) = x^{\frac{1}{3}}$ at $x = 0$ $(f(x) = x^{\frac{1}{3}} \Rightarrow f'(x) = \frac{1}{3\sqrt[3]{x^2}})$.

Endpoints: The derivative is defined using a two-sided limit. At the endpoints of a function's domain, you can only approach the point from one side, so the full limit cannot be taken. Therefore, the derivative does not exist at the endpoints.

The exact number of points where a derivative does not exist depends entirely on the specific function in question. A function can have zero, one, a finite number, or even an infinite number of points where its derivative does not exist. The function shown in Fig. 7.1 lacks a derivative at =0, 1, 2, 3, and 4.

Example 7.2 Find the angle θ that the tangent line to $y = x^2$ at $x = 1$ makes with the x-axis (Fig. 7.2).

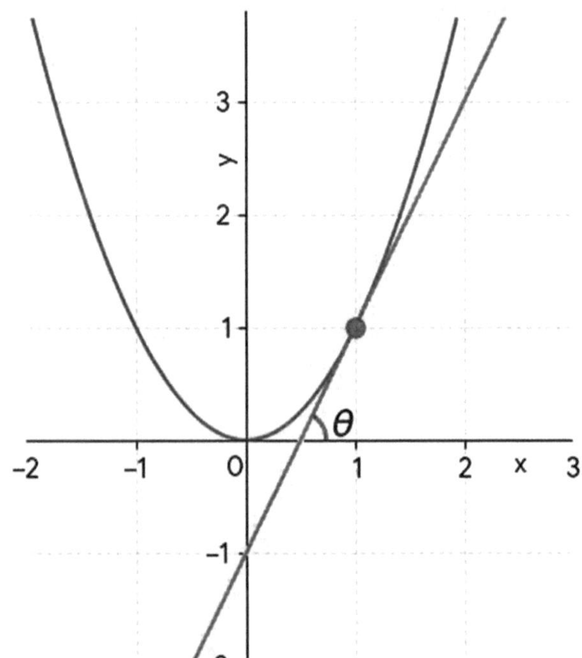

Fig. 7.2 Tangent to $y = x^2$ at $x = 1$

Solution The slope of the tangent line to $y = x^2$ at $x = 1$ is found by evaluating the derivative, $y' = 2x$, at $x = 1$. This gives a slope of $m = 2 \times 1 = 2$. The angle this line makes with the positive x-axis (θ in Fig. 7.2) is given by $\theta = \tan^{-1}(m) = \tan^{-1}(2) \approx 63.43°$.

Example 7.3 Find the angle θ that the tangent line to $y = x^2$ at $x = -1$ makes with the x-axis (Fig. 7.3).

Fig. 7.3 Tangent to $y = x^2$ at $x = -1$

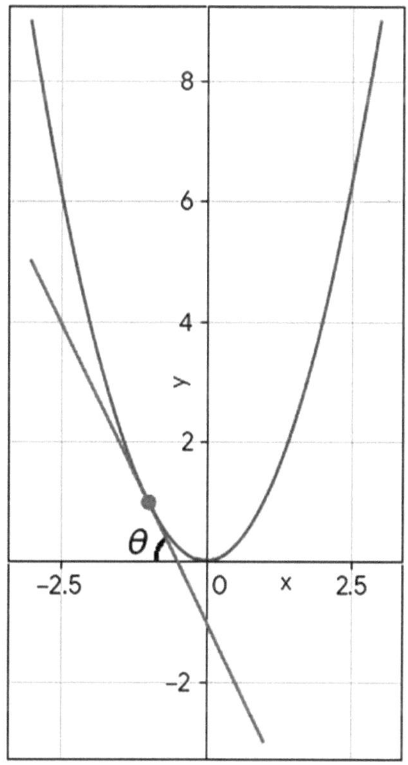

Solution The slope of the tangent line to $y = x^2$ at $x = -1$ is found by evaluating the derivative, $y' = 2x$, at $x = -1$. This gives a slope of $m = 2 \times -1 = -2$. The angle this line makes with the positive x-axis (φ in Fig. 7.4) is given by $\varphi = \tan^{-1}(m) = \tan^{-1}(-2) \approx -63.43°$. Therefore, $\theta = |\varphi| = 63.43°$.

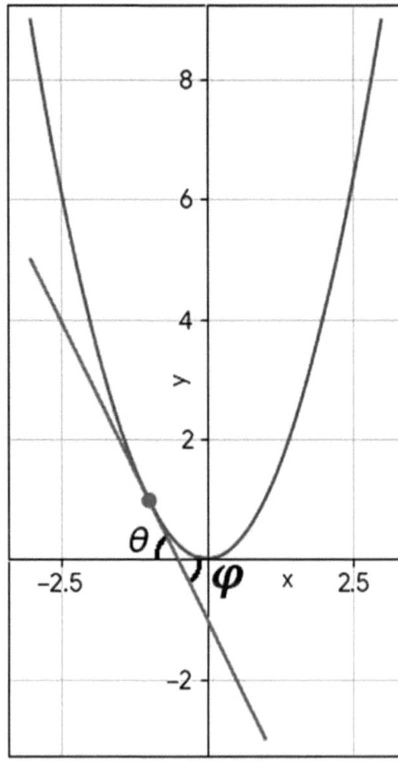

Fig. 7.4 φ represents the angle measured from the positive x-axis

Exercise 7.1 Find the angle θ that the tangent line to $y = x^3 - 5x$ at $x = -1$ makes with the x-axis (Fig. 7.5). (Ans. $\theta = 2.035$ rad $= 116.56°$)

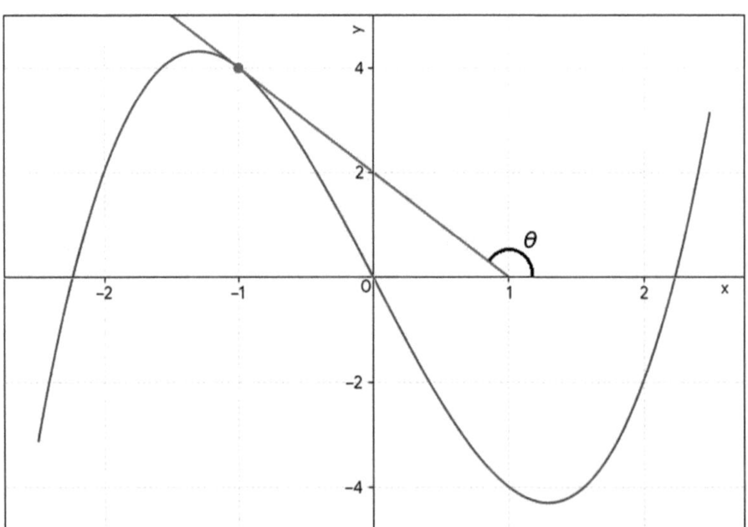

Fig. 7.5 Tangent to $y = x^3 - 5x$ at $x = -1$

Example 7.4 Find the angle between $y = -2x + 1$ and $y = \sqrt{x+1}$.

Solution Finding the angle between a line and a curve requires determining the angle between the line and the tangent to the curve at the point of intersection.

Let's set the two equations equal to each other to find the x-coordinate of the intersection point.

$$-2x+1=\sqrt{x+1}\Rightarrow 4x^2-4x+1=x+1\Rightarrow 4x^2-5x=0\Rightarrow x(4x-5)=0\Rightarrow \begin{cases} x=0 \\ x=\frac{5}{4} \end{cases}$$

We must check both in the original equation, as squaring can introduce extraneous solutions.

$\begin{cases} y=-2x+1^{x=0}{}_{x=0}{}^{y=1} \\ y=\sqrt{x+1}^{x=0}{}_{x=0}{}^{y=1} \end{cases}$ $\Rightarrow$ The point (0, 1) is a valid intersection point (Fig. 7.6).

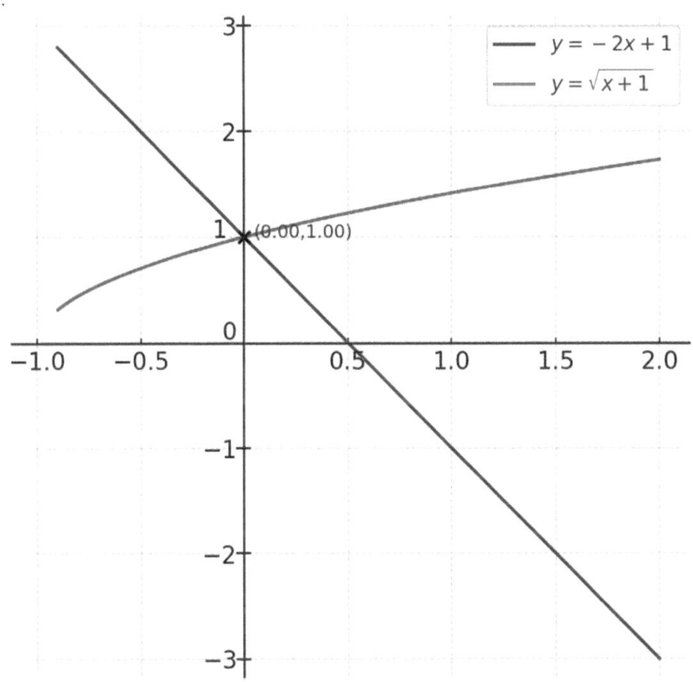

Fig. 7.6 Plot of $y = -2x + 1$ and $y = \sqrt{x+1}$

$\begin{cases} y=-2x+1 \overset{x=\frac{5}{4}}{\Rightarrow} y=-\frac{5}{2}+1=-\frac{3}{2} \\ y=\sqrt{x+1} \overset{x=\frac{5}{4}}{\Rightarrow} y=\sqrt{\frac{5}{4}+1}=\frac{3}{2} \end{cases}$ $\Rightarrow$ The y-values do not match, so this is an extraneous solution.

The slope of the line $y = -2x + 1$ is its derivative, which is a constant $m_1 = -2$.

To find the slope of the curve $y = \sqrt{x+1}$ at the intersection point, we need its derivative. $y' = \frac{1}{2\sqrt{x+1}}$. Now, evaluate this derivative at the intersection point, $x = 0$. $m_2 = \frac{1}{2\sqrt{0+1}} = \frac{1}{2}$.

The angle θ between two lines with slopes m_1 and m_2 is given by the formula: $\tan(\theta) = \left|\frac{m_1 - m_2}{1+m_1 m_2}\right|$. Therefore, $\tan(\theta) = \left|\frac{m_1 - m_2}{1+m_1 m_2}\right| = \left|\frac{-2-\frac{1}{2}}{1+(-2)\times\frac{1}{2}}\right| = \left|\frac{-2.5}{1-1}\right|$ $= \infty \Rightarrow \theta = \tan^{-1}(\infty) = \frac{\pi}{2}$. This makes sense because the product of the slopes is $m_1 \times m_2 = -2 \times \frac{1}{2} = -1$. Recall that two lines are perpendicular if and only if the product of their slopes is -1.

Exercise 7.2 Find the angle between $y = x^2$ and $y = \frac{32}{\sqrt{x}}$. (Ans. $\tan^{-1}\left(\frac{2}{3}\right) = 33.69° = 0.5880$ rad)

Example 7.5 Find the equation of the line tangent to $y = x^3 + 5x^2 - x - 7$ at $x = 1$.

Solution

$$y(1) = 1^3 + 5 \times 1^2 - 1 - 7 = 1 + 5 - 1 - 7 = -2$$

$$y' = 3x^2 + 10x - 1 \Rightarrow y'(1) = 3 \times 1^2 + 10 \times 1 - 1 \Rightarrow y'(1) = 12$$

The equation of the tangent line is given by:

$$y - y(1) = y'(1)(x - 1) \Rightarrow y - (-2) = 12(x - 1) \Rightarrow y = 12x - 14$$

The tangent line is shown in Fig. 7.7.

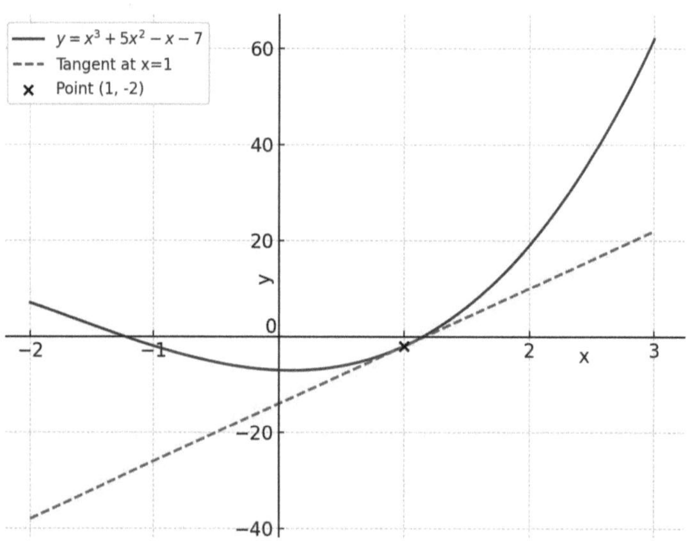

Fig. 7.7 Tangent to $y = x^3 + 5x^2 - x - 7$ at $x = 1$

Example 7.6 Find the equation of the line tangent to $y = x^2 + 6x + 7$ at $x = -3$.

Solution

$$y(-3) = (-3)^2 + 6 \times -3 + 7 = 9 - 18 + 7 = -2$$

$$y' = 2x + 6 \Rightarrow y'(-3) = 2 \times -3 + 6 \Rightarrow y'(-3) = 0$$

The equation of the tangent line is given by:

$$y - y(-3) = y'(-3)(x - (-3)) \Rightarrow y - (-2) = 0(x + 3) \Rightarrow y = -2$$

The tangent line is shown in Fig. 7.8.

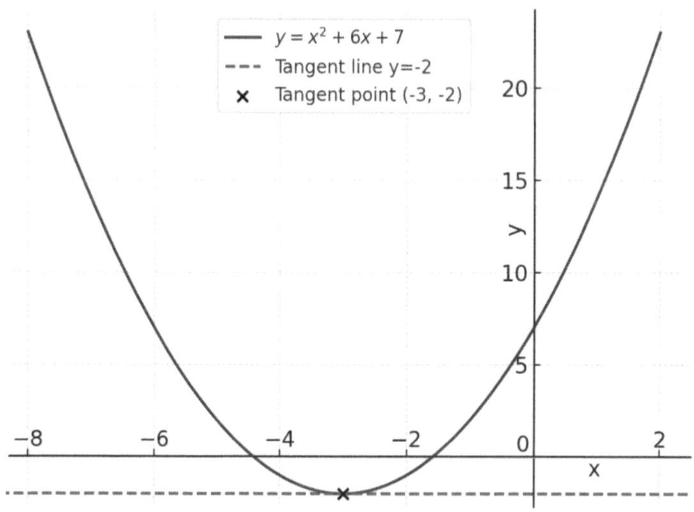

Fig. 7.8 Tangent to $y = x^2 + 6x + 7$ at $x = -3$

Example 7.7 Find the equation of the line tangent to $y = \sqrt[3]{(x-5)^2}$ at $x = 5$.

Solution

$$y(5) = \sqrt[3]{(5-5)^2} = 0$$

$$y' = \left((x-5)^{\frac{2}{3}}\right)' = \frac{2}{3}(x-5)^{\frac{2}{3}-1} = \frac{2}{3}(x-5)^{-\frac{1}{3}} = \frac{2}{3\sqrt[3]{x-5}}$$

$$y'(5) = \frac{2}{3\sqrt[3]{5-5}} = \frac{2}{3 \times 0} = \frac{2}{0} = \infty$$

Because the derivative is undefined (approaches infinity), the tangent line is vertical, and its equation is: $x = 5$. The tangent line is shown in Fig. 7.9.

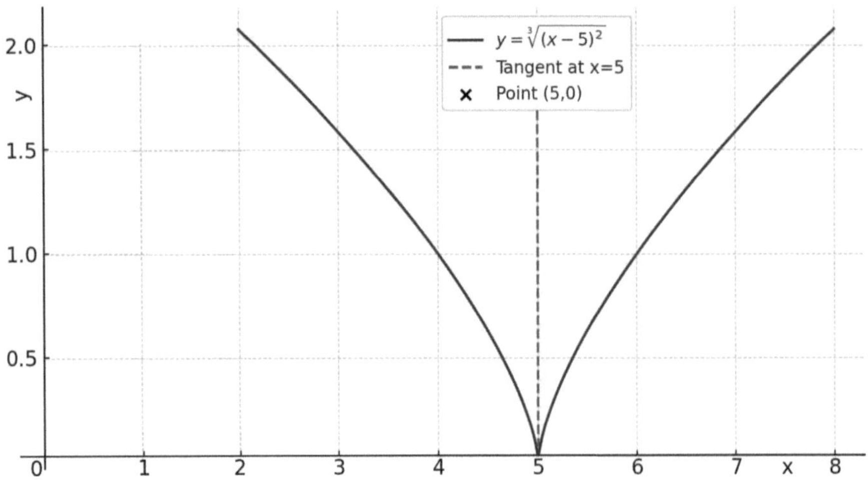

Fig. 7.9 Tangent to $y = \sqrt[3]{(x-5)^2}$ at $x = 5$

Example 7.8 Find the equation of the line perpendicular (normal line) to the function $y = x^3 + \sin(3x)$ at $x = 0$.

Solution

$$y = x^3 + \sin(3x) \Rightarrow y(0) = 0^3 + \sin(3 \times 0) \Rightarrow y(0) = 0$$

$$y' = 3x^2 + 3\cos(3x) \Rightarrow y'(0) = 3 \times 0^2 + 3\cos(3 \times 0) = 3$$

The slope of the tangent line at $x = 0$ is 3. Therefore, the slope of the perpendicular line is $-\frac{1}{3}$. The equation of the perpendicular (normal) line is:

$$y - y(0) = -\frac{1}{3}(x - 0) \Rightarrow y = -\frac{1}{3}x$$

The perpendicular line is shown in Fig. 7.10.

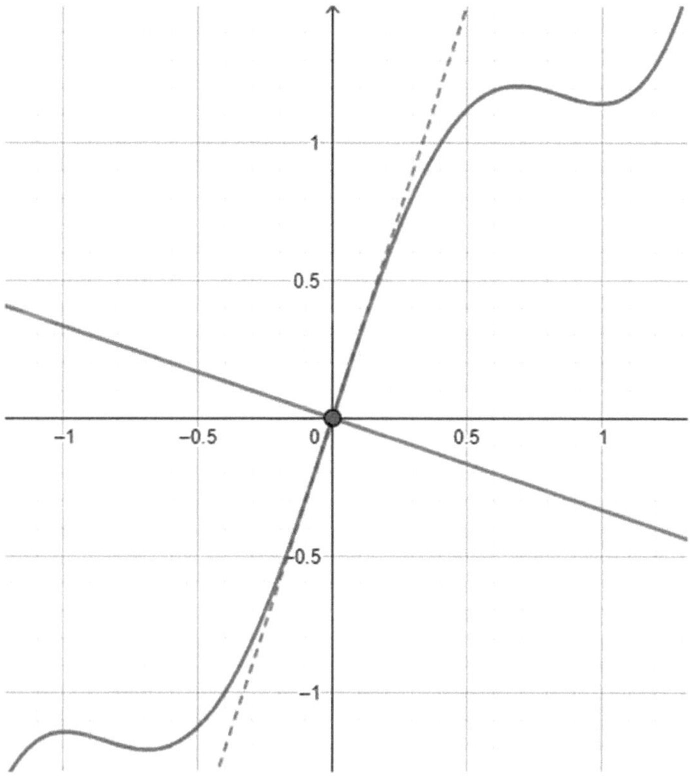

Fig. 7.10 Tangent (dotted red) and perpendicular (solid blue) lines

Exercise 7.3 Find the equation of the line tangent to $y = \sqrt{5x}e^{x+2}$ at $x = 0$. (Ans. $x = 0$)

Exercise 7.4 Find the equation of the line tangent to $y = \sqrt{5x}e^{x+2}$ at $x = 1$. (Ans. $y = \frac{3\sqrt{5}e^3}{2}x - \frac{\sqrt{5}e^3}{2}$)

Example 7.9 For what values of m, $y = 2x - 4$ is tangent to $y = (m + 3)x^2 + mx$.

Solution For the line $y = 2x - 4$ to be tangent to the parabola $y = (m + 3)x^2 + mx$, the two equations must have exactly one point of intersection. We can find this by setting the equations equal to each other and solving for x. The resulting quadratic equation must have a discriminant equal to zero.

$$2x - 4 = (m + 3)x^2 + mx \Rightarrow (m + 3)x^2 + (m - 2)x + 4 = 0 \Rightarrow \Delta = (m - 2)^2 - 4(m + 3) \times 4 \Rightarrow \Delta = m^2 - 4m + 4 - 16m - 48 \Rightarrow \Delta = m^2 - 20m - 44$$

$$\Delta=0\Rightarrow m^2-20m-44=0\Rightarrow m=\frac{10\pm\sqrt{10^2-(1\times-44)}}{1}\Rightarrow m$$
$$=\frac{10\pm 12}{1}\Rightarrow m=22,\ -2$$

The values of m for which the line is tangent to the parabola are $m = 22$ and $m = -2$.

Example 7.10 For what values of m, $y = 3x + m$ is tangent to $y = x^3 + mx^2 + 2x$.

Solution

$$x^3+mx^2+2x=3x+m\Rightarrow x^3+mx^2-x-m=0\Rightarrow x^2(x+m)-(x+m)$$
$$=0\Rightarrow(x^2-1)(x+m)=0\Rightarrow\begin{cases}x=1\\x=-1\\x=-m\end{cases}$$

The roots are repeated when $m = \pm 1$. Therefore, the values of m for which the line $y = 3x + m$ is tangent to the curve $y = x^3 + mx^2 + 2x$ are $m = -1$ and $m = 1$.

Example 7.11 Show that $f(x) = (x - 1)[x]$ is continuous at $x = 1$ but not differentiable there. $[x]$ shows the floor function.

Solution

$$f(x)=(x-1)[x]\Rightarrow f(1)=(1-1)[1]=0.$$
$$\lim_{x\to 1^+}(x-1)[x]=\lim_{x\to 1^+}0\times 1=\lim_{x\to 1^+}0=0$$
$$\lim_{x\to 1^-}(x-1)[x]=\lim_{x\to 1^-}0\times 0=\lim_{x\to 1^-}0=0$$
$$\lim_{x\to 1^+}(x-1)[x]=\lim_{x\to 1^-}(x-1)[x]=f(1)\Rightarrow \text{The function is continuous at } x=1.$$

$$y'(1)=\lim_{x\to 1}\frac{f(x)-f(1)}{x-1}=\lim_{x\to 1}\frac{(x-1)[x]-0}{x-1}=\lim_{x\to 1}\frac{(x-1)[x]}{x-1}=\lim_{x\to 1}[x]=\begin{cases}\lim_{x\to 1^+}[x]=1\\ \lim_{x\to 1^-}[x]=0\end{cases}$$

Therefore, since the right-hand and left-hand derivatives are not equal, the function is not differentiable at $x = 1$.

Example 7.12 Calculate the left-hand derivative of $y=|x-3|+\sqrt{3x}$ at $x = 3$.

Solution

$$y(3)=|3-3|+\sqrt{3\times 3}=3$$

$$y'(3^-)=\lim_{h\to 0^-}\frac{y(3+h)-y(3)}{h}=\lim_{h\to 0^-}\frac{|3+h-3|+\sqrt{3(3+h)}-3}{h}$$
$$=\lim_{h\to 0^-}\frac{|h|+\sqrt{9+3h}-3}{h}=\lim_{h\to 0^-}\frac{-h+\sqrt{9+3h}-3}{h}$$

The limit is in the indeterminate form $\frac{0}{0}$. So we can use L'Hôpital's Rule:

$$y'(3^-)=\lim_{h\to 0^-}\frac{-h+\sqrt{9+3h}-3}{h}=\lim_{h\to 0^-}\frac{-1+\frac{1}{2}\times 3\times \frac{1}{\sqrt{9+3h}}}{1}=\lim_{h\to 0^-}-1+\frac{3}{2\sqrt{9+3h}}=-\frac{1}{2}$$

Example 7.13 Calculate the left-hand derivative of $y=|x-3|+\sqrt{3x}$ at $x=3$.

Solution This problem was previously solved. Here, we will calculate the limit without using L'Hôpital's Rule.

$$y(3)=|3-3|+\sqrt{3\times 3}=3$$

$$y'(3^-)=\lim_{h\to 0^-}\frac{y(3+h)-y(3)}{h}=\lim_{h\to 0^-}\frac{|3+h-3|+\sqrt{3(3+h)}-3}{h}$$
$$=\lim_{h\to 0^-}\frac{|h|+\sqrt{9+3h}-3}{h}=\lim_{h\to 0^-}\frac{-h+\sqrt{9+3h}-3}{h}=\lim_{h\to 0^-}\frac{-h-3+\sqrt{9+3h}}{h}$$
$$=\lim_{h\to 0^-}\frac{\left(-h-3+\sqrt{9+3h}\right)\left(-h-3-\sqrt{9+3h}\right)}{h\left(-h-3-\sqrt{9+3h}\right)}=\lim_{h\to 0^-}\frac{(-h-3)^2-\left(\sqrt{9+3h}\right)^2}{h\left(-h-3-\sqrt{9+3h}\right)}$$
$$=\lim_{h\to 0^-}\frac{(h+3)^2-(9+3h)}{-h\left(h+3+\sqrt{9+3h}\right)}=\lim_{h\to 0^-}\frac{h^2+6h+9-9-3h}{-h\left(h+3+\sqrt{9+3h}\right)}$$
$$=\lim_{h\to 0^-}\frac{h^2+3h}{-h\left(h+3+\sqrt{9+3h}\right)}=\lim_{h\to 0^-}\frac{h+3}{-\left(h+3+\sqrt{9+3h}\right)}=\frac{3}{-6}=-\frac{1}{2}$$

Exercise 7.5 Calculate the right-hand derivative of $y=|x-3|+\sqrt{3x}$ at $x=3$. (Ans. 1)

Exercise 7.6 Show that $f(x)=(x-1)[x]$ is discontinuous at $x=0$. $[x]$ shows the floor function.

Example 7.14 Calculate the right-hand derivative of $y=([x]-|x|)\sqrt[3]{4x^2}$ at $x=-4$.

Solution As $x\to -4^+, y=([x]-|x|)\sqrt[3]{4x^2}=(-4-(-x))\sqrt[3]{4x^2}=\sqrt[3]{4}(x-4)x^{\frac{2}{3}}$. Therefore, $y'=\sqrt[3]{4}x^{\frac{2}{3}}+\sqrt[3]{4}(x-4)\times\frac{2}{3}x^{-\frac{1}{3}}$ and $y'(-4)=\sqrt[3]{4}(-4)^{\frac{2}{3}}+\sqrt[3]{4}(-4-4)\times\frac{2}{3}(-4)^{-\frac{1}{3}}=4+\frac{16}{3}=\frac{28}{3}$.

Example 7.15 Calculate the left-hand derivative of $y=([x]-|x|)\sqrt[3]{4x^2}$ at $x=-4$.

Solution As $x\to -4^-, y=([x]-|x|)\sqrt[3]{4x^2}=(-5-(-x))\sqrt[3]{4x^2}=\sqrt[3]{4}(x-5)x^{\frac{2}{3}}$. Therefore, $y'=\sqrt[3]{4}x^{\frac{2}{3}}+\sqrt[3]{4}(x-5)\times\frac{2}{3}x^{-\frac{1}{3}}$ and $y'(-4)=\sqrt[3]{4}(-4)^{\frac{2}{3}}+\sqrt[3]{4}\times(-4-5)\times\frac{2}{3}(-4)^{-\frac{1}{3}}=4-6\sqrt[3]{4}(-4)^{-\frac{1}{3}}=4-\frac{6\sqrt[3]{4}}{\sqrt[3]{-4}}=4-(-6)=4+6=10$.

Example 7.16 Calculate the derivative of $y=x^2+5x+6$.

Solution

$$y'=2x+5$$

Example 7.17 Calculate the derivative of $y=\frac{7}{x^2}+\frac{6}{x}+x^4+5x-6$ at $x=-1$.

Solution

$$y = \frac{7}{x^2} + \frac{6}{x} + x^4 + 5x - 6 \Rightarrow y = 7x^{-2} + 6x^{-1} + x^4 + 5x - 6 \Rightarrow y' = -14x^{-3}$$
$$-6x^{-2} + 4x^3 + 5 \Rightarrow y' = -\frac{14}{x^3} - \frac{6}{x^2} + 4x^3 + 5 \Rightarrow y' = -\frac{14}{(-1)^3} - \frac{6}{(-1)^2}$$
$$+4(-1)^3 + 5 = 9$$

Example 7.18 Calculate the derivative of $y = (x-1)^2 \sqrt[3]{(x+1)^4}$.

Solution

$$y = (x-1)^2 \sqrt[3]{(x+1)^4} = (x-1)^2 (x+1)^{\frac{4}{3}} \Rightarrow y' = 2(x-1)(x+1)^{\frac{4}{3}} + (x-1)^2$$
$$\times \frac{4}{3}(x+1)^{\frac{1}{3}} = 2(x-1)(x+1)^{\frac{1}{3}}\left((x+1) + \frac{2}{3}(x-1)\right)$$
$$= 2(x-1)(x+1)^{\frac{1}{3}}\left(\frac{5}{3}x + \frac{1}{3}\right)$$

Example 7.19 Determine whether $y = \begin{cases} 6x - 4 & x \geq 2 \\ x^2 + 4x - 4 & x < 2 \end{cases}$ is differentiable at $x = 2$.

Solution A function is differentiable at a point if the derivative exists and is the same from both the left and right sides. If a function is differentiable at $x = a$, it must be continuous at that point. However, the reverse is NOT true: A function can be continuous at a point but not differentiable there. For example, $y = |x|$ is continuous at $x = 0$, but not differentiable at $x = 0$.

The plot of $y = \begin{cases} 6x - 4 & x \geq 2 \\ x^2 + 4x - 4 & x < 2 \end{cases}$ is shown in Fig. 7.11.

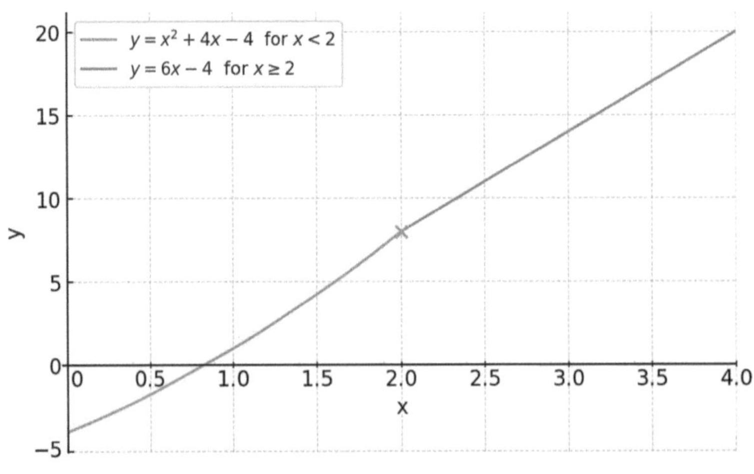

Fig. 7.11 Plot of $y = \begin{cases} 6x - 4 & x \geq 2 \\ x^2 + 4x - 4 & x < 2 \end{cases}$

From Fig. 7.11, we can see that the function is continuous at $x = 2$. Determining if a function is continuous at a point doesn't require a plot. You can simply calculate the left- and right-hand limits and check if they are equal to each other and to the function's value at that point:

$$y = \begin{cases} 6x - 4 & x \geq 2 \\ x^2 + 4x - 4 & x < 2 \end{cases} \Rightarrow \begin{cases} \lim\limits_{x \to 2^+} 6x - 4 = 8 \\ y(2) = 6 \times 2 - 4 = 8 \\ \lim\limits_{x \to 2^-} x^2 + 4x - 4 = 8 \end{cases}$$

Let's calculate the left- and right-hand derivatives:

$$y = \begin{cases} 6x - 4 & x \geq 2 \\ x^2 + 4x - 4 & x < 2 \end{cases} \Rightarrow y' = \begin{cases} 6 & x \geq 2 \\ 2x + 4 & x < 2 \end{cases} \Rightarrow \begin{cases} y'(2^+) = 6 \\ y'(2^-) = 8 \end{cases}$$

The right-hand derivative is 6, while the left-hand derivative is 8. Because these derivatives are not equal, the function is not differentiable at $x = 2$.

Example 7.20 Show that $y = \begin{cases} x \sin\left(\frac{1}{x}\right) & x \neq 0 \\ 0 & x = 0 \end{cases}$ is continuous at $x = 0$ but not differentiable at $x = 0$.

Solution To show that y is continuous at $x = 0$, we must prove that the limit of the function as x approaches 0 is equal to the value of the function at $x = 0$. $y(0) = 0$ and $\lim\limits_{x \to 0} y = \lim\limits_{x \to 0} x \sin\left(\frac{1}{x}\right) = 0$. Therefore, the function is continuous at $x = 0$.

To show that y is not differentiable at $x = 0$, we must prove that the limit of the difference quotient does not exist. The definition of the derivative at $x = 0$ is: $y'(0) = \lim\limits_{h \to 0} \frac{y(0+h) - y(0)}{h}$. From the function definition, we have $y(0) = 0$ and $y(h) = h sin\left(\frac{1}{h}\right)$ for $h = 0$. Substituting these into the limit expression, we get:

$$y'(0) = \lim_{h \to 0} \frac{y(0+h) - y(0)}{h} = \lim_{h \to 0} \frac{y(h) - 0}{h} = \lim_{h \to 0} \frac{h \sin\left(\frac{1}{h}\right)}{h} = \lim_{h \to 0} \sin\left(\frac{1}{h}\right)$$

As $h \to 0$, the value of $\frac{1}{h}$ approaches $\pm\infty$. The sine of a value approaching infinity oscillates between -1 and 1, never settling on a single value. Therefore, the limit does not exist and the function is not differentiable at $x = 0$.

Example 7.21 Determine whether $y = \begin{cases} 6x - 4 & x \geq 2 \\ -x^2 + 4x + 4 & x < 2 \end{cases}$ is differentiable at $x = 2$.

Solution The plot of $y = \begin{cases} 6x - 4 & x \geq 2 \\ -x^2 + 4x + 4 & x < 2 \end{cases}$ is shown in Fig. 7.12.

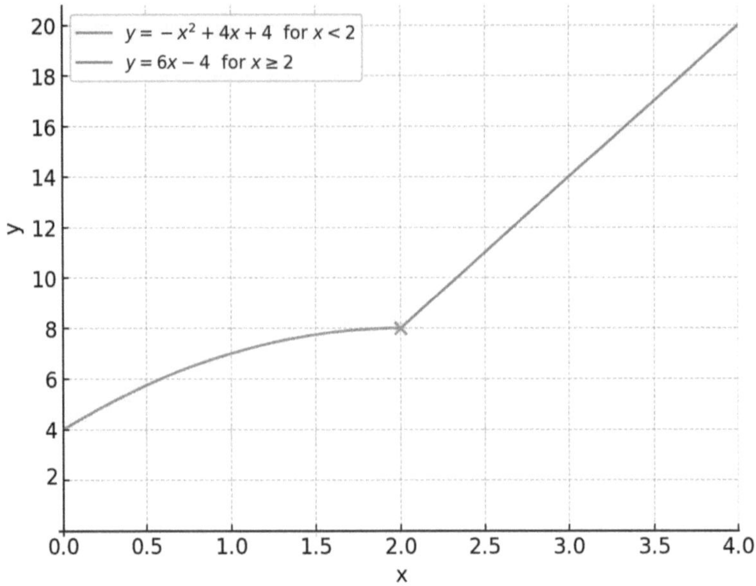

Fig. 7.12 Plot of $y = \begin{cases} 6x - 4 & x \geq 2 \\ -x^2 + 4x + 4 & x < 2 \end{cases}$

From Fig. 7.12, we can see that the function is continuous at $x = 2$. The equality of the function's value and its right- and left-hand limits at $x = 2$ proves continuity at that point, as well:

$$y = \begin{cases} 6x - 4 & x \geq 2 \\ -x^2 + 4x + 4 & x < 2 \end{cases} \Rightarrow \begin{cases} \lim\limits_{x \to 2^+} 6x - 4 = 8 \\ y(2) = 6 \times 2 - 4 = 8 \\ \lim\limits_{x \to 2^-} -x^2 + 4x + 4 = 8 \end{cases}$$

Let's calculate the left- and right-hand derivatives:

$$y = \begin{cases} 6x - 4 & x \geq 2 \\ -x^2 + 4x + 4 & x < 2 \end{cases} \Rightarrow y' = \begin{cases} 6 & x \geq 2 \\ -2x + 4 & x < 2 \end{cases} \Rightarrow \begin{cases} y'(2^+) = 6 \\ y'(2^-) = 0 \end{cases}$$

Since the left-hand and right-hand derivatives are not equal, the function is not differentiable at $x = 2$.

Example 7.22 Calculate the derivative of $y = |x^2 - 3x|$.

Solution To find the general derivative, we define the function as a piecewise function based on where the expression inside the absolute value is positive or negative. The roots of $x^2 - 3x$ are $x = 0$ and $x = 3$. The sign table is shown in Fig. 7.13.

Fig. 7.13 Sign table for $x^2 - 3x$

x	$-\infty$		0		3		$+\infty$
$x^2 - 3x$		$+$	0	$-$	0	$+$	

The function can be written as:

$$y = \begin{cases} x^2 - 3x & x \leq 0 \; or \; x \geq 3 \\ -(x^2 - 3x) & 0 < x < 3 \end{cases}$$

The derivative, for $x \neq 0$ and $x \neq 3$, is:

$$y' = \begin{cases} 2x - 3 & x < 0 \; or \; x > 3 \\ -(2x - 3) & 0 < x < 3 \end{cases}$$

The derivative does not exist at $x = 0$ and $x = 3$ because the graph has sharp corners at these points (Fig. 7.14).

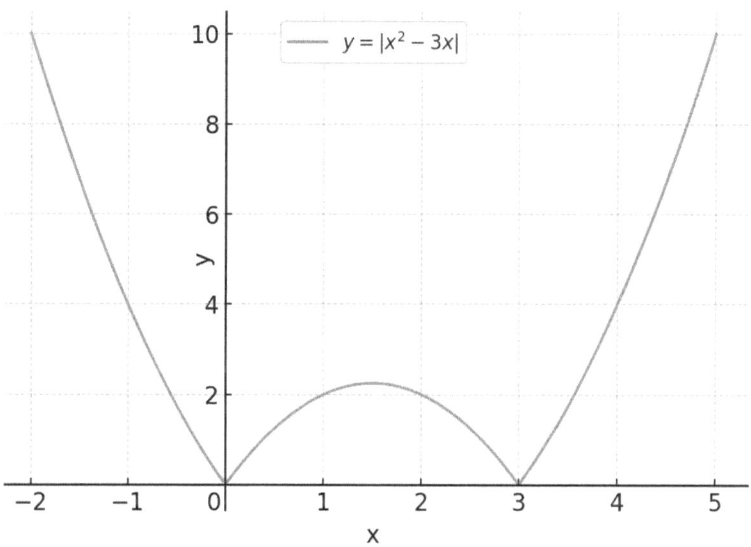

Fig. 7.14 Plot of $y = |x^2 - 3x|$

Example 7.23 Calculate the derivative of $y = |x^2 - 3x|$.

Solution This problem was solved previously. Here, we'll solve it again using the formula: $\frac{d}{dx}|f(x),| = f'(x)\frac{f(x)}{|f(x)|} = f'(x).sgn(f(x))$. This formula holds true for all values of x where $f(x) \neq 0$ and $f(x)$ is differentiable.

$$y=\left|x^2-3x\right| \Rightarrow y'=(2x-3)sgn\left(x^2-3x\right)=\begin{cases}(2x-3)\times 1 & x^2-3x>0\\(2x-3)\times -1 & x^2-3x<0\end{cases}$$

The sign table is shown in Fig. 7.15.

Fig. 7.15 Sign table for $x^2 - 3x$

x	$-\infty$		0		3		$+\infty$
x^2-3x		+	0	−	0	+	

Based on Fig. 7.15, we derive the following piecewise function for the derivative of $y = |x^2 - 3x|$:

$$y=\left|x^2-3x\right| \Rightarrow y'=(2x-3)sgn\left(x^2-3x\right)=\begin{cases}2x-3 & x<0 \text{ or } x>3\\-(2x-3) & 0<x<3\end{cases}$$

This result matches the one from the previous example.

Example 7.24 Calculate the derivative of $y = |x^2 - 3x|$ at $x = 2$.

Solution Let $f(x) = x^2 - 3x$. Next, we evaluate the expression inside the absolute value at $x = 2$: $f(2) = 2^2 - 3 \times 2 = 4 - 6 = -2$.

This means that in the neighborhood of $x = 2$, the absolute value function changes the sign of the inner expression. Specifically, for values of x close to 2, we can write $|x^2 - 3x|$ as $-(x^2 - 3x)$, which simplifies to $-x^2 + 3x$.

Now, we can differentiate this simplified expression: $-2x + 3$. Finally, we substitute $x = 2$ into the derivative we just found: $-2 \times 2 + 3 = -4 + 3 = -1$.

Therefore, the derivative of $y = |x^2 - 3x|$ at $x = 2$ is -1.

Example 7.25 Calculate the right-hand derivative of $y = |x^2 - 3x|$ at $x = 3$.

Solution The absolute value function $y = |x^2 - 3x|$ needs to be simplified based on the domain. The expression inside the absolute value, $x^2 - 3x$, changes its sign at its roots, $x = 0$ and $x = 3$.

Since we are calculating the right-hand derivative at $x = 3$, we need to consider the domain where $x > 3$. In this domain, both factors, x and $(x - 3)$, are positive, so their product is also positive. Therefore, for $x > 3$, the function simplifies to: $y = x^2 - 3x$.

Now, we calculate the derivative of the simplified function $y' = 2x - 3$. To find the right-hand derivative at $x = 3$, we evaluate the derivative we just found at that point: $y'(3^+) = 2 \times 3 - 3 = 3$.

The right-hand derivative of $|x^2 - 3x|$ at $x = 3$ is 3.

Example 7.26 Calculate the left-hand derivative of $y = |x^2 - 3x|$ at $x = 3$.

Solution The absolute value function $y = |x^2 - 3x|$ needs to be simplified based on the domain. The expression inside the absolute value, $x^2 - 3x$, changes its sign at its roots, $x = 0$ and $x = 3$.

Since we are calculating the left-hand derivative at $x = 3$, we need to consider the domain where $x < 3$. In this domain, x is positive while $(x - 3)$ is negative, so their product is negative. Therefore, for $x < 3$, the function simplifies to: $y = -(x^2 - 3x) = -x^2 + 3x$.

Now, we calculate the derivative of the simplified function $y' = -2x + 3$. To find the left-hand derivative at $x = 3$, we evaluate the derivative we just found at that point: $y'(3^-) = -2 \times 3 + 3 = -3$.

The left-hand derivative of $|x^2 - 3x|$ at $x = 3$ is -3.

Exercise 7.7 Calculate the left-hand and right-hand derivatives of $y = |x^2 + 5x + 6|$ at $x = -2$. (Ans. The left-hand and right-hand derivatives are -1 and 1, respectively)

Exercise 7.8 Calculate the left-hand and right-hand derivatives of $y = |x^2 + 5x + 6|$ at $x = 2$. (Ans. 9)

Example 7.27 Calculate the derivative of $y = |x(x - 2)(1 + \sin(x))|$ at $x = -\frac{\pi}{2}$.

Solution The expression $x(x - 2)(1 + \sin(x))$ is non-negative in the neighborhood of $x = -\frac{\pi}{2}$. This is due to the fact that the product of x and $x - 2$ is positive, while the term $1 + \sin(x)$ is always greater than or equal to zero. Therefore, in neighborhood of $x = -\frac{\pi}{2}$, $y = |x(x - 2)(1 + \sin(x))| = x(x - 2)(1 + \sin(x))$.

The value of y at $x = 0$ is 0 since $y\left(-\frac{\pi}{2}\right) = \left|-\frac{\pi}{2}\left(-\frac{\pi}{2} - 2\right)\left(1 + \sin\left(-\frac{\pi}{2}\right)\right)\right| = 0$. The derivative at $x = -\frac{\pi}{2}$ is:

$$y'\left(-\frac{\pi}{2}\right) = \lim_{x \to -\frac{\pi}{2}} \frac{|x(x-2)(1 + \sin(x))| - y\left(-\frac{\pi}{2}\right)}{x - \left(-\frac{\pi}{2}\right)} = \lim_{x \to -\frac{\pi}{2}} \frac{|x(x-2)(1 + \sin(x))| - 0}{x + \frac{\pi}{2}}$$

$$= \lim_{x \to -\frac{\pi}{2}} \frac{x(x-2)(1 + \sin(x))}{x + \frac{\pi}{2}}$$

The limit is in the indeterminate form $\frac{0}{0}$. So we can use L'Hôpital's Rule:

$$y'\left(-\frac{\pi}{2}\right) = \lim_{x \to -\frac{\pi}{2}} \frac{x(x-2)(1 + \sin(x))}{x + \frac{\pi}{2}}$$

$$= \lim_{x \to -\frac{\pi}{2}} \frac{(x-2)(1 + \sin(x)) + x(1 + \sin(x)) + x(x-2)\cos(x)}{1}$$

$$= \frac{\left(-\frac{\pi}{2} - 2\right)\left(1 + \sin\left(-\frac{\pi}{2}\right)\right) + x\left(1 + \sin\left(-\frac{\pi}{2}\right)\right) + -\frac{\pi}{2}\left(-\frac{\pi}{2} - 2\right)\cos\left(-\frac{\pi}{2}\right)}{1}$$

$$= \frac{0 + 0 + 0}{1} = 0$$

The derivative of $y = |x(x - 2)(1 + \sin(x))|$ at $x = -\frac{\pi}{2} = -1.5708$ is 0 (Fig. 7.16).

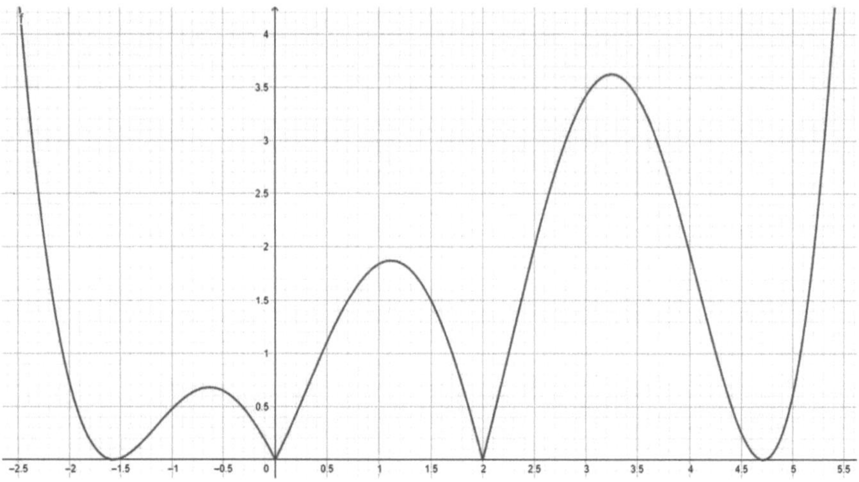

Fig. 7.16 Plot of $y = |x(x - 2)(1 + \sin(x))|$

Example 7.28 Calculate the derivative of $y = |x(x - 2)^2|$ at $x = 2$.

Solution $x = 2$ is a root of multiplicity 2 for the function inside the absolute value. The derivative at $x = 2$ is:

$$y'(2) = \lim_{x \to 2} \frac{|x(x-2)^2| - y(2)}{x-2} = \lim_{x \to 2} \frac{|x(x-2)^2| - 0}{x-2} = \lim_{x \to 2} \frac{x(x-2)^2}{x-2} = \lim_{x \to 2} x(x-2) = 0$$

The derivative of $y = |x(x - 2)^2|$ at $x = 2$ is 0 (Fig. 7.17).

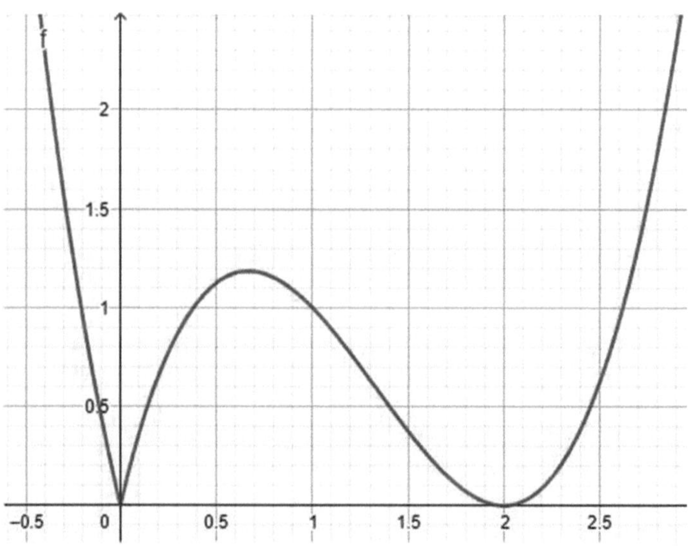

Fig. 7.17 Plot of $y = |x(x - 2)^2|$

Example 7.29 Calculate the derivative of $y = |x(x-2)^3|$ at $x = 2$.

Solution $x = 2$ is a root of multiplicity 3 for the function inside the absolute value. The derivative at $x = 2$ is:

$$y'(2) = \lim_{x\to 2} \frac{|x(x-2)^3| - y(2)}{x-2} = \lim_{x\to 2} \frac{|x(x-2)^3| - 0}{x-2} = \lim_{x\to 2} \frac{|x(x-2)^3|}{x-2}$$
$$= \begin{cases} \lim\limits_{x\to 2+} \dfrac{|x(x-2)^3|}{x-2} = \lim\limits_{x\to 2^+} \dfrac{x(x-2)^3}{x-2} = \lim\limits_{x\to 2^+} x(x-2)^2 = 0 \\ \lim\limits_{x\to 2^-} \dfrac{|x(x-2)^3|}{x-2} = \lim\limits_{x\to 2^-} \dfrac{-x(x-2)^3}{x-2} = \lim\limits_{x\to 2^-} -x(x-2)^2 = 0 \end{cases}$$

Since the right-hand and left-hand limits are equal, the overall limit exists and is equal to that value: $y'(2) = \lim\limits_{x\to 2} \frac{|x(x-2)^3| - y(2)}{x-2} = 0$. Therefore, the derivative of $y = |x(x-2)^3|$ at $x = 2$ is 0 (Fig. 7.18).

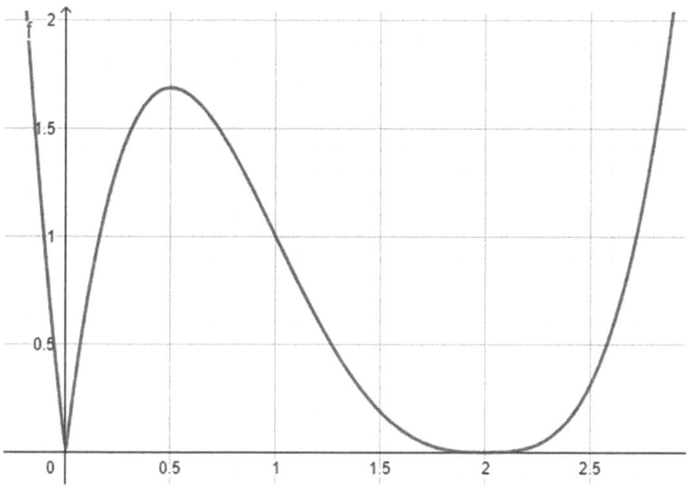

Fig. 7.18 Plot of $|x(x-2)^3|$

Example 7.30 Determine if $y = |x(x-2)^3|$ is differentiable at $x = 0$.

Solution $x = 0$ is a simple root for the function inside the absolute value. The derivative at $x = 0$ is:

$$y'(0) = \lim_{x\to 0} \frac{\left|x(x-2)^3\right| - y(0)}{x-0} = \lim_{x\to 0} \frac{\left|x(x-2)^3\right| - 0}{x} = \lim_{x\to 0} \frac{\left|x(x-2)^3\right|}{x}$$

$$= \begin{cases} \lim\limits_{x\to 0^+} \dfrac{\left|x(x-2)^3\right|}{x} = \lim\limits_{x\to 0^+} \dfrac{-x(x-2)^3}{x} = \lim\limits_{x\to 0^+} -(x-2)^3 = 8 \\ \lim\limits_{x\to 0^-} \dfrac{\left|x(x-2)^3\right|}{x} = \lim\limits_{x\to 0^-} \dfrac{x(x-2)^3}{x} = \lim\limits_{x\to 0^-} (x-2)^3 = -8 \end{cases}$$

Since the right-hand and left-hand limits are not equal, the overall limit does not exist and the function $y = |x(x-2)^3|$ is not differentiable at $x = 0$ (Fig. 7.19).

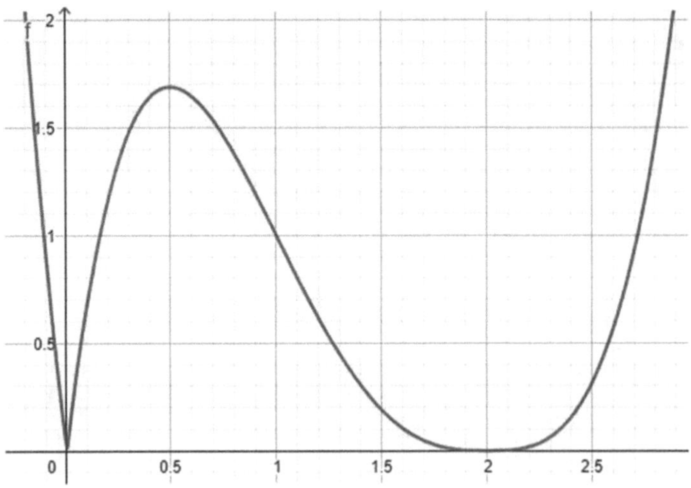

Fig. 7.19 Plot of $y = |x(x-2)^3|$

Example 7.31 In how many points is the function $f(x) = 1 - |1 - |x||$ is not differentiable?

Solution We need to begin with the inner absolute value and count the simple, i.e., non-repeated, roots of the function within it.

The inner absolute value function is $|x|$. The function inside it, x, has a simple root at $x = 0$. The outer absolute value function is $|1 - |x||$. The function inside this one, $1 - |x|$, has two simple roots at $x = 1$ and $x = -1$. Therefore, we expect the full function to be non-differentiable at these three points. A plot of this function (Fig. 7.20) clearly shows that it is not differentiable at $x = 0$, $x = -1$, and $x = 1$ (Recall that a function is not differentiable at a point where its graph has a sharp corner or a cusp because the left-hand and right-hand derivatives are not equal at that point).

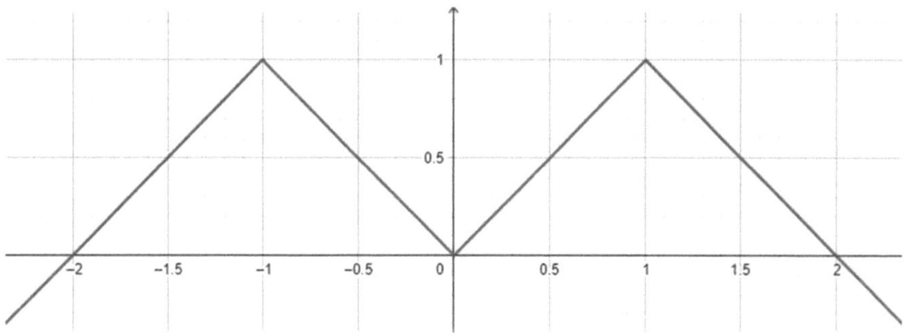

Fig. 7.20 Plot of $f(x) = 1 - |1 - |x||$

Example 7.32 In how many points is the function $f(x) = |(x^2 + 5x + 6)(x + 2)|$ is not differentiable?

Solution

$$x^2+5x+6=0 \Rightarrow x_{1,2}=\frac{-5\pm\sqrt{5^2-4(1)(6)}}{2\times 1} \Rightarrow x_{1,2}=\frac{-5\pm 1}{2} \Rightarrow x_1=-2, x_2=-3$$

$$x^2+5x+6=(x-(-2))(x-(-3))=(x+2)(x+3)$$

$$f(x)=\left|\left(x^2+5x+6\right)(x+2)\right|=\left|(x+2)^2(x+3)\right|$$

The function inside the absolute value is $g(x) = (x + 2)^2(x + 3)$. This function has a simple root at $x = -3$ and a repeated root at $x = -2$. Therefore, $f(x) = |(x + 2)^2(x + 3)|$ is not differentiable at $x = -3$. A plot of this function (Fig. 7.21) clearly shows that it is not differentiable at $x = -3$.

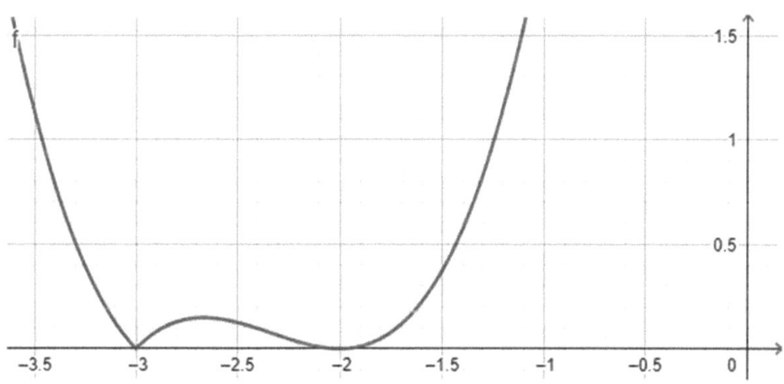

Fig. 7.21 Plot of $f(x) = |(x + 2)^2(x + 3)|$

Example 7.33 Show that the roots of $f(x) = 1 - \sin(x)$ are repeated.

Solution

$$1 - \sin(x) = 0 \Rightarrow \sin(x) = 1 \Rightarrow x = \frac{\pi}{2} + 2k\pi \quad k = 0, \pm 1, \pm 2, \ldots$$

A repeated root means that not only $f(x_0) = 0$, but also $f'(x_0) = 0$.

$$f'(x) = (1 - \sin(x))' = -\cos(x)$$

At $x = \frac{\pi}{2} + 2k\pi : f'\left(\frac{\pi}{2} + 2k\pi\right) = -\cos\left(\frac{\pi}{2} + 2k\pi\right) = -\cos\left(\frac{\pi}{2}\right) = 0$

The first derivative vanishes at the root. So each root has multiplicity **2** (a double root).

Example 7.34 In how many points is the function $f(x) = |x(1 - \sin(x))|$ is not differentiable?

Solution The function inside the absolute value is $g(x) = x(1 - \sin(x))$. This function has a simple root at $x = 0$ and repeated roots at $x = \frac{\pi}{2} + 2k\pi \quad k = 0, \pm 1, \pm 2, \ldots$. Therefore, $f(x) = |x(1 - \sin(x))|$ is not differentiable at $x = 0$. A plot of this function (Fig. 7.22) clearly shows that it is not differentiable at $x = 0$.

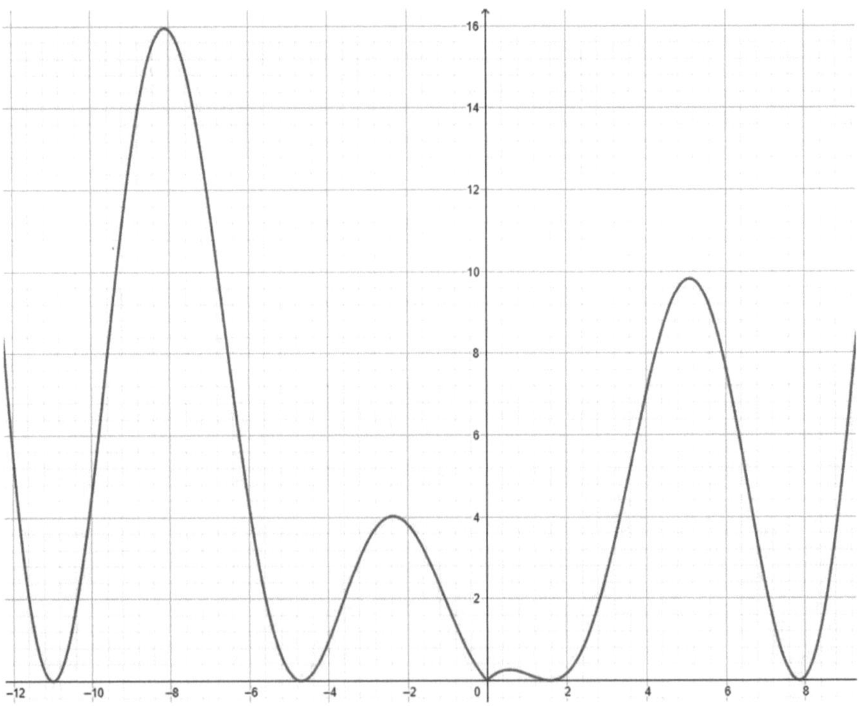

Fig. 7.22 Plot of $f(x) = |x(1 - \sin(x))|$

Example 7.35 Calculate the derivative of $y=\sqrt[3]{(x^2+5x+6)^2}$.

Solution

$$y=\sqrt[3]{(x^2+5x+6)^2}=\left(x^2+5x+6\right)^{\frac{2}{3}}\Rightarrow y'=\frac{2}{3}(2x+5)\left(x^2+5x+6\right)^{\frac{2}{3}-1}$$
$$=\frac{4x+10}{3}\left(x^2+5x+6\right)^{-\frac{1}{3}}=\frac{4x+10}{3\sqrt[3]{x^2+5x+6}}\qquad \forall x\neq -2,\,-3$$

Example 7.36 Show that the right-hand derivative of $y=\sqrt[3]{(x^2+5x+6)^2}$ does not exist at $x=\ -2$.

Solution

$$y(-2)=\sqrt[3]{\left((-2)^2+5\times -2+6\right)^2}=0$$

$$y'(-2^+)=\lim_{h\to 0^+}\frac{y(-2+h)-y(-2)}{h}=\lim_{h\to 0^+}\frac{y(-2+h)-0}{h}=\lim_{h\to 0^+}\frac{y(-2+h)}{h}$$
$$=\lim_{h\to 0^+}\frac{\sqrt[3]{\left((-2+h)^2+5(-2+h)+6\right)^2}}{h}=\lim_{h\to 0^+}\frac{\sqrt[3]{\left(4+h^2-4h-imh+6\right)^2}}{h}$$
$$=\lim_{h\to 0^+}\frac{\left(h^2+h\right)^{\frac{2}{3}}}{h}=\lim_{h\to 0^+}\frac{\left(h(h+1)\right)^{\frac{2}{3}}}{h}=\lim_{h\to 0^+}\frac{h^{\frac{2}{3}}(h+1)^{\frac{2}{3}}}{h}$$
$$=\lim_{h\to 0^+}\frac{(h+1)^{\frac{2}{3}}}{h^{\frac{1}{3}}}=\lim_{h\to 0^+}\frac{1}{0}=\infty$$

Exercise 7.9 Show that the left-hand derivative of $y=\sqrt[3]{(x^2+5x+6)^2}$ does not exist at $x=\ -2$.

Example 7.37 Calculate the derivative of $y=\frac{x^2+5}{3x+7}$.

Solution

$$y=\frac{x^2+5}{3x+7}\Rightarrow y'=\frac{(2x)(3x+7)-(3)(x^2+5)}{(3x+7)^2}=\frac{6x^2+14x-3x^2-15}{(3x+7)^2}$$
$$=\frac{3x^2+14x-15}{(3x+7)^2}\qquad \forall x\neq -\frac{7}{3}$$

Example 7.38 Calculate the derivative of $y=\frac{\sqrt[3]{x^2+5}}{3x+7}$.

Solution

$$y=\frac{\sqrt[3]{x^2+5}}{3x+7}=y=\frac{(x^2+5)^{\frac{1}{3}}}{3x+7}=\frac{\frac{1}{3}(2x)(x^2+5)^{\frac{1}{3}-1}(3x+7)-3(x^2+5)^{\frac{1}{3}}}{(3x+7)^2}$$

$$=\frac{2x(x^2+5)^{-\frac{2}{3}}(3x+7)-9(x^2+5)^{\frac{1}{3}}}{3(3x+7)^2}=\frac{2x(3x+7)-9(x^2+5)}{3(3x+7)^2(x^2+5)^{\frac{2}{3}}}$$

$$=\frac{6x^2+14x-9x^2-45}{3(3x+7)^2(x^2+5)^{\frac{2}{3}}}=\frac{-3x^2+14x-45}{3(3x+7)^2(x^2+5)^{\frac{2}{3}}}$$

Example 7.39 Calculate the derivative of $y=(3x^4+5x-3)^3$.

Solution

$$y'=3\times\left(12x^3+5\right)\times\left(3x^4+5x-3\right)^{3-1}=3\left(12x^3+5\right)\left(3x^4+5x-3\right)^2$$

Example 7.40 Calculate the derivative of $y=\frac{(3x+2)^2}{4x+5}$.

Solution

$$y'=\frac{2\times 3\times(3x+2)(4x+5)-4\times(3x+2)^2}{(4x+5)^2}$$

$$=\frac{72x^2+138x+60-36x^2-48x-16}{(4x+5)^2}=\frac{36x^2+90x+44}{(4x+5)^2}$$

Example 7.41 Calculate the derivative of $y=\frac{1}{(3x+2)^2}$.

Solution

$$y=\frac{1}{(3x+2)^2}\Rightarrow y=(3x+2)^{-2}\Rightarrow y'=-2\times 3\times(3x+2)^{-3}\Rightarrow y'=-\frac{6}{(3x+2)^3}$$

Example 7.42 Calculate the derivative of $y=\frac{1}{(3x+2)^2}$.

Solution

$$y'=\frac{0\times(3x+2)^2-1\times(2\times 3\times(3x+2))}{\left((3x+2)^2\right)^2}=\frac{-6(3x+2)}{(3x+2)^4}=-\frac{6}{(3x+2)^3}$$

Example 7.43 Calculate the derivative of $y=\frac{1}{\left(x+\frac{1}{3x+2}\right)^2}$.

Solution

$$y=\frac{1}{\left(x+\frac{1}{3x+2}\right)^2}\Rightarrow y=\left(x+\frac{1}{3x+2}\right)^{-2}\Rightarrow y=\left(x+(3x+2)^{-1}\right)^{-2}\Rightarrow y'=-2$$

$$\times\left(1-3(3x+2)^{-2}\right)\times\left(x+(3x+2)^{-1}\right)^{-3}\Rightarrow y'=\frac{2\left(\frac{3}{(3x+2)^2}-1\right)}{\left(x+\frac{1}{3x+2}\right)^3}$$

Example 7.44 Calculate the derivative of $y = (2x + 3)(4x + 5)$.

Solution

$$y' = 2 \times (4x + 5) + 4 \times (2x + 3) = 8x + 10 + 8x + 12 = 16x + 22$$

Example 7.45 Calculate the derivative of $y = (2x + 3)^3(4x + 5)^7$.

Solution

$$\begin{aligned} y' &= 3 \times 2 \times (2x+3)^2(4x+5)^7 + (2x+3)^3 \times 7 \times 4 \times (4x+5)^6 \\ &= 6(2x+3)^2(4x+5)^7 + 28(2x+3)^3(4x+5)^6 \\ &= (2x+3)^2(4x+5)^6(6(4x+5) + 28(2x+3)) \\ &= (2x+3)^2(4x+5)^6(80x+114) \end{aligned}$$

Example 7.46 Calculate the derivative of $y = \frac{x^2+1}{x^3-2}$.

Solution

$$y = \frac{x^2+1}{x^3-2} \Rightarrow y' = \frac{2x(x^3-2) - 3x^2(x^2+1)}{(x^3-2)^2} \Rightarrow y' = \frac{2x^4 - 4x - 3x^4 - 3x^2}{(x^3-2)^2} \Rightarrow y'$$
$$= \frac{-x^4 - 3x^2 - 4x}{(x^3-2)^2}$$

Example 7.47 Calculate the derivative of $y = \sqrt[8]{\left(\frac{3x+4}{5x+6}\right)^7}$.

Solution

$$y = \sqrt[8]{\left(\frac{3x+4}{5x+6}\right)^7} \Rightarrow y = \left(\frac{3x+4}{5x+6}\right)^{\frac{7}{8}} \Rightarrow y' = \frac{7}{8}\left(\frac{3 \times (5x+6) - 5 \times (3x+4)}{(5x+6)^2}\right)\left(\frac{3x+4}{5x+6}\right)^{\frac{7}{8}-1}$$
$$\Rightarrow y' = \frac{7}{8}\left(\frac{-2}{(5x+6)^2}\right)\left(\frac{3x+4}{5x+6}\right)^{-\frac{1}{8}} \Rightarrow y' = \frac{-7(3x+4)^{-\frac{1}{8}}}{4(5x+6)^{\frac{15}{8}}} \Rightarrow y' = \frac{-7}{4(5x+6)^{\frac{15}{8}}(3x+4)^{\frac{1}{8}}}$$

Example 7.48 Find the derivative of $y(3x + 4)$, where $y = x^2 + 5$.

Solution

$$y(3x+4) = (3x+4)^2 + 5 \Rightarrow (y(3x+4))' = 2 \times 3 \times (3x+4) = 18x + 24$$

Example 7.49 Find the derivative of $y(3x + 4)$, where $y = x^2 + 5$.

Solution

$$y = x^2 + 5 \Rightarrow y' = 2x$$

$$(y(3x+4))' = 3 \times y'(3x+4) = 3 \times 2(3x+4) = 18x + 24$$

Example 7.50 Calculate the derivative of $y =\ \sin(3x^2 + 5x)$.

Solution

$$y' = \left(3x^2 + 5x\right)' \cos\left(3x^2 + 5x\right) = (6x+5)\cos\left(3x^2 + 5x\right)$$

Example 7.51 Calculate the derivative of $y = \cos\left(\sqrt{3x^2 + 5x}\right)$.

Solution

$$y' = \left(\sqrt{3x^2 + 5x}\right)' \times -\sin\left(\sqrt{3x^2 + 5x}\right) = \frac{1}{2} \times (6x+5) \times \frac{1}{\sqrt{3x^2 + 5x}} \times$$

$$-\sin\left(\sqrt{3x^2 + 5x}\right) = -\frac{6x+5}{2\sqrt{3x^2 + 5x}} \sin\left(\sqrt{3x^2 + 5x}\right)$$

Example 7.52 Calculate the derivative of $y =\ \cos(x^2)$.

Solution

$$y' = \left(x^2\right)' \times -\sin\left(x^2\right) = -2x\sin\left(x^2\right)$$

Example 7.53 Calculate the derivative of $y =\ \tan(\cos(x^2))$.

Solution

$$y' = \left(\cos\left(x^2\right)\right)' \times \left(1 +\ \tan^2\left(\cos\left(x^2\right)\right)\right) = -2x\sin\left(x^2\right)\left(1 +\ \tan^2\left(\cos\left(x^2\right)\right)\right)$$

Example 7.54 Calculate the derivative of $y = \cot\left(\sqrt{3x^2 + 5x}\right)$.

Solution

$$y' = \left(\sqrt{3x^2 + 5x}\right)' \times -\left(1 + \cot^2\left(\sqrt{3x^2 + 5x}\right)\right) = \frac{1}{2} \times (6x+5) \times \frac{1}{\sqrt{3x^2 + 5x}}$$

$$\times -\left(1 + \cot^2\left(\sqrt{3x^2 + 5x}\right)\right) = -\frac{6x+5}{2\sqrt{3x^2 + 5x}}\left(1 + \cot^2\left(\sqrt{3x^2 + 5x}\right)\right)$$

Example 7.55 Calculate the derivative of $y =\ \cos(\tan(x^2))$.

Solution

$$y' = \left(\tan\left(x^2\right)\right)' \times -\sin\left(\tan\left(x^2\right)\right) = -2x\left(1 +\ \tan^2\left(x^2\right)\right)\sin\left(\tan\left(x^2\right)\right)$$

Example 7.56 Calculate the derivative of $y = 7\cos^4(3x^2 + 5x + 8)$.

Solution The derivative of $y = 7u^4$ is $y' = 7 \times 4u'u^3$. Therefore,

$$y' = 7 \times \left(\cos\left(3x^2 + 5x + 8\right)\right)' \times 4\cos^3\left(3x^2 + 5x + 8\right) = 7 \times$$
$$-(6x + 5)\sin\left(3x^2 + 5x + 8\right) \times 4\cos^3\left(3x^2 + 5x + 8\right) =$$
$$-28(6x + 5)\sin\left(3x^2 + 5x + 8\right)\cos^3\left(3x^2 + 5x + 8\right)$$

Example 7.57 Calculate the derivative of $y = 7\tan^3(4x^2 + 6x)$.

Solution The derivative of $y = 7u^3$ is $y' = 7 \times 3u'u^2$. Therefore,

$$y' = 7 \times 3\left(\tan\left(4x^2 + 6x\right)\right)' \tan^2\left(4x^2 + 6x\right) = 21(8x + 6)$$
$$\times \left(1 + \tan^2\left(4x^2 + 6x\right)\right)\tan^2\left(4x^2 + 6x\right)$$

Example 7.58 Calculate the derivative of $y = \cot(4x^2 + 6x)$.

Solution

$$y' = \left(4x^2 + 6x\right)' \times -\left(1 + \cot^2\left(4x^2 + 6x\right)\right) = -(8x + 6)\left(1 + \cot^2\left(4x^2 + 6x\right)\right)$$

Exercise 7.10 Show that $(\cos^2(x))' = -\sin(2x)$.

Exercise 7.11 Show that $(\sin^2(x))' = \sin(2x)$.

Example 7.59 Calculate the derivative of $y = \frac{\cos^2(x)}{1 + \sin^2(x)}$ at $x = \frac{\pi}{4}$.

Solution

$$y' = \frac{-\sin(2x)\left(1 + \sin^2(x)\right) - \sin(2x)\cos^2(x)}{\left(1 + \sin^2(x)\right)^2}$$
$$= \frac{-\sin(2x) - \sin(2x)\sin^2(x) - \sin(2x)\cos^2(x)}{\left(1 + \sin^2(x)\right)^2}$$
$$= \frac{-\sin(2x) - \sin(2x)\left(\sin^2(x) + \cos^2(x)\right)}{\left(1 + \sin^2(x)\right)^2} = \frac{-\sin(2x) - \sin(2x)}{\left(1 + \sin^2(x)\right)^2}$$
$$= \frac{-2\sin(2x)}{\left(1 + \sin^2(x)\right)^2}$$

$$y'\left(\frac{\pi}{4}\right) = \frac{-2\sin\left(\frac{2\pi}{4}\right)}{\left(1 + \sin^2\left(\frac{\pi}{4}\right)\right)^2} = \frac{-2\sin\left(\frac{\pi}{2}\right)}{\left(1 + \sin^2\left(\frac{\pi}{4}\right)\right)^2} = \frac{-2}{\left(1 + \left(\frac{\sqrt{2}}{2}\right)^2\right)^2} = \frac{-2}{\left(1 + \frac{1}{2}\right)^2} = -\frac{2}{\frac{9}{4}} = -\frac{8}{9}$$

Example 7.60 Calculate the derivative of $y = e^{4x^2 - 7x}$.

Solution

$$y' = \left(4x^2 - 7x\right)' e^{4x^2 - 7x} = (8x - 7)e^{4x^2 - 7x}$$

Example 7.61 Calculate the derivative of $y = e^{\sqrt{4x^2 - 7x}}$.

Solution

$$y' = \left(\sqrt{4x^2 - 7x}\right)' e^{\sqrt{4x^2 - 7x}} = \frac{8x - 7}{2\sqrt{4x^2 - 7x}} e^{\sqrt{4x^2 - 7x}}$$

Example 7.62 Calculate the derivative of $y = e^{\cos(3x)}$.

Solution

$$y' = (\cos(3x))' e^{\cos(3x)} = -3\sin(3x)e^{\cos(3x)}$$

Example 7.63 Calculate the derivative of $y = 6^{4x^2 - 7x}$.

Solution

$$y' = \left(4x^2 - 7x\right)' 6^{4x^2 - 7x}\ln(6) = (8x - 7)6^{4x^2 - 7x}\ln(6)$$

Example 7.64 Calculate the derivative of $y = 6^{\sqrt{4x^2 - 7x}}$.

Solution

$$y' = \left(\sqrt{4x^2 - 7x}\right)' 6^{\sqrt{4x^2 - 7x}}\ln(6) = \frac{8x - 7}{2\sqrt{4x^2 - 7x}} 6^{\sqrt{4x^2 - 7x}}\ln(6)$$

Example 7.65 Calculate the derivative of $y = 6^{\cos(3x)}$.

Solution

$$y' = (\cos(3x))' 6^{\cos(3x)}\ln(6) = -3\sin(3x)6^{\cos(3x)}\ln(6)$$

Example 7.66 Calculate the derivative of $y = 2^{2x}$.

Solution

$$y' = (2x)' 2^{2x}\ln(2) = 2 \times 2^{2x}\ln(2) = 2^{2x+1}\ln(2)$$

Example 7.67 Calculate the derivative of $y = \ln\left(3x^2 - 7x\right)$.

Solution

$$y' = \frac{\left(3x^2 - 7x\right)'}{3x^2 - 7x} = \frac{6x - 7}{3x^2 - 7x}$$

Example 7.68 Calculate the derivative of $y = \ln\left(\frac{3x+4}{5x+6}\right)$.

Solution

$$y' = \frac{\left(\frac{3x+4}{5x+6}\right)'}{\frac{3x+4}{5x+6}} = \frac{\frac{3(5x+6)-5(3x+4)}{(5x+6)^2}}{\frac{3x+4}{5x+6}} = \frac{\frac{-2}{(5x+6)^2}}{\frac{3x+4}{5x+6}} = -\frac{2}{(3x+4)(5x+6)}$$

Exercise 7.12 Calculate the derivative of $y = \ln\left(\sqrt{\frac{\sin(x)}{1+\cos(x)}}\right)$ at $x = \frac{\pi}{2}$. (Ans. $\frac{1}{2}$)

Example 7.69 Calculate the derivative of $y = \ln(\cos(\tan(3x)))$.

Solution

$$y' = \frac{(\cos(\tan(3x)))'}{\cos(\tan(3x))} = \frac{(\tan(3x))' \times -\sin(\tan(3x))}{\cos(\tan(3x))}$$

$$= \frac{3(1+\tan^2(3x)) \times -\sin(\tan(3x))}{\cos(\tan(3x))} = \frac{-3\sin(\tan(3x))(1+\tan^2(3x))}{\cos(\tan(3x))} =$$

$$-3\tan(\tan(3x))\left(1+\tan^2(3x)\right)$$

Example 7.70 Calculate the derivative of $y = ln\left(e^{\sqrt{3x+4}}\right)$ at $x = 0$.

Solution

$$y = \ln\left(e^{\sqrt{3x+4}}\right) = \sqrt{3x+4} \Rightarrow y' = \left(\sqrt{3x+4}\right)' = \frac{1}{2} \times 3 \times \frac{1}{\sqrt{3x+4}} \Rightarrow y'$$

$$= \frac{3}{2\sqrt{3x+4}} \Rightarrow y'(0) = \frac{3}{2\sqrt{4}} \Rightarrow y'(0) = \frac{3}{4}$$

Exercise 7.13 Find the equation of the line tangent to $y = \ln\left(\sqrt{\frac{\sin(x)}{1+\cos(x)}}\right)$ at $x = \frac{\pi}{2}$. (Ans. $y = \frac{1}{2}x - \frac{\pi}{4}$)

Exercise 7.14 Find the equation of the line tangent to $y = \ln\left(\frac{\sqrt{4x+1}}{x^2-2x+3}\right)$ at $x = 2$. (Ans. $y = -\frac{4}{9}x + \frac{8}{9}$)

Example 7.71 Calculate the derivative of $y = \frac{ax+b}{cx+d}$ where a, b, c, and d are constants.

Solution

$$y' = \frac{a(cx+d) - c(ax+b)}{(cx+d)^2} - \frac{ad-bc}{(cx+d)^2}$$

Example 7.72 Calculate the derivative of $y = \frac{au(x)+b}{cu(x)+d}$ where a, b, c, and d are constants and u is a function of x.

Solution

$$y' = \frac{au'(cu+d) - cu'(au+b)}{(cu+d)^2} = \frac{(ad-bc)u'}{(cu+d)^2}$$

Example 7.73 Calculate the derivative of $y = \frac{2\sin(7x)+3}{4\sin(7x)+5}$.

Solution

$$y' = \frac{2\times 7\cos(7x)(4\sin(7x)+5) - 4\times 7\cos(7x)(2\sin(7x)+3)}{(4\sin(7x)+5)^2}$$

$$= \frac{70\cos(7x) - 84\cos(7x)}{(4\sin(7x)+5)^2} = \frac{-14\cos(7x)}{(4\sin(7x)+5)^2}$$

Example 7.74 Calculate the second derivative of $y = \frac{2\sin(7x)+3}{4\sin(7x)+5}$.

Solution

$$y' = \frac{2\times 7\cos(7x)(4\sin(7x)+5) - 4\times 7\cos(7x)(2\sin(7x)+3)}{(4\sin(7x)+5)^2}$$

$$= \frac{70\cos(7x) - 84\cos(7x)}{(4\sin(7x)+5)^2} = \frac{-14\,cos(7x)}{(4sin(7x)+5)^2}$$

$$y'' = (y')' = \left(\frac{-14\cos(7x)}{(4\sin(7x)+5)^2}\right)'$$

$$= \frac{-14\times 7\times -\sin(7x)\times(4\sin(7x)+5)^2 - 2(4\times 7\cos(7x))(4\sin(7x)+5)(-14\cos(7x))}{(4\sin(7x)+5)^4}$$

$$= \frac{98\sin(7x)(4\sin(7x)+5)^2 + 784\cos^2(7x)(4\sin(7x)+5)}{(4\sin(7x)+5)^4}$$

$$= \frac{98\sin(7x)(4\sin(7x)+5) + 784\cos^2(7x)}{(4\sin(7x)+5)^3}$$

Example 7.75 Calculate the derivative of $y = \frac{2x+\sqrt{x}}{3x-2}\cot\left(\frac{\pi}{x}\right)$.

Solution

$$y' = \left(\frac{2x+\sqrt{x}}{3x-2}\right)' \cot\left(\frac{\pi}{x}\right) + \frac{2x+\sqrt{x}}{3x-2}\left(\cot\left(\frac{\pi}{x}\right)\right)' = \left(\frac{2x+\sqrt{x}}{3x-2}\right)' \cot\left(\frac{\pi}{x}\right)$$

$$+\frac{2x+\sqrt{x}}{3x-2}\left(\frac{\pi}{x}\right)' \times -\left(1+\cot^2\left(\frac{\pi}{x}\right)\right)$$

$$= \frac{\left(2+\frac{1}{2\sqrt{x}}\right)(3x-2)-3(2x+\sqrt{x})}{(3x-2)^2}\cot\left(\frac{\pi}{x}\right) + \frac{2x+\sqrt{x}}{3x-2} \times -\frac{\pi}{x^2} \times$$

$$-\left(1+\cot^2\left(\frac{\pi}{x}\right)\right) = \frac{6x-4+\frac{3}{2}\sqrt{x}-\frac{1}{\sqrt{x}}-6x-3\sqrt{x}}{(3x-2)^2}\cot\left(\frac{\pi}{x}\right) + \frac{\pi(2x+\sqrt{x})}{(3x-2)x^2}$$

$$\times\left(1+\cot^2\left(\frac{\pi}{x}\right)\right) = \frac{-4-\frac{3}{2}\sqrt{x}-\frac{1}{\sqrt{x}}}{(3x-2)^2}\cot\left(\frac{\pi}{x}\right) + \frac{\pi(2x+\sqrt{x})}{(3x-2)x^2}\left(1+\cot^2\left(\frac{\pi}{x}\right)\right)$$

$$= \frac{-4-\frac{3}{2}\sqrt{x}-\frac{1}{\sqrt{x}}}{(3x-2)^2}\cot\left(\frac{\pi}{x}\right) + \frac{\pi(2x+\sqrt{x})}{(3x-2)x^2}\frac{1}{\sin^2\left(\frac{\pi}{x}\right)} = \frac{-4-\frac{3}{2}\sqrt{x}-\frac{1}{\sqrt{x}}}{(3x-2)^2}\cot\left(\frac{\pi}{x}\right)$$

$$+\frac{\pi(2x+\sqrt{x})}{(3x-2)x^2}\csc^2\left(\frac{\pi}{x}\right)$$

Example 7.76 Find the derivative of the function $g(x) = f\left(\sqrt{|x|+3}\right)$ at $x = -1$, given that $\lim_{x\to 2}\frac{f(x)-f(2)}{x-2} = -\frac{1}{3}$.

Solution

$$\lim_{x\to 2}\frac{f(x)-f(2)}{x-2} = -\frac{1}{3} \Rightarrow f'(2) = -\frac{1}{3}$$

In neighborhood of $x = -1$, $g(x) = f\left(\sqrt{|x|+3}\right) = f\left(\sqrt{-x+3}\right)$. Therefore,

$$g'(x) = \left(\sqrt{-x+3}\right)' f'\left(\sqrt{-x+3}\right) = \left((-x+3)^{\frac{1}{2}}\right)' f'\left(\sqrt{-x+3}\right)$$

$$= \frac{1}{2} \times -1 \times \frac{1}{\sqrt{-x+3}} \times f'\left(\sqrt{-x+3}\right) = \frac{-1}{2\sqrt{-x+3}} f'\left(\sqrt{-x+3}\right)$$

$$g'(x) = \frac{-1}{2\sqrt{-x+3}} f'\left(\sqrt{-x+3}\right) \overset{x=-1}{\Rightarrow} g'(-1) = \frac{-1}{2\sqrt{-(-1)+3}} f'\left(\sqrt{-(-1)+3}\right)$$

$$\Rightarrow g'(-1) = \frac{-1}{4} f'(2) \Rightarrow g'(-1) = \frac{-1}{4} \times \frac{-1}{3} \Rightarrow g'(-1) = \frac{1}{12}$$

Example 7.77 Calculate the derivative of $y = \ln(|\ln(x^2)|)$ for $-1 < x < 1$ excluding $x = 0$.

Solution

$$\left.\begin{array}{r}-1<x<1\\ x\neq 0\end{array}\right\}\Rightarrow 0<x^2<1\Rightarrow \ln\left(x^2\right)\leq 0\Rightarrow \left|\ln\left(x^2\right)\right| = -\ln\left(x^2\right)$$

$$y=\ln(|\ln(x^2)|)\Rightarrow y=\ln(-\ln(x^2))\Rightarrow y'=\frac{(-\ln(x^2))'}{-\ln(x^2)}\Rightarrow y'=\frac{-\frac{(x^2)'}{x^2}}{-\ln(x^2)}\Rightarrow y'$$

$$=\frac{\frac{2x}{x^2}}{\ln(x^2)}\Rightarrow y'=\frac{2}{x\ln(x^2)}\Rightarrow y'=\frac{2}{2x\ln(|x|)}\Rightarrow y'=\frac{1}{x\ln(|x|)}$$

Exercise 7.15 Calculate the derivative of $y = \ln(|\ln(x^2)|)$ for $|x| > 1$. (Ans. $y' = \frac{1}{x\ln(|x|)}$)

Example 7.78 Find the derivative of y with respect to x, where $y = \cos^3(u)$ and $u = \tan(x)$.

Solution

$$y'=\frac{dy}{dx}=\frac{dy}{du}\times\frac{du}{dx}=-3\sin(u)\cos^2(u)\times\left(1+\tan^2(x)\right)=$$
$$-3\sin(\tan(x))\cos^2(\tan(x))\left(1+\tan^2(x)\right)$$

Example 7.79 Find the derivative of y with respect to x at $x = 0$, where $y=\sqrt{u}-\frac{1}{u}$ and $u = \cos^2(x) - \sin(2x)$.

Solution

$$y'=\frac{dy}{dx}=\frac{dy}{du}\times\frac{du}{dx}=\left(\frac{1}{2\sqrt{u}}+\frac{1}{u^2}\right)\times(-\sin(2x)-2\cos(2x))$$

$$=\left(\frac{1}{2\sqrt{\cos^2(x)-\sin(2x)}}+\frac{1}{\left(\cos^2(x)-\sin(2x)\right)^2}\right)\times(-\sin(2x)-2\cos(2x))$$

$$\Rightarrow y'(0)=\left(\frac{1}{2\sqrt{\cos^2(0)-\sin(2\times 0)}}+\frac{1}{\left(\cos^2(0)-\sin(2\times 0)\right)^2}\right)$$
$$\times(-\sin(2\times 0)-2\cos(2\times 0))\Rightarrow y'(0)=\left(\frac{1}{2}+\frac{1}{1}\right)\times -2\Rightarrow y'(0)=-3$$

Example 7.80 Calculate $\frac{d}{dx}f(f(x))$, where $f(x)=\sin^2(\sqrt{x})-\frac{1}{x}$.

Solution

$$\frac{d}{dx}f(f(x))=f'(x)\times f'(f(x))\Rightarrow\frac{d}{dx}f(f(x))=\left(\frac{1}{2\sqrt{x}}\sin\left(2\sqrt{x}\right)+\frac{1}{x^2}\right)$$
$$\times\left(\frac{1}{2\sqrt{f(x)}}\sin\left(2\sqrt{f(x)}\right)+\frac{1}{(f(x))^2}\right)=\left(\frac{1}{2\sqrt{x}}\sin\left(2\sqrt{x}\right)+\frac{1}{x^2}\right)$$
$$\times\left(\frac{1}{2\sqrt{\sin^2(\sqrt{x})-\frac{1}{x}}}\sin\left(2\sqrt{\sin^2(\sqrt{x})-\frac{1}{x}}\right)+\frac{1}{\left(\sin^2(\sqrt{x})-\frac{1}{x}\right)^2}\right)$$

Example 7.81 Calculate $\frac{d}{dx}fog(x)$, where $f(x) = \cos^2(x)$ and $g(x) = \frac{1}{2}\sqrt{3x+4}$.

Solution

$$f' = -\sin(2x)$$

$$g' = \frac{1}{2} \times \frac{1}{2} \times 3 \times \frac{1}{\sqrt{3x+4}} = \frac{3}{4\sqrt{3x+4}}$$

$$\frac{d}{dx}fog(x) = g'(x) \times f'(g(x)) = \frac{3}{4\sqrt{3x+4}} \times -\sin(2g(x)) = \frac{3}{4\sqrt{3x+4}} \times$$

$$-\sin\left(2 \times \frac{1}{2}\sqrt{3x+4}\right) = \frac{-3\sin\left(\sqrt{3x+4}\right)}{4\sqrt{3x+4}}$$

Example 7.82 Calculate $(fog(x))'$, where $f(x) = \frac{4}{5}x - \frac{1}{5}|x|$ and $g(x) = 4x + |x|$.

Solution

$$f(x) = \frac{4}{5}x - \frac{1}{5}|x| = \begin{cases} \frac{4}{5}x - \frac{1}{5}x & x \geq 0 \\ \frac{4}{5}x + \frac{1}{5}x & x < 0 \end{cases}, = \begin{cases} \frac{3}{5}x & x \geq 0 \\ x & x < 0 \end{cases},$$

$$g(x) = 4x + |x| = \begin{cases} 5x & x \geq 0 \\ 3x & x < 0 \end{cases}$$

$$fog(x) = \begin{cases} \frac{3}{5}g(x) & g(x) \geq 0 \\ g(x) & g(x) < 0 \end{cases} \xRightarrow{\begin{cases} x > 0 \Rightarrow \begin{cases} g(x) = 5x \\ g(x) \geq 0 \end{cases} \\ x < 0 \Rightarrow \begin{cases} g(x) = 3x \\ g(x) < 0 \end{cases} \end{cases}} fog(x)$$

$$= \begin{cases} \frac{3}{5} \times 5x & x \geq 0 \\ 3x & x < 0 \end{cases} \Rightarrow fog(x) = \begin{cases} 3x & x \geq 0 \\ 3x & x < 0 \end{cases} \Rightarrow fog(x) = 3x$$

$$(fog(x),)' = (3x)' = 3$$

Example 7.83 Find the derivative $\frac{dy}{dx}$ of the implicit equation $x^2 + y^2 = 25$.

Solution

$$f(x,y) = 0 \Rightarrow y' = -\frac{f'_x}{f'_y}$$

$$x^2 + y^2 = 25 \Rightarrow x^2 + y^2 - 25 = 0 \Rightarrow y' = -\frac{f'_x}{f'_y} = -\frac{2x}{2y} = -\frac{x}{y}$$

Example 7.84 Find the derivative $\frac{dy}{dx}$ of the implicit equation $x^2 + y^2 = 25$.

Solution

$$x^2+y^2=25 \Rightarrow \frac{d}{dx}\left(x^2+y^2\right)=\frac{d}{dx}(25) \Rightarrow \frac{d}{dx}x^2+\frac{d}{dx}y^2=0 \Rightarrow 2x+2y'y=0 \Rightarrow x$$
$$+y'y=0 \Rightarrow y'=-\frac{x}{y}$$

Example 7.85 Find the derivative $\frac{dy}{dx}$ of the implicit equation $x^3y^2+\sin(xy)=5$.

Solution

$$x^3y^2+\sin(xy)=5 \Rightarrow x^3y^2+\sin(xy)-5=0 \Rightarrow y'=-\frac{f'_x}{f'_y}=-\frac{3x^2y^2+y\cos(xy)}{2x^3y+x\cos(xy)}$$

Example 7.86 Find the derivative $\frac{dy}{dx}$ of the implicit equation $x^3y^2+\sin(xy)=5$.

Solution

$$x^3y^2+\sin(xy)=5 \Rightarrow x^3y^2+\sin(xy)-5=0 \Rightarrow \frac{d}{dx}\left(x^3y^2+\sin(xy)-5\right)=\frac{d}{dx}$$
$$\times(0) \Rightarrow 3x^2y^2+x^3\times 2y'y+(xy)'\cos(xy)=0 \Rightarrow 3x^2y^2+2x^3y'y$$
$$+(y+xy')\cos(xy)=0 \Rightarrow 3x^2y^2+2x^3y'y+y\cos(xy)+xy'\cos(xy)=0 \Rightarrow 3x^2y^2$$
$$+y\cos(xy)+\left(2x^3y+x\cos(xy)\right)y'=0 \Rightarrow y'=-\frac{3x^2y^2+y\cos(xy)}{2x^3y+x\cos(xy)}$$

Example 7.87 Find the derivative $\frac{dy}{dx}$ of the implicit equation $\frac{\sqrt{2y}}{x}+y\sqrt{3x}-7=0$ at $(3, 2.105)$.

Solution

$$\frac{\sqrt{2y}}{x}+y\sqrt{3x}-7=0 \Rightarrow \sqrt{2}y^{\frac{1}{2}}x^{-1}+\sqrt{3}yx^{\frac{1}{2}}-7=0 \Rightarrow y'=-\frac{f'_x}{f'_y}$$

$$=-\frac{-\frac{\sqrt{2y}}{x^2}+\sqrt{3}y\times\frac{1}{2\sqrt{x}}}{\frac{\sqrt{2}}{x}\times\frac{1}{2\sqrt{y}}+\sqrt{3x}} \Rightarrow y'=-\frac{-\frac{\sqrt{2y}}{x^2}+\frac{\sqrt{3}y}{2\sqrt{x}}}{\frac{1}{x\sqrt{2y}}+\sqrt{3x}} \overset{\substack{x=3\\y=2.105}}{\Rightarrow} y'=-0.261$$

Example 7.88 Find the derivative $\frac{dy}{dx}$ of the implicit equation $\frac{\sqrt{2y}}{x}+y\sqrt{3x}-7=0$ at $(3, 2.105)$.

Solution

$$\frac{\sqrt{2y}}{x}+y\sqrt{3x}-7=0\Rightarrow\sqrt{2}y^{\frac{1}{2}}x^{-1}+\sqrt{3}yx^{\frac{1}{2}}-7=0\Rightarrow\sqrt{2}\times\frac{1}{2}\frac{y'}{\sqrt{y}}\times x^{-1}$$

$$-\sqrt{2}y^{\frac{1}{2}}x^{-2}+\sqrt{3}y'x^{\frac{1}{2}}+\frac{\sqrt{3}}{2}yx^{-\frac{1}{2}}=0\Rightarrow\frac{y'}{x\sqrt{2y}}+\sqrt{3}y'x^{\frac{1}{2}}-\sqrt{2}y^{\frac{1}{2}}x^{-2}$$

$$+\frac{\sqrt{3}}{2}yx^{-\frac{1}{2}}=0\Rightarrow\left(\frac{1}{x\sqrt{2y}}+\sqrt{3}x^{\frac{1}{2}}\right)y'-\sqrt{2}y^{\frac{1}{2}}x^{-2}+\frac{\sqrt{3}}{2}yx^{-\frac{1}{2}}=0\Rightarrow y'$$

$$=\frac{\sqrt{2}y^{\frac{1}{2}}x^{-2}-\frac{\sqrt{3}}{2}yx^{-\frac{1}{2}}}{\frac{1}{x\sqrt{2y}}+\sqrt{3}x^{\frac{1}{2}}}$$

$$\overset{\substack{x=3\\y=2.105}}{\Rightarrow}\quad y'=-0.261$$

Example 7.89 Find the derivative $\frac{dy}{dx}$ of the implicit equation $e^{xy}+\ln(x^2+y^2)=\tan^{-1}\left(\frac{x}{y}\right)$.

Solution

$$e^{xy}+\ln\left(x^2+y^2\right)=\tan^{-1}\left(\frac{x}{y}\right)\Rightarrow e^{xy}+\ln\left(x^2+y^2\right)-\tan^{-1}\left(\frac{x}{y}\right)=0\Rightarrow e^{xy}$$

$$+\ln\left(x^2+y^2\right)-\tan^{-1}\left(xy^{-1}\right)=0\Rightarrow y'=-\frac{f'_x}{f'_y}=$$

$$-\frac{ye^{xy}+\frac{2x}{x^2+y^2}-y^{-1}\times\frac{1}{1+(xy^{-1})^2}}{xe^{xy}+\frac{2y}{x^2+y^2}-x\times-\frac{1}{y^2}\times\frac{1}{1+(xy^{-1})^2}}=-\frac{ye^{xy}+\frac{2x}{x^2+y^2}-\frac{1}{y+x^2y^{-1}}}{xe^{xy}+\frac{2y}{x^2+y^2}+\frac{x}{y^2+x^2}}=$$

$$-\frac{\frac{(x^2+y^2)ye^{xy}}{x^2+y^2}+\frac{2x}{x^2+y^2}-\frac{y}{y^2+x^2}}{\frac{xe^{xy}(x^2+y^2)}{x^2+y^2}+\frac{2y}{x^2+y^2}+\frac{x}{y^2+x^2}}=-\frac{(x^2+y^2)ye^{xy}+2x-y}{(x^2+y^2)xe^{xy}+2y+x}$$

Example 7.90 Find the value of $y'(1)$ for the implicit equation $x^2y+5y=3x+9$.

Solution

$$x^2y+5y=3x+9\overset{x=1}{\Rightarrow}1^2y+5y=3\times1+9\Rightarrow6y=12\Rightarrow y=2$$

$$x^2y+5y=3x+9\Rightarrow x^2y+5y-3x-9=0\Rightarrow y'=-\frac{f'_x}{f'_y}=-\frac{2xy-3}{x^2+5}=\frac{3-2xy}{x^2+5}\overset{\substack{x=1\\y=2}}{\Rightarrow}y'=-\frac{1}{6}$$

Exercise 7.16 Find the derivative $\frac{dy}{dx}$ of the implicit equation $y=y^2e^{\sin(2x)}+\sin(x)$ at (0, 1). (Ans. −3)

Exercise 7.17 Find the derivative $\frac{dy}{dx}$ of the implicit equation $y = \sin(x - 2y) + \sqrt{x - y}$ at (2, 1). (Ans. $\frac{3}{7}$)

Exercise 7.18 Find the equation of the line tangent to $\ln(x^2 - y) - \sqrt{y + 1} + x = 0$ at (2, 3). (Ans. $y = 4x - 5$)

Example 7.91 Find the 3rd derivative of the function $y = 8x^4 + 5x^3 + 7x^2 + 6x + 2$ with respect to x.

Solution

$$y = 8x^4 + 5x^3 + 7x^2 + 6x + 2$$

$$y' = 8 \times 4 \times x^3 + 5 \times 3 \times x^2 + 7 \times 2 \times x + 6 = 32x^3 + 15x^2 + 14x + 6$$

$$y'' = 8 \times 4 \times 3 \times x^2 + 5 \times 3 \times 2 \times x + 7 \times 2 = 96x^2 + 30x + 14$$

$$y''' = 8 \times 4 \times 3 \times 2 \times x + 5 \times 3 \times 2 = 192x + 30$$

Example 7.92 Find the 7th derivative of the function $y = 4x^7$ with respect to x.

Solution

$$y^{(7)} = 4 \times 7! = 20160$$

Example 7.93 Find the 10th derivative of the function $y = 4x^7$ with respect to x.

Solution

$$y^{(10)} = 0$$

Example 7.94 Find the value of 6th and 7th derivatives of the function $y = 4x^4|x^3|$ with respect to x at $x = 0$.

Solution

$$y = 4x^4|x^3| = \begin{cases} 4x^7 & x \geq 0 \\ -4x^7 & x < 0 \end{cases}.$$

$$y' = \begin{cases} 4 \times 7 \times x^6 & x \geq 0 \\ -4 \times 7 \times x^6 & x < 0 \end{cases}$$

$$y'' = \begin{cases} 4 \times 7 \times 6 \times x^5 & x \geq 0 \\ -4 \times 7 \times 6 \times x^5 & x < 0 \end{cases}$$

$$y''' = \begin{cases} 4 \times 7 \times 6 \times 5 \times x^4 & x \geq 0 \\ -4 \times 7 \times 6 \times 5 \times x^4 & x < 0 \end{cases}$$

$$y^{(4)} = \begin{cases} 4 \times 7 \times 6 \times 5 \times 4 \times x^3 & x \geq 0 \\ -4 \times 7 \times 6 \times 5 \times 4 \times x^3 & x < 0 \end{cases}$$

$$y^{(5)} = \begin{cases} 4\times7\times6\times5\times4\times3\times x^2 & x\geq0 \\ -4\times7\times6\times5\times4\times3\times x^2 & x<0 \end{cases}$$

$$y^{(6)} = \begin{cases} 4\times7\times6\times5\times4\times3\times2\times x & x\geq0 \\ -4\times7\times6\times5\times4\times3\times2\times x & x<0 \end{cases} \Rightarrow y^{(6)}$$

$$= \begin{cases} 4\times7! & x\geq0 \\ -4\times7! & x<0 \end{cases} \Rightarrow y^{(6)}(0^+) = y^{(6)}(0^-) = 0 \Rightarrow y^{(6)}(0) = 0$$

$$y^{(7)} = \begin{cases} 4\times7\times6\times5\times4\times3\times2 & x\geq0 \\ -4\times7\times6\times5\times4\times3\times2 & x<0 \end{cases} \Rightarrow y^{(7)} = \begin{cases} 4\times7! & x\geq0 \\ -4\times7! & x<0 \end{cases} \Rightarrow \begin{cases} y^{(7)}(0^+) = 4\times7! \\ y^{(7)}(0^-) = -4\times7! \end{cases}$$

Therefore, the 6th derivative of $y = 4x^4|x^3|$ at $x = 0$ is 0, but the 7th derivative does not exist at $x = 0$.

Example 7.95 Find the 5th derivative of the function $y = 8x^4 + 5x^3 + 7x^2 + 6x + 2$ with respect to x.

Solution

$$y^{(5)} = 0$$

Example 7.96 Find the 8th derivative of the function $y = (x^2 + 2)^4 + x^8$ with respect to x.

Solution

$$y = \left(x^2 + 2\right)^4 + x^8 = x^8 + \ldots + x^8 = 2x^8 + \ldots \Rightarrow y^{(8)} = 2\times8!$$

Example 7.97 Find the 100th derivative of the function $y = \sin(x) + \cos(x)$ with respect to x.

Solution The derivatives of $\sin(x)$ follow a cycle of four:

1st derivative: $\cos(x)$
2nd derivative: $-\sin(x)$
3rd derivative: $-\cos(x)$
4th derivative: $\sin(x)$

The pattern repeats every four derivatives. To find the 100th derivative, we divide 100 by 4: $100 \div 4 = 25$ with a remainder of 0. A remainder of 0 means the 100th derivative of $\sin(x)$ is the same as the 4th derivative, which is $\sin(x)$.

The derivatives of $\cos(x)$ also follow a cycle of four:

1st derivative: $-\sin(x)$
2nd derivative: $-\cos(x)$
3rd derivative: $\sin(x)$
4th derivative: $\cos(x)$

The pattern also repeats every four derivatives. To find the 100th derivative, we divide 100 by 4: $100 \div 4 = 25$ with a remainder of 0. A remainder of 0 means the 100th derivative of cos(x) is the same as the 4th derivative, which is cos(x).

Since the 100th derivative of the sum is the sum of the 100th derivatives, we have: $y^{(100)} = \sin^{(100)}(x) + \cos^{(100)}(x) = \sin(x) + \cos(x)$.

Example 7.98 Find the 101th derivative of the function $y = \sin(3x) + \cos(4x)$ with respect to x.

Solution The derivatives of sin(x) and cos(x) are cyclical, repeating every four differentiations. The derivatives of sin($3x$) follow this pattern, with a factor of 3 from the chain rule for each step.

1st derivative: $3\cos(3x)$
2nd derivative: $-3^2 \sin(3x)$
3rd derivative: $-3^3 \cos(3x)$
4th derivative: $3^4 \sin(3x)$

To find the 101st derivative, we find the remainder of $101 \div 4$, which is 1. This means the 101st derivative is the same as the 1st derivative in the cycle, but with a factor of 3^{101}. Therefore, $\frac{d^{101}}{dx^{101}}(\sin(3x)) = 3^{101}\cos(3x)$.

Similarly, the derivatives of cos($4x$) follow a cycle of four, with a factor of 4 from the chain rule.

1st derivative: $-4\sin(4x)$
2nd derivative: $-4^2\cos(4x)$
3rd derivative: $4^3 \sin(4x)$
4th derivative: $4^4 \cos(4x)$

Since the remainder of $101 \div 4$ is 1, the 101st derivative is the same as the 1st derivative in the cycle, with a factor of 4^{101}. Therefore, $\frac{d^{101}}{dx^{101}}(\cos(3x)) = -4^{101}\sin(4x)$.

Therefore, $y^{(101)} = \sin^{(101)}(3x) + \cos^{(101)}(4x) = 3^{101}\cos(3x) - 4^{101}\sin(4x)$.

Exercise 7.19 Given the functions $y_1 = \sin(x)\cos(x)$ and $y_2 = \cos(2x)$, calculate the value of $\frac{y_1^{(10)}\left(\frac{\pi}{4}\right)}{y_2^{(7)}\left(\frac{\pi}{4}\right)}$. (Ans. $-\frac{512}{128} = -4$)

Example 7.99 Given the function $f(x) = \begin{cases} x - \sin(x) & x \geq 0 \\ ax^n & x < 0 \end{cases}$ and given that $f'''(0)$ exists, calculate the value of a.

Solution We find the derivatives for both parts of the piecewise function.

$$f(x) = \begin{cases} x - \sin(x) & x \geq 0 \\ ax^n & x < 0 \end{cases}$$

$$f'(x) = \begin{cases} 1 - \cos(x) & x \geq 0 \\ nax^{n-1} & x < 0 \end{cases}$$

Let's find the condition for $f'(x)$ to exist at $x = 0$.

$$\lim_{x \to 0^+} f'(x) = \lim_{x \to 0^-} f'(x) \Rightarrow \lim_{x \to 0^+} 1 - \cos(x) = \lim_{x \to 0^-} nax^{n-1} \Rightarrow 0$$
$$= na \lim_{x \to 0^+} x^{n-1} \Rightarrow 0 = \lim_{x \to 0^+} x^{n-1}$$

Let's now calculate the $\lim_{x \to 0^+} x^{n-1}$, for different values of n.

$$\lim_{x \to 0^+} x^{n-1} = \begin{cases} 0 & n > 1 \\ \lim_{x \to 0^+} x^0 = 1 & n = 1 \\ \infty & n < 1 \end{cases}$$

Therefore, $0 = \lim_{x \to 0^+} x^{n-1}$ holds when $n > 1$. For $f'(x)$ to exist at $x = 0$, the condition is: $n > 1$.

$$f''(x) = \begin{cases} \sin(x) & x \geq 0 \\ n(n-1)ax^{n-2} & x < 0 \end{cases}$$

Let's find the condition for $f''(x)$ to exist at $x = 0$.

$$\lim_{x \to 0^+} f''(x) = \lim_{x \to 0^-} f''(x) \Rightarrow \lim_{x \to 0^+} \sin(x) = \lim_{x \to 0^-} n(n-1)ax^{n-2} \Rightarrow 0$$
$$= n(n-1)a \lim_{x \to 0^+} x^{n-2} \Rightarrow 0 = \lim_{x \to 0^+} x^{n-2}$$

Let's now calculate the $\lim_{x \to 0^+} x^{n-2}$, for different values of n.

$$\lim_{x \to 0^+} x^{n-2} = \begin{cases} 0 & n > 2 \\ \lim_{x \to 0^+} x^0 = 1 & n = 2 \\ \infty & n < 2 \end{cases}$$

Therefore, $0 = \lim_{x \to 0^+} x^{n-2}$ holds when $n > 2$. For $f''(x)$ to exist at $x = 0$, the condition is: $n > 2$.

$$f'''(x) = \begin{cases} \cos(x) & x \geq 0 \\ n(n-1)(n-2)ax^{n-3} & x < 0 \end{cases}$$

Let's find the condition for $f'''(x)$ to exist at $x = 0$.

$$\lim_{x \to 0^+} f'''(x) = \lim_{x \to 0^-} f'''(x) \Rightarrow \lim_{x \to 0^+} \cos(x) = \lim_{x \to 0^-} n(n-1)(n-2)ax^{n-3} \Rightarrow 1$$
$$= n(n-1)(n-2)a \lim_{x \to 0^+} x^{n-3}$$

Let's now calculate the $\lim_{x \to 0^+} x^{n-3}$, for different values of n.

$$\lim_{x \to 0^+} x^{n-3} = \begin{cases} 0 & n > 3 \\ \lim_{x \to 0^+} x^0 = 1 & n = 3 \\ \infty & n < 3 \end{cases}$$

Therefore, $1 = n(n-1)(n-2)a \lim_{x \to 0^+} x^{n-3}$ is satisfied only when $n = 3$. For $f'''(x)$ to exist at $x = 0$, the condition is: $n = 3$ and a suitable value of a. The value of a is calculated as follows:

$$1 = n(n-1)(n-2)a \lim_{x \to 0^+} x^{n-3} \overset{n=3}{\Rightarrow} 1 = 3(3-1)(3-2)a \times 1 \Rightarrow 1 = 6a \Rightarrow a = \frac{1}{6}$$

Example 7.100 Given the function $f(x) = \begin{cases} 3\sqrt{4x+5} & x \geq 1 \\ ax^2 + bx & x < 1 \end{cases}$ is differentiable at all points of its domain, calculate the value of a and b.

Solution The function needs to be continuous at $x = 1$. Therefore,

$$\lim_{x \to 1^-} f(x) = \lim_{x \to 1^+} f(x) \Rightarrow \lim_{x \to 1^-} ax^2 + bx = \lim_{x \to 1^+} 3\sqrt{4x+5} \Rightarrow a + b = 3\sqrt{4+5} \Rightarrow a$$
$$+ b = 9$$

For the function to be differentiable at $x = 1$, the left-hand and right-hand derivatives must be equal at that point. First, we find the derivative of each piece of the function.

$$f(x) = \begin{cases} 3\sqrt{4x+5} & x \geq 1 \\ ax^2 + bx & x < 1 \end{cases} \Rightarrow f'(x) = \begin{cases} \dfrac{6}{\sqrt{4x+5}} & x \geq 1 \\ 2ax + b & x < 1 \end{cases} \Rightarrow \lim_{x \to 1^-} 2ax + b$$
$$= \lim_{x \to 1^+} \frac{6}{\sqrt{4x+5}} \Rightarrow 2a + b = \frac{6}{\sqrt{4+5}} \Rightarrow 2a + b = 2$$

$$\begin{cases} a+b=9 \\ 2a+b=2 \end{cases} \Rightarrow \begin{cases} -2a-2b=-18 \\ 2a+b=2 \end{cases} \Rightarrow -2a-2b+2a+b=-18+2 \Rightarrow$$

$$-b=-16 \Rightarrow b=16$$

$$a+b=9 \Rightarrow a+16=9 \Rightarrow a=9-16 \Rightarrow a=-7$$

Exercise 7.20 Show that there are no values of a and b for which the function $f(x)$ $= \begin{cases} ax+\ln\left(\frac{1}{\sqrt{2x}}\right) & x \geq 1 \\ x^3+bx & x<1 \end{cases}$ is differentiable at $x = 1$.

Exercise 7.21 Given the function $f(x) = \begin{cases} 3\sin^2(x)-1 & 0<x\leq \frac{\pi}{4} \\ a.\tan(x)+b.\sin(2x) & \frac{\pi}{4}<x<\frac{\pi}{2} \end{cases}$ is differentiable at all points of its domain, calculate the value of a and b. (Ans. $a = \frac{3}{2}$ and $b = -1$)

Example 7.101 Determine whether $y = [3x + 4]$ is differentiable at $x = 0.5$ and $x = 1$.

Solution The floor function $y = [f(x)]$ is not differentiable at any value of x for which $f(x)$ is an integer.

At $x = 0.5$, the expression inside the floor function is not an integer: $3x + 4 = 3 \times 0.5 + 4 = 5.5 \notin \mathbb{Z}$. Therefore, the function $[3x + 4]$ is continuous and differentiable at this point.

At $x = 1$, the expression inside the floor function is an integer: $3x + 4 = 3 \times 1 + 4 = 7 \in \mathbb{Z}$. Therefore, the function $[3x + 4]$ is neither continuous nor differentiable at this point. Let's show that $y = [3x + 4]$ is not differentiable at $x = 1$.

$$y'(1) = \lim_{h\to 0}\frac{y(1+h)-y(1)}{h} = \lim_{h\to 0}\frac{[3(1+h)+4]-[3\times 1+4]}{h}$$

$$= \lim_{h\to 0}\frac{[7+3h]-7}{h}$$

$$= \begin{cases} \lim_{h\to 0^+}\frac{[7+3h]-7}{h} = \lim_{h\to 0^+}\frac{7-7}{h} = \lim_{h\to 0^+} 0 = 0 \\ \lim_{h\to 0^-}\frac{[7+3h]-7}{h} = \lim_{h\to 0^-}\frac{6-7}{h} = \lim_{h\to 0}\frac{-1}{h} = +\infty \end{cases}$$

The left-hand and right-hand derivatives are not equal. Therefore, the function is not differentiable at $x = 1$.

Example 7.102 Determine whether $y = (x - 1)[3x + 4]$ is differentiable at $x = 1$.

Solution The function is continuous at $x = 1$ since:

$$y(1)=(1-1)[3\times 1+4]=0\times 7=0$$
$$\lim_{x\to 1^+}(x-1)[3x+4]=\lim_{x\to 1^-}(x-1)\times 7=0$$
$$\lim_{x\to 1^-}(x-1)[3x+4]=\lim_{x\to 1^-}(x-1)\times 6=0$$

Let's determine if the function is differentiable at $x = 1$.

$$y'(1)=\lim_{h\to 0}\frac{y(1+h)-y(1)}{h}$$
$$=\lim_{h\to 0}\frac{(1+h-1)[3\times(1+h)+4]-(1-1)[3\times 1+4]}{h}=\lim_{h\to 0}\frac{h[7+3h]-0}{h}$$
$$=\lim_{h\to 0}[7+3h]=\begin{cases}\lim\limits_{h\to 0^+}[7+3h]=7\\ \lim\limits_{h\to 0^-}[7+3h]=6\end{cases}$$

The left-hand and right-hand derivatives are not equal. Therefore, the function is not differentiable at $x = 1$.

Example 7.103 Determine whether $y = (x - 1)^2[3x + 4]$ is differentiable at $x = 1$.

Solution The function is continuous at $x = 1$ since:

$$y(1)=(1-1)^2[3\times 1+4]=0\times 7=0$$
$$\lim_{x\to 1^+}(x-1)^2[3x+4]=\lim_{x\to 1^-}(x-1)^2\times 7=0$$
$$\lim_{x\to 1^-}(x-1)^2[3x+4]=\lim_{x\to 1^-}(x-1)^2\times 6=0$$

Since the left-hand limit, right-hand limit, and function value are all equal to 0, the function is indeed continuous at $x = 1$. Let's determine if the function is differentiable at $x = 1$.

$$y'(1)=\lim_{h\to 0}\frac{y(1+h)-y(1)}{h}$$
$$=\lim_{h\to 0}\frac{(1+h-1)^2[3\times(1+h)+4]-(1-1)^2[3\times 1+4]}{h}$$
$$=\lim_{h\to 0}\frac{h^2[7+3h]-0}{h}=\lim_{h\to 0}h[7+3h]=\begin{cases}\lim\limits_{h\to 0^+}h\times 7=0\\ \lim\limits_{h\to 0^-}h\times 6=0\end{cases}$$

The left-hand and right-hand derivatives are equal. Therefore, the function is differentiable at $x = 1$.

Example 7.104 Determine whether $y = (x^2 + 6x + 9)[3x + 2]$ is differentiable at $x = -3$.

Solution $y = (x^2 + 6x + 9)[3x + 2] = (x + 3)^2[3x + 2]$. The function is continuous at $x = -3$ since:

$$y(-3) = \left((-3)^2 + 6\times -3 + 9\right)[3\times -3 + 2] = 0\times[-7] = 0$$

$$\lim_{x\to -3^+} (x+3)^2[3x+2] = \lim_{x\to -3^+} (x+3)^2 \times -7 = 0$$

$$\lim_{x\to -3^-} (x+3)^2[3x+2] = \lim_{x\to 3^-} (x+3)^2 \times -8 = 0$$

Since the left-hand limit, right-hand limit, and function value are all equal to 0, the function is indeed continuous at $x = -3$. Let's determine if the function is differentiable at $x = -3$.

$$y'(-3) = \lim_{h\to 0} \frac{y(-3+h) - y(-3)}{h} = \lim_{h\to 0} \frac{(-3+h+3)^2[3\times(-3+h)+2] - 0}{h}$$

$$= \lim_{h\to 0} \frac{h^2[-7+3h]}{h} = \lim_{h\to 0} h[-7+3h] = \begin{cases} \lim\limits_{h\to 0^+} h\times -7 = 0 \\ \lim\limits_{h\to 0^-} h\times -8 = 0 \end{cases}$$

The left-hand and right-hand derivatives are equal. Therefore, the function is differentiable at $x = 1$.

Example 7.105 Calculate the average rate of change for the function $y = x^2 + \sqrt{x}$ over the interval $[5, 6]$.

Solution

$$\frac{f(b) - f(a)}{b - a} = \frac{f(6) - f(5)}{6 - 5} = \frac{6^2 + \sqrt{6} - (5^2 + \sqrt{5})}{6 - 5} = 11.2134$$

Example 7.106 Given the function $y = \sin(2x) - 3x^2$, find the differential, dy.

Solution The complete differential for a function of a single variable, $y = f(x)$, is given by the formula $dy = y'dx$. Therefore, $dy = y'dx \Rightarrow dy = (2\cos(2x) - 6x)dx$.

Example 7.107 Given the function $y = \sin(2x) - 3x^2$. (A) Find the exact value of $y(0)$. (B) Use your answer from part A and differentials to approximate $y(0.05)$. (C) Find the exact value of $y(0.05)$.

Solution

(A) $y = \sin(2x) - 3x^2 \Rightarrow y(0) = \sin(2\times 0) - 3\times 0^2 = 0 - 0 = 0$

(B) $dy = y'dx \Rightarrow dy = (2\cos(2x) - 6x)dx \overset{x=0}{\Rightarrow} dy = 2$

$$\begin{aligned} y(x_0 + \Delta x) &= y(x_0) + y'(x_0)\Delta x \Rightarrow y(0.05) = y(0 + 0.05) \\ &= y(0) + y'(0)\times 0.05 = 0 + 2\times 0.05 = 0.1 \end{aligned}$$

(C) $y = \sin(2x) - 3x^2 \Rightarrow y(0.05) = \sin(2\times 0.05) - 3\times 0.05^2 = 0.0923$

Example 7.108 Given that the derivative of $f\left(\sqrt[3]{25x+2}\right)$ at $x=1$ is -7, calculate $f'(3)$.

Solution

$$\left(\sqrt[3]{25x+2}\right)'=\left((25x+2)^{\frac{1}{3}}\right)'=\frac{1}{3}\times 25\times(25x+2)^{-\frac{2}{3}}=\frac{25}{3\sqrt[3]{(25x+2)^2}}$$

$$\left(f\left(\sqrt[3]{25x+2}\right)\right)'=\left(\sqrt[3]{25x+2}\right)'f'\left(\sqrt[3]{25x+2}\right)=\frac{25}{3\sqrt[3]{(25x+2)^2}}$$
$$\times f'\left(\sqrt[3]{25x+2}\right)\overset{x=1}{\Rightarrow}\frac{25}{3\sqrt[3]{(25\times1+2)^2}}\times f'\left(\sqrt[3]{25\times1+2}\right)$$
$$=-7\Rightarrow\frac{25}{3\times9}\times f'(3)=-7\Rightarrow f'(3)=-\frac{7\times3\times9}{25}\Rightarrow f'(3)=-\frac{189}{25}$$

Example 7.109 Show that for $u(x)>0$, the derivative of $u(x)^{v(x)}$ is given by

$$\left(u(x)^{v(x)}\right)'=u(x)^{v(x)}\left(v'(x)\ln(u(x))+v(x)\frac{u'(x)}{u(x)}\right)$$

Solution Let y be the function we want to differentiate: $y=u(x)^{v(x)}$. Since $u(x)>0$ is given, we can take the natural logarithm of both sides:

$$y=u(x)^{v(x)}\Rightarrow\ln(y)=v(x)\ln(u(x))$$

We now differentiate both sides with respect to x:

$$y=u(x)^{v(x)}\Rightarrow\ln(y)=v(x)\ln(u(x))\Rightarrow\frac{y'}{y}=v'(x)\ln(u(x))+v(x)\frac{u'(x)}{u(x)}\Rightarrow y'$$
$$=y\left(v'(x)\ln(u(x))+v(x)\frac{u'(x)}{u(x)}\right)\Rightarrow y'=u(x)^{v(x)}\left(v'(x)\ln(u(x))+v(x)\frac{u'(x)}{u(x)}\right)$$

Example 7.110 Calculate the derivative of $y=\ln(x)^{\ln(x)}$ for $x\in(1,\infty)$.

Solution

$$y=\ln(x)^{\ln(x)}\Rightarrow\ln(y)=\ln\left(\ln(x)^{\ln(x)}\right)\Rightarrow\ln(y)=\ln(x)\times\ln(\ln(x))$$
$$\overset{\frac{d}{dx}}{\Rightarrow}\frac{y'}{y}=\frac{1}{x}\times\ln(\ln(x))+\ln(x)\times\frac{\frac{1}{x}}{\ln(x)}\Rightarrow\frac{y'}{y}=\frac{1}{x}\ln(\ln(x))+\frac{1}{x}\Rightarrow\frac{y'}{y}=\frac{1}{x}$$
$$\times(\ln(\ln(x))+1)\Rightarrow y'=y\left(\frac{1}{x}(\ln(\ln(x))+1)\right)\Rightarrow y'$$
$$=\ln(x)^{\ln(x)}\left(\frac{1}{x}(\ln(\ln(x))+1)\right)$$

Alternatively, the formula $u(x) > 0 \Rightarrow \left(u(x)^{v(x)}\right)' = u(x)^{v(x)}\left(v'(x)\,ln(u(x)) + v(x)\frac{u'(x)}{u(x)}\right)$ can be used to solve this problem.

Example 7.111 Calculate the derivative of $y = (x^2 + 1)^{\sin(x)}$.

Solution

$$y = \left(x^2+1\right)^{\sin(x)} \Rightarrow \ln(y) = \ln\left(\left(x^2+1\right)^{\sin(x)}\right) \Rightarrow \ln(y) = \sin(x)\ln\left(x^2+1\right)$$

$$\overset{\frac{d}{dx}}{\Rightarrow} \frac{y'}{y} = \cos(x)\ln\left(x^2+1\right) + \sin(x)\frac{2x}{x^2+1} \Rightarrow y'$$

$$= y\left(\cos(x)\ln\left(x^2+1\right) + \sin(x)\frac{2x}{x^2+1}\right) \Rightarrow y'$$

$$= \left(x^2+1\right)^{\sin(x)}\left(\cos(x)\ln\left(x^2+1\right) + \sin(x)\frac{2x}{x^2+1}\right)$$

Alternatively, the formula $u(x) > 0 \Rightarrow \left(u(x)^{v(x)}\right)' = u(x)^{v(x)}\left(v'(x)\ln(u(x)) + v(x)\frac{u'(x)}{u(x)}\right)$ can be used to solve this problem.

Example 7.112 The code in Fig. 7.23 evaluates the function $y = \frac{x+1}{(x+3)^2(x+4)}$ at $x = 17$.

```
Command Window
>> syms x
>> y=(x+1)/(x+3)^2/(x+4);
>> subs(y,17)

ans =

3/1400

>> eval(ans)

ans =

    0.0021

fx >>
```

Fig. 7.23 Evaluation of $y = \frac{x+1}{(x+3)^2(x+4)}$ at $x = 17$

Example 7.113 The code in Fig. 7.24 computes $\frac{d}{dx}\left(\frac{x+1}{(x+3)^2(x+4)}\right)$.

```
Command Window
>> syms x
>> dydx=diff((x+1)/(x+3)^2/(x+4),x)

dydx =

1/((x + 3)^2*(4 + x)) - (1 + x)/((x + 3)^2*(x + 4)^2) - (2*(1 + x))/((x + 3)^3*(4 + x))

>> simplify(dydx)

ans =

-(- 1 + 7*x + 2*x^2)/((x + 3)^3*(x + 4)^2)

>> pretty(ans)
                   2
   - 1 + 7 x + 2 x
- -----------------
         3        2
  (3 + x)  (4 + x)

fx >>
```

Fig. 7.24 Calculation of $\frac{d}{dx}\left(\frac{x+1}{(x+3)^2(x+4)}\right)$

Example 7.114 The code in Fig. 7.25 computes $\frac{d^3}{dx^3}\left(\frac{x+1}{(x+3)^2(x+4)}\right)$ at $x = -2$.

```
Command Window
>> syms x
>> dy3dx3=diff((x+1)/(x+3)^2/(x+4),x,3);
>> subs(dy3dx3,x,-2)

ans =

249/8

>> eval(ans)

ans =

   31.1250

fx >>
```

Fig. 7.25 Calculation of $\frac{d^3}{dx^3}\left(\frac{x+1}{(x+3)^2(x+4)}\right)$ at $x = -2$

Example 7.115 The code shown in Fig. 7.26 plots the graph of $(x) = \frac{x+1}{(x+3)^2(x+4)}$. The resulting plot is displayed in Fig. 7.27.

Fig. 7.26 Plotting $f(x) = \frac{x+1}{(x+3)^2(x+4)}$ on the interval $[-6, 0]$

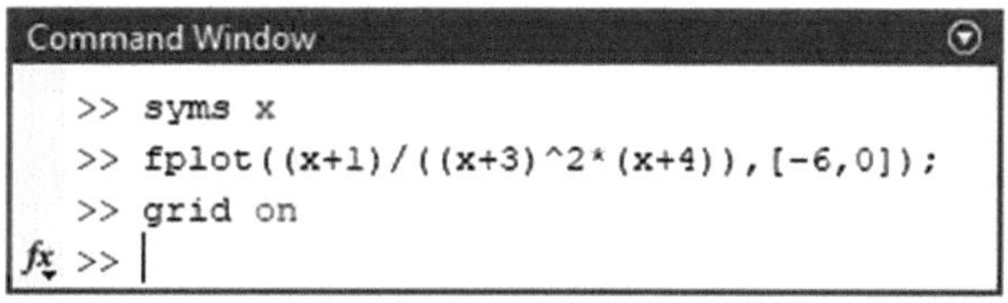

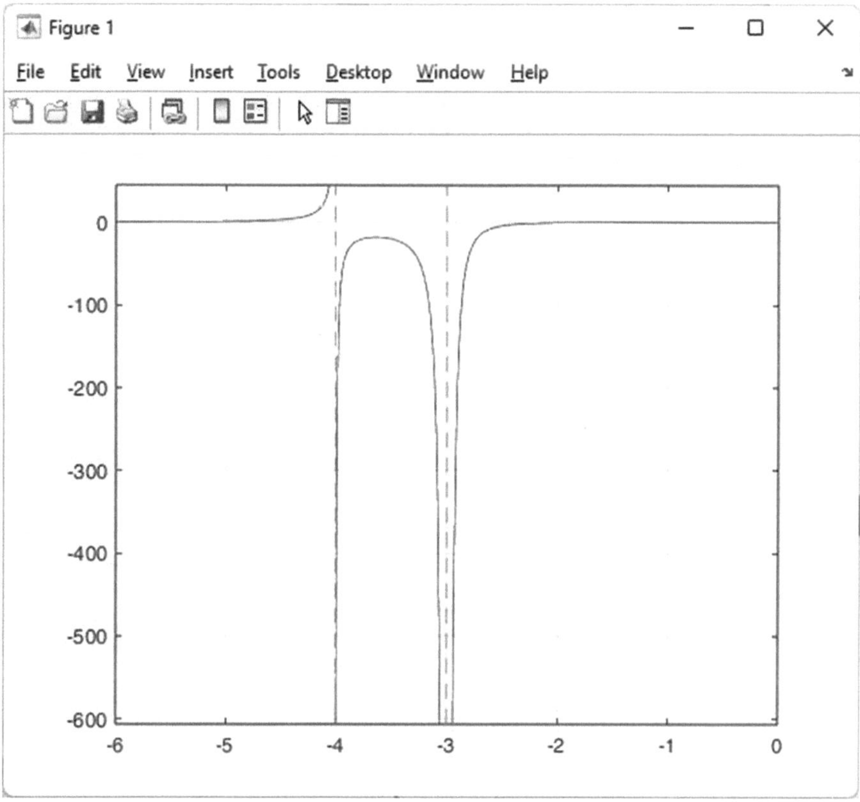

Fig. 7.27 Output of the code in Fig. 7.26

Hovering the mouse cursor over the graph displays the menu shown in Fig. 7.28, which provides options to zoom in and out.

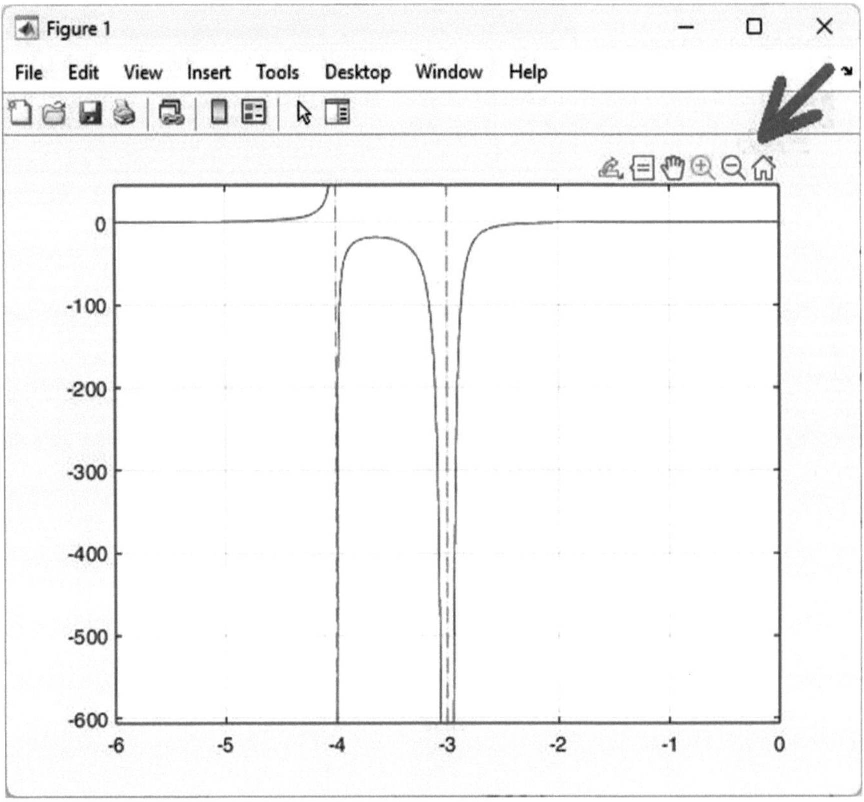

Fig. 7.28 Interactive toolbar with zoom and pan options

Example 7.116 The code shown in Fig. 7.29 calculates the $\frac{d}{dx}\left(\left|x^2-5x+6\right|\right)$, assuming that $2<x<3$.

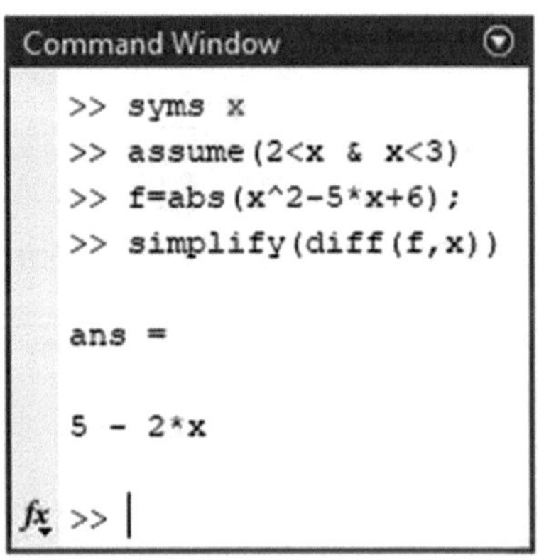

Fig. 7.29 Calculation of the $\frac{d}{dx}\left(\left|x^2-5x+6\right|\right)$ for $2<x<3$

We will now confirm the result presented in Fig. 7.29 through hand calculation.

$$f(x) = |x^2 - 5x + 6| \overset{2<x<3}{\Rightarrow} f(x) = |x^2 - 5x + 6| = -(x^2 - 5x + 6)$$

$$2 < x < 3 \Rightarrow \frac{d}{dx}(|x^2 - 5x + 6|) = \frac{d}{dx}(-(x^2 - 5x + 6)) = -(2x - 5) = 5 - 2x$$

The obtained result agrees with the one shown in Fig. 7.29.

Example 7.117 The code shown in Fig. 7.30 calculates the $\frac{d}{dx}(|x^2 - 5x + 6|)$, assuming that $3 < x$.

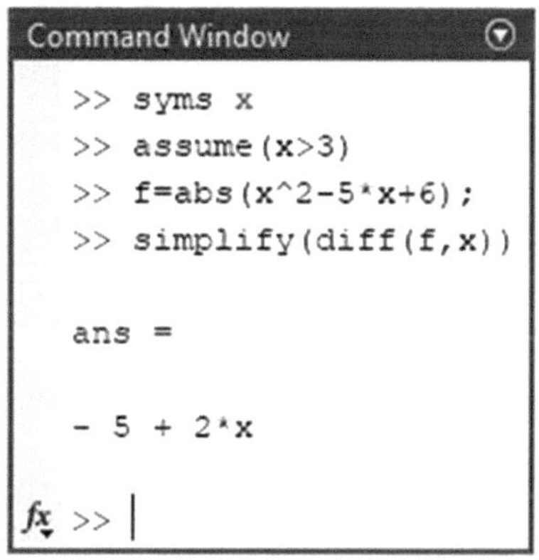

Fig. 7.30 Calculation of the $\frac{d}{dx}(|x^2 - 5x + 6|)$ for $3 < x$

We will now confirm the result presented in Fig. 7.30 through hand calculation.

$$f(x) = |x^2 - 5x + 6| \overset{x>3}{\Rightarrow} f(x) = |x^2 - 5x + 6| = x^2 - 5x + 6$$

$$x > 3 \Rightarrow \frac{d}{dx}(|x^2 - 5x + 6|) = \frac{d}{dx}(x^2 - 5x + 6) = 2x - 5$$

The obtained result agrees with the one shown in Fig. 7.30.

Example 7.118 The code shown in Fig. 7.31 calculates the $\frac{d}{dx}(|x^2 - 5x + 6|)$, assuming that $0 < x$.

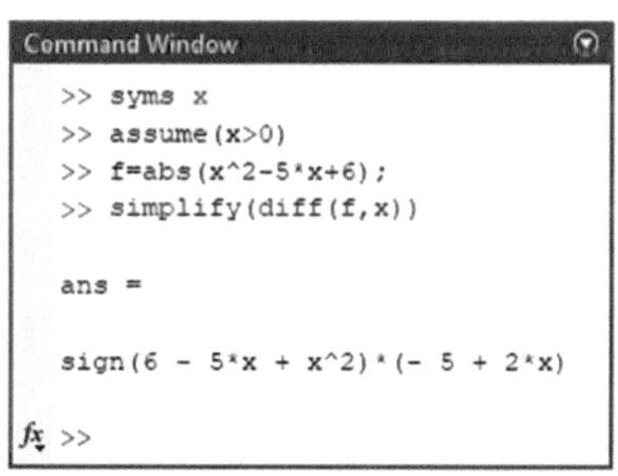

Fig. 7.31 Calculation of the $\frac{d}{dx}(|x^2 - 5x + 6|)$ for $0 < x$

The expression presented in Fig. 7.31 may be simplified to the following form:

$$sign\left(x^2-5x+6\right)(2x-5)=\begin{cases} 2x-5 & 0<x<2 \\ 5-2x & 2<x<3 \\ 2x-5 & 3<x \end{cases}$$

We will now confirm this result through manual calculation.

$$f(x)=\left|x^2-5x+6\right| \overset{0<x<2}{\Rightarrow} f(x)=\left|x^2-5x+6\right|=x^2-5x+6 \overset{\frac{d}{dx}}{\Rightarrow} 2x-5$$

$$f(x)=\left|x^2-5x+6\right| \overset{2<x<3}{\Rightarrow} f(x)=\left|x^2-5x+6\right|=-x^2+5x-6 \overset{\frac{d}{dx}}{\Rightarrow} -2x+5$$

$$f(x)=\left|x^2-5x+6\right| \overset{3<x}{\Rightarrow} f(x)=\left|x^2-5x+6\right|=x^2-5x+6 \overset{\frac{d}{dx}}{\Rightarrow} 2x-5$$

Example 7.119 The code shown in Fig. 7.32 draws the plot of $|x^2-5x+6|$ over the $[0,5]$ interval. The output of this code is shown in Fig. 7.33. Figure 7.33 clearly shows that the derivative does not exist at $x=2$ and $x=3$.

Fig. 7.32 Code for drawing the plot of $|x^2-5x+6|$ over the $[0,5]$ interval

```
Command Window
>> syms x
>> f=abs(x^2-5*x+6);
>> fplot(f,[0,5])
>> grid on
>>
```

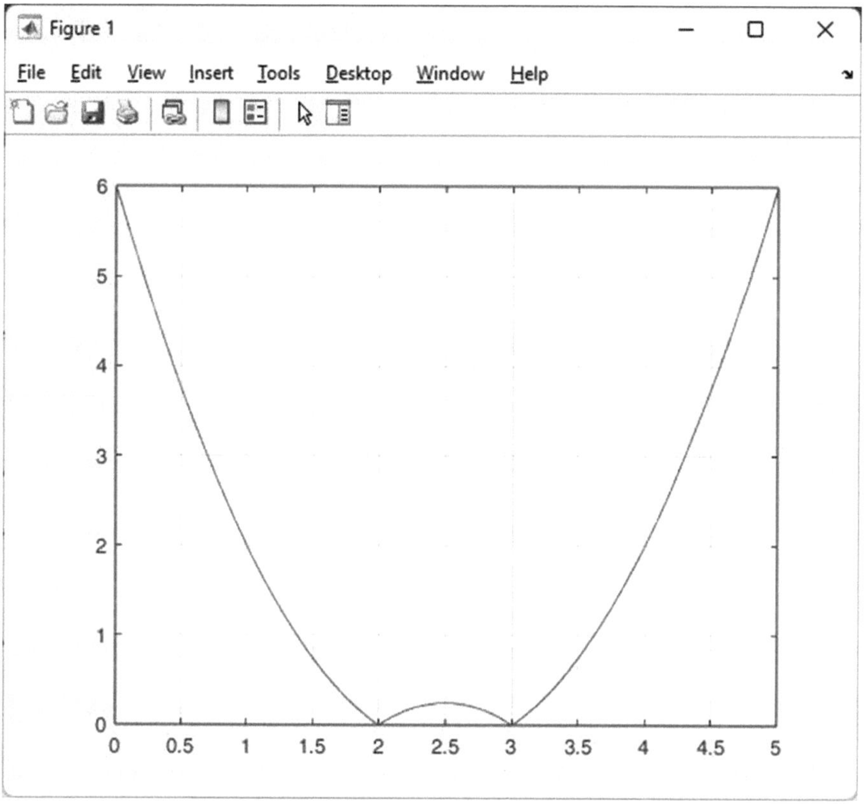

Fig. 7.33 Output of the code in Fig. 7.32

Example 7.120 The code in Fig. 7.34 calculates the derivative y' for the implicit equation $y\sin(x) - x\cos(y) = 0$.

```
Command Window
>> syms x
>> syms y(x)
>> syms Dydx
>> F=y*sin(x)-x*cos(y)==0;%define the implicit equation
>> dF_dx = diff(F,x);
>> substituted_eq = subs(dF_dx, diff(y, x), Dydx);
>> dy_dx_sol = solve(substituted_eq, Dydx);
>> pretty(dy_dx_sol)
cos(y(x)) - cos(x) y(x)
-----------------------
  sin(x) + x sin(y(x))

fx >>
```

Fig. 7.34 Calculation of y' for the implicit equation $y\sin(x) - x\cos(y) = 0$

Example 7.121 The code in Fig. 7.35 calculates the derivative y' for the implicit equation $x^2 + y^2 + xy + \cos(y^2) = 0$.

```
Command Window
>> syms x
>> syms y(x)
>> syms Dydx
>> F=x^2+y^2+x*y+cos(y^2)==0;%define the implicit equation
>> dF_dx = diff(F,x);
>> substituted_eq = subs(dF_dx, diff(y, x), Dydx);
>> dy_dx_sol = solve(substituted_eq, Dydx);
>> pretty(dy_dx_sol)
                 2 x + y(x)
- ------------------------------
                          2
  x + 2 y(x) - 2 sin(y(x) ) y(x)

fx >>
```

Fig. 7.35 Calculation of y' for the implicit equation $x^2 + y^2 + xy + \cos(y^2) = 0$

References for Further Study

1. Engineering Mathematics with MATLAB® and Simulink®, Farzin Asadi, Springer, 2025.
2. Applied Numerical Analysis with MATLAB®/Simulink®: For Engineers and Scientists, Farzin Asadi, Springer, 2023.
3. Advanced Engineering Mathematics, Erwin Kreyszig, Wiley, 2020.
4. A Problem-Solving Approach to Single-Variable Calculus (Volume II), Farzin Asadi, Springer, 2026.
5. A Problem-Solving Approach to Multi-Variable Calculus (Volume I), Farzin Asadi, Springer, 2026.
6. A Problem-Solving Approach to Multi-Variable Calculus (Volume II), Farzin Asadi, Springer, 2026.
7. A Problem-Solving Approach to Multi-Variable Calculus (Volume III), Farzin Asadi, Springer, 2026.
8. A Problem-Solving Approach to Ordinary Differential Equations (Volume I), Farzin Asadi, Springer, 2026.
9. A Problem-Solving Approach to Ordinary Differential Equations (Volume II), Farzin Asadi, Springer, 2026.
10. A Problem-Solving Approach to Ordinary Differential Equations (Volume III), Farzin Asadi, Springer, 2026.
11. A Problem-Solving Approach to Ordinary Differential Equations (Volume IV), Farzin Asadi, Springer, 2026.

Chapter 8
Applications of the Differentiation

8.1 Introduction

Let $f(x)$ be a function defined on an interval I. Then $f(x)$ is classified as follows:

Non-decreasing Function: For any two numbers x_1 and x_2 in I, if $x_1 < x_2$, then $f(x_1) \leq f(x_2)$. For example, $f(x) = [x]$ is non-decreasing for all real numbers (Fig. 8.1).

Increasing Function: For any two numbers x_1 and x_2 in I, if $x_1 < x_2$, then $f(x_1) < f(x_2)$. For example, $f(x) = x^3$ is increasing for all real numbers (Fig. 8.2). Some books use the term "strictly increasing" to mean increasing.

Non-increasing Function: For any two numbers x_1 and x_2 in I, if $x_1 < x_2$, then $f(x_1) \geq f(x_2)$. For example, $f(x) = -[x]$ is non-increasing for all real numbers (Fig. 8.3).

Decreasing Function: For any two numbers x_1 and x_2 in I, if $x_1 < x_2$, then $f(x_1) > f(x_2)$. For example, $f(x) = -x$ is decreasing for all real numbers (Fig. 8.4). Some books use the term "strictly decreasing" to mean decreasing.

F. Asadi, *A Problem-Solving Approach to Single-Variable Calculus*,
https://doi.org/10.1007/978-3-032-13505-6_8

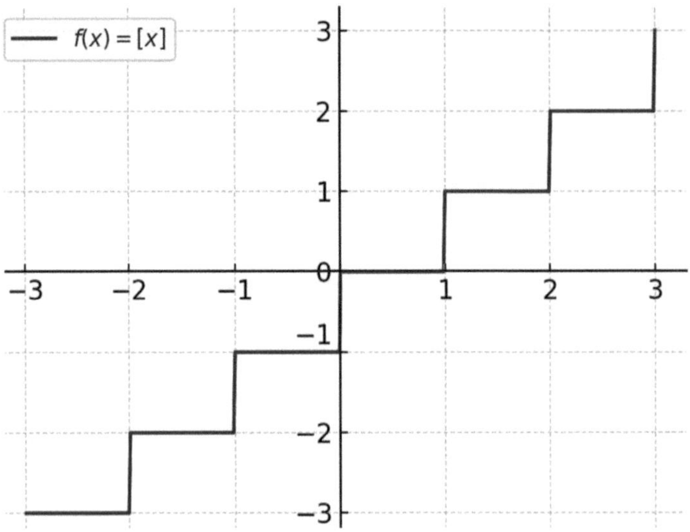

Fig. 8.1 Plot of $f(x) = [x]$

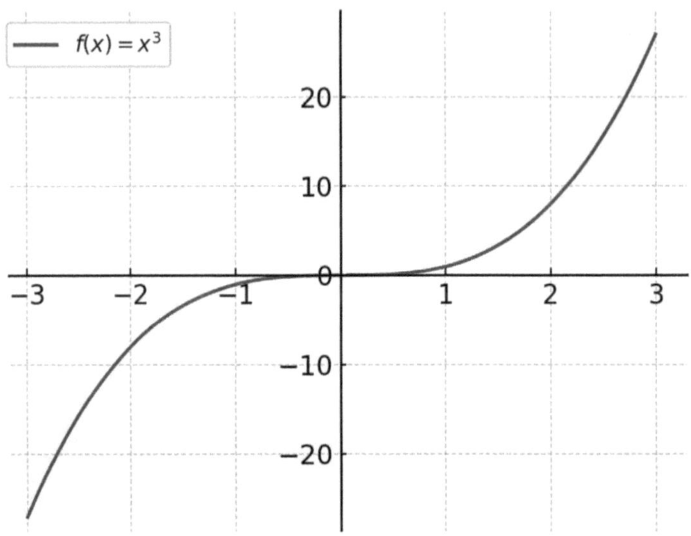

Fig. 8.2 Plot of $f(x) = x^3$

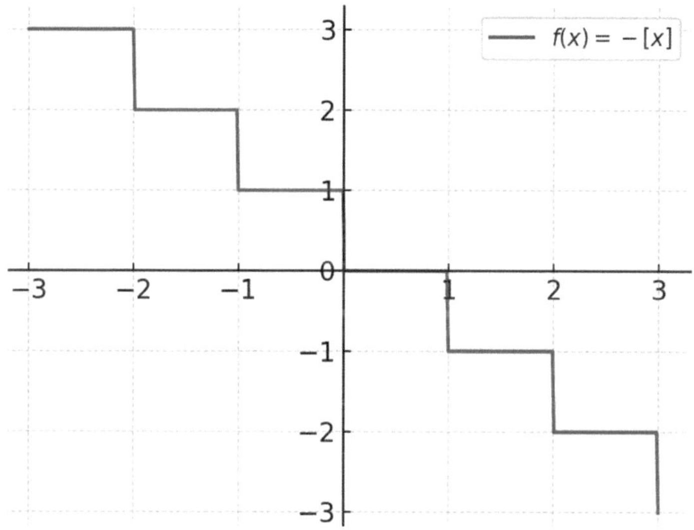

Fig. 8.3 Plot of $f(x) = -[x]$

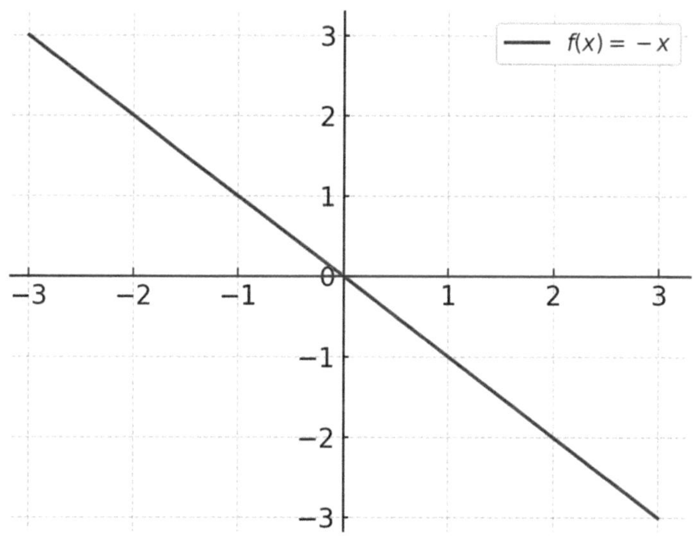

Fig. 8.4 Plot of $f(x) = -x$

The behavior of a function's derivative indicates its monotonicity:

If $f'(x) > 0$ for all x in an interval (with the derivative allowed to vanish at isolated points), then $f(x)$ is increasing there.
If $f'(x) \geq 0$ for all x in the interval, then $f(x)$ is non-decreasing (Here the function may "flatten out" on subintervals where the derivative is zero, without decreasing).
If $f'(x) < 0$ for all x in an interval (with the derivative allowed to vanish at isolated points), then $f(x)$ is decreasing there.
If $f'(x) \leq 0$ for all x in an interval, then $f(x)$ is non-increasing (Here the function may "flatten out" on subintervals where the derivative is zero, without increasing).

Throughout this chapter, $\nearrow$ indicates an increase, $\searrow$ indicates a decrease, $\uparrow$ indicates upward concavity, and $\downarrow$ indicates downward concavity.

8.1.1 Some Useful Definitions

A critical point of a function is a point in the domain where the first derivative is either zero or does not exist.

An extremum is a point where a function has a maximum or minimum value. Every extremum (local maximum or minimum) that occurs in the interior of a function's domain must be a critical point. However, not all critical points are extrema. For example, a point where the derivative is zero could be an inflection point (like the function $f(x) = x^3$ at $x = 0$), not a maximum or minimum.

A **relative maximum** (or local maximum) of a function is a point where the function takes a value larger than at all nearby points, even if higher values exist elsewhere. Similarly, a **relative minimum** (or local minimum) is a point where the function takes a value smaller than at nearby points, even if lower values exist elsewhere (Fig. 8.5).

A **global maximum** is the absolute highest value of the function over the entire domain (or over a given interval), while a **global minimum** is the absolute lowest value. Unlike relative extrema, which depend on local neighborhoods, global extrema are determined by comparing the function's values everywhere in the domain (Fig. 8.5).

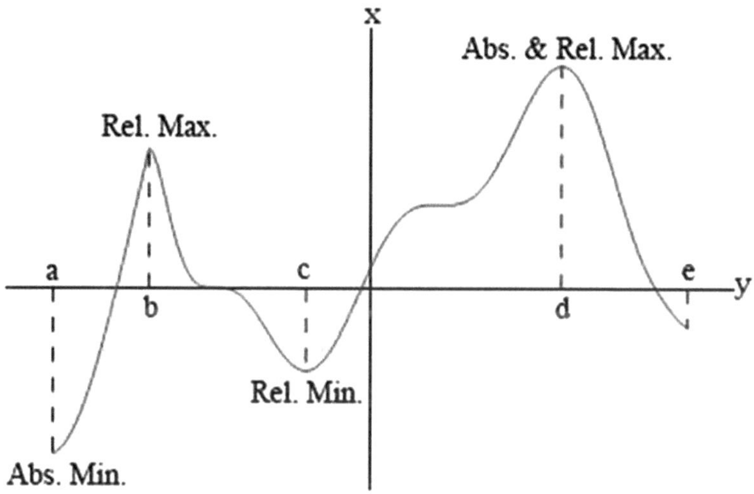

Fig. 8.5 Local and glob extrema

An **inflection point** is a point on the graph of a function where the **concavity changes**—that is, the function changes from concave up to concave down, or vice versa. Potential inflection points occur where the second derivative is **zero** or **does not exist**. A point is an **inflection point only if the concavity changes** across it. For example, inflection point of $f(x) = x^3 - 3x^2 + 2$ is $(1, 0)$ (Fig. 8.6).

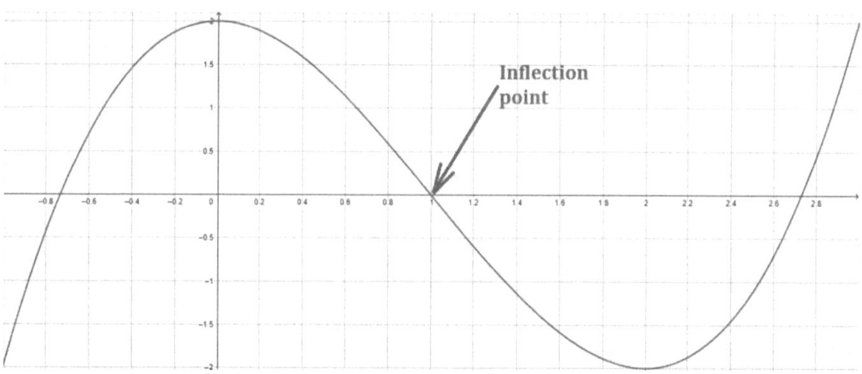

Fig. 8.6 Inflection point

A point where the second derivative, $f''(x)$, has a repeated root (even multiplicity) is typically not an inflection point.

8.1.2 The Procedure for Finding Extrema

The procedure for finding local and global extrema is as follows:

First, compute the **first derivative** $f'(x)$ of the function. Critical points occur where $f'(x) = 0$ or where $f'(x)$ is undefined. These are the candidates for local maxima or minima. Next, apply the **second derivative test**: if $f''(x) > 0$ at a critical point, the function has a local minimum there; if $f''(x) < 0$, it has a local maximum; if $f''(x) = 0$, the test is inconclusive and further analysis is needed.

To find **global extrema** on a closed interval $[a, b]$, evaluate the function at all critical points inside the interval and at the endpoints a and b. The largest of these values gives the global maximum, and the smallest gives the global minimum.

Let $f(x)$ and $g(x)$ be two functions differentiable at a point c, with $g(c) \neq 0$ and $g'(c) \neq 0$. If the quotient function $h(x) = \frac{f(x)}{g(x)}$ has a relative extremum (maximum or minimum) at $x = c$, then the following identity holds:

$$\frac{f(c)}{g(c)} = \frac{f'(c)}{g'(c)}$$

The following is the proof:

$$h(x) = \frac{f(x)}{g(x)} \Rightarrow h'(x) = \frac{f'(x)g(x) - g'(x)f(x)}{(g(x))^2}$$

$$h'(x) = 0 \Rightarrow \frac{f'(x)g(x) - g'(x)f(x)}{(g(x))^2} = 0 \Rightarrow f'(x)g(x) - g'(x)f(x) = 0$$

$$\Rightarrow f'(c)g(c) - g'(c)f(c) = 0 \Rightarrow \frac{f(c)}{g(c)} = \frac{f'(c)}{g'(c)}$$

8.1.3 The Mean Value Theorem

The Mean Value Theorem states that for a continuous function $f(x)$ on a closed interval $[a, b]$ and differentiable on the open interval (a, b), there exists at least one number c in (a, b) such that the instantaneous rate of change at c, $f'(c)$, equals the average rate of change over the interval $[a, b]$. In mathematical terms: $f'(c) = \frac{f(b) - f(a)}{b - a}$. In simpler terms, there's a point on the curve where the tangent line is parallel to the secant line connecting the endpoints.

8.1.4 Newton-Raphson Method

The Newton-Raphson method is a powerful iterative technique used to find the roots of nonlinear equations, including polynomials. It's based on the idea of linear approximation. In each step the guess is updated with the aid of following formula:

$$x_{n+1} = x_n - \frac{f(x_n)}{f'(x_n)}$$

8.2 Solved Problems

Example 8.1 Determine whether the function $f(x) = e^{-x}$ is increasing, decreasing, non-increasing, or non-decreasing on $(-\infty, \infty)$.

Solution

$$f(x) = e^{-x} \Rightarrow f'(x) = -e^{-x} \Rightarrow \forall x : f'(x) < 0 \Rightarrow \text{strictly decreasing}$$

Example 8.2 Determine whether the function $f(x) = \sqrt{x}$ is increasing, decreasing, non-increasing, or non-decreasing $[0, \infty)$.

Solution

$$f(x) = \sqrt{x} \Rightarrow f'(x) = \frac{1}{2\sqrt{x}} \Rightarrow \forall x : f'(x) > 0 \Rightarrow \text{strictly increasing}$$

Example 8.3 Determine whether the function $f(x) = \sin(x)$ is increasing, decreasing, non-increasing, or non-decreasing $[0, \pi]$.

Solution

$$f(x) = \sin(x)$$

$$f'(x) = \cos(x) \Rightarrow \begin{cases} x \in \left[0, \frac{\pi}{2}\right] \Rightarrow f'(x) = \cos(x) \geq 0 \Rightarrow \text{strictly increasing} \\ x \in \left[\frac{\pi}{2}, \pi\right] \Rightarrow f'(x) = \cos(x) \leq 0 \Rightarrow \text{strictly decreasing} \end{cases}$$

$$f'\left(\frac{\pi}{2}\right) = 0 \Rightarrow x = \frac{\pi}{2} \text{ is a critical point}$$

Figure 8.7 shows the graph of sin(x), visually representing its behavior on the interval $[0, \pi]$.

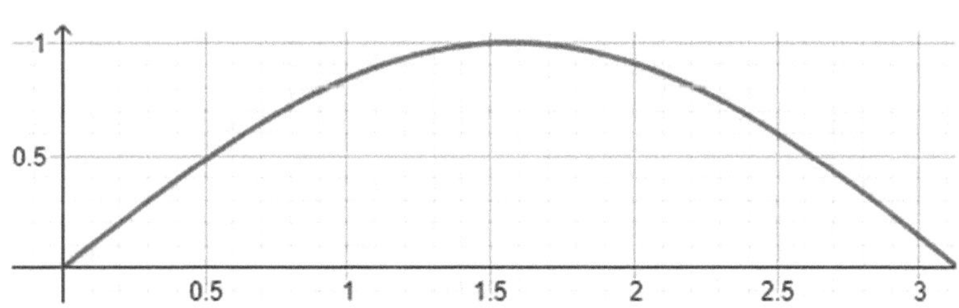

Fig. 8.7 Plot of $y = \sin(x)$ on the interval $[0, \pi]$

Example 8.4 Determine whether the function $f(x) = x + \sin(2x)$ is increasing, decreasing, non-increasing, or non-decreasing on $[0, \pi]$.

Solution

$$f(x) = x + \sin(2x) \Rightarrow f'(x) = 1 + 2\cos(2x)$$

To find where the function might change from increasing to decreasing, we find the values of x where $f'(x) = 0$:

$$f'(x) = 0 \Rightarrow 1 + 2\cos(2x) = 0 \Rightarrow \cos(2x) = -\frac{1}{2}$$

The general solutions for this equation $2x = \frac{2\pi}{3} + 2k\pi$ and $2x = \frac{4\pi}{3} + 2k\pi$, where k is an integer (see Chap. 4). For the interval $x \in [0, \pi]$, the corresponding interval for $2x$ is $[0, 2\pi]$. The solutions for $2x$ in this range are $\frac{2\pi}{3}$ and $\frac{4\pi}{3}$. Solving for x, we get the critical points:

$$2x = \frac{2\pi}{3} \Rightarrow x = \frac{\pi}{3}$$

$$2x = \frac{4\pi}{3} \Rightarrow x = \frac{2\pi}{3}$$

The critical points $\frac{\pi}{3}$ and $\frac{2\pi}{3}$ divide the interval $[0, \pi]$ into three subintervals. We choose a test point in each interval to determine the sign of $f'(x)$.

Interval $\left(0, \frac{\pi}{3}\right)$: Let's choose $x = \frac{\pi}{4}$:

$$f'\left(\frac{\pi}{4}\right) = 1 + 2\cos\left(2 \times \frac{\pi}{4}\right) = 1 + 2\cos\left(\frac{\pi}{2}\right) = 1 + 2(0) = 1$$

Since $f'\left(\frac{\pi}{4}\right) > 0$, the function is increasing on this interval.

Interval $\left(\frac{\pi}{3}, \frac{2\pi}{3}\right)$: Let's choose $x = \frac{\pi}{2}$:

$$f'\left(\frac{\pi}{2}\right) = 1 + 2\cos\left(2 \times \frac{\pi}{2}\right) = 1 + 2\cos(\pi) = 1 + 2(-1) = -1.$$

Since $f'\left(\frac{\pi}{2}\right) < 0$, the function is decreasing on this interval.

Interval $\left(\frac{2\pi}{3}, \pi\right)$: Let's choose $x = \frac{3\pi}{4}$:

$$f'\left(\frac{3\pi}{4}\right) = 1 + 2\cos\left(2 \times \frac{3\pi}{4}\right) = 1 + 2\cos\left(\frac{3\pi}{2}\right) = 1 + 2(0) = 1.$$

Since $f'\left(\frac{3\pi}{4}\right) > 0$, the function is increasing on this interval.

Because the sign of the derivative changes on the interval $[0, \pi]$, the function is neither increasing nor decreasing on the entire interval. Instead, its behavior is as follows:

The function is increasing on the intervals $\left[0, \frac{\pi}{3}\right]$ and $\left[\frac{2\pi}{3}, \pi\right]$ (Fig. 8.8).
The function is decreasing on the interval $\left[\frac{\pi}{3}, \frac{2\pi}{3}\right]$ (Fig. 8.8).

Fig. 8.8 Sign table for $f'(x)$

x	0		$\frac{\pi}{3}$		$\frac{2\pi}{3}$		π
$f'(x)$		$+$	0	$-$	0	$+$	
$f(x)$		$\nearrow$		$\searrow$		$\nearrow$	

Figure 8.9 shows the graph of $x + \sin(2x)$, visually representing its behavior on the interval $[0, \pi]$.

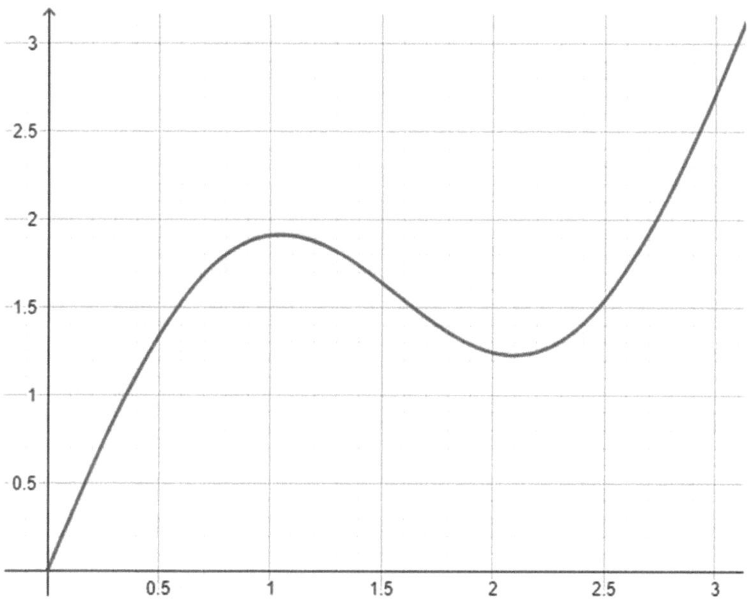

Fig. 8.9 Plot of $x + \sin(2x)$

Example 8.5 Determine whether the function $f(x) = x\sqrt{9 - x^2}$ is increasing, decreasing, non-increasing, or non-decreasing on $[-3, 3]$.

Solution

$$f(x) = x\sqrt{9 - x^2} \Rightarrow f'(x) = \sqrt{9 - x^2} + x \times \frac{1}{2} \times -2x \times \frac{1}{\sqrt{9 - x^2}}$$

$$= \sqrt{9 - x^2} - \frac{x^2}{\sqrt{9 - x^2}} = \frac{9 - 2x^2}{\sqrt{9 - x^2}}$$

$$f'(x) = 0 \Rightarrow \frac{9 - 2x^2}{\sqrt{9 - x^2}} = 0 \Rightarrow 9 - 2x^2 = 0 \Rightarrow 2x^2 = 9 \Rightarrow x^2 = \frac{9}{2} \Rightarrow x = \pm \frac{3}{\sqrt{2}}$$

$f'(x) = \frac{9 - 2x^2}{\sqrt{9 - x^2}}$ is undefined at the endpoints of the domain, $x = \pm 3$.

These critical points divide the interval $[-3, 3]$ into three subintervals. We test a value within each subinterval to determine the sign of $f'(x)$:

On the interval $\left(-3, -\frac{3}{\sqrt{2}} = -2.12\right)$: Test $x = -2.5$. $f'(-2.5) = \frac{9 - 2(-2.5)^2}{\sqrt{9 - (-2.5)^2}} = -2.1106 < 0$.

Since $f'(x) < 0$, the function is decreasing on this interval (Fig. 8.10).

On the interval $\left(-\frac{3}{\sqrt{2}}, \frac{3}{\sqrt{2}}\right)$: Test $x = 0$. $f'(0) = \frac{9 - 2(0)^2}{\sqrt{9 - 0^2}} = \frac{9}{3} = 3 > 0$.

Since $f'(x) > 0$, the function is increasing on this interval (Fig. 8.10).

On the interval $\left(\frac{3}{\sqrt{2}}, 3\right)$: Test $x = 2.5$. $f'(2.5) = \frac{9 - 2(2.5)^2}{\sqrt{9 - (2.5)^2}} = 2.1106 < 0$.

Since $f'(x) < 0$, the function is decreasing on this interval (Fig. 8.10).

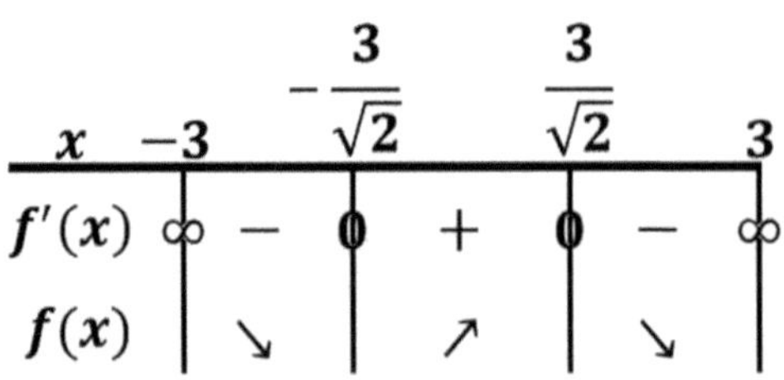

Fig. 8.10 Sign table for $f'(x)$

Figure 8.11 shows the graph of $x\sqrt{9 - x^2}$, visually representing its behavior on the interval $[-3, 3]$.

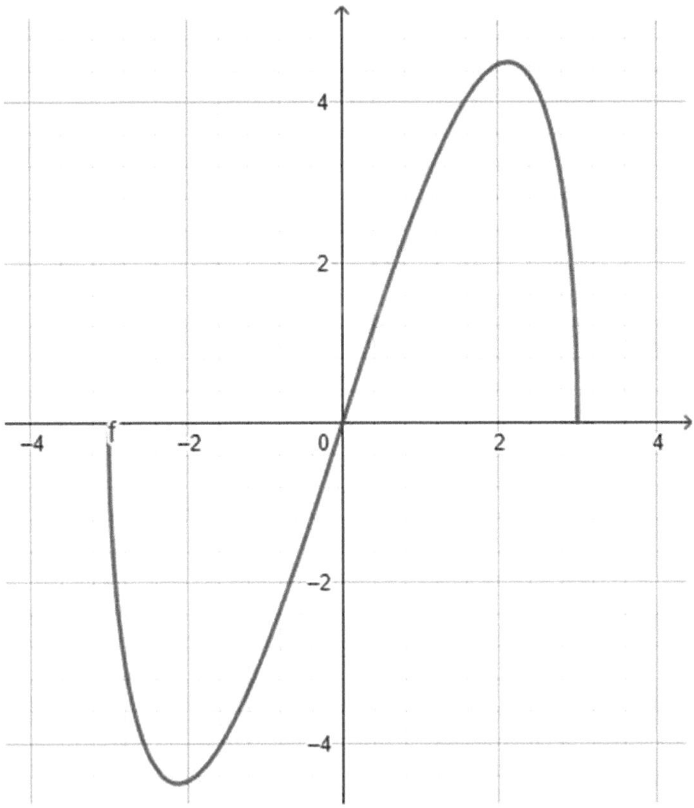

Fig. 8.11 Graph of $x\sqrt{9-x^2}$

Example 8.6 Determine whether the function $f(x)=x\sqrt{x^2-9}$ is increasing, decreasing, non-increasing, or non-decreasing on $(-\infty,3]\bigcup[3,+\infty)$.

Solution First, let's find the domain of the function. For the square root to be a real number, the term inside must be non-negative: $x^2-9\geq 0\Rightarrow x^2\geq 9\Rightarrow\begin{cases}x\geq 3\\ x\leq -3\end{cases}$. Therefore, the domain of the function is: $(-\infty,3]\bigcup[3,+\infty)$.

$$f(x)=x\sqrt{x^2-9}\Rightarrow f'(x)=\sqrt{x^2-9}+x\times\frac{1}{2}\times 2x\times\frac{1}{\sqrt{x^2-9}}=\sqrt{x^2-9}+\frac{x^2}{\sqrt{x^2-9}}=\frac{2x^2-9}{\sqrt{x^2-9}}$$

$$f'(x)=0\Rightarrow\frac{2x^2-9}{\sqrt{x^2-9}}=0\Rightarrow 2x^2-9=0\Rightarrow 2x^2=9\Rightarrow x^2=\frac{9}{2}\Rightarrow x=\pm\frac{3}{\sqrt{2}}\approx\pm 2.1213$$

However, these critical values are not in the domain of the function, so the sign of the derivative will not change within the two intervals of the domain.

$f'(x) = \frac{2x^2-9}{\sqrt{x^2-9}}$ is undefined at $x = \pm 3$.

On the interval $(-\infty, -3)$: Choose a test value, for example, $x = -4$.

$$f'(-4) = \frac{2(-4)^2-9}{\sqrt{(-4)^2-9}} = \frac{32-9}{\sqrt{7}} = \frac{23}{\sqrt{7}} > 0$$

Since $f'(x) > 0$, the function is increasing on $(-\infty, -3)$ (Fig. 8.12).

On the interval $(3, \infty)$: Choose a test value, for example, $x = 4$.

$$f'(4) = \frac{2 \times 4^2 - 9}{\sqrt{4^2-9}} = \frac{32-9}{\sqrt{7}} = \frac{23}{\sqrt{7}} > 0$$

Since $f'(x) > 0$, the function is increasing on $(3, \infty)$ (Fig. 8.12).

Fig. 8.12 Sign table for $f'(x)$

x	$-\infty$		-3		3		$+\infty$
$f'(x)$		$+$	⫴	Not allowed	⫴	$+$	
$f(x)$		↗		Not allowed		↗	

Figure 8.13 shows the graph of $x\sqrt{x^2-9}$, visually representing its behavior on the interval $(-\infty, 3] \bigcup [3, +\infty)$.

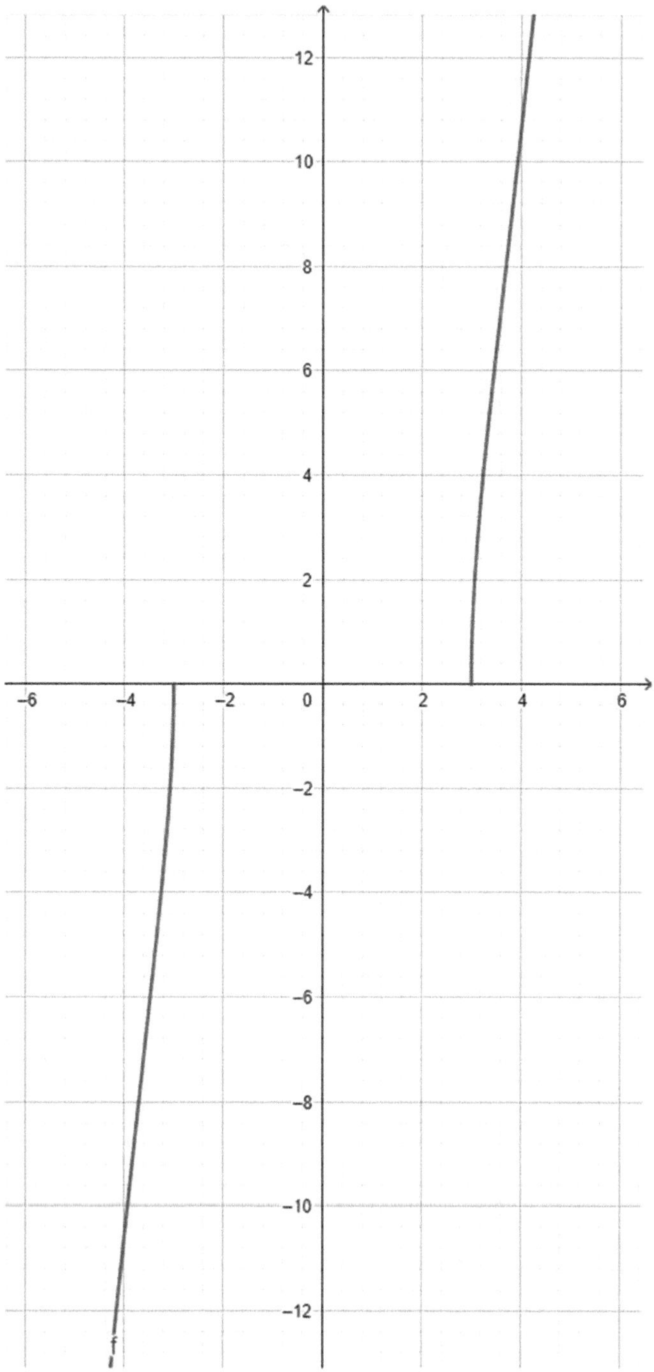

Fig. 8.13 Plot of $x\sqrt{x^2-9}$

Example 8.7 Determine whether the function $f(x) = \cos^2(x) - \cos(x)$ is increasing, decreasing, non-increasing, or non-decreasing on $\left[\frac{\pi}{2}, \pi\right]$.

Solution

$$f(x) = \cos^2(x) - \cos(x) \Rightarrow f'(x) = -2\sin(x)\cos(x) + \sin(x) = \sin(x)(1 - 2\cos(x))$$

$$f'(x) = 0 \Rightarrow \sin(x)(1 - 2\cos(x)) = 0 \Rightarrow \begin{cases} \sin(x) = 0 \Rightarrow x = k\pi & k \in \mathbb{Z} \\ \cos(x) = \frac{1}{2} \Rightarrow x = 2k\pi \pm \frac{\pi}{3} & k \in \mathbb{Z} \end{cases}$$

$$k = 1 \Rightarrow x = \pi \in \left[\frac{\pi}{2}, \pi\right]$$

Now, let's analyze the signs of the factors $\sin(x)$ and $(1 - 2\cos(x))$ on the given interval.

Sign of $\sin(x)$ on $\left[\frac{\pi}{2}, \pi\right]$: In the second quadrant, where $x \in \left[\frac{\pi}{2}, \pi\right]$, the value of $\sin(x)$ is positive or zero (at the endpoint $x = \pi$). So, $\sin(x) \geq 0$.

Sign of $(1 - 2\cos(x))$ on $\left[\frac{\pi}{2}, \pi\right]$: In the second quadrant, the value of $\cos(x)$ is negative or zero (at the endpoints). So, for $x \in \left[\frac{\pi}{2}, \pi\right]$, we have $-1 \leq \cos(x) \leq 0$. Multiplying by -2, we get $0 \leq -2\cos(x) \leq 2$. Adding 1, we get $1 \leq 1 - 2\cos(x) \leq 3$. Therefore, the term $(1 - 2\cos(x))$ is positive on this interval.

$f'(x) = \sin(x)(1 - 2\cos(x)) \geq 0$ for all $x \in \left[\frac{\pi}{2}, \pi\right]$. More specifically, $f'(x) > 0$ for all $x \in \left[\frac{\pi}{2}, \pi\right]$ and $f'(x) = 0$ only at the endpoint $x = \pi$ (Fig. 8.14). Therefore, $f(x)$ is increasing on the interval $\left[\frac{\pi}{2}, \pi\right]$.

x	$\frac{\pi}{2}$		π
$f'(x)$		+	0
$f(x)$		↗	

Fig. 8.14 Sign table for $f'(x)$ on the interval $\left[\frac{\pi}{2}, \pi\right]$

Figure 8.15 shows the graph of $\cos^2(x) - \cos(x)$, visually representing its behavior on the interval $\left[\frac{\pi}{2}, \pi\right]$.

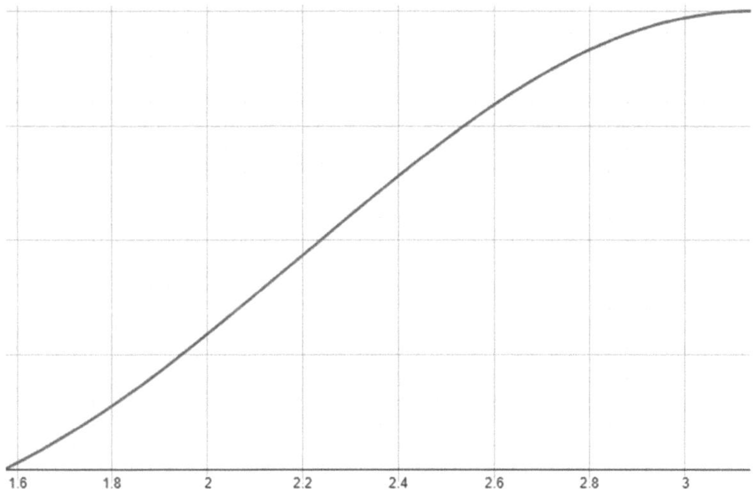

Fig. 8.15 Plot of $\cos^2(x) - \cos(x)$ on the interval $\left[\frac{\pi}{2}, \pi\right]$

Example 8.8 Determine the interval(s) on $[0, 2\pi]$, where the function $f(x) = \frac{\cos(x)}{2+\sin(x)}$ is increasing.

Solution To determine the interval where the function $f(x) = \frac{\cos(x)}{2+\sin(x)}$ is increasing on $[0, 2\pi]$, we need to find the derivative and analyze its sign.

$$f(x) = \frac{\cos(x)}{2+\sin(x)} \Rightarrow f'(x) = \frac{-\sin(x)(2+\sin(x)) - \cos(x)\times\cos(x)}{(2+\sin(x))^2}$$
$$= \frac{-2\sin(x) - \sin^2(x) - \cos^2(x)}{(2+\sin(x))^2} = \frac{-2\sin(x) - 1}{(2+\sin(x))^2}$$

A function is increasing when its derivative is positive, so we need to solve the inequality $f'(x) > 0$. The denominator, $(2+\sin(x))^2$, is always positive since $-1 \le \sin(x) \le 1$. Therefore, the sign of $f'(x)$ depends only on the sign of the numerator:

$$-2\sin(x) - 1 > 0 \Rightarrow -2\sin(x) > 1 \Rightarrow \sin(x) < -\frac{1}{2}$$

On the interval $[0, 2\pi]$, the values of x for which $\sin(x) = -\frac{1}{2}$ are $x = \frac{7\pi}{6}$ and $x = \frac{11\pi}{6}$. The sine function is less than $-\frac{1}{2}$ in the third and fourth quadrants (Fig. 8.16). Thus, the function $f(x)$ is increasing on the interval $\left(\frac{7\pi}{6}, \frac{11\pi}{6}\right)$.

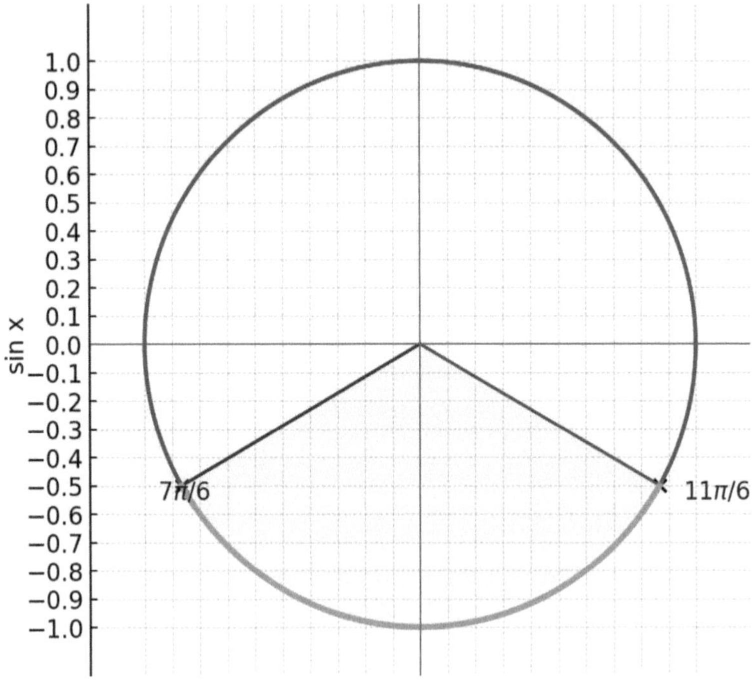

Fig. 8.16 Regions where $\sin(x) < -\frac{1}{2}$

Example 8.9 Determine the value(s) of a for which the function $f(x) = x^3 + ax^2 + x$ is always increasing.

Solution To determine the values of a for which the function $f(x) = x^3 + ax^2 + x$ is always increasing, we need to analyze its first derivative. A function is always increasing if its derivative is ≥ 0 for all real numbers x.

First, we find the derivative of $f(x)$: $f'(x) = 3x^2 + 2ax + 1$.

For $f(x)$ to be always increasing, we must have $f'(x) \geq 0$ for all x. This means the quadratic expression $3x^2 + 2ax + 1$ must be non-negative for all real values of x.

The graph of the quadratic $3x^2 + 2ax + 1$ is an upward-opening parabola (since the leading coefficient, 3, is positive). For the parabola to be always above or on the x-axis, it must have either one real root or no real roots. This condition is met when the discriminant of the quadratic formula is ≤ 0.

We set the discriminant to be ≤ 0:

$$\Delta \leq 0 \Rightarrow (2a)^2 - 4(3)(1) \leq 0 \Rightarrow 4a^2 - 12 \leq 0 \Rightarrow 4a^2 \leq 12 \Rightarrow a^2 \leq 3 \Rightarrow -\sqrt{3} \leq a \leq \sqrt{3}$$

Therefore, the function $f(x) = x^3 + ax^2 + x$ is always increasing for values of a in the closed interval $\left[-\sqrt{3}, \sqrt{3}\right]$.

Example 8.10 Determine the interval for m such that the function $f(x) = x^3 - (m + 2)x^2 + 3x$ is always increasing.

Solution The function $f(x) = x^3 - (m + 2)x^2 + 3x$ is always increasing if its first derivative, $f'(x)$, is non-negative for all real numbers x, i.e., $f'(x) \geq 0$.

Since $f(x)$ is a cubic polynomial, its derivative $f'(x)$ will be a quadratic function. For a quadratic function to be always non-negative, two conditions must be met:

1. The quadratic must have its leading coefficient (the coefficient of the x^2 term) be positive.
2. The quadratic must have at most one real root, meaning its discriminant (Δ) must be ≤ 0 ($\Delta \leq 0$).

$$f(x)=x^3-(m+2)x^2+3x \Rightarrow f'(x)=3x^2-2(m+2)x+3 \Rightarrow \begin{cases} 3>0 \\ \Delta=4(m+2)^2-4(3)(3)\leq 0 \end{cases}$$

$$\Delta\leq 0 \Rightarrow 4(m+2)^2-4(3)(3)\leq 0 \Rightarrow (m+2)^2\leq 9 \Rightarrow -3\leq m+2\leq 3 \Rightarrow -5\leq m\leq 1$$

Therefore, the function $f(x)$ is always increasing when the parameter m is in the interval: $m \in [-5, 1]$.

Example 8.11 Determine the interval(s) for which the function $f(x) = \frac{x-1}{4-x^2}$ is increasing.

Solution To determine where the function $f(x) = \frac{x-1}{4-x^2}$ is increasing, we need to find the interval(s) where its first derivative, $f'(x)$, is positive. The function is a rational expression, so it is undefined where the denominator is zero:

$$4-x^2=0 \Rightarrow x^2=4 \Rightarrow x=\pm 2$$

The domain of the function is all real numbers except $x = 2$ and $x = -2$.

$$f(x)=\frac{x-1}{4-x^2} \Rightarrow f'(x)=\frac{1\times(4-x^2)-(-2x)\times(x-1)}{(4-x^2)^2} \Rightarrow f'(x)$$
$$=\frac{4-x^2+2x^2-2x}{(4-x^2)^2} \Rightarrow f'(x)=\frac{x^2-2x+4}{(4-x^2)^2}$$

We need to determine where $f'(x) > 0$. We can analyze the signs of the numerator and the denominator separately:

1. The numerator is a quadratic expression, $x^2 - 2x + 4$. To determine its sign, we can check its discriminant: $\Delta = (-2)^2 - 4(1)(4) \Rightarrow \Delta = 4 - 16 = -12 < 0$. Since the discriminant is negative ($\Delta < 0$) and the leading coefficient (the

coefficient of x^2) is positive, the quadratic $x^2 - 2x + 4$ is always positive for all real values of x.

2. The denominator is $(4 - x^2)^2$. Since this term is squared, it is always positive for all values in the function's domain (it is only zero at the excluded values $x = \pm 2$).

Since the numerator is always positive and the denominator is always positive, the entire derivative, $f'(x)$, is always positive for all values in the domain of $f(x)$:

$$f'(x) = \frac{\text{positive}}{\text{positive}} > 0$$

Therefore, the function is increasing on every interval in its domain: $(-\infty, -2) \cup (-2, 2) \cup (2, \infty)$.

Example 8.12 Determine the critical point(s) of the function $f(x) = x^2(x - 3)^2$.

Solution To determine the critical points of a function, we must find the values of x where the first derivative, $f'(x)$, is either equal to zero or is undefined.

$$f(x)=x^2(x-3)^2 \Rightarrow f'(x)=2x(x-3)^2+2x^2(x-3) \Rightarrow f'(x)=(x-3)\left(2x(x-3)+2x^2\right)$$
$$\Rightarrow f'(x)=(x-3)\left(4x^2-6x\right) \Rightarrow f'(x)=x(x-3)(4x-6)$$

$$f'(x)=0 \Rightarrow x(x-3)(4x-6)=0 \Rightarrow \begin{cases} x=0 \\ x=3 \\ x=\dfrac{6}{4}=\dfrac{3}{2} \end{cases}$$

The critical points are $x = 0$, $x = 3$, and $x = \frac{3}{2}$.

Example 8.13 Determine the critical point(s) of the function $f(x) = |x^3 - x|$.

Solution We must find all values of x where the derivative is either zero or undefined. For $f(x) = |x^3 - x|$, the derivative will be undefined when the expression inside the absolute value is zero, as this indicates a sharp corner:

$$x^3-x=0 \Rightarrow x\left(x^2-1\right)=0 \Rightarrow x(x+1)(x-1)=0 \Rightarrow \begin{cases} x=0 \\ x=-1 \\ x=1 \end{cases}$$

These are three of the critical points.

$$f'(x)=\left(\left|x^3-x\right|\right)'=\begin{cases} 3x^2-1 & x^3-x>0 \\ -\left(3x^2-1\right) & x^3-x<0 \end{cases}$$

$$f'(x)=0 \Rightarrow 3x^2-1=0 \Rightarrow x^2=\frac{1}{3} \Rightarrow x=\pm\frac{1}{\sqrt{3}}$$

Combining the results from both steps, the critical points of the function $f(x) = |x^3 - x|$ are:

$$x=0, x=-1, x=1, x=\frac{1}{\sqrt{3}}, x=-\frac{1}{\sqrt{3}}$$

Exercise 8.1 Determine whether the function $f(x)=\frac{\sqrt{x^2+9}}{x}$ has any critical points, and if so, find them.

Example 8.14 Determine the critical point(s) of the function $f(x)=|x-3|\sqrt[3]{x^2}$.

Solution To find the critical points of a function, we must identify all values in its domain where the first derivative is either zero or undefined.

The function is a product of two terms, $f(x)=|x-3|\sqrt[3]{x^2}$, which can be written as $f(x)=|x-3|x^{\frac{2}{3}}$. The domain of this function is all real numbers.

We need to consider two cases to find the derivative, based on the sign of $(x-3)$.

Case 1

$$x-3>0 \Rightarrow f(x)=|x-3|x^{\frac{2}{3}}=(x-3)x^{\frac{2}{3}} \Rightarrow f(x)=x^{\frac{5}{3}}-3x^{\frac{2}{3}} \Rightarrow f'(x)=\frac{5}{3}x^{\frac{2}{3}}-2x^{-\frac{1}{3}}$$

$$f'(x)=0 \Rightarrow \frac{5}{3}x^{\frac{2}{3}}-2x^{-\frac{1}{3}}=0 \Rightarrow \frac{5}{3}x^{\frac{2}{3}}=2x^{-\frac{1}{3}} \Rightarrow \frac{5}{3}x=2 \Rightarrow x=\frac{6}{5}$$

Since $\frac{6}{5}<3$, this solution is not in the interval $x>3$. Therefore, there are no critical points in this case.

Case 2

$$x-3<0 \Rightarrow f(x)=|x-3|x^{\frac{2}{3}}=-(x-3)x^{\frac{2}{3}} \Rightarrow f(x)=-x^{\frac{5}{3}}+3x^{\frac{2}{3}} \Rightarrow f'(x)=-\frac{5}{3}x^{\frac{2}{3}}+2x^{-\frac{1}{3}}$$

$$f'(x)=0 \Rightarrow -\frac{5}{3}x^{\frac{2}{3}}+2x^{-\frac{1}{3}}=0 \Rightarrow -\frac{5}{3}x^{\frac{2}{3}}=-2x^{-\frac{1}{3}} \Rightarrow \frac{5}{3}x=2 \Rightarrow x=\frac{6}{5}$$

Since $\frac{6}{5}<3$, this solution is in the interval $x<3$. Therefore, $x=\frac{6}{5}$ is a critical point.

Let's determine the points where the derivative is undefined. In this function, this occurs at two points:

1. The term $|x-3|$ has a sharp corner where its argument is zero, at $x=3$.
2. The term $\sqrt[3]{x^2}=\frac{x^2}{3}$ has a vertical tangent where its derivative, $\frac{2}{3}x^{-\frac{1}{3}}$, is undefined, at $x=0$.

Therefore, $x=0$ and $x=3$ are also critical points.

Combining the results, the function has three critical points: $x = \frac{6}{5}, x = 0$, and $x = 3$.

Example 8.15 Find the maximum value of the function $f(x) = x^3 - 3x^2 - 9x + 5$ on interval $[-2, 2]$.

Solution To find the maximum value of a function on a closed interval, we must evaluate the function at its critical points within the interval and at the endpoints of the interval.

$$f(x) = x^3 - 3x^2 - 9x + 5 \Rightarrow f'(x) = 3x^2 - 6x - 9 \Rightarrow f'(x) = 3\left(x^2 - 2x - 3\right) \Rightarrow f'(x) = 3(x + 1)(x - 3)$$

$$f'(x) = 0 \Rightarrow 3(x + 1)(x - 3) = 0 \Rightarrow \begin{cases} x = -1 \\ x = 3 \end{cases}$$

The critical points are $x = 3$ and $x = -1$.The given interval is $[-2, 2]$. We only need to consider the critical points that lie within this interval, and the endpoints.

The critical point $x = 3$ is outside the interval, so we can ignore it.

The critical point $x = -1$ is within the interval.

The endpoints are $x = -2$ and $x = 2$.

Now, we evaluate the function at these three points:

$$f(x) = x^3 - 3x^2 - 9x + 5 \overset{x=-1}{\Rightarrow} f(-1) = (-1)^3 - 3(-1)^2 - 9(-1) + 5 \Rightarrow f(-1) = 10$$

$$f(x) = x^3 - 3x^2 - 9x + 5 \overset{x=-2}{\Rightarrow} f(-2) = (-2)^3 - 3(-2)^2 - 9(-2) + 5 \Rightarrow f(-2) = 3$$

$$f(x) = x^3 - 3x^2 - 9x + 5 \overset{x=2}{\Rightarrow} f(2) = (2)^3 - 3(2)^2 - 9(2) + 5 \Rightarrow f(2) = -17$$

By comparing the function values at the critical point and the endpoints, we find the maximum value. The maximum value of the function on the interval $[-2, 2]$ is 10, which occurs at $x = -1$.

The graph of $f(x) = x^3 - 3x^2 - 9x + 5$ is shown in Fig. 8.17.

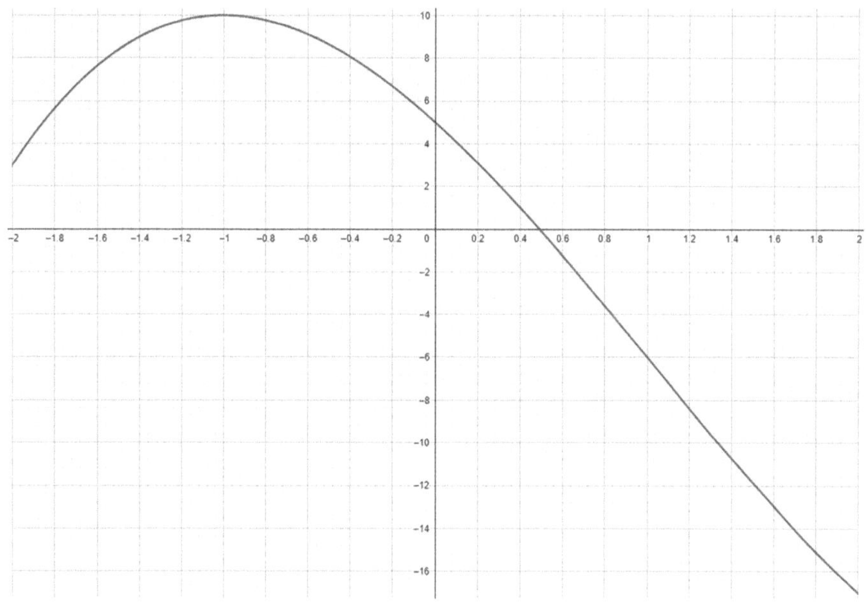

Fig. 8.17 Plot of $f(x) = x^3 - 3x^2 - 9x + 5$

Exercise 8.2 Find the minimum value of the function $f(x) = \frac{1}{4}x^4 - x^3 - x^2 + 2$. (Ans. $f\left(\frac{3+\sqrt{17}}{2}\right) = \frac{-55-17\sqrt{17}}{8} = -15.6366$).

Example 8.16 For what value of k are the maximum and minimum values of the function $f(x) = x^3 - 3x^2 + k$ on the interval $[1, 3]$ opposites of each other?

Solution The maximum and minimum values of a continuous function on a closed interval must occur at the critical points within the interval or at the endpoints.

First, we find the derivative of the function $f(x) = x^3 - 3x^2 + k$ and set it to zero:

$$f(x) = x^3 - 3x^2 + k \Rightarrow f'(x) = 3x^2 - 6x \Rightarrow f'(x) = 3x(x-2)$$

$$f'(x) = 0 \Rightarrow 3x(x-2) = 0 \Rightarrow \begin{cases} x = 0 \\ x = 2 \end{cases}$$

The critical points are $x = 0$ and $x = 2$.

The given interval is $[1, 3]$. The critical point $x = 0$ is outside this interval, so we only need to evaluate the function at the critical point $x = 2$ and the endpoints $x = 1$ and $x = 3$.

$$f(x) = x^3 - 3x^2 + k \overset{x=1}{\Rightarrow} f(1) = 1^3 - 3(1)^2 + k \Rightarrow f(1) = k - 2$$

$$f(x)=x^3-3x^2+k\overset{x=2}{\Rightarrow}f(2)=2^3-3(2)^2+k\Rightarrow f(2)=k-4$$

$$f(x)=x^3-3x^2+k\overset{x=3}{\Rightarrow}f(3)=3^3-3(3)^2+k\Rightarrow f(3)=k$$

By comparing the three values, we find the maximum and minimum: Maximum value = k and minimum value = $k-4$.

The problem states that the maximum and minimum values are opposites of each other.

$$y_{\max}=-y_{\min}\Rightarrow k=-(k-4)\Rightarrow 2k=4\Rightarrow k=2$$

The value of k for which the maximum and minimum values are opposites is 2.

The graph of $f(x)=x^3-3x^2+k$ for $k=2$ is shown in Fig. 8.18. As the figure shows, the maximum value is 2, while the minimum value is -2.

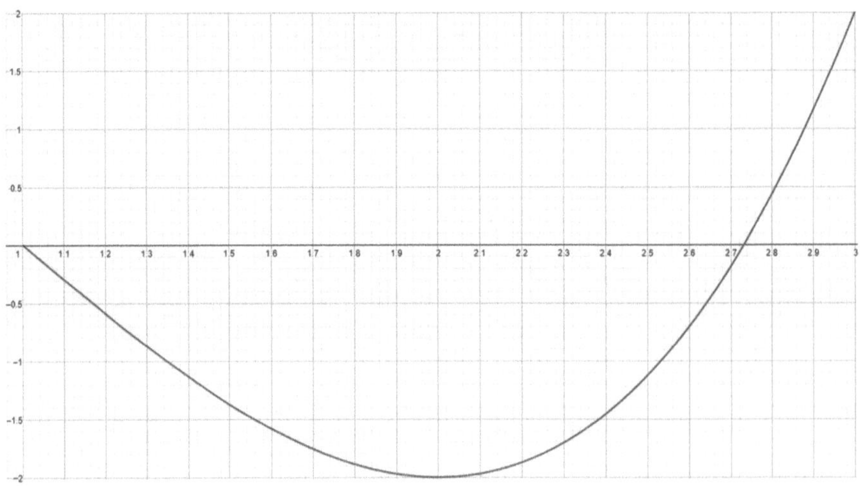

Fig. 8.18 Plot of $f(x)=x^3-3x^2+2$

Example 8.17 A wire of length 16 m is to be bent into the shape of a rectangle. Determine the dimensions of the rectangle that will maximize its area.

Solution Let the length of the rectangle be l and the width be w. The perimeter of the rectangle is equal to the length of the wire, which is 16 m: $2l+2w=16\Rightarrow l+w=8\Rightarrow w=8-l$.

The area of the rectangle, A, is given by the formula: $A=l\times w=l\times(8-l)=-l^2+8l$.

To find the maximum area, we need to find the critical points of the area function, by taking its derivative with respect to l and setting it equal to zero:

$$A = A(l) = -l^2 + 8l \Rightarrow \frac{dA(l)}{dl} = -2l + 8$$

$$\frac{dA(l)}{dl} = 0 \Rightarrow -2l + 8 = 0 \Rightarrow l = 4$$

To confirm that this value of l corresponds to a maximum, we can use the second derivative test:

$$A(l) = -l^2 + 8l \Rightarrow \frac{dA(l)}{dl} = -2l + 8 \Rightarrow \frac{d^2A(l)}{dl^2} = -2$$

Since $A''(4) = -2 < 0$, the area is indeed a maximum at $l = 4$. Now, we can find the corresponding width using the perimeter equation: $w = 8 - l = 8 - 4 = 4$.

Therefore, the dimensions that maximize the area are a length of 4 m and a width of 4 m. This means the rectangle with the maximum area is a square.

Example 8.18 A box with an open top is to be constructed from a 12-inch by 12-inch piece of cardboard by cutting out equal squares from each corner and folding up the sides (Fig. 8.19). What size should the squares be to maximize the volume of the box?

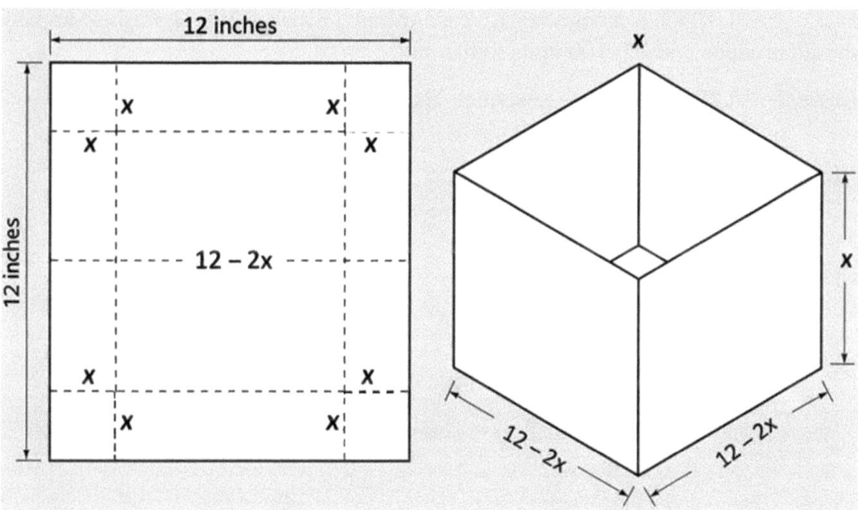

Fig. 8.19 Figure for Example 8.18

Solution The volume, V, is a function of x: $V(x) = x(12 - 2x)^2$.

To find the maximum volume, we take the derivative of $V(x)$ and set it to zero. Using the product rule:

$$V(x) = x(12-2x)^2 \Rightarrow V'(x) = 1(12-2x)^2 + x \times 2(12-2x) \times -2 \Rightarrow V'(x)$$
$$= (12-2x)(12-2x-4x) \Rightarrow V'(x) = (12-2x)(12-6x)$$

$$V'(x) = 0 \Rightarrow (12-2x)(12-6x) = 0 \Rightarrow \begin{cases} x = 6 \\ x = 2 \end{cases}$$

The value $x = 6$ would result in a base of size 0×0, so the volume would be 0, which is a minimum.

The value $x = 2$ gives the maximum volume.

The size of the squares cut from the corners should be 2 inches by 2 inches.

Example 8.19 A company sells x units of a product. The price per unit is given by the demand function $p(x) = 500 - 2x$. The cost to produce x units is $C(x) = 100x + 1000$. How many units should the company sell to maximize its profit?

Solution

$$\text{Revenue function}: R(x) = \text{price} \times \text{quantity} = p(x) \times x.$$

$$\text{Objective (profit)}: P(x) = R(x) - C(x) = -2x^2 + 400x - 1000.$$

$$P(x) = -2x^2 + 400x - 1000 \Rightarrow P'(x) = -4x + 400$$

$$P'(x) = 0 \Rightarrow -4x + 400 = 0 \Rightarrow 4x = 400 \Rightarrow x = 100$$

$P''(x) = -4$, which is always negative, confirming a maximum. Thus, the company should produce and sell 100 units to maximize profit.

Example 8.20 Find the absolute maximum value of the function $f(x) = \sin(2x) + \cos(x)$ on the interval $[0, \pi]$.

Solution

$$f(x) = \sin(2x) + \cos(x) \Rightarrow f'(x) = 2\cos(2x) - \sin(x)$$

Let's simplify this expression:

$$f'(x) = 2\cos(2x) - \sin(x) \Rightarrow f'(x) = 2\left(\cos^2(x) - \sin^2(x)\right) - \sin(x)$$
$$\Rightarrow f'(x) = 2\left(1 - \sin^2(x) - \sin^2(x)\right) - \sin(x)$$
$$\Rightarrow f'(x) = 2\left(1 - 2\sin^2(x)\right) - \sin(x) \Rightarrow f'(x) = -4\sin^2(x) - \sin(x) + 2$$

$$f'(x) = 0 \Rightarrow -4\sin^2(x) - \sin(x) + 2 = 0 \xrightarrow{u = \sin(x)} -4u^2 - u + 2$$

$$= 0 \Rightarrow u_{1,2} = \frac{1 \pm \sqrt{(-1)^2 - 4(-4)(2)}}{-8} \Rightarrow u_{1,2} = \frac{1 \pm \sqrt{33}}{-8} \Rightarrow \begin{cases} u_1 \approx 0.593 \\ u_2 \approx -0.843 \end{cases}$$

We find the values of x in the interval $[0, \pi]$ that satisfy these conditions:

$\sin(x) \approx 0.593 \Rightarrow x \approx 0.636$ rad.

$\sin(x) \approx -0.843 \Rightarrow$ No solution in the interval $[0, \pi]$ since $\sin(x)$ is non-negative on this interval.

The only critical point in the interval is $x \approx 0.636$. We evaluate $f(x)$ at the critical point and the endpoints of the interval, $x = 0$, $x \approx 0.636$, and $x = \pi$:

$$f(0) = \sin(2(0)) + \cos(0) = 0 + 1 = 1$$

$$f(\pi) = \sin(2\pi) + \cos(\pi) = 0 + (-1) = -1$$

$$f(0.636) \approx \sin(2(0.636)) + \cos(0.636) \approx \sin(1.272) + \cos(0.636) \approx 0.956 + 0.806 = 1.762$$

Comparing the values, the absolute maximum value is the largest value we found. Thus, the absolute maximum value is approximately 1.762, occurring at $x \approx 0.636$.

Figure 8.20 shows the graph of $f(x) = \sin(2x) + \cos(x)$ on $[0, \pi]$.

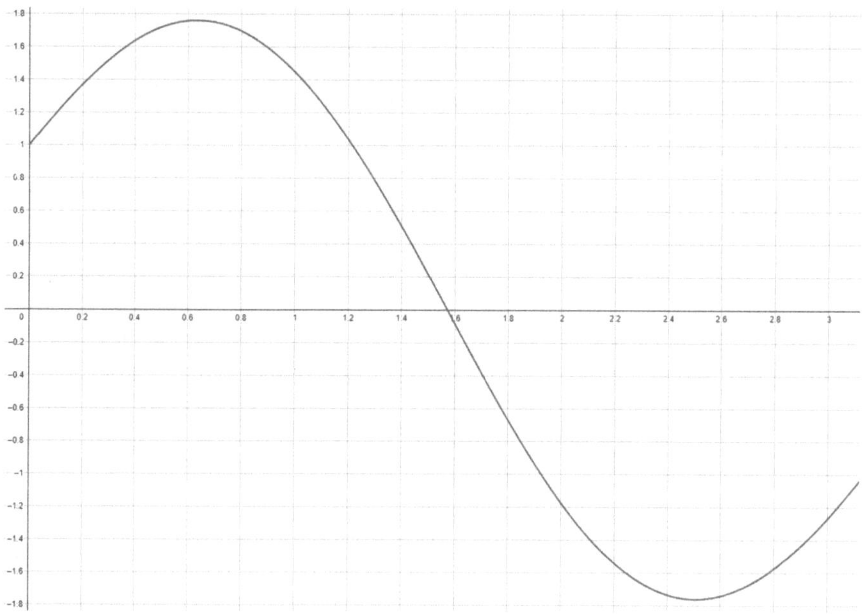

Fig. 8.20 Plot of $f(x) = \sin(2x) + \cos(x)$ on $[0, \pi]$

Example 8.21 Find the absolute maximum and minimum values of the function $f(x) = \frac{x^2-1}{x^2+1}$ on the interval $[-2, 2]$.

Solution

$$f(x) = \frac{x^2-1}{x^2+1} \Rightarrow f'(x) = \frac{2x(x^2+1) - 2x(x^2-1)}{(x^2+1)^2} = \frac{4x}{(x^2+1)^2}$$

The critical points occur where the derivative is zero or undefined.

$f'(x) = 0$ when the numerator is zero: $4x = 0 \Rightarrow x = 0$.

$f'(x)$ is never undefined, as the denominator $(x^2 + 1)^2$ is always positive.

So, the only critical point is $x = 0$. To find the absolute maximum and minimum values on the closed interval $[-2, 2)$, we must evaluate the function at the critical point and at the endpoints of the interval:

$$f(-2) = \frac{(-2)^2 - 1}{(-2)^2 + 1} = \frac{3}{5} = 0.6$$

$$f(0) = \frac{0^2 - 1}{0^2 + 1} = -1$$

$$f(2) = \frac{2^2 - 1}{2^2 + 1} = \frac{3}{5} = 0.6$$

By comparing the values found, we can determine the absolute maximum and minimum: Maximum value: $\frac{3}{5}$ and minimum value: -1. The absolute maximum value is $\frac{3}{5}$, and the absolute minimum value is -1.

Figure 8.21 shows the graph of $f(x) = \frac{x^2-1}{x^2+1}$ on $[-2, 2]$.

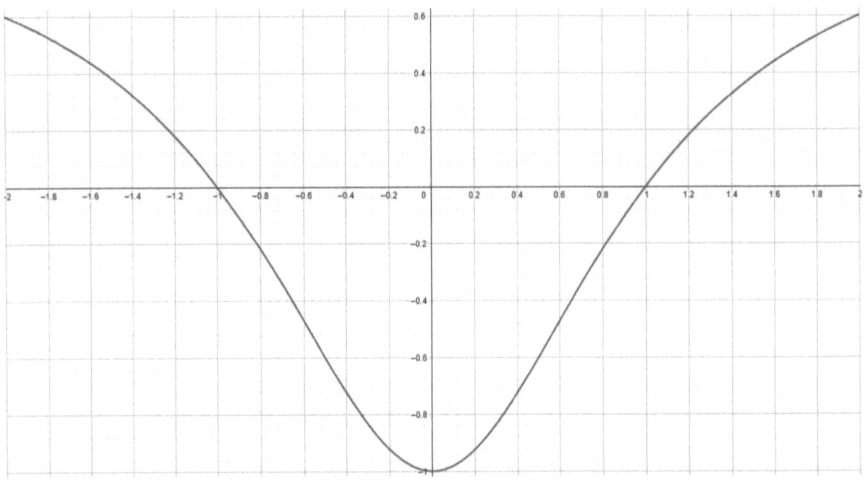

Fig. 8.21 Plot of $f(x) = \frac{x^2-1}{x^2+1}$ on $[-2, 2]$

Example 8.22 A car travels 180 miles in 3 h. At some point during the trip, what must the car's instantaneous speed have been?

Solution The average speed of the car is the total distance divided by the total time: $\frac{180}{3} = 60$ miles per hour.

According to the Mean Value Theorem, since the position function of a car is continuous and differentiable (assuming no teleportation or instantaneous changes in speed), there must exist a time c between 0 and 3 h such that the instantaneous speed, is equal to the average speed.

Example 8.23 Given the function $f(x) = \sqrt{x-1}$ on the interval $[2, 5]$, find the value (s) of c that satisfy the Mean Value Theorem.

Solution First, we need to verify that the conditions for the MVT are met. The function $f(x) = x - 1$ is continuous on $[2, 5]$ and differentiable on $(2, 5)$.

Next, we calculate the average rate of change:

$$f(a) = f(2) = \sqrt{2-1} = \sqrt{1} = 1$$
$$f(b) = f(5) = \sqrt{5-1} = \sqrt{4} = 2$$

Average rate of change $= \frac{f(5)-f(2)}{5-2} = \frac{2-1}{3} = \frac{1}{3.}$

Now, we find the derivative of the function, $f'(x)$:

$$f'(x) = \frac{1}{2\sqrt{x-1}}$$

Finally, we set the derivative equal to the average rate of change to find c:

$$f'(c) = \frac{1}{2\sqrt{c-1}} = \frac{1}{3}$$

Solving for c:

$$\frac{1}{2\sqrt{c-1}} = \frac{1}{3} \Rightarrow \frac{1}{4(c-1)} = \frac{1}{9} \Rightarrow 4(c-1) = 9 \Rightarrow 4c = 13 \Rightarrow c = \frac{13}{4}$$

Since $\frac{13}{4} = 3.25$ which is within the interval $(2, 5)$, the value of c is $\frac{13}{4}$.

Example 8.24 Verify that the function $f(x) = x^2 - 4x - 3$ satisfies the Mean Value Theorem on the interval $[1, 4]$. If it does, find the value of c that is guaranteed by the theorem.

Solution First, we must check if the conditions of the MVT are met. Since $f(x) = x^2 - 4x - 3$ is a polynomial, it is continuous on the closed interval $[1, 4]$ and differentiable on the open interval $(1, 4)$. The conditions are met.

Next, we calculate the average rate of change:

$$f(a) = f(1) = 1^2 - 4 \times 1 - 3 = 1 - 4 - 3 = -6$$
$$f(b) = f(4) = 4^2 - 4 \times 4 - 3 = 16 - 16 - 3 = -3$$
$$\text{Average rate of change} = \frac{f(4) - f(1)}{4-1} = \frac{-3-(-6)}{3} = 1.$$

Now, we find the derivative of the function: $f'(x) = 2x - 4$.

Finally, we set the derivative equal to the average rate of change and solve for c:

$$f'(c) = 1 \Rightarrow 2c - 4 = 1 \Rightarrow 2c = 5 \Rightarrow c = 2.5$$

Since $c = 2.5$ is in the interval $(1, 4)$, the Mean Value Theorem is satisfied.

Example 8.25 Find the extrema of $y = \tan(x) + 3\cot(x)$.

Solution Extrema is a general term for maximum and minimum values. First, we find the derivative of the function with respect to x: $y' = \sec^2(x) - 3\mathrm{cosec}^2(x)$. To find the critical points, we set the derivative equal to zero and solve for x:

$$y' = 0 \Rightarrow \sec^2(x) - 3\mathrm{cosec}^2(x) = 0 \Rightarrow \sec^2(x) = 3\mathrm{cosec}^2(x) \Rightarrow \frac{1}{\cos^2(x)}$$
$$= \frac{3}{\sin^2(x)} \Rightarrow \frac{\sin^2(x)}{\cos^2(x)} = 3 \Rightarrow \tan^2(x) = 3 \Rightarrow \tan(x) = \pm\sqrt{3}$$

This gives us two sets of critical points for x (k is an integer $k \in \mathbb{Z}$):

$$\tan(x) = \sqrt{3} \Rightarrow \tan(x) = \tan\left(\frac{\pi}{3}\right) \Rightarrow x = \frac{\pi}{3} + k\pi$$
$$\tan(x) = -\sqrt{3} \Rightarrow \tan(x) = \tan\left(-\frac{\pi}{3}\right) \Rightarrow x = -\frac{\pi}{3} + k\pi$$

We only need to evaluate the function at these two distinct values within a single period, for example, at $x = \frac{\pi}{3}$ and $x = -\frac{\pi}{3}$. We use the second derivative test to determine the type of these extrema. Recall that:

1. If the second derivative is positive at a critical point, the function is concave up, and the point is a local minimum.
2. If the second derivative is negative at a critical point, the function is concave down, and the point is a local maximum.
3. If the second derivative is zero, the test is inconclusive, and you must use another method (like the first derivative test, see next example) to determine the nature of the extremum.

$$y = \tan(x) + 3\cot(x) \Rightarrow y' = \sec^2(x) - 3\mathrm{cosec}^2(x) \Rightarrow y' = 1 + \tan^2(x)$$
$$-3\left(1 + \cot^2(x)\right) \Rightarrow y' = \tan^2(x) - 3\cot^2(x) - 2 \Rightarrow y'' = 2\tan(x)\left(1 + \tan^2(x)\right)$$
$$+ 3 \times 2\cot(x)\left(1 + \cot^2(x)\right) \Rightarrow y'' = 2\tan(x)\left(1 + \tan^2(x)\right) + 6\cot(x)$$
$$\times \left(1 + \cot^2(x)\right)$$

$$y''\left(\frac{\pi}{3}\right) = 2\tan\left(\frac{\pi}{3}\right)\left(1 + \tan^2\left(\frac{\pi}{3}\right)\right) + 6\cot\left(\frac{\pi}{3}\right)\left(1 + \cot^2\left(\frac{\pi}{3}\right)\right)$$
$$= 2\sqrt{3}(1 + 3) + 6 \times \frac{\sqrt{3}}{3}\left(1 + \frac{3}{9}\right) = 8\sqrt{3} + \frac{8\sqrt{3}}{3} > 0$$

The second derivative at $\frac{\pi}{3}$ is positive. According to the second derivative test, the function has a local minimum at these points. The minimum value is $y\left(\frac{\pi}{3}\right) = \tan\left(\frac{\pi}{3}\right) + 3\cot\left(\frac{\pi}{3}\right) = 2\sqrt{3} = 3.4641$.

$$y''\left(-\frac{\pi}{3}\right) = 2\tan\left(-\frac{\pi}{3}\right)\left(1 + \tan^2\left(-\frac{\pi}{3}\right)\right) + 6\cot\left(-\frac{\pi}{3}\right)\left(1 + \cot^2\left(-\frac{\pi}{3}\right)\right)$$
$$= 2\times -\sqrt{3}(1+3) + 6\times -\frac{\sqrt{3}}{3}\left(1+\frac{3}{9}\right) = -8\sqrt{3} - \frac{8\sqrt{3}}{3} < 0$$

The second derivative at $-\frac{\pi}{3}$ is negative. According to the second derivative test, the function has a local maximum at these points. The maximum value is $y\left(-\frac{\pi}{3}\right) = \tan\left(-\frac{\pi}{3}\right) + 3\cot\left(-\frac{\pi}{3}\right) = -2\sqrt{3} = -3.4641$.

Figure 8.22 shows the graph of $f(x) = \tan(x) + 3\cot(x)$ on $\left[-\frac{\pi}{2}, \frac{\pi}{2}\right]$.

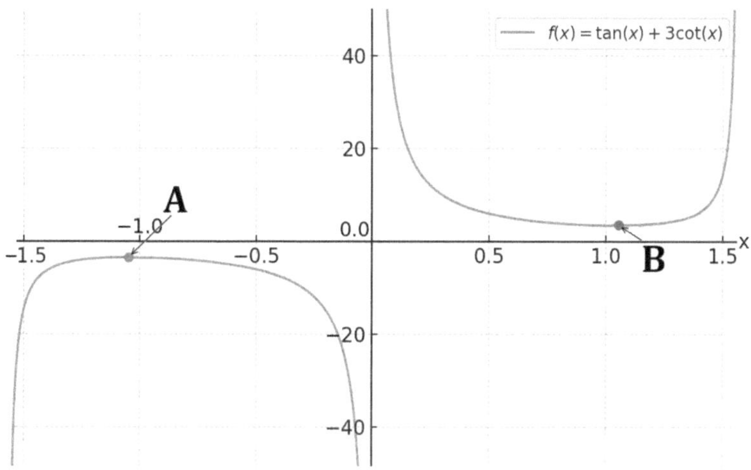

Fig. 8.22 Point *A* is a local maximum, whereas point *B* is a local minimum

Example 8.26 Determine the extremum of the function $y = -x^4 + 6x^3 - 7$.

Solution

$$y = -x^4 + 6x^3 - 7 \Rightarrow y' = -4x^3 + 18x^2$$

$$y' = 0 \Rightarrow -4x^3 + 18x^2 = 0 \Rightarrow 2x^2(-2x+9) = 0 \Rightarrow \begin{cases} x_1 = 0 \\ x_2 = 0 \\ x_3 = \dfrac{9}{2} \end{cases}$$

Figure 8.23 presents the sign table for y'.

Fig. 8.23 Sign table for y'

x	$-\infty$		0		$\frac{9}{2}$		$+\infty$
$y' = -4x^3 + 18x^2$		$+$	0	$+$	0	$-$	

Figure 8.24 illustrates the function's increasing and decreasing behavior. $x = \frac{9}{2}$ is the point where the function attains its local maximum. $y = -x^4 + 6x^3 - 7 \Rightarrow y\left(\frac{9}{2}\right) = 129.6875$.

Fig. 8.24 The function's behavior

x	$-\infty$		0		$\frac{9}{2}$		$+\infty$
y		↗	0	↗	0	↘	

Figure 8.25 shows the graph of $y = -x^4 + 6x^3 - 7$ on $[-1, 5]$.

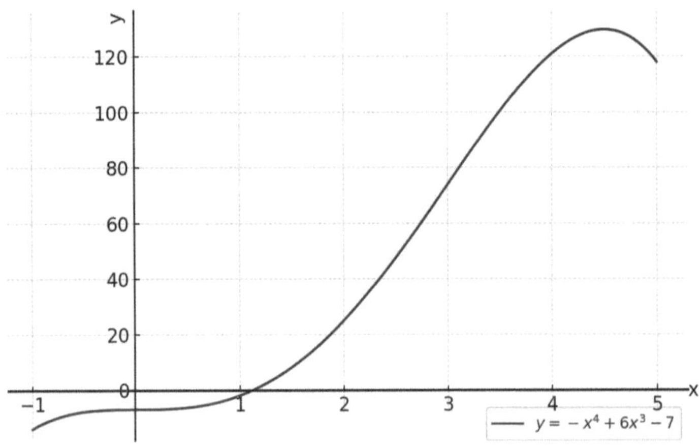

Fig. 8.25 Plot of $y = -x^4 + 6x^3 - 7$ on $[-1, 5]$

Example 8.27 Find the minimum value of $y = x\sin(x) + \cos(x)$ on the interval $[0, 2\pi]$.

Solution

$$y = x\sin(x) + \cos(x) \Rightarrow y' = \sin(x) + x\cos(x) - \sin(x) \Rightarrow y' = x\cos(x)$$

$$y' = 0 \Rightarrow x\cos(x) = 0 \Rightarrow \begin{cases} x = 0 \\ \cos(x) = 0 \Rightarrow x = \frac{\pi}{2}, \frac{3\pi}{2} \end{cases}$$

The minimum value must occur at one of the critical points or at the endpoints of the interval, $x = 0$ and $x = 2\pi$:

$$y = x\sin(x) + \cos(x) \overset{x=0}{\Rightarrow} y(0) = 1$$

$$y = x\sin(x) + \cos(x) \overset{x=\frac{\pi}{2}}{\Rightarrow} y\left(\frac{\pi}{2}\right) = \frac{\pi}{2} \approx 1.571$$

$$y = x\sin(x) + \cos(x) \overset{x=\frac{3\pi}{2}}{\Rightarrow} y\left(\frac{3\pi}{2}\right) = \frac{-3\pi}{2} \approx -4.712$$

$$y = x\sin(x) + \cos(x) \overset{x=2\pi}{\Rightarrow} y(2\pi) = 1$$

The smallest value is approximately -4.712. Therefore, the minimum value of the function on the interval $[0, 2\pi]$ is $-\frac{3\pi}{2}$ occurring at $x = \frac{3\pi}{2}$.

Example 8.28 The function $f(x) = \frac{a}{x} + bx^2$ has a relative extremum at the point $(1, -3)$. Determine the value of a and b.

Solution

$$f(x) = \frac{a}{x} + bx^2 \Rightarrow f(1) = -3 \Rightarrow f(1) = \frac{a}{1} + b(1)^2 \Rightarrow a + b = -3$$

$$f(x) = \frac{a}{x} + bx^2 \Rightarrow f'(x) = -\frac{a}{x^2} + 2bx \Rightarrow f'(1) = -\frac{a}{1^2} + 2b \times 1 \Rightarrow -a + 2b = 0$$

$$\begin{cases} a + b = -3 \\ -a + 2b = 0 \end{cases} \Rightarrow a + b - a + 2b = -3 + 0 \Rightarrow 3b = -3 \Rightarrow b = -1$$

$$a + b = -3 \Rightarrow a - 1 = -3 \Rightarrow a = -2$$

Example 8.29 The function $f(x) = \frac{a}{x} + bx^2$ has a relative extremum at the point $(1, -3)$. Determine the value of a and b.

Solution Let $f(x)$ and $g(x)$ be two functions differentiable at a point c, with $g(c) \neq 0$ and $g'(c) \neq 0$. If the quotient function $h(x) = \frac{f(x)}{g(x)}$ has a relative extremum (maximum or minimum) at $x = c$, then $\frac{f(c)}{g(c)} = \frac{f'(c)}{g'(c)}$.

$$f(x) = \frac{a}{x} + bx^2 \Rightarrow f(x) = \frac{a + bx^3}{x} \Rightarrow f(1) = \frac{a + b \times 1^3}{1} \Rightarrow -3 = a + b \Rightarrow a + b = -3$$

$$f(x) = \frac{a + bx^3}{x} \Rightarrow f(1) = \frac{3b \times 1^?}{1} \Rightarrow -3 = 3b \Rightarrow b = -1$$

$$a + b = -3 \Rightarrow a - 1 = -3 \Rightarrow a = -2$$

The results obtained are consistent with those of the previous example.

Exercise 8.3 The function $f(x) =$ sin $(x) + ax^2 + b$ has a relative extremum at the point $(\pi, 4)$. Determine the values of a and b. (Ans. $a = \frac{1}{2\pi}, b = 4 - \frac{\pi}{2}$)

Exercise 8.4 The function $f(x) = \frac{ax+b}{x^2+1}$ has a relative extremum at the point $(1, 1)$. Determine the values of a and b. (Ans. $a = 2$, $b = 0$)

Example 8.30 Given that $(1, 3)$ is a local minimum of $y = \frac{ax^2+1}{x+b}$, find the local maximum.

Solution Since the point $(1, 3)$ is on the graph of $y = \frac{ax^2+1}{x+b}$, substituting $x = 1$ and $y = 3$ gives:

$$y = \frac{ax^2 + 1}{x + b} \overset{\substack{x=1 \\ y=3}}{\Rightarrow} 3 = \frac{a \times 1^2 + 1}{1 + b} \Rightarrow 3 = \frac{a + 1}{1 + b} \Rightarrow 3 + 3b = a + 1 \Rightarrow a - 3b = 2$$

$$y = \frac{ax^2 + 1}{x + b} \Rightarrow y' = \frac{2ax(x + b) - 1(ax^2 + 1)}{(x + b)^2} \Rightarrow y'$$
$$= \frac{2ax^2 + 2abx - ax^2 - 1}{(x + b)^2} \Rightarrow y' = \frac{ax^2 + 2abx - 1}{(x + b)^2}$$

Since the local minimum is at $x = 1$, we set $y' = 0$ when $x = 1$:

$$y'(1) = 0 \Rightarrow \frac{a \times 1^2 + 2ab - 1}{(1 + b)^2} = 0 \Rightarrow a + 2ab - 1 = 0 \Rightarrow a(1 + 2b) = 1$$

$$\begin{cases} a - 3b = 2 \\ a(1 + 2b) = 1 \end{cases} \Rightarrow \begin{cases} a = 3b + 2 \\ a(1 + 2b) = 1 \end{cases} \Rightarrow (3b + 2)(1 + 2b) = 1 \Rightarrow 6b^2 + 7b + 2$$
$$= 1 \Rightarrow 6b^2 + 7b + 1 = 0 \Rightarrow \begin{cases} b_1 = -1 \\ b_2 = -\frac{1}{6} \end{cases}$$

Case 1

$$b = b_1 = -1 \overset{a-3b=2}{\Rightarrow} a = -1 \Rightarrow y = \frac{ax^2 + 1}{x + b} = \frac{-x^2 + 1}{x - 1} = \frac{(1 + x)(1 - x)}{x - 1} =$$
$$-(1 + x)$$

This is a line, which has no local maximum or minimum. So we discard this case.

Case 2

$$b = b_2 = -\frac{1}{6} \overset{a-3b=2}{\Rightarrow} a = \frac{3}{2} \Rightarrow y = \frac{ax^2 + 1}{x + b} = \frac{\frac{3}{2}x^2 + 1}{x - \frac{1}{6}} = \frac{9x^2 + 6}{6x - 1}$$

The function is $y = \frac{9x^2+6}{6x-1}$. This must be the correct function.

$$y = \frac{9x^2+6}{6x-1} \Rightarrow y' = \frac{18x(6x-1)-6(9x^2+6)}{(6x-1)^2}$$

$$y' = 0 \Rightarrow \frac{18x(6x-1)-6(9x^2+6)}{(6x-1)^2} = 0 \Rightarrow 18x(6x-1)-6\left(9x^2+6\right)$$

$$= 0 \Rightarrow 18x(6x-1) = 6\left(9x^2+6\right) \Rightarrow 3x(6x-1) = 9x^2+6 \Rightarrow 18x^2-3x = 9x^2$$

$$+6 \Rightarrow 9x^2-3x-6=0 \Rightarrow \begin{cases} x_1 = 1 \\ x_2 = -\dfrac{6}{9} = -\dfrac{2}{3} \end{cases}$$

The critical points are $x = 1$ and $x = -\frac{2}{3}$. $x = 1$ is the local minimum (as given). $x = -\frac{2}{3}$ must be the local maximum.

Now, find the y-coordinate of the local maximum by substituting $x = -\frac{2}{3}$ into the original function:

$$y = \frac{9x^2+6}{6x-1} \overset{x=-\frac{2}{3}}{\Rightarrow} y = \frac{9\left(-\frac{2}{3}\right)^2+6}{6\left(-\frac{2}{3}\right)-1} = \frac{4+6}{-4-1} = \frac{10}{-5} = 2$$

The local maximum is at the point $\left(-\frac{2}{3}, -2\right)$.

Example 8.31 In which interval is the graph of the function $y = -x^4 + 6x^3 - 7$ increasing and concave down?

Solution

$$y = -x^4 + 6x^3 - 7$$

$$y' = -4x^3 + 18x^2$$

$$y'' = -12x^2 + 36x$$

$$y' = 0 \Rightarrow -4x^3 + 18x^2 = 0 \Rightarrow 2x^2(-2x+9) = 0 \Rightarrow \begin{cases} x_1 = 0 \\ x_2 = 0 \\ x_3 = \dfrac{9}{2} \end{cases}$$

$$y'' = 0 \Rightarrow -12x^2 + 36x = 0 \Rightarrow -12x(x-3) = 0 \Rightarrow \begin{cases} x_1 = 0 \\ x_2 = 3 \end{cases}$$

Figure 8.26 presents the sign table for y' and y''.

x	$-\infty$		0		3		$\frac{9}{2}$		$+\infty$
$y' = -4x^3 + 18x^2$		+	0	+		+	0	−	
$y'' = -12x^2 + 36x$		−	0	+	0	−		−	

Fig. 8.26 Sign table for y' and y''

Figure 8.27 illustrates the function's increasing and decreasing behavior, as well as its concavity.

Fig. 8.27 The function's behavior

x	$-\infty$		0		3		$\frac{9}{2}$		$+\infty$
y		↗	0	↗		↗	0	↘	
y		↓	0	↑	0	↓		↓	

According to Fig. 8.27, the function is increasing on the interval $\left(-\infty, \frac{9}{2}\right)$.

According to Fig. 8.27, the function is concave down on the intervals $(-\infty, 0) \cup (3, \infty)$.

We find the intersection of these two sets: $(-\infty, 0) \cup \left(3, \frac{9}{2}\right)$.

Exercise 8.5 Assume $y = x\sqrt{x^2 + 2}$ is concave up for all $x > a$. Determine the least value of a for which this holds. (Ans. $a = 0$)

Example 8.32 $y = (x + 3)\sqrt{x}$ is concave down in the interval (a, b). Determine the maximum value of $b - a$.

Solution The function is $y = (x + 3)\sqrt{x}$. Since the square root is involved, the domain of y is $x \geq 0$.

A function is concave down when its second derivative, y'', is negative ($y'' < 0$).

$$y = (x + 3)\sqrt{x}$$

$$y' = \sqrt{x} + \frac{(x + 3)}{2\sqrt{x}}$$

$$y'' = \frac{1}{2\sqrt{x}} + \frac{2\sqrt{x} - \frac{x+3}{\sqrt{x}}}{(2\sqrt{x})^2} = \frac{1}{2\sqrt{x}} + \frac{2\sqrt{x} - \frac{x+3}{\sqrt{x}}}{(2\sqrt{x})^2} = \frac{1}{2\sqrt{x}} + \frac{x - 3}{4x\sqrt{x}} = \frac{3x - 3}{4x\sqrt{x}} = \frac{3}{4}\frac{x - 1}{x\sqrt{x}}$$

We must restrict the domain to $x > 0$ for the derivatives to be defined.

$$y'' < 0 \Rightarrow \frac{3}{4}\frac{x - 1}{x\sqrt{x}} < 0 \Rightarrow \frac{x - 1}{x\sqrt{x}} < 0 \xRightarrow{x > 0 \Rightarrow x\sqrt{x} > 0} x - 1 < 0 \Rightarrow x < 1$$

Considering the domain restriction $x > 0$, the function is concave down on the interval $(0, 1)$. The largest open interval (a, b) on which the function is concave down is $(0, 1)$. Thus, $a = 0$ and $b = 1$. The maximum value of $b - a$ is $1 - 0 = 1$.

Example 8.33 Find the point(s) of inflection for the function $f(x) = x^3 - 6x^2 + 5x + 1$.

Solution

$$f(x) = x^3 - 6x^2 + 5x + 1 \Rightarrow f'(x) = 3x^2 - 12x + 5 \Rightarrow f''(x) = 6x - 12$$

$$f''(x) = 0 \Rightarrow 6x - 12 = 0 \Rightarrow x = 2$$

The sign table for $f''(x) = 6x - 12$ is shown in Fig. 8.28. Since the concavity changes at $x = 2$, it is an inflection point.

Fig. 8.28 Sign table for $f''(x)$

x	$-\infty$		2		$+\infty$
$f''(x) = 6x - 12$		$-$	0	$+$	

Example 8.34 Find the point(s) of inflection for the function $f(x) = x\sqrt{x+2}$, if any exist.

Solution The function $f(x) = x\sqrt{x+2}$ is defined when $x + 2 \geq 0$, so the domain is $x \geq -2$. For the derivatives to be defined, we need $x + 2 > 0$, so we will look for inflection points in the interval $(-2, \infty)$.

$$f(x) = x\sqrt{x+2}$$

$$f'(x) = \sqrt{x+2} + \frac{x}{2\sqrt{x+2}} = \frac{3x+4}{2\sqrt{x+2}}$$

$$f''(x) = \frac{3x+8}{4(x+2)^{\frac{3}{2}}}$$

Potential inflection points occur where $f''(x) = 0$ or where $f''(x)$ is undefined.

1. $f''(x)$ **is undefined:** The denominator is zero when $x + 2 = 0$, or $x = -2$. Since $x = -2$ is an endpoint of the domain and $f''(x)$ is not defined there, it is not an inflection point.
2. $f''(x) = 0$**:** Set the numerator to zero: $3x + 8 = 0 \Rightarrow x = -\frac{8}{3} \approx -2.67$. This value is outside the domain of the original function ($x \geq -2$).

Since $f''(x) > 0$ for all $x > -2$, the function is concave up throughout its domain. Therefore, the function has no point(s) of inflection.

Exercise 8.6 Find the point(s) of inflection for the function $f(x) = x\sqrt{x^2+2}$. (Ans. $(0, 0)$)

Example 8.35 Find the point(s) of inflection for the function $f(x) = \frac{x}{1+|x|}$.

Solution The function can be written as a piecewise function:

$$f(x) = \frac{x}{1+|x|} = \begin{cases} \frac{x}{1+x} & x \geq 0 \\ \frac{x}{1-x} & x < 0 \end{cases}$$

Case 1 For $x > 0$, the function is $f(x) = \frac{x}{1+x}$. Therefore,

$$f'(x) = \frac{1}{(1+x)^2} = (1+x)^{-2}$$

$$f''(x) = \frac{-2}{(1+x)^3} = -2(1+x)^{-3}$$

Since $x > 0$, $(1 + x)^3$ is always positive. Therefore, $f''(x)$ is always negative on the interval $(0, +\infty)$. $f''(x) < 0$ implies that the function is concave down on $(0, +\infty)$.

Case 2 For $x < 0$, the function is $f(x) = \frac{x}{1-x}$. Therefore,

$$f'(x) = \frac{1}{(1-x)^2} = (1-x)^{-2}$$

$$f''(x) = \frac{2}{(1-x)^3} = 2(1-x)^{-3}$$

Since $x < 0$, $(1 - x)^3$ is always positive. Therefore, $f''(x)$ is always positive on the interval $(-\infty, 0)$. $f''(x) > 0$ implies that the function is concave up on $(-\infty, 0)$.

An inflection point occurs where the concavity changes.

On $(-\infty, 0)$, the function is concave up ($f''(x) > 0$).

On $(0, +\infty)$, the function is concave down ($f''(x) < 0$).

The concavity changes at $x = 0$. Therefore, $x = 0$ is an inflection point, provided the function has a well-defined tangent there (i.e., $f'(0)$ exists). We check the left-hand and right-hand derivatives at $x = 0$:

From the left ($x \to 0^-$): $f'(x) = \frac{1}{(1-x)^2} \Rightarrow f'(0^-) = \frac{1}{(1-0^-)^2} = 1$.

From the left ($x \to 0^+$): $f'(x) = \frac{1}{(1+x)^2} \Rightarrow f'(0^+) = \frac{1}{(1+0^+)^2} = 1$.

Since $f'(0^-) = f'(0^+) = 1$, the first derivative exists at $x = 0$, and the slope of the tangent line is 1.

Since the concavity changes at $x = 0$ and the function is continuous and differentiable at this point, the point of inflection is at $x = 0$. $f(x) = \frac{x}{1+|x|} \Rightarrow f(0) = 0$.

The function has one point of inflection at $(0, 0)$.

Example 8.36 Find the point(s) of inflection for the function $f(x) = x^2e^x$.

Solution To find the points of inflection for the function $f(x) = x^2e^x$, we must find the second derivative, $f''(x)$, set it to zero, and determine where its sign changes. The domain of $f(x)$ is all real numbers, $\mathbb{R}$.

$$f(x) = x^2e^x \Rightarrow f'(x) = 2xe^x + x^2e^x$$

$$f''(x) = 2e^x + 2xe^x + 2xe^x + x^2e^x = 2e^x + 4xe^x + x^2e^x = e^x\left(x^2 + 4x + 2\right)$$

Potential inflection points occur where $f''(x) = 0$ or where $f''(x)$ is undefined. Since $f''(x)$ is defined for all real numbers, we only need to solve $f''(x) = 0$:

$$f''(x) = 0 \Rightarrow e^x\left(x^2 + 4x + 2\right) = 0$$

Since e^x is never zero, we only need to set the quadratic factor to zero: $x^2 + 4x + 2 = 0$. We use the quadratic formula:

$$x^2 + 4x + 2 = 0 \Rightarrow x_{1,2} = \frac{-4 \pm \sqrt{4^2 - 4(1)(2)}}{2} = -2 \pm \sqrt{2} \Rightarrow \begin{cases} x_1 = -2 - \sqrt{2} = -3.414 \\ x_2 = -2 + \sqrt{2} = -0.586 \end{cases}$$

The sign table for $f''(x) = e^x(x^2 + 4x + 2)$ is shown in Fig. 8.29. Since e^x is always positive, the sign of $f''(x)$ is determined by the quadratic factor, $x^2 + 4x + 2$, which is a parabola opening upward.

x	$-\infty$		-3.414		-0.586		$+\infty$
$f''(x) = e^x(x^2 + 4x + 2)$		$+$	|	$-$	|	$+$	

Fig. 8.29 Sign table for $f''(x)$

Figure 8.30 illustrates the function's concavity.

Fig. 8.30 The function's concavity

x	$-\infty$		-3.414		-0.586		$+\infty$
f		↑	|	↓	|	↑	

The concavity changes at both -3.414 and -0.586, so both are points of inflection. Let's substitute the x-values into the original function $f(x) = x^2e^x$:

$$f(x) = x^2e^x \Rightarrow \begin{cases} f(-3.414) = 0.384 \\ f(-0.586) = 0.191 \end{cases}$$

The points of inflection are: $(-3.414, 0.384)$ and $(-0.586, 0.191)$. Fig. 8.31 shows the graph of $f(x) = x^2e^x$.

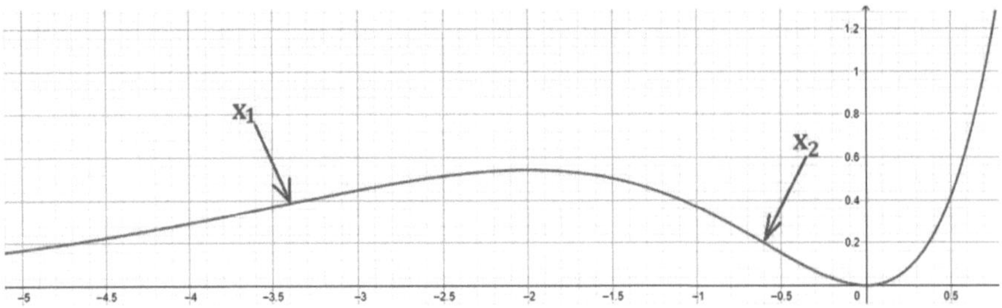

Fig. 8.31 Plot of $f(x) = x^2e^x$. x_1 and x_2 are inflection points

Example 8.37 Find the point(s) of inflection for the function $f(x) = \begin{cases} x^3 - 3x^2 & x \geq -1 \\ -13 - \frac{9}{x} & x < -1 \end{cases}$.

Solution Finding the point(s) of inflection for a piecewise function requires analyzing the second derivative, $f''(x)$, for each piece separately and also checking the potential point where the definition changes, $x = -1$.

Piece 1 $f(x) = x^3 - 3x^2$ for $x \geq -1$

$$f(x) = x^3 - 3x^2 \Rightarrow f'(x) = 3x^2 - 6x \Rightarrow f''(x) = 6x - 6$$

$$f''(x) = 0 \Rightarrow 6x - 6 = 0 \Rightarrow x = 1$$

Check for sign change around $x = 1$ (within the domain $x \geq -1$):

Test $x = 0$ (to the left of $x = 1$): $f''(0) = 6(0) - 6 = -6 < 0$ (concave down)

Test $x = 2$ (to the right of $x = 1$): $f''(2) = 6(2) - 6 = 6 > 0$ (concave up)

Since the concavity changes, $x = 1$ is an inflection point. $f(1) = -2$.

Piece 2 $f(x) = -13 - \frac{9}{x} = -13 - 9x^{-1}$ for $x < -1$

$$f(x) = -13 - 9x^{-1} \Rightarrow f'(x) = 9x^{-2} \Rightarrow f''(x) = -18x^{-3} = -\frac{18}{x^3}$$

$f''(x) = 0 \Rightarrow -\frac{18}{x^3} = 0 \Rightarrow$ This equation has no solution.

$f''(x)$ is undefined at $x = 0$. However, $x = 0$ is outside the domain of this piece $(x < -1)$.

Since $f''(x)$ is never zero and is defined everywhere in its domain $(x < -1)$, there are no inflection points from this piece. We can confirm the concavity:

For any $x < -1$, x^3 is negative, so $-\frac{18}{x^3}$ is $\frac{\text{negative}}{\text{negative}}$, which is positive ($f''(x) > 0$). This piece is always concave up.

Check the boundary point $x = -1$: For $x = -1$ to be an inflection point, the concavity must change as we cross it, and the function must be continuous at that point.

Limit from the right ($x \geq -1$): $\lim_{x \to -1^+} x^3 - 3x^2 = -4$.

Limit from the left ($x < -1$): $\lim_{x \to -1^-} -13 - \frac{9}{x} = -4$.

Since, $\lim_{x \to -1^+} f(x) = \lim_{x \to -1^-} f(x)$, the function is continuous at $x = -1$.

Let's check concavity change across $x = -1$:

Concavity to the left ($x < -1$): We found $f''(x) > 0$ (concave up).

Concavity to the right (e.g., $x = 0$): We found $f''(0) = -6 < 0$ (concave down).

The concavity changes from up (left) to down (right) across $x = -1$. Therefore, $x = -1$ is an inflection point. $f(-1) = -4$.

Conclusion The function $f(x)$ has two points of inflection: $(1, -2)$ and $(-1, -4)$.

Exercise 8.7 Find the point(s) of inflection for the function $f(x) = \begin{cases} \frac{1}{x} & x \geq 1 \\ \frac{3 - x^2}{2} & x < 1 \end{cases}$. (Ans. $x = 1$)

Example 8.38 Find the roots of $f(x) = x^2 - 4$ with the aid of Newton-Raphson method (initial guess is 3).

Solution We know that roots of $(x) = x^2 - 4$ are ± 2 since $f(x) = 0 \Rightarrow x^2 - 4 = 0 \Rightarrow x = \sqrt{4} \Rightarrow x = \pm 2$. Following iterations are converged to +2.0000.

$$f'(x) = \frac{df}{dx} = 2x \Rightarrow x_{n+1} = x_n - \frac{f(x_n)}{f'(x_n)} = x_n - \frac{x_n{}^2 - 4}{2x_n}$$

$$x_0 = 3, x_{n+1} = x_n - \frac{x_n{}^2 - 4}{2x_n}$$

$$x_1 = x_0 - \frac{f(x_0)}{f'(x_0)} = 3 - \frac{3^2 - 4}{2 \times 3} = 2.1667 \Rightarrow x_1 = 2.1667$$

$$x_2 = x_1 - \frac{f(x_1)}{f'(x_1)} = 2.1667 - \frac{2.1667^2 - 4}{2 \times 2.1667} = 2.0064 \Rightarrow x_2 = 2.0064$$

$$x_3 = x_2 - \frac{f(x_2)}{f'(x_2)} = 2.0064 - \frac{2.0064^2 - 4}{2 \times 2.0064} = 2.0000 \Rightarrow x_3 = 2.0000$$

Example 8.39 Find the roots of $f(x) = x^2 - 4$ with the aid of Newton-Raphson method (initial guess is -4).

Solution Newton-Raphson iterations with initial guess of -4 are shown below. This time the result converged to -2.000.

$$x_0 = -4$$

$$x_1 = x_0 - \frac{f(x_0)}{f'(x_0)} = -4 - \frac{(-4)^2 - 4}{2 \times -4} = -2.5000 \Rightarrow x_1 = -2.5000$$

$$x_2 = x_1 - \frac{f(x_1)}{f'(x_1)} = -2.5000 - \frac{(-2.5000)^2 - 4}{2 \times -2.5000} = -2.0500 \Rightarrow x_2 = -2.0500$$

$$x_3 = x_2 - \frac{f(x_2)}{f'(x_2)} = -2.0500 - \frac{(-2.0500)^2 - 4}{2 \times -2.0500} = 2.0006 \Rightarrow x_3 = -2.0006$$

$$x_4 = x_3 - \frac{f(x_3)}{f'(x_3)} = -2.0006 - \frac{(-2.0006)^2 - 4}{2 \times -2.0006} = 2.0000 \Rightarrow x_4 = -2.0000$$

Exercise 8.8 Use Newton-Raphson method to find the two roots of $x + e^{-x} - 2 = 0$. Use −1 and +1 as initial guess.

Example 8.40 Use MATLAB to find the minimum of $f(x) = x^3 - 4x^2 + 5$ on the interval $[-1, 3]$.

Solution The code shown in Fig. 8.32 draws the plot of $f(x) = x^3 - 4x^2 + 5$ on the interval $[-1, 3]$. The output of this code is shown in Fig. 8.33.

```
Command Window
>> f=@(x) x.^3-4*x.^2+5;
>> fplot(f,[-1 3])
>> grid on
>>
```

Fig. 8.32 The code for drawing the plot of $f(x) = x^3 - 4x^2 + 5$ on $[-1, 3]$

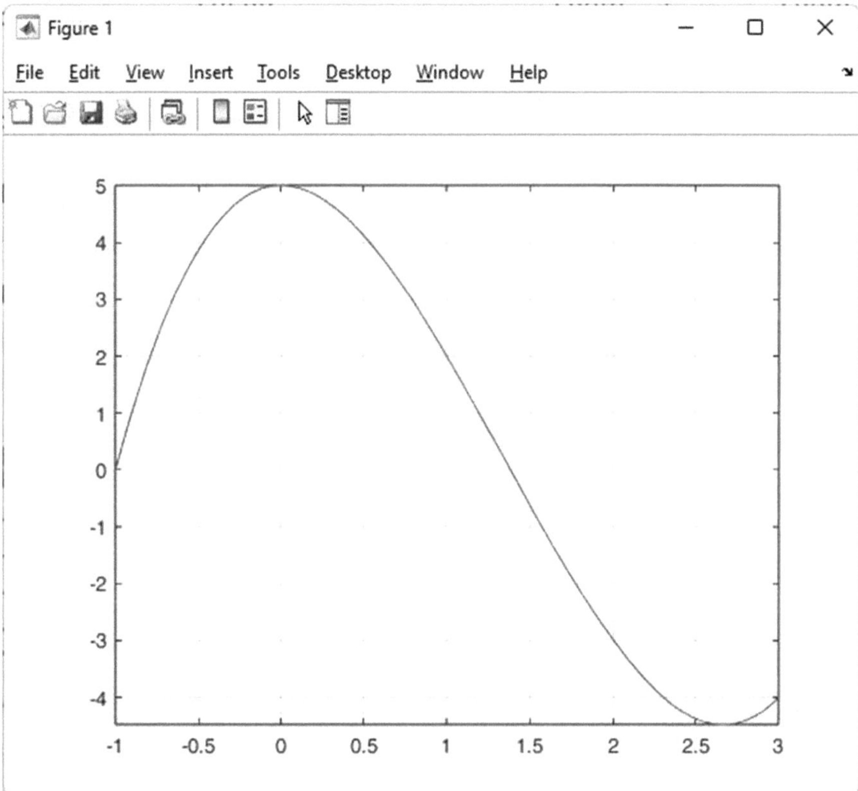

Fig. 8.33 Output of the code in Fig. 8.32

The coordinates of the minimum point can be viewed by clicking on it. As shown in Fig. 8.34, the minimum is approximately at $(x, y) = (2.67067, -4.48142)$.

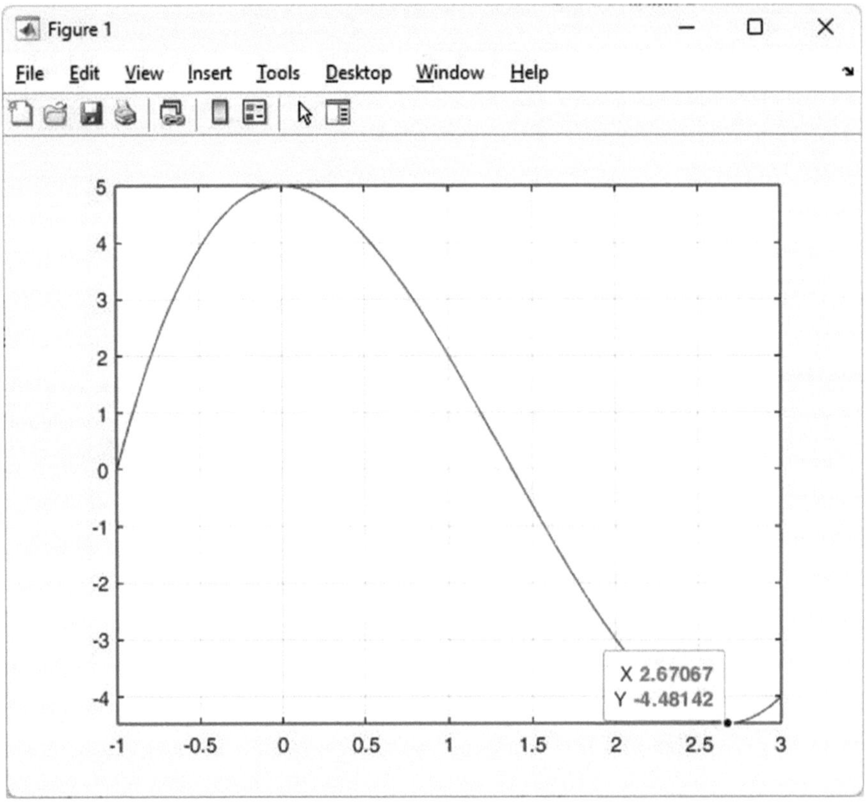

Fig. 8.34 The minimum occurs approximately at $(x, y) = (2.67067, -4.48142)$

Example 8.41 Use MATLAB to find the maximum of $f(x) = \sin(x) - \frac{x^2}{4}$ on the interval $[0, 2\pi]$.

Solution The code shown in Fig. 8.35 draws the plot of $f(x) = \sin(x) - \frac{x^2}{4}$ on the interval $[0, 2\pi]$. The output of this code is shown in Fig. 8.36.

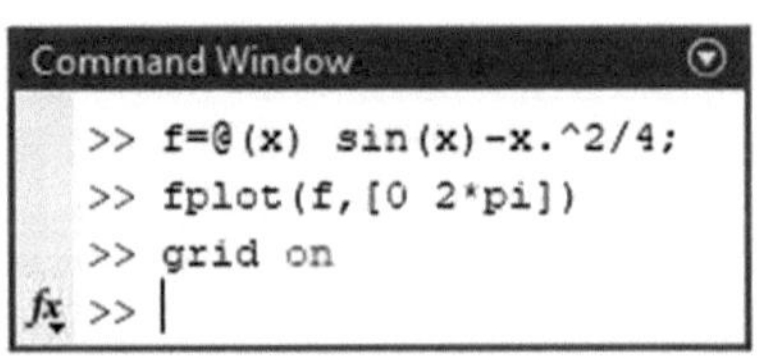

Fig. 8.35 The code for drawing the plot of $f(x) = \sin(x) - \frac{x^2}{4}$ on $[0, 2\pi]$

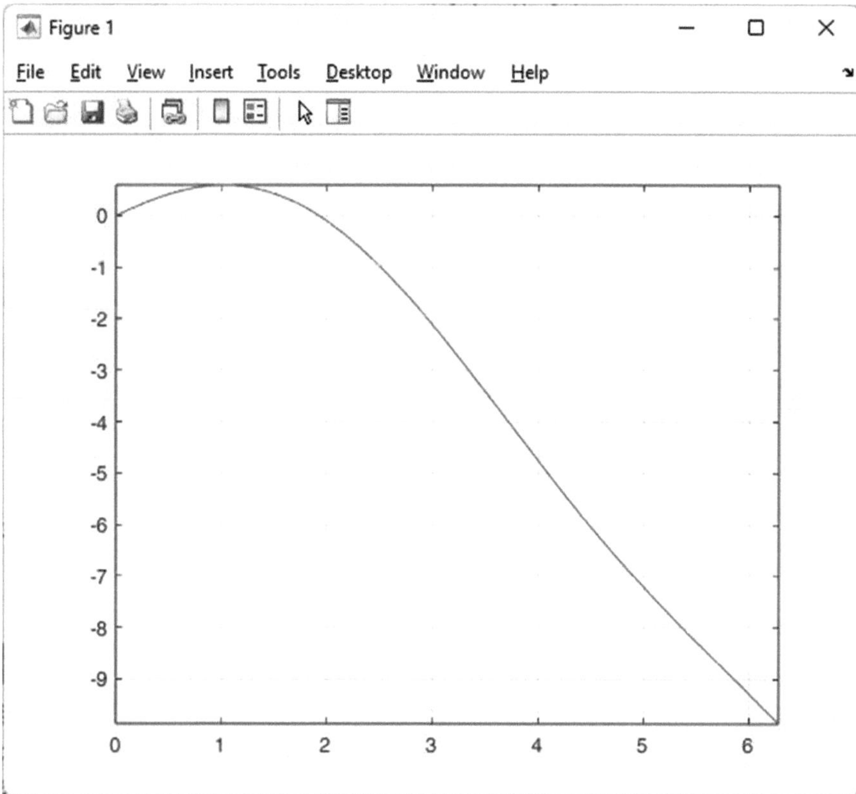

Fig. 8.36 Output of the code in Fig. 8.35

The coordinates of the maximum point can be viewed by clicking on it. As shown in Fig. 8.37, the maximum is approximately at $(x, y) = (1.01436, 0.591911)$.

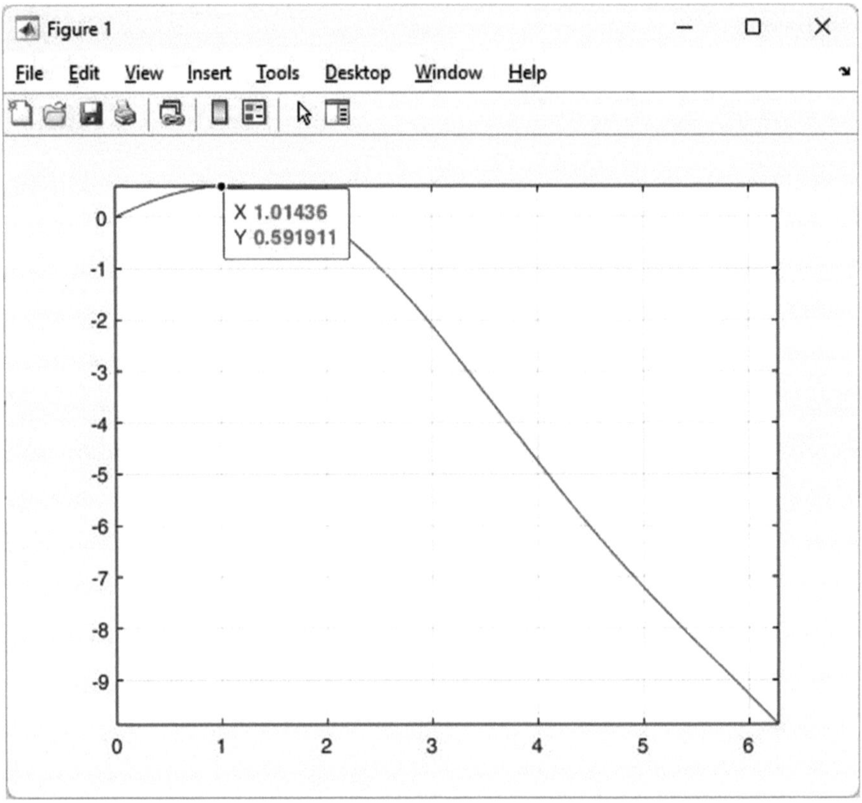

Fig. 8.37 The maximum occurs approximately at $(x, y) = (1.01436, 0.591911)$

Example 8.42 Find the minimum of the function $f(x) = x \sin(x) + \cos(x)$ on the interval $[-10, 10]$.

Solution The code shown in Fig. 8.38 draws the plot of $f(x) = x \sin(x) + \cos(x)$ on the interval $[-10, 10]$. The output of this code is shown in Fig. 8.39.

```
Command Window
>> f = @(x) x.*sin(x) + cos(x);
>> fplot(f, [-10, 10]);
>> title('plot of x*sin(x) + cos(x)');
>> xlabel('x');
>> ylabel('f(x)');
>> grid on;
>>
```

Fig. 8.38 The code for drawing the plot of $f(x) = x \sin(x) + \cos(x)$ on $[-10, 10]$

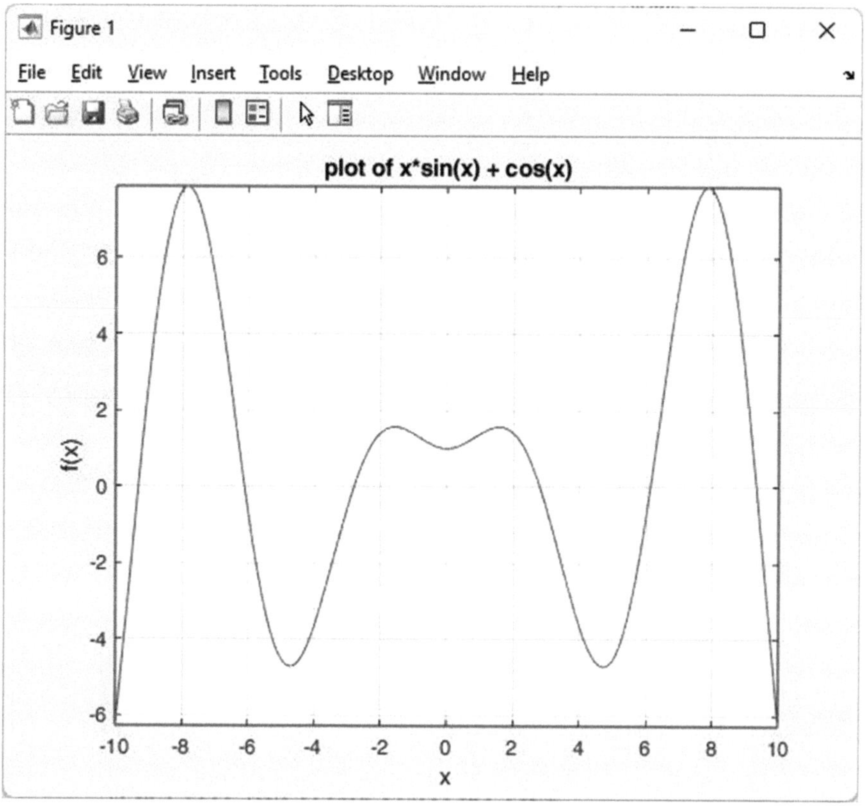

Fig. 8.39 Output of the code in Fig. 8.38

The coordinates of the minimum points can be viewed by clicking on them. As shown in Fig. 8.40, the minimum value is approximately -6.27928 at $x = \pm 10$.

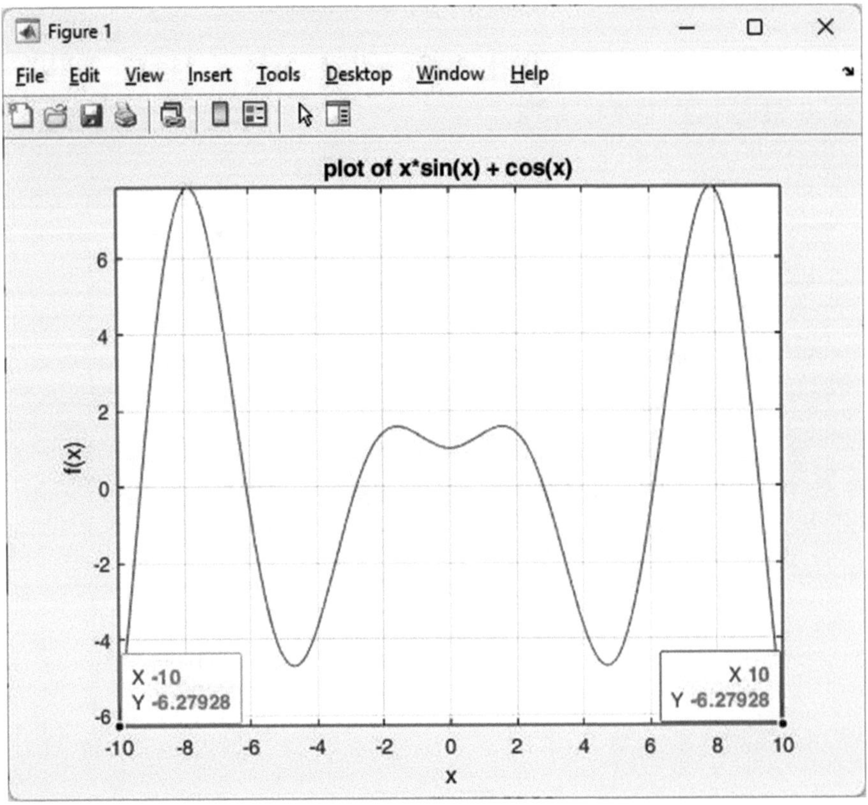

Fig. 8.40 The minimum occurs approximately at $(x, y) = (\pm 10, -6.27928)$

Example 8.43 Find the roots of the equation $e^{-x} - x = 0$.

Solution The code shown in Fig. 8.41 draws the plot of $f(x) = e^{-x} - x$ on the interval $[-1, 1]$. The output of this code is shown in Fig. 8.42.

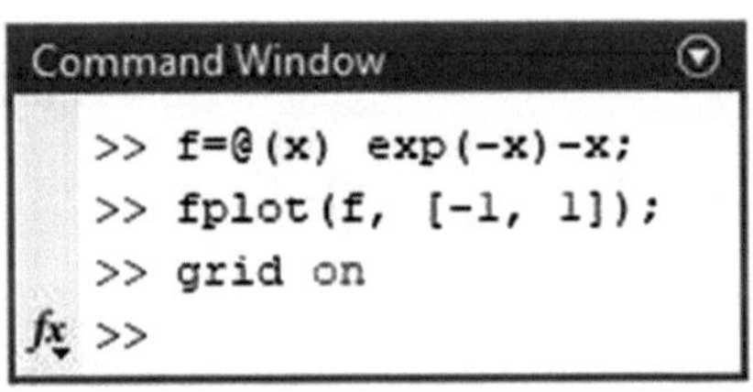

Fig. 8.41 The code for drawing the plot of $f(x) = e^{-x} - x$ on $[-1, 1]$

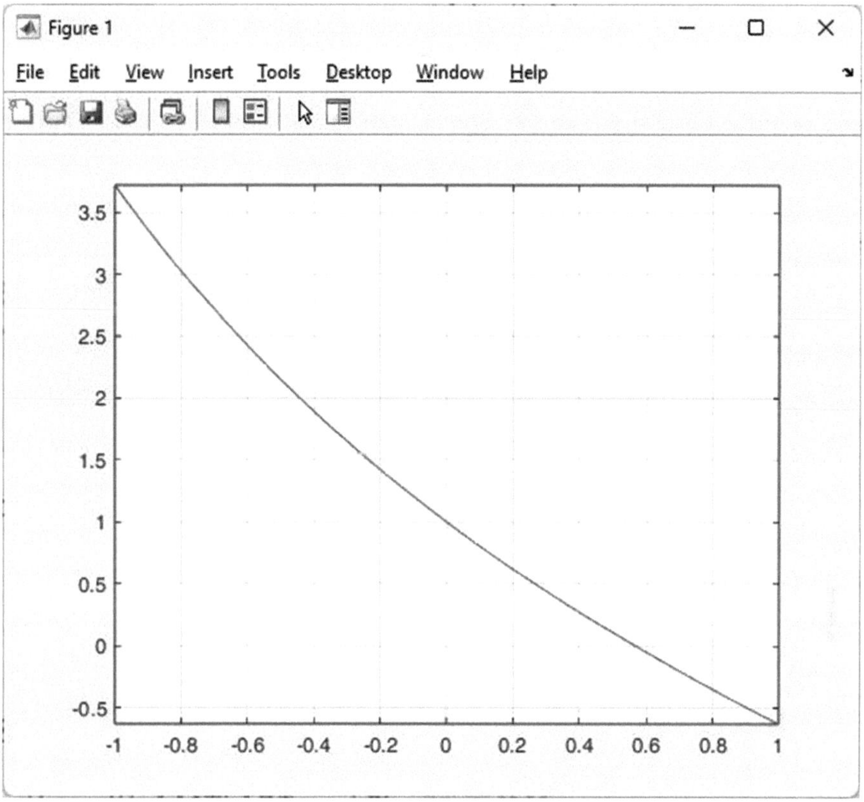

Fig. 8.42 Output of the code in Fig. 8.41

The root is located at approximately $x = 0.594225$ (Fig. 8.43).

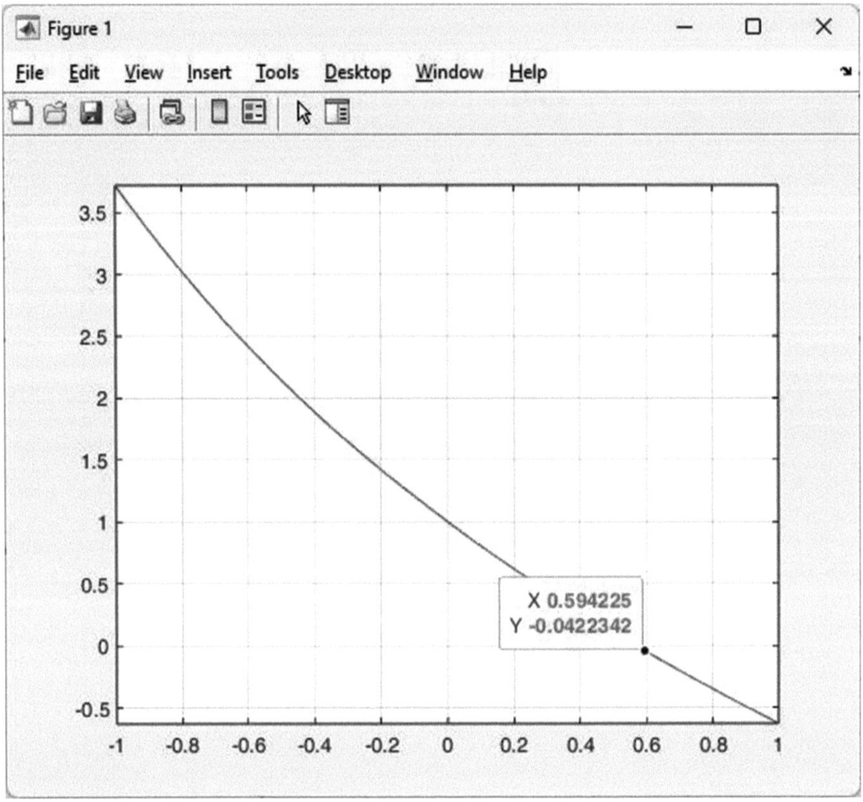

Fig. 8.43 The approximate coordinates of the root

To find the precise root value, the approximation from Fig. 8.43 is used as the starting point for the fzero command. Figure 8.44 shows that the precise value of the root is 0.5671.

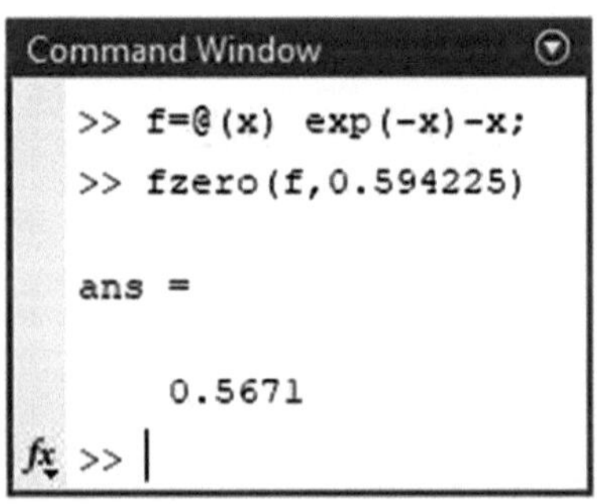

Fig. 8.44 The precise coordinates of the root

Example 8.44 Find the roots of the equation $x^4 + 2x^3 + 6x + 8 = 0$.

Solution Figure 8.45 displays the code used to find the roots of $x^4 + 2x^3 + 6x + 8 = 0$. The output in Fig. 8.45 shows the roots are $0.7902 + 1.5070i$, $0.7902 - 1.5070i$, -2.4549 and -1.1255.

Fig. 8.45 The roots of $x^4 + 2x^3 + 6x + 8 = 0$

```
Command Window

>> roots([1 2 0 6 8])

ans =

   0.7902 + 1.5070i
   0.7902 - 1.5070i
  -2.4549 + 0.0000i
  -1.1255 + 0.0000i

fx >>
```

Example 8.45 The code in Fig. 8.46 calculates the $\frac{d}{dx}(x^3 - 2\sin(6x))$.

Fig. 8.46 Calculation of $\frac{d}{dx}(x^3 - 2\sin(6x))$

```
Command Window

>> syms x
>> f=x^3-2*sin(6*x);
>> diff(f,x)

ans =

3*x^2 - 12*cos(6*x)

fx >>
```

Example 8.46 The code in Fig. 8.47 calculates the $\frac{d^2}{dx^2}(x^3 - 2\sin(6x))$.

Fig. 8.47 Calculation of $\frac{d^2}{dx^2}(x^3 - 2\sin(6x))$

```
Command Window

>> syms x
>> f=x^3-2*sin(6*x);
>> diff(f,x,3)

ans =

432*cos(6*x) + 6

fx >>
```

Example 8.47 The code in Fig. 8.48 calculates the value of the $\frac{d}{dx}\left(\left|x^2-5x+6\right|\right)$ at $x = 2.0001$.

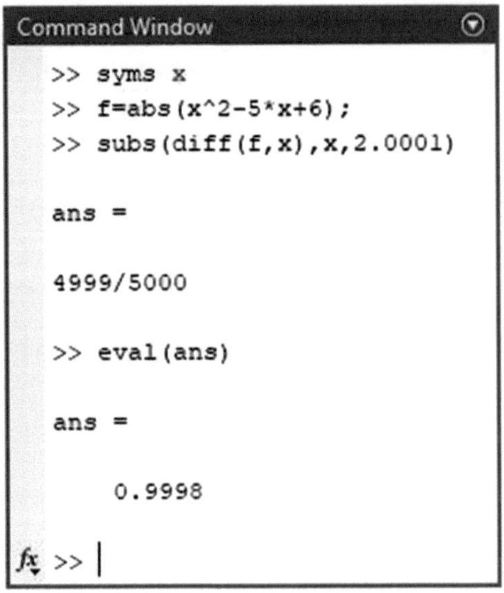

```
>> syms x
>> f=abs(x^2-5*x+6);
>> subs(diff(f,x),x,2.0001)

ans =

4999/5000

>> eval(ans)

ans =

    0.9998

>>
```

Fig. 8.48 Calculation of $\frac{d}{dx}\left(\left|x^2-5x+6\right|\right)$ at $x = 2.0001$

Example 8.47 The code in Fig. 8.49 calculates the value of the left-hand derivative of $|\ x^2 - 5x + 6\ |$ at $x = 2$ (i.e., $\frac{d}{dx}\left(\left|x^2-5x+6\right|\right)$ at $x = 2^-$).

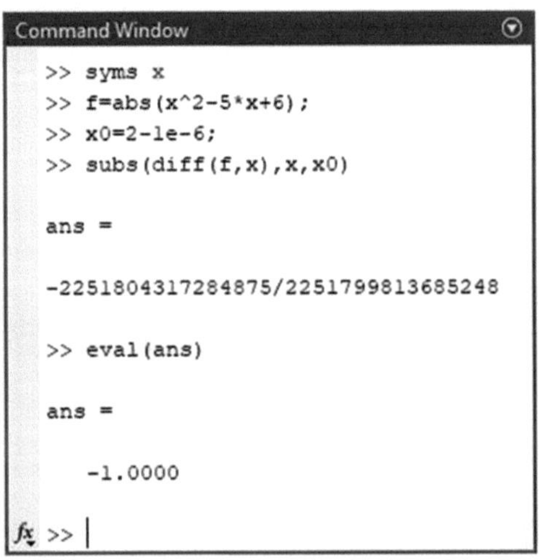

```
>> syms x
>> f=abs(x^2-5*x+6);
>> x0=2-1e-6;
>> subs(diff(f,x),x,x0)

ans =

-2251804317284875/2251799813685248

>> eval(ans)

ans =

   -1.0000

>>
```

Fig. 8.49 Calculation of $\frac{d}{dx}\left(\left|x^2-5x+6\right|\right)$ at $x = 2^-$

References for Further Study

1. Engineering Mathematics with MATLAB® and Simulink®, Farzin Asadi, Springer, 2025.
2. Applied Numerical Analysis with MATLAB®/Simulink®: For Engineers and Scientists, Farzin Asadi, Springer, 2023.
3. Advanced Engineering Mathematics, Erwin Kreyszig, Wiley, 2020.
4. A Problem-Solving Approach to Single-Variable Calculus (Volume II), Farzin Asadi, Springer, 2026.
5. A Problem-Solving Approach to Multi-Variable Calculus (Volume I), Farzin Asadi, Springer, 2026.
6. A Problem-Solving Approach to Multi-Variable Calculus (Volume II), Farzin Asadi, Springer, 2026.
7. A Problem-Solving Approach to Multi-Variable Calculus (Volume III), Farzin Asadi, Springer, 2026.
8. A Problem-Solving Approach to Ordinary Differential Equations (Volume I), Farzin Asadi, Springer, 2026.
9. A Problem-Solving Approach to Ordinary Differential Equations (Volume II), Farzin Asadi, Springer, 2026.
10. A Problem-Solving Approach to Ordinary Differential Equations (Volume III), Farzin Asadi, Springer, 2026.
11. A Problem-Solving Approach to Ordinary Differential Equations (Volume IV), Farzin Asadi, Springer, 2026.

Chapter 9
Partial Fraction Decomposition

9.1 Introduction

Partial fraction decomposition is a technique used in algebra and calculus to break down a complex rational expression into simpler fractions. This process involves expressing a rational function, which is a fraction where both the numerator and denominator are polynomials, as a sum of simpler fractions with simpler denominators.

By breaking down a complex rational expression into these simpler components, it becomes easier to integrate, differentiate, or analyze the function. This chapter is dedicated to partial fraction decomposition.

9.2 Addition of Rational Expressions

This section reviews the addition of rational expressions with a few examples.

Example 9.1 Add the expression $\frac{x+1}{x^2+5x+6}$ to $\frac{x+4}{x+5}$.

$$\begin{aligned}\frac{x+1}{x^2+5x+6}+\frac{x+4}{x+5}&=\frac{(x+1)(x+5)+(x+4)(x^2+5x+6)}{(x^2+5x+6)(x+5)}\\&=\frac{(x+1)(x+5)+(x+4)(x^2+5x+6)}{(x^2+5x+6)(x+5)}\\&=\frac{x^3+10x^2+32x+29}{x^3+10x^2+31x+30}\end{aligned}$$

Example 9.2 Add the expression $\frac{x+1}{(x+2)(x+3)}$ to $\frac{x+4}{(x+3)(x+5)}$.

F. Asadi, *A Problem-Solving Approach to Single-Variable Calculus*,
https://doi.org/10.1007/978-3-032-13505-6_9

$$\frac{x+1}{(x+2)(x+3)}+\frac{x+4}{(x+3)(x+5)}=\frac{(x+1)(x+5)+(x+4)(x+2)}{(x+2)(x+3)(x+5)}$$
$$=\frac{2x^2+12x+13}{x^3+10x^2+31x+30}$$

Example 9.3 Add the expression $\frac{5}{x+2}$ to $\frac{6}{x+3}$.

$$\frac{5}{x+2}+\frac{6}{x+3}=\frac{5x+15+6x+12}{(x+2)(x+3)}=\frac{11x+27}{x^2+5x+6}$$

9.3 Partial Fraction Decomposition

Partial fraction decomposition is a technique used in algebra to break down a complex rational expression into simpler fractions. Let's study some numeric examples.

Example 9.4 Decompose the rational function $\frac{7x-23}{x^2-7x+10}$ into partial fractions.

$$x^2-7x+10=0\Rightarrow x_{1,2}=\frac{+7\pm\sqrt{7^2-4(1)(10)}}{2}=\frac{7\pm 3}{2}=\begin{cases}\frac{7-3}{2}=2\\ \frac{7+3}{2}=5\end{cases}$$

$$\frac{7x-23}{x^2-7x+10}=\frac{7x-23}{(x-2)(x-5)}=\frac{A}{(x-2)}+\frac{B}{(x-5)}$$

Cover the $(x-2)$ term with your finger and substitute $x=2$ (the root of covered section) into the remaining expression (Fig. 9.1). The value of A is $\left.\frac{7x-23}{(x-5)}\right|_{x=2}=\frac{7\times 2-23}{(2-5)}=3$.

Fig. 9.1 $(x-2)$ term is covered

Now cover the $(x - 5)$ term with your finger and substitute $x = 5$ (the root of covered section) into the remaining expression (Fig. 9.2). The value of B is $\left.\frac{7x-23}{(x-2)}\right|_{x=5} = \frac{7\times 5-23}{(5-2)} = 4.$

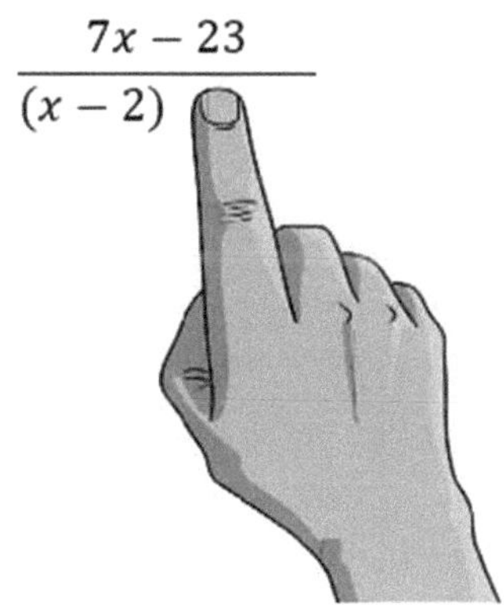

Fig. 9.2 $(x - 5)$ term is covered

Therefore, $\frac{7x-23}{x^2-7x+10} = \frac{3}{x-2} + \frac{4}{x-5}.$

Example 9.5 Decompose the rational function $\frac{4x^2-21x+17}{x^2-7x+10}$ into partial fractions.

When the degree of the numerator is greater than or equal to the degree of the denominator, we have an improper rational function. To apply partial fraction decomposition, we first need to perform polynomial division to divide the numerator by the denominator.

$$\frac{4x^2-21x+17}{x^2-7x+10} = \frac{4(x^2-7x+10)+7x-23}{x^2-7x+10} = 4 + \frac{7x-23}{x^2-7x+10} = 4 + \frac{3}{x-2} + \frac{4}{x-5}$$

Example 9.6 Decompose the rational function $\frac{x^3+2x^2+x+1}{x^2+1}$ into partial fractions.

$$\frac{x^3+2x^2+x+1}{x^2+1} = \frac{x(x^2+1)+2x^2+1}{x^2+1} = \frac{x(x^2+1)+2(x^2+1)-1}{x^2+1} = x+2-\frac{-1}{x^2+1}$$

Example 9.7 Decompose the rational function $\frac{1}{x^2(x+1)}$ into partial fractions.

$$\frac{1}{x^2(x+1)} = \frac{A}{x^2} + \frac{B}{x} + \frac{C}{x+1}$$

Cover the x^2 term with your finger and substitute $x = 0$ (the root of covered section) into the remaining expression (Fig. 9.3). The value of A is $\left.\frac{1}{x+1}\right|_{x=0} = \frac{1}{0+1} = 1.$

Fig. 9.3 x^2 term is covered

We'll need to use derivatives to find the value of B. $B = \frac{d}{dx}\left(\frac{1}{x+1}\right)\Big|_{x=0} = -\frac{1}{(x+1)^2}\Big|_{x=0} = -1.$

Now cover the $x + 1$ term with your finger and substitute $x = -1$ (the root of covered section) into the remaining expression (Fig. 9.4). The value of C is $\frac{1}{x^2}\big|_{x=-1} = \frac{1}{1} = 1.$

Fig. 9.4 $(x + 1)$ term is covered

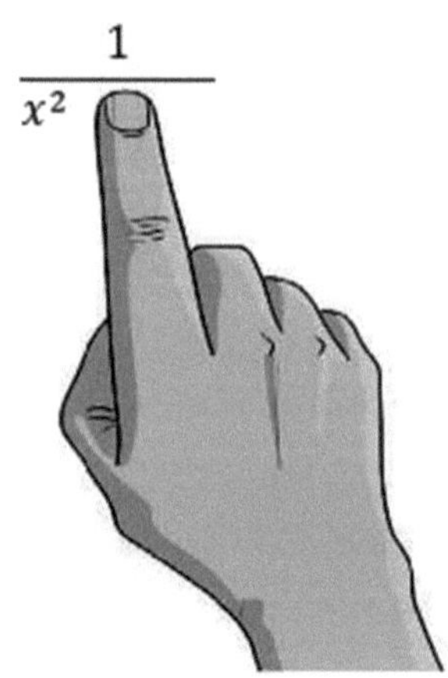

Therefore, $\frac{1}{x^2(x+1)} = \frac{1}{x^2} + \frac{-1}{x} + \frac{1}{x+1}.$

Example 9.8 Decompose the rational function $\frac{2x+1}{(x+1)(x+2)^2}$ into partial fractions.

$$\frac{2x+1}{(x+1)(x+2)^2} = \frac{A}{(x+2)^2} + \frac{B}{x+2} + \frac{C}{x+1}$$

Cover the $(x + 2)^2$ term with your finger and substitute $x = -2$ (the root of covered section) into the remaining expression (Fig. 9.5). The value of A is $\left.\frac{2x+1}{(x+1)}\right|_{x=-2} = \frac{2\times -2+1}{-2+1} = 3.$

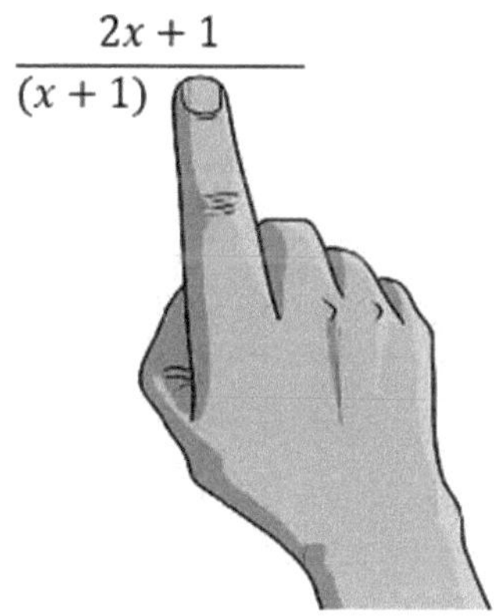

Fig. 9.5 $(x + 2)^2$ term is covered

We'll need to use derivatives to find the value of B:

$$B=\frac{d}{dx}\left(\frac{2x+1}{x+1}\right)\Big|_{x=-2}=\left.\frac{2(x+1)-1(2x+1)}{(x+1)^2}\right|_{x=-2}=\frac{2(-2+1)-1(-4+1)}{(-2+1)^2}=1$$

Now cover the $(x + 1)$ term with your finger and substitute $x = -1$ (the root of covered section) into the remaining expression (Fig. 9.6). The value of C is $\left.\frac{2x+1}{(x+2)^2}\right|_{x=-1} = \frac{-2+1}{(-1+2)^2} = -1.$

Fig. 9.6 $(x + 1)$ term is covered

Therefore, $\frac{2x+1}{(x+1)(x+2)^2} = \frac{3}{(x+2)^2} + \frac{1}{x+2} + \frac{-1}{x+1}$.

Example 9.9 Decompose the rational function $\frac{-2x^2-5x-1}{(x+2)^3}$ into partial fractions.

$$\frac{-2x^2-5x-1}{(x+2)^3} = \frac{A}{(x+2)^3} + \frac{B}{(x+2)^2} + \frac{C}{(x+2)}$$

Cover the $(x + 2)^3$ term with your finger and substitute $x = -2$ (the root of covered section) into the remaining expression (Fig. 9.7). The value of A is $-2x^2 - 5x - 1|_{x=-2} \Rightarrow -2(-2)^2 + 10 - 1 = 1$.

Fig. 9.7 $(x + 2)^3$ term is covered

We'll need to use derivatives to find the value of B and C:

$$B = \frac{d}{dx}(-2x^2-5x-1)\Big|_{x=-2} = -4x-5|_{x=-2} = -4(-2)-5 = 3$$

$$C = \frac{1}{2}\frac{d^2}{dx^2}(-2x^2-5x-1)\Big|_{x=-2} = \frac{1}{2} \times -4|_{x=-2} = -2$$

Example 9.10 Decompose the rational function $\frac{x+1}{(x+2)^3(x+3)^2(x+4)}$ into partial fractions.

$$\frac{x+1}{(x+2)^3(x+3)^2(x+4)} = \frac{A}{(x+2)^3} + \frac{B}{(x+2)^2} + \frac{C}{(x+2)} + \frac{D}{(x+3)^2} + \frac{E}{(x+3)} + \frac{F}{(x+4)}$$

Cover the $(x + 2)^3$ term with your finger and substitute $x = -2$ (the root of covered section) into the remaining expression (Fig. 9.8). The value of A is $\left.\frac{x+1}{(x+3)^2(x+4)}\right|_{x=-2} = \frac{-1}{2}$.

Fig. 9.8 $(x + 2)^3$ term is covered

We'll need to use derivatives to find the value of B and C:

$$B = \frac{d}{dx}\left(\frac{x+1}{(x+3)^2(x+4)}\right)\Bigg|_{x=-2} = \frac{7}{4}$$

$$C = \frac{1}{2}\frac{d^2}{dx^2}\left(\frac{x+1}{(x+3)^2(x+4)}\right)\Bigg|_{x=-2} = -\frac{27}{8}$$

Cover the $(x + 3)^2$ term with your finger and substitute $x = -3$ (the root of covered section) into the remaining expression (Fig. 9.9). The value of D is $\left.\frac{x+1}{(x+2)^3(x+4)}\right|_{x=-3} = \frac{-2}{-1^3 \times 1} = 2.$

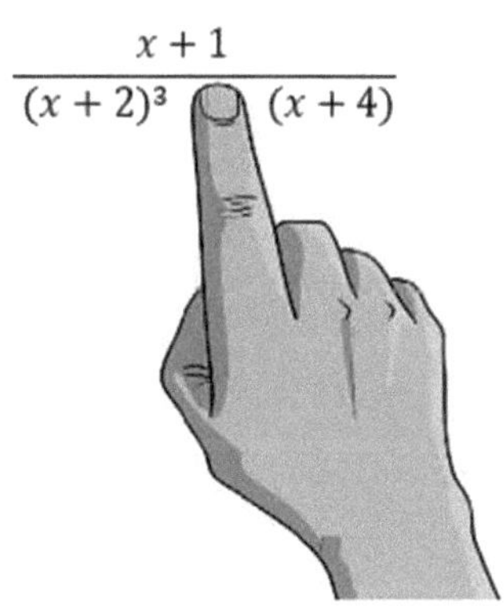

Fig. 9.9 $(x + 3)^2$ term is covered

We'll need to use derivatives to find the value of E:

$$E = \frac{d}{dx}\left(\frac{x+1}{(x+2)^3(x+4)}\right)\Bigg|_{x=-3} = 3$$

Cover the $(x + 4)$ term with your finger and substitute $x = -4$ (the root of covered section) into the remaining expression (Fig. 9.10). The value of F is $\frac{x+1}{(x+2)^3(x+3)^2}\Big|_{x=-4} = \frac{-3}{-8\times 1} = \frac{3}{8}$.

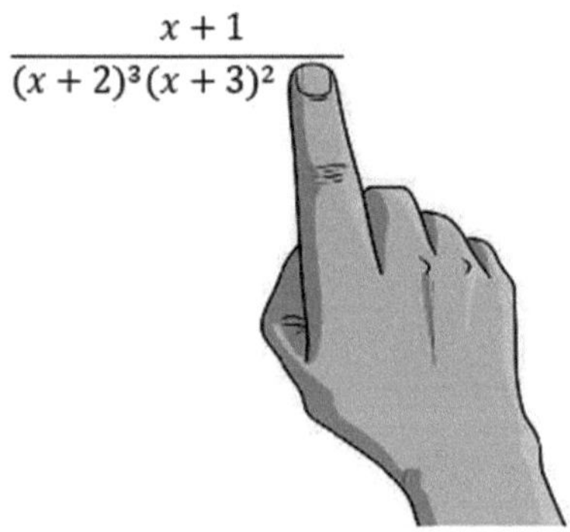

Fig. 9.10 $(x + 4)$ term is covered

Therefore, $\frac{x+1}{(x+2)^3(x+3)^2(x+4)} = \frac{\frac{-1}{2}}{(x+2)^3} + \frac{\frac{7}{4}}{(x+2)^2} + \frac{\frac{-27}{8}}{(x+2)} + \frac{2}{(x+3)^2} + \frac{3}{(x+3)} + \frac{\frac{3}{8}}{(x+4)}$.

Example 9.11 Decompose the rational function $\frac{5s^2+s+4}{(s+4)(s^2+4)}$ into partial fractions.

$$\frac{5s^2+s+4}{(s+4)(s^2+4)} = \frac{5s^2+s+4}{(s+4)(s+2j)(s-2j)} = \frac{A}{s+4} + \frac{B}{s+2j} + \frac{C}{s-2j}$$

Cover the $(s + 4)$ term with your finger and substitute $s = -4$ (the root of covered section) into the remaining expression (Fig. 9.11). The value of A is $\frac{5(-4)^2-4+4}{(-4+2j)(-4-2j)} = 4$.

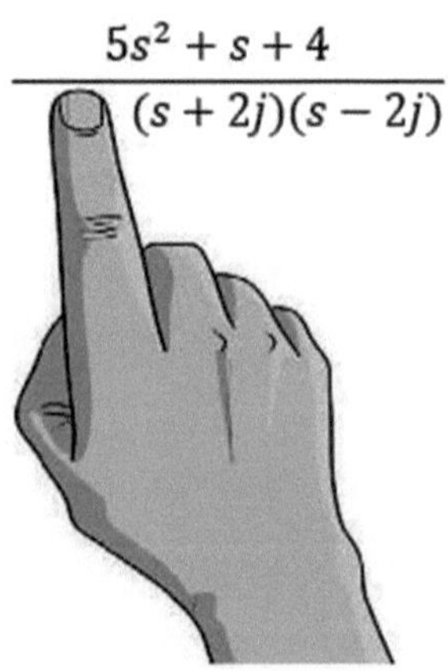

Fig. 9.11 $(s + 4)$ term is covered

Cover the $(s + 2j)$ term with your finger and substitute $s = -2j$ (the root of covered section) into the remaining expression (Fig. 9.12). The value of B is $\frac{5(-2j)^2 - 2j + 4}{(-2j + 4)(-2j - 2j)} = \frac{-16 - 2j}{-8 - 16j} = 0.5 - 0.75j.$

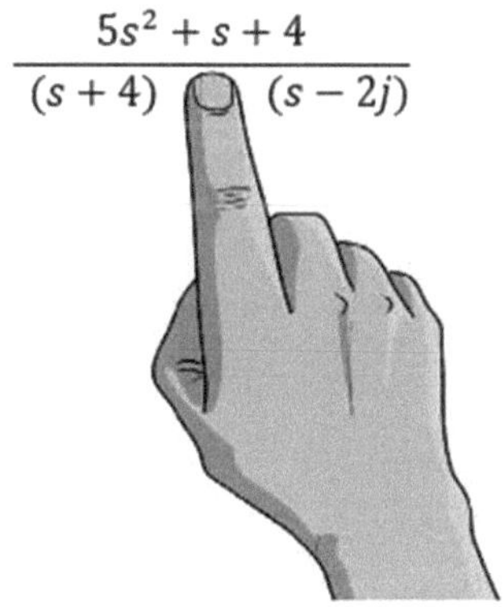

Fig. 9.12 $(s + 2j)$ term is covered

Cover the $(s - 2j)$ term with your finger and substitute $s = +2j$ (the root of covered section) into the remaining expression (Fig. 9.13). The value of C is $\frac{5(2j)^2 + 2j + 4}{(2j + 4)(2j + 2j)} = \frac{-16 + 2j}{-8 + 16j} = 0.5 + 0.75j.$

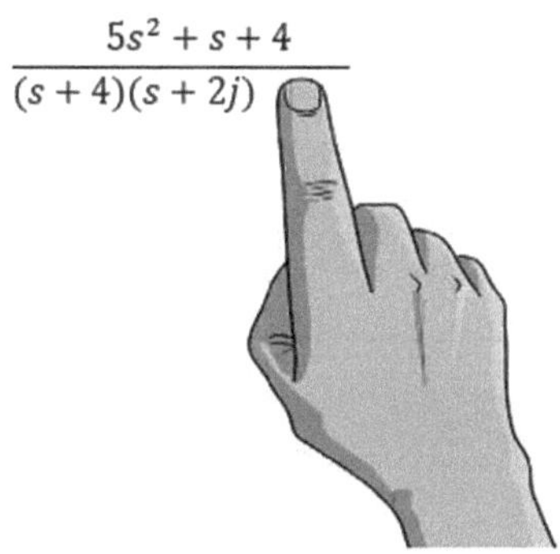

Fig. 9.13 $(s - 2j)$ term is covered

Therefore, $\frac{5s^2+s+4}{(s+4)(s^2+4)} = \frac{4}{s+4} + \frac{0.5-0.75j}{s+2j} + \frac{0.5+0.75j}{s-2j} = \frac{4}{s+4} + \frac{(0.5-0.75j)(s-2j)+(0.5+0.75j)(s+2j)}{(s-2j)(s+2j)} = \frac{4}{s+4} + \frac{(0.5-0.75j)(s-2j)+(0.5+0.75j)(s+2j)}{s^2+4} = \frac{4}{s+4} + \frac{s-3}{s^2+4}.$

Alternatively, you can solve this problem as follows:

$$\frac{5s^2+s+4}{(s+4)(s^2+4)}=\frac{A}{s+4}+\frac{Bs+C}{s^2+4}$$

Cover the $(s + 4)$ term with your finger and substitute $s = -4$ (the root of covered section) into the remaining expression (Fig. 9.14). The value of A is $\frac{5(-4)^2-4+4}{(-4)^2+4}=4.$

Fig. 9.14 $(s + 4)$ term is covered

Value of B and C can be found as follows:

$$\frac{5s^2+s+4}{(s+4)(s^2+4)}=\frac{4}{s+4}+\frac{Bs+C}{(s^2+4)}\Rightarrow\frac{5s^2+s+4}{(s+4)(s^2+4)}-\frac{4}{s+4}$$
$$=\frac{Bs+C}{(s^2+4)}\Rightarrow\frac{s^2+s-12}{(s+4)(s^2+4)}=\frac{Bs+C}{(s^2+4)}\Rightarrow\frac{s^2+s-12}{(s+4)}$$
$$=Bs+C\Rightarrow\frac{(s+4)(s-3)}{(s+4)}$$
$$=Bs+C\Rightarrow s-3=Bs+C\Rightarrow B=1,C=-3$$

Therefore, $\frac{5s^2+s+4}{(s+4)(s^2+4)}=\frac{4}{s+4}+\frac{s-3}{(s^2+4)}.$

Example 9.12 Decompose the rational function $\frac{3}{(s-1)(s-5)}$ into partial fractions.

$$\frac{k}{(s+a)(s+b)}=\frac{A}{(s+a)}+\frac{B}{(s+b)}=\frac{\frac{k}{b-a}}{(s+a)}+\frac{\frac{k}{a-b}}{(s+b)}$$

$$\frac{3}{(s-1)(s-5)}=\frac{\frac{-3}{4}}{(s-1)}+\frac{\frac{3}{4}}{(s-5)}$$

Exercise 9.1 Show that $\frac{4x}{x^3-x^2-x+1}=\frac{1}{x-1}+\frac{2}{(x-1)^2}+\frac{-1}{x+1}.$

9.4 Polynomial Division

This section provides a review of polynomial division with a few examples.

Example 9.13 Perform polynomial division on $\frac{3x^2-10}{x^2-4x+4}$.

$$\frac{3x^2-10}{x^2-4x+4}=\frac{3x^2-12x+12+12x-12-10}{x^2-4x+4}=3+\frac{12x+22}{x^2-4x+4}$$

You can use the method shown in Fig. 9.15 as well.

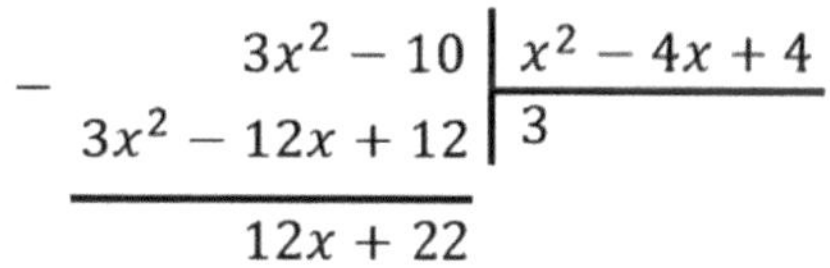

Fig. 9.15 Long division of $3x^2-10$ by x^2-4x+4

Example 9.14 Perform polynomial division on $\frac{x^3+x}{x-1}$.

According to Fig. 9.16, $\frac{x^3+x}{x-1}=x^2+x+2+\frac{2}{x-1}$.

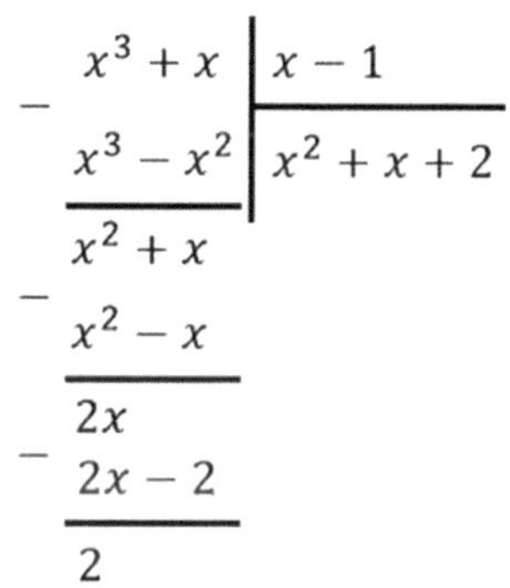

Fig. 9.16 Long division of x^3+x by $x-1$

Exercise 9.2 Perform polynomial division on $\frac{x^4-2x^2+4x+1}{x^3-x^2-x+1}$ (Ans. $\frac{x^4-2x^2+4x+1}{x^3-x^2-x+1}=x+1+\frac{4x}{x^3-x^2-x+1}$).

9.5 Problem Solving with MATLAB®

This section offers MATLAB examples relevant to the problems in this chapter. If you are new to MATLAB, begin with Chap. 11.

Example 9.15 The code in Fig. 9.17 computes the partial fraction decomposition of the rational function $\frac{7x-23}{x^2-7x+10}$. From Fig. 9.17, $\frac{7x-23}{x^2-7x+10}=\frac{3}{x-2}+\frac{4}{x-5}$.

Fig. 9.17 Partial fraction decomposition of $\frac{7x-23}{x^2-7x+10}$

```
Command Window
>> [r,p,k]=residue([7 -23],[1 -7 10])

r =

     4
     3


p =

     5
     2


k =

     []

fx >>
```

Example 9.16 The code in Fig. 9.18 computes the partial fraction decomposition of $\frac{x+1}{(x+2)^3(x+3)^2(x+4)}$. Note that $(x + 2)^3(x + 3)^2(x + 4) = x^6 + 16x^5 + 105x^4 + 362x^3 + 692x^2 + 696x + 288$ (Fig. 9.19). From Fig. 9.18, $\frac{x+1}{(x+2)^3(x+3)^2(x+4)} = \frac{0.375}{(x+4)} + \frac{3}{(x+3)} + \frac{2}{(x+3)^2} + \frac{-3.375}{(x+2)} + \frac{1.75}{(x+2)^2} + \frac{-0.5}{(x+2)^3}$.

```
Command Window
>> [r,p,k]=residue([1 1],[1 16 105 362 692 696 288])

r =

    0.3750
    3.0000
    2.0000
   -3.3750
    1.7500
   -0.5000


p =

   -4.0000
   -3.0000
   -3.0000
   -2.0000
   -2.0000
   -2.0000


k =

     []

fx >>
```

Fig. 9.18 Partial fraction decomposition of $\frac{x+1}{(x+2)^3(x+3)^2(x+4)}$

```
Command Window
>> syms x
>> expand((x+2)^3*(x+3)^2*(x+4))

ans =

x^6 + 16*x^5 + 105*x^4 + 362*x^3 + 692*x^2 + 696*x + 288

fx >>
```

Fig. 9.19 Expansion of $(x + 2)^3(x + 3)^2(x + 4)$

Example 9.17 The code in Fig. 9.20 computes the partial fraction decomposition of $\frac{x+7}{x^3+4x^2+13x}$. From Fig. 9.20, $\frac{x+7}{x^3+4x^2+13x}=\frac{0.5385}{x}+\frac{-0.2692+0.0128j}{x-(-2+3j)}+\frac{-0.2692-0.0128j}{x-(-2-3j)}$. Note that $\frac{-0.2692+0.0128j}{x-(-2+3j)}+\frac{-0.2692-0.0128j}{x-(-2-3j)}=-\frac{\frac{673}{1250}x+\frac{1442}{1250}}{x^2+4x+13}=-\frac{0.5384x+1.1536}{x^2+4x+13}$ (Fig. 9.21). Therefore, $\frac{x+7}{x^3+4x^2+13x}=\frac{0.5385}{x}-\frac{0.5384x+1.1536}{x^2+4x+13}$.

Fig. 9.20 Partial fraction decomposition of $\frac{x+7}{x^3+4x^2+13x}$

```
Command Window

>> [r,p,k]=residue([1 7],[1 4 13 0]

r =

  -0.2692 + 0.0128i
  -0.2692 - 0.0128i
   0.5385 + 0.0000i


p =

  -2.0000 + 3.0000i
  -2.0000 - 3.0000i
   0.0000 + 0.0000i


k =

     []

fx >>
```

```
Command Window
>> syms x
>> simplify((-.2692+0.0128i)/(x+2-3i)+(-.2692-0.0128i)/(x+2+3i))

ans =

-(673*x + 1442)/(1250*(x^2 + 4*x + 13))

>> pretty(ans)
      673 x + 1442
- --------------------
     2
  (x  + 4 x + 13) 1250

fx >>
```

Fig. 9.21 Simplification of $\frac{-0.2692+0.0128j}{x-(-2+3j)}+\frac{-0.2692-0.0128j}{x-(-2-3j)}$

Example 9.18 The code in Fig. 9.22 divides $x^3 + x$ by $x - 1$. From Fig. 9.22, $x^3 + x = (x^2 + x + 2)(x - 1) + 2$.

```
Command Window
>> dividend=[1 0 1 0];
>> divisor=[1 -1];
>> [quotient, remainder] = deconv(dividend, divisor)

quotient =

     1     1     2


remainder =

     0     0     0     2

fx >>
```

Fig. 9.22 Dividing $x^3 + x$ by $x - 1$

Example 9.19 The code in Fig. 9.23 divides $x^4 - 2x^2 + 4x + 1$ by $x^3 - x^2 - x + 1$. From Fig. 9.23, $x^4 - 2x^2 + 4x + 1 = (x + 1)(x^3 - x^2 - x + 1) + 4x$.

```
Command Window
>> dividend=[1 0 -2 4 1];
>> divisor=[1 -1 -1 1];
>> [quotient, remainder] = deconv(dividend, divisor)

quotient =

     1     1


remainder =

     0     0     0     4     0

fx >>
```

Fig. 9.23 Dividing $x^4 - 2x^2 + 4x + 1$ by $x^3 - x^2 - x + 1$

References for Further Study

1. Engineering Mathematics with MATLAB® and Simulink®, Farzin Asadi, Springer, 2025.
2. Applied Numerical Analysis with MATLAB®/Simulink®: For Engineers and Scientists, Farzin Asadi, Springer, 2023.
3. Advanced Engineering Mathematics, Erwin Kreyszig, Wiley, 2020.
4. A Problem-Solving Approach to Single-Variable Calculus (Volume II), Farzin Asadi, Springer, 2026.
5. A Problem-Solving Approach to Multi-Variable Calculus (Volume I), Farzin Asadi, Springer, 2026.
6. A Problem-Solving Approach to Multi-Variable Calculus (Volume II), Farzin Asadi, Springer, 2026.
7. A Problem-Solving Approach to Multi-Variable Calculus (Volume III), Farzin Asadi, Springer, 2026.
8. A Problem-Solving Approach to Ordinary Differential Equations (Volume I), Farzin Asadi, Springer, 2026.
9. A Problem-Solving Approach to Ordinary Differential Equations (Volume II), Farzin Asadi, Springer, 2026.
10. A Problem-Solving Approach to Ordinary Differential Equations (Volume III), Farzin Asadi, Springer, 2026.
11. A Problem-Solving Approach to Ordinary Differential Equations (Volume IV), Farzin Asadi, Springer, 2026.

Chapter 10
Indefinite Integrals

10.1 Introduction

The indefinite integral of a function $f(x)$ is the entire family of functions $F(x)$ whose derivative is $f(x)$: $\frac{dF(x)}{dx} = f(x) \Leftrightarrow \int f(x)dx = F(x) + C$. The constant C must be included because the derivative of any constant is zero. For example:

The derivative of x^2 is $2x$.
The derivative of $x^2 + 5$ is $2x$.
The derivative of $x^2 - 100$ is $2x$.

Since all these functions have the same derivative, their indefinite integral must represent all of them. The $+C$ accounts for every possible constant that could be present.

10.1.1 Properties of Indefinite Integrals

Here are the important properties of indefinite integrals:

1. $\int k_1 f(x)dx = k_1 \int f(x)dx \quad k_1 \in \mathbb{R}$
2. $\int (k_1 f(x) + k_2 g(x) - k_3 h(x))dx = k_1 \int f(x)dx + k_2 \int g(x)dx - k_3 \int h(x)dx \quad k_1, k_2, k_3 \in \mathbb{R}$
3. $\int f(x)dx = F(x) + C \Rightarrow \int f(k_1 x + k_2)dx = \frac{1}{k_1} F(k_1 x + k_2) + C \quad k_1, k_2, C \in \mathbb{R}$
4. $\int \frac{df(x)}{dx} dx = f(x) + C$
5. $\frac{d}{dx}\left(\int f(x)dx\right) = f(x)$

F. Asadi, *A Problem-Solving Approach to Single-Variable Calculus*,
https://doi.org/10.1007/978-3-032-13505-6_10

10.1.2 Important Integration Formulas

Important integral formulas are listed below.

1. $\int k_1 dx = k_1 x + C, \quad k_1 \in \mathbb{R}$
2. $\int x^n dx = \frac{1}{n+1} x^{n+1} + C, \quad n \in \mathbb{R}, n \neq -1$
3. $\int x^{-1} dx = \int \frac{1}{x} dx = \ln(|x|) + C$
4. $\int \frac{u(x)'}{u(x)} dx = \ln(|u(x)|) + C$
5. $\int \frac{1}{ax+b} dx = \frac{1}{a} \ln(|ax+b|) + C, \quad a, b \in \mathbb{R}, a \neq 0$
6. $\int (ax+b)^n dx = \frac{1}{a} \frac{1}{n+1} (ax+b)^{n+1} + C, \quad a, b, n \in \mathbb{R}, \quad a \neq 0, \quad n \neq -1$
7. $\int \frac{u'(x)}{(u(x))^2 + a^2} dx = \frac{1}{a} \tan^{-1}\left(\frac{u(x)}{a}\right) + C$
8. $\int \frac{1}{x^2+a^2} dx = \frac{1}{a} \tan^{-1}\left(\frac{x}{a}\right) + C$
9. $\int \frac{u'(x)}{\sqrt{a^2 - u(x)^2}} dx = \sin^{-1}\left(\frac{u(x)}{a}\right) + C, \quad |u(x)| < a$
10. $\int \frac{u'(x)}{\sqrt{1 - u(x)^2}} dx = \sin^{-1}(u(x)) + C, \quad |u(x)| < 1$
11. $\int \frac{u'(x)}{\sqrt{a^2 - u(x)^2}} dx = -\cos^{-1}\left(\frac{u(x)}{a}\right) + C$
12. $\int \frac{u'(x)}{\sqrt{1 - u(x)^2}} dx = -\cos^{-1}(u(x)) + C$
13. $\int \frac{u'(x)}{1+u(x)^2} dx = \tan^{-1}(u(x)) + C$
14. $\int \frac{u'(x)}{a^2+u(x)^2} dx = -\frac{1}{a} \cot^{-1}\left(\frac{u(x)}{a}\right) + C$
15. $\int \frac{u'(x)}{1+u(x)^2} dx = -\cot^{-1}(u(x)) + C$
16. $\int \frac{u'(x)}{1-u(x)^2} dx = \frac{1}{2} \ln\left(\left|\frac{1+u(x)}{1-u(x)}\right|\right) + C, \quad |u(x)| < 1$
17. $\int \frac{u'(x)}{|u(x)|\sqrt{(u(x))^2 - a^2}} dx = \frac{1}{a} \sec^{-1}\left(\left|\frac{u(x)}{a}\right|\right) + C, \quad |u(x)| > a$
18. $\int \frac{u'(x)}{|u(x)|\sqrt{(u(x))^2 - a^2}} dx = -\frac{1}{a} \csc^{-1}\left(\left|\frac{u(x)}{a}\right|\right) + C, \quad |u(x)| > a$
19. $\int u'(x) \sin(u(x)) dx = -\cos(u(x)) + C$
20. $\int u'(x) \cos(u(x)) dx = \sin(u(x)) + C$
21. $\int u'(x)(1 + \tan^2(u(x))) dx = \tan(u(x)) + C$
22. $\int u'(x)(1 + \cot^2(u(x))) dx = -\cot(u(x)) + C$
23. $\int \sin(ax+b) dx = -\frac{1}{a} \cos(ax+b) + C, \quad a, b \in \mathbb{R}, \quad a \neq 0$
24. $\int \cos(ax+b) dx = \frac{1}{a} \sin(ax+b) + C, \quad a, b \in \mathbb{R}, \quad a \neq 0$
25. $\int (1 + \tan^2(ax+b)) dx = \frac{1}{a} \tan(ax+b) + C, \quad a, b \in \mathbb{R}, \quad a \neq 0$
26. $\int (1 + \cot^2(ax+b)) dx = -\frac{1}{a} \cot(ax+b) + C, \quad a, b \in \mathbb{R}, \quad a \neq 0$
27. $\int u'(x) e^{u(x)} dx = e^{u(x)} + C$
28. $\int e^{ax} dx = \frac{1}{a} e^{ax} + C$
29. $\int u'(x) a^{u(x)} dx = \frac{a^{u(x)}}{\ln(a)} + C$
30. $\int u'(x) (u(x))^n dx = \frac{1}{n+1} u^{n+1}(x) + C$

31. $\int u'(x) \sinh(u(x))dx = \cosh(u(x)) + C$
32. $\int u'(x) \cosh(u(x))dx = \sinh(u(x)) + C$
33. $\int u'(x) \operatorname{sech}^2(u(x))dx = \tanh(u(x)) + C$
34. $\int u'(x) \operatorname{csch}^2(u(x))dx = -\coth(u(x)) + C$
35. $\int \frac{u'(x)}{\sqrt{u(x)^2+1}}dx = \sinh^{-1}(u(x)) + C = \ln\left(u(x) + \sqrt{u(x)^2+1}\right) + C$
36. $\int \frac{u'(x)}{\sqrt{u(x)^2 - 1}}dx = \cosh^{-1}(u(x)) + C = \ln\left(u(x) + \sqrt{u(x)^2-1}\right) + C, \quad u(x) > 1$
37. $\int \frac{u'(x)}{1-u(x)^2}dx = \tanh^{-1}(u(x)) + C = \frac{1}{2}\ln\left(\frac{1+u(x)}{1-u(x)}\right) + C, \quad |u(x)| < 1$
38. $\int \tanh(x)dx = \ln(\cosh(x)) + C$
39. $\int \coth(x)dx = \ln(|\sinh(x)|) + C$
40. $\int \operatorname{sech}(x)dx = 2\tan^{-1}(e^x) + C = \sin^{-1}(\tanh(x)) + C = \tan^{-1}(\sinh(x)) + C$

10.1.3 Some of the Important Trigonometric Identities

Some of the important trigonometric identities are listed below.

1. $\sin^2(x) = \frac{1-\cos(2x)}{2}$
2. $\cos^2(x) = \frac{1+\cos(2x)}{2}$
3. $\sin(2x) = 2\sin(x)\cos(x)$
4. $\cos(2x) = \cos^2(x) - \sin^2(x)$
5. $1 \pm \sin(2x) = (\sin(x) \pm \cos(x))^2$
6. $\frac{1}{\cos^2(x)} = 1 + \tan^2(x)$
7. $\frac{1}{\sin^2(x)} = 1 + \cot^2(x)$
8. $\tan^{-1}(x) + \cot^{-1}(x) = \frac{\pi}{2}$
9. $\sin^2(x) + \cos^2(x) = 1$
10. $\frac{d}{dx}\left(\sin^2(u(x))\right) = 2u'(x)\sin(u(x))\cos(u(x)) = u'(x)\sin(2u(x))$, u is a function of x
11. $\frac{d}{dx}\left(\cos^2(u(x))\right) = -2u'(x)\sin(u(x))\cos(u(x)) = -u'(x)\sin(2u(x))$, u is a function of x.

10.1.4 Trigonometric Substitutions

1. For $\sqrt{a^2 - x^2}$, the substitution is $x = a.\sin(\theta)$. The standard range for θ is $\left[-\frac{\pi}{2}, \frac{\pi}{2}\right]$. This range ensures that the substitution is a one-to-one function. In this interval, $\cos(\theta) \geq 0$, which is important for simplifying the term $\sqrt{a^2 - x^2}$ to $a.\cos(\theta)$.

2. For $\sqrt{a^2 + x^2}$, the substitution is $x = a.\tan(\theta)$. The standard range for θ is $\left(-\frac{\pi}{2}, \frac{\pi}{2}\right)$. This range covers all possible real values of x since the tangent function has a range of $(-\infty, \infty)$ in this interval. In this interval, $\sec(\theta) > 0$, which is necessary for the simplification of the term $\sqrt{a^2 + x^2}$ to $a.\sec(\theta)$.
3. For $\sqrt{x^2 - a^2}$, the substitution is $x = a.\sec(\theta)$. The standard range for θ is $\left[0, \frac{\pi}{2}\right] \cup \left[\pi, \frac{3\pi}{2}\right]$. This range ensures that the substitution is one-to-one. It also ensures that $\tan(\theta) \geq 0$ for the relevant simplification of $\sqrt{x^2 - a^2}$ to $a.\tan(\theta)$.

10.2 Solved Problems

Example 10.1 Evaluate the integral $\int 7dx$.

Solution $\int 7dx = 7x + C$

Example 10.2 Evaluate the integral $\int \frac{1}{7}dx$.

Solution $\int \frac{1}{7}dx = \frac{1}{7}x + C$

Example 10.3 Evaluate the integral $\int (t^3 + t)dx$.

Solution $\int (t^3 + t)dx = (t^3 + t)x + C$

Example 10.4 Evaluate the integral $\int (t^3 + t)dt$.

Solution $\int (t^3 + t)dt = \frac{t^4}{4} + \frac{t^2}{2} + C$

Example 10.5 Evaluate the integral $\int (x^4 - 5x^2 + 7x - 3)dx$.

Solution $\int (x^4 - 5x^2 + 7x - 3)dx = \frac{x^{4+1}}{4+1} - 5\frac{x^{2+1}}{2+1} + 7\frac{x^{1+1}}{1+1} - 3x + C = \frac{1}{5}x^5 - \frac{5}{3}x^3 + \frac{7}{2}x^2 - 3x + C$

Example 10.6 Evaluate the integral $\int \frac{1}{\sqrt[5]{x^2}}dx$.

Solution $\int \frac{1}{\sqrt[5]{x^2}}dx = \int x^{-\frac{2}{5}}dx = \frac{x^{-\frac{2}{5}+1}}{-\frac{2}{5}+1} + C = \frac{x^{\frac{3}{5}}}{\frac{3}{5}} + C = \frac{5}{3}x^{\frac{3}{5}} + C = \frac{5}{3}\sqrt[5]{x^3} + C$

Example 10.7 Evaluate the integral $\int x^{\frac{2}{7}}dx$.

Solution $\int x^{\frac{2}{7}}dx = \frac{x^{\frac{2}{7}+1}}{\frac{2}{7}+1} + C = \frac{x^{\frac{9}{7}}}{\frac{9}{7}} + C = \frac{7}{9}x^{\frac{9}{7}} + C$

Example 10.8 Evaluate the integral $\int x^{-\frac{2}{7}}dx$.

Solution $\int x^{-\frac{2}{7}}dx = \frac{x^{-\frac{2}{7}+1}}{-\frac{2}{7}+1} + C = \frac{x^{\frac{5}{7}}}{\frac{5}{7}} + C = \frac{7}{5}x^{\frac{5}{7}} + C$

Example 10.9 Evaluate the integral $\int \frac{2x^4+6}{x^3}dx$.

Solution $\int \frac{2x^4+6}{x^3}dx = \int(2x+6x^{-3})dx = 2\frac{x^{1+1}}{1+1} + 6\frac{x^{-3+1}}{-3+1} = x^2 - 3x^{-2} + C$

Example 10.10 Evaluate the integral $\int \frac{2(5x-3)^5+6}{(5x-3)^2}dx$.

Solution Recall that $\int(ax+b)^n dx = \frac{1}{a}\frac{1}{n+1}(ax+b)^{n+1} + C\ a,b,n \in \mathbb{R}, a \neq 0, n \neq -1$. Therefore,

$$\int \frac{2(5x-3)^5+6}{(5x-3)^2}dx = \int\left(2(5x-3)^3 + 6(5x-3)^{-2}\right)dx$$
$$= 2\times\frac{1}{5}\times\frac{1}{4}\times(5x-3)^4 + 6\times\frac{1}{5}\times\frac{1}{-1}\times(5x-3)^{-1} +$$
$$C = \frac{1}{10}(5x-3)^4 - \frac{6}{5}(5x-3)^{-1} + C$$

Example 10.11 Evaluate the integral $\int(2x+3)^4 dx$.

Solution Recall that $\int(ax+b)^n dx = \frac{1}{a}\frac{1}{n+1}(ax+b)^{n+1} + C\ a,b,n \in \mathbb{R}, a \neq 0, n \neq -1$. Therefore,

$$\int(2x+3)^4 dx = \frac{1}{2}\times\frac{1}{4+1}\times(2x+3)^{4+1} + C = \frac{1}{10}(2x+3)^5 + C$$

Example 10.12 Evaluate the integral $\int(5x+6)^7 dx$.

Solution Recall that $\int(ax+b)^n dx = \frac{1}{a}\frac{1}{n+1}(ax+b)^{n+1} + C\ a,b,n \in \mathbb{R}, a \neq 0, n \neq -1$. Therefore,

$$\int(5x+6)^7 dx = \frac{1}{5}\times\frac{1}{7+1}\times(5x+6)^{7+1} + C = \frac{1}{40}(5x+6)^8 + C$$

Example 10.13 Show that $\int(5x+6)^7 dx = \frac{1}{40}(5x+6)^8 + C$.

Solution In this example, we need to demonstrate that the derivative of the proposed solution is equal to the original integrand.

$$\frac{d}{dx}\left(\frac{1}{40}(5x+6)^8 + C\right) = \frac{d}{dx}\left(\frac{1}{40}(5x+6)^8\right) = \frac{1}{40}\frac{d}{dx}\left((5x+6)^8\right)$$
$$= \frac{1}{40}\times 8\times 5\times(5x+6)^7 = (5x+6)^7$$

Example 10.14 Evaluate the integral $\int \frac{7}{2x^2-12x+18}dx$.

Solution $2x^2 - 12x + 18 = 2(x^2 - 6x + 9) = 2(x - 3)^2$. Therefore,

$$\int \frac{7}{2x^2-12x+18}dx = \int \frac{7}{2(x-3)^2}dx = \frac{7}{2}\int \frac{1}{(x-3)^2}dx = \frac{7}{2}\int (x-3)^{-2}dx.$$

Recall that $\int (ax+b)^n dx = \frac{1}{a}\frac{1}{n+1}(ax+b)^{n+1} + C \quad a,b,n \in \mathbb{R}, a \neq 0, n \neq -1$. Therefore,

$$\int \frac{7}{2x^2-12x+18}dx = \frac{7}{2}\int (x-3)^{-2}dx$$
$$= \frac{7}{2} \times \frac{1}{1} \times \frac{1}{-2+1}(x-3)^{-2+1} +$$
$$C = -\frac{7}{2}(x-3)^{-1} + C = \frac{7}{6-2x} + C$$

Example 10.15 Show that $\int \frac{7}{2x^2-12x+18}dx = \frac{7}{6-2x} + C$.

Solution

$$\frac{d}{dx}\left(\frac{7}{6-2x} + C\right) = \frac{d}{dx}\left(\frac{7}{6-2x}\right) = \frac{0 \times (6-2x) - (-2) \times 7}{(6-2x)^2} = \frac{14}{(6-2x)^2}$$
$$= \frac{14}{4(3-x)^2} = \frac{7}{2(3-x)^2} = \frac{7}{2(x^2-6x+9)}$$
$$= \frac{7}{2x^2-12x+18}$$

Example 10.16 Evaluate the integral $\int \frac{7}{x-3}dx$.

Solution Recall that $\int \frac{u'}{u}dx = \ln(|u|) + C$. Therefore, $\int \frac{7}{x-3}dx = 7\int \frac{1}{x-3}dx = 7\ln(|x-3|) + C$

Example 10.17 Evaluate the integral $\int \frac{7x}{x^2-3}dx$.

Solution Recall that $\int \frac{u'}{u}dx = \ln(|u|) + C$. Therefore, $\int \frac{7x}{x^2-3}dx = \int \frac{7}{2}\frac{2x}{x^2-3}dx = \frac{7}{2}\int \frac{2x}{x^2-3}dx = \frac{7}{2}\ln\left(\left|x^2-3\right|\right) + C$

Example 10.18 Evaluate the integral $\int \frac{(\sqrt{x}+1)^2}{\sqrt{x}}dx$.

Solution

$$\int \frac{(\sqrt{x}+1)^2}{\sqrt{x}}dx = \int \frac{x+2\sqrt{x}+1}{\sqrt{x}}dx = \int \left(x^{\frac{1}{2}} + 2 + x^{-\frac{1}{2}}\right)dx$$
$$= \frac{x^{\frac{1}{2}+1}}{\frac{1}{2}+1} + 2x + \frac{x^{-\frac{1}{2}+1}}{-\frac{1}{2}+1} + C = \frac{2}{3}x^{\frac{3}{2}} + 2x + 2x^{\frac{1}{2}} + C$$

Example 10.19 Evaluate the integral $\int \frac{4x^3-3}{x^4-3x}dx$.

Solution Recall that $\int \frac{u'}{u}dx = \ln(|u|) + C$. Therefore,

$$\int \frac{4x^3 - 3}{x^4 - 3x}dx = \ln\left(\left|x^4 - 3x\right|\right) + C$$

Example 10.20 Evaluate the integral $\int \tan(x)dx$.

Solution Recall that $\int \frac{u'}{u}dx = \ln(|u|) + C$. Therefore, $\int \tan(x)dx = \int \frac{\sin(x)}{\cos(x)}dx = -\int \frac{-\sin(x)}{\cos(x)}dx = -\ln(|\cos(x)|) + C = \ln\left(\frac{1}{|\cos(x)|}\right) + C = \ln(\sec(x)) + C$

Example 10.21 Evaluate the integral $\int \cot(x)dx$.

Solution Recall that $\int \frac{u'}{u}dx = \ln(|u|) + C$. Therefore,

$$\int \cot(x)dx = \int \frac{\cos(x)}{\sin(x)}dx = \ln(|\sin(x)|) + C$$

Example 10.22 Evaluate the integral $\int \cot(3x)dx$.

Solution Recall that $\int \frac{u'}{u}dx = \ln(|u|) + C$. Therefore,

$$\int \cot(3x)dx = \int \frac{\cos(3x)}{\sin(3x)}dx = \frac{1}{3}\int \frac{3\cos(3x)}{\sin(3x)}dx = \frac{1}{3}\ln(|\sin(3x)|) + C$$
$$= \ln\left(\sqrt[3]{|\sin(3x)|}\right) + C$$

Example 10.23 Evaluate the integral $\int \frac{x}{x^2+1}dx$.

Solution Recall that $\int \frac{u'}{u}dx = \ln(|u|) + C$. Therefore,

$$\int \frac{x}{x^2 + 1}dx = \int \frac{1}{2}\frac{2x}{x^2 + 1}dx = \frac{1}{2}\int \frac{2x}{x^2 + 1}dx = \frac{1}{2}\ln\left(\left|x^2 + 1\right|\right) + C$$
$$= \frac{1}{2}\ln\left(x^2 + 1\right) + C = \ln\left(\sqrt{x^2 + 1}\right) + C$$

Example 10.24 Evaluate the integral $\int \sin(4x)dx$.

Solution $\int \sin(4x)dx = -\frac{1}{4}\cos(4x) + C$

Example 10.25 Evaluate the integral $\int \sin\left(4x + \frac{\pi}{3}\right)dx$.

Solution $\int \sin(4x)dx = -\frac{1}{4}\cos\left(4x + \frac{\pi}{3}\right) + C$

Example 10.26 Evaluate the integral $\int \cos(4x)dx$.

Solution $\int \cos(4x)dx = \frac{1}{4}\sin(4x) + C$

Example 10.27 Evaluate the integral $\int \cos\left(4x + \frac{\pi}{3}\right)dx$.

Solution

$$\int \cos\left(4x + \frac{\pi}{3}\right)dx = \frac{1}{4}\sin\left(4x + \frac{\pi}{3}\right) + C$$

Example 10.28 Evaluate the integral $\int(1 + \tan^2(3x))dx$.

Solution Recall that $\int(1 + \tan^2(ax + b))dx = \frac{1}{a}\tan(ax + b) + C$. Therefore,

$$\int \left(1 + \tan^2(3x)\right)dx = \frac{1}{3}\tan(3x) + C$$

Example 10.29 Evaluate the integral $\int\left(1 + \cot^2\left(3x + \frac{\pi}{4}\right)\right)dx$.

Solution Recall that $\int(1 + \cot^2(ax + b))dx = -\frac{1}{a}\cot(ax + b) + C$. Therefore,

$$\int \left(1 + \cot^2\left(3x + \frac{\pi}{4}\right)\right)dx = \frac{1}{3}\cot\left(3x + \frac{\pi}{4}\right) + C$$

Example 10.30 Evaluate the integral $\int(\cos(3x)e^{\sin(3x)})dx$.

Solution Recall that $\int u'e^u dx = e^u + C$. Therefore,

$$\int \left(\cos(3x)e^{\sin(3x)}\right)dx = \frac{1}{3}\int \left(3\cos(3x)e^{\sin(3x)}\right)dx = \frac{1}{3}e^{\sin(3x)} + C$$

Example 10.31 Evaluate the integral $\int(\sin(3x)e^{\cos(3x)})dx$.

Solution Recall that $\int u'e^u dx = e^u + C$. Therefore,

$$\int \left(\sin(3x)e^{\cos(3x)}\right)dx = \frac{1}{-3}\int \left(-3\sin(3x)e^{\cos(3x)}\right)dx = -\frac{1}{3}e^{\cos(3x)} + C$$

Example 10.32 Evaluate the integral $\int\left((x + 3)e^{x^2+3x}\right)dx$.

Solution Recall that $\int u'e^u dx = e^u + C$. Therefore,

$$\int \left((x + 3)e^{x^2+3x}\right)dx = e^{x^2+3x} + C$$

Example 10.33 Evaluate the integral $\int \frac{x^5 - 4x^3}{(x^6 - 6x^4)^7}dx$.

Solution Recall that $\int u'(x)u^n(x)dx = \frac{1}{n+1}u^{n+1}(x) + C$. Therefore,

$$\int \frac{x^5-4x^3}{(x^6-6x^4)^7}dx = \frac{1}{6}\int (6x^5-24x^3)(x^6-6x^4)^{-7}dx = \frac{1}{6}\frac{(x^6-6x^4)^{-7+1}}{-7+1}+C$$
$$= \frac{1}{-36}(x^6-6x^4)^{-6}+C$$

Example 10.34 Evaluate the integral $\int (3x+4)^5dx$.

Solution Let's use a substitution and set $u = 3x + 4$. Thus,

$$u = 3x+4 \Longrightarrow du = 3dx \Longrightarrow dx = \frac{1}{3}du$$

$$\int (3x+4)^5dx = \int u^5 \times \frac{1}{3}du = \frac{1}{3}\int u^5du = \frac{1}{3}\frac{u^6}{6}+C = \frac{1}{18}u^6+C$$
$$= \frac{1}{18}(3x+4)^6+C$$

Example 10.35 Evaluate the integral $\int x^2\sqrt{x+2}dx$.

Solution Let's use a substitution and set $u = x + 2$. Thus,

$$u = x+2 \Longrightarrow du = dx$$

$$\int x^2\sqrt{x+2}dx = \int (u-2)^2\sqrt{u}du = \int (u^2-4u+4)\sqrt{u}du$$
$$= \int \left(u^{\frac{5}{2}}-4u^{\frac{3}{2}}+4u^{\frac{1}{2}}\right)du = \frac{u^{\frac{5}{2}+1}}{\frac{5}{2}+1} - \frac{4u^{\frac{3}{2}+1}}{\frac{3}{2}+1} + \frac{4u^{\frac{1}{2}+1}}{\frac{1}{2}+1}+C$$
$$= \frac{2}{7}u^{\frac{7}{2}} - \frac{8}{5}u^{\frac{5}{2}} + \frac{8}{3}u^{\frac{3}{2}}+C = \frac{2}{7}(x+2)^{\frac{7}{2}} - \frac{8}{5}(x+2)^{\frac{5}{2}} + \frac{8}{3}(x+2)^{\frac{3}{2}}+C$$

Example 10.36 Evaluate the integral $\int \frac{(\sqrt{x}+1)^2}{\sqrt{x}}dx$.

Solution Let's use a substitution and set $u = \sqrt{x} + 1$. Thus,

$$u = \sqrt{x}+1 \Longrightarrow du = \frac{1}{2\sqrt{x}}dx \Longrightarrow du = \frac{1}{2(u-1)}dx \Longrightarrow dx = 2(u-1)du$$

$$\int \frac{(\sqrt{x}+1)^2}{\sqrt{x}}dx = \int \frac{u^2}{u-1} \times 2(u-1)du = \int 2u^2du = \frac{2}{3}u^3+C$$
$$= \frac{2}{3}(\sqrt{x}+1)^3+C = \frac{2}{3}x^{\frac{3}{2}} + \frac{2}{3}\times 3x + \frac{2}{3}\times 3\sqrt{x} + \frac{2}{3}+C$$
$$= \frac{2}{3}x^{\frac{3}{2}} + 2x + 2\sqrt{x} + \frac{2}{3} + C = \frac{2}{3}x^{\frac{3}{2}} + 2x + 2\sqrt{x} + C'$$

The result obtained is consistent with that of Example 10.18.

Example 10.37 Evaluate the integral $\int x\sqrt{7-x^2}dx$.

Solution

$$\int x\sqrt{7-x^2}dx=\frac{1}{-2}\int -2x\left(7-x^2\right)^{\frac{1}{2}}dx=\frac{1}{-2}\times\frac{\left(7-x^2\right)^{\frac{1}{2}+1}}{\frac{1}{2}+1}+C=-\frac{1}{3}\left(7-x^2\right)^{\frac{3}{2}}+C$$

Example 10.38 Evaluate the integral $\int x(3x^2+4)^5dx$.

Solution

$$\int x\left(3x^2+4\right)^5dx=\frac{1}{6}\int 6x\left(3x^2+4\right)^5dx=\frac{1}{6}\frac{\left(3x^2+4\right)^{5+1}}{5+1}+C=\frac{1}{36}\left(3x^2+4\right)^6+C$$

Example 10.39 Evaluate the integral $\int\frac{x-4}{(x^2-8x+6)^7}dx$.

Solution

$$\int\frac{x-4}{\left(x^2-8x+6\right)^7}dx=\frac{1}{2}\int (2x-8)\left(x^2-8x+6\right)^{-7}dx=\frac{1}{2}\frac{\left(x^2-8x+6\right)^{-7+1}}{-7+1}+C$$

$$=\frac{\left(x^2-8x+6\right)^{-6}}{-12}+C=\frac{1}{-12\left(x^2-8x+6\right)^6}+C$$

Example 10.40 Evaluate the integral $\int\frac{\sqrt{4-x^2}}{x^4}dx$.

Solution Let's use a substitution and set $x=2\sin(\theta)$. Thus,

$$x=2\sin(\theta)\Rightarrow dx=2\cos(\theta)d\theta$$

$$\int\frac{\sqrt{4-x^2}}{x^4}dx=\int\frac{\sqrt{4-4\sin^2(\theta)}}{16\sin^4(\theta)}\times 2\cos(\theta)d\theta=\int\frac{\sqrt{\cos^2(\theta)}}{4\sin^4(\theta)}\times\cos(\theta)d\theta$$

$$=\frac{1}{4}\int\frac{|\cos(\theta)|}{\sin^4(\theta)}\times\cos(\theta)d\theta$$

Let's assume $\cos(\theta)>0$, thus,

$$\int\frac{\sqrt{4-x^2}}{x^4}dx=\frac{1}{4}\int\frac{|\cos(\theta)|}{\sin^4(\theta)}\times\cos(\theta)d\theta=\frac{1}{4}\int\frac{\cos^2(\theta)}{\sin^4(\theta)}d\theta$$

$$=\frac{1}{4}\int\frac{\cos^2(\theta)}{\sin^2(\theta)}\times\frac{1}{\sin^2(\theta)}d\theta=\frac{1}{4}\int\cot^2(\theta)\times\left(1+\cot^2(\theta)\right)d\theta$$

$$=-\frac{1}{4}\int\cot^2(\theta)\times-\left(1+\cot^2(\theta)\right)d\theta=-\frac{1}{4}\frac{\cot^3(\theta)}{3}+C=-\frac{1}{12}\cot^3(\theta)+C$$

Finally, use the initial substitution $x = 2\sin(\theta)$ to find $\cot(\theta)$ in terms of x. Draw a right triangle where $\sin(\theta) = \frac{x}{2}$ (Fig. 10.1). According to Fig. 10.1 and Pythagorean theorem, $\cot(\theta) = \frac{AB}{AC} = \frac{\sqrt{2^2 - x^2}}{2} = \frac{\sqrt{4-x^2}}{x}$. Therefore, $\int \frac{\sqrt{4-x^2}}{x^4} dx = -\frac{1}{12}\cot^3(\theta) + C = -\frac{1}{12}\left(\frac{\sqrt{4-x^2}}{x}\right)^3 + C$.

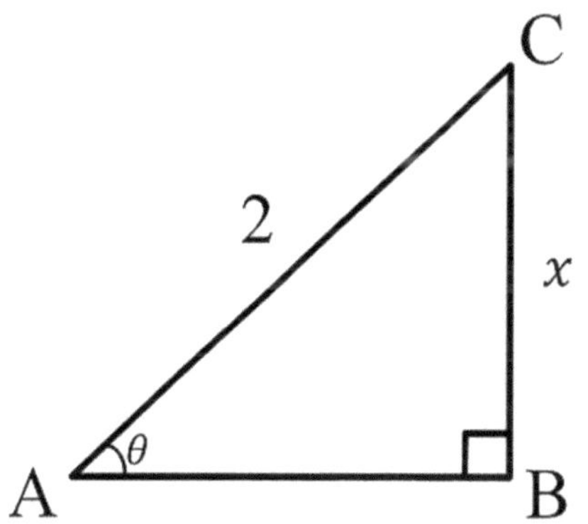

Fig. 10.1 The right triangle for which $\sin(\theta) = \frac{x}{2}$

Exercise 10.1 Evaluate the integral $\int \frac{\sqrt{4-x^2}}{x^4} dx$. Use the substitution $x = 2\cos(\theta)$.

Example 10.41 Evaluate the integral $\int \frac{6\cos(\sqrt{x})}{\sqrt{x}} dx$.

Solution Let's use a substitution and set $u = \sqrt{x}$. Thus,

$$u = \sqrt{x} \Rightarrow du = \frac{1}{2\sqrt{x}} dx \Rightarrow du = \frac{1}{2u} dx \Rightarrow dx = 2u du$$

$$\int \frac{6\cos(\sqrt{x})}{\sqrt{x}} dx = 6\int \frac{\cos(u)}{u} 2u du = 12\int \cos(u) du = 12\sin(u) + C = 12\sin(\sqrt{x}) + C$$

Example 10.42 Evaluate the integral $\int \frac{\cos(x)}{(1+\sin(x))^2} dx$.

Solution Let's use a substitution and set $u = 1 + \sin(x)$. Thus,

$$u = 1 + \sin(x) \Rightarrow du = \cos(x) dx$$

$$\int \frac{\cos(x)}{(1+\sin(x))^2} dx = \int \frac{du}{u^2} = \int u^{-2} du = -u^{-1} + C = -\frac{1}{1+\sin(x)} + C$$

Example 10.43 Evaluate the integral $\int \frac{\cos^3(3x)}{\sqrt[3]{\sin(3x)}} dx$.

Solution Let's use a substitution and set $u = \sin(3x)$. Thus,

$$u = \sin(3x) \Rightarrow du = 3\cos(3x) dx$$

$$\int \frac{\cos^3(3x)}{\sqrt[3]{\sin(3x)}}dx=\frac{1}{3}\int \frac{\cos^2(3x)}{\sqrt[3]{\sin(3x)}}3\cos(3x)dx$$
$$=\frac{1}{3}\int \frac{1-\sin^2(3x)}{\sqrt[3]{\sin(3x)}}3\cos(3x)dx=\frac{1}{3}\int \frac{1-u^2}{\sqrt[3]{u}}du=\frac{1}{3}\int \left(u^{-\frac{1}{3}}-u^{\frac{5}{3}}\right)du$$
$$=\frac{1}{3}\left(\frac{u^{-\frac{1}{3}+1}}{-\frac{1}{3}+1}-\frac{u^{\frac{5}{3}+1}}{\frac{5}{3}+1}\right)+C=$$
$$\frac{u^{\frac{2}{3}}}{2}-\frac{u^{\frac{8}{3}}}{8}+C=\frac{(\sin(3x))^{\frac{2}{3}}}{2}-\frac{(\sin(3x))^{\frac{8}{3}}}{8}+C$$

Example 10.44 Evaluate the integral $\int \frac{dx}{\sqrt{x}\cos^2(\sqrt{x})}$.

Solution Let's use a substitution and set $u=\sqrt{x}$. Thus,

$$u=\sqrt{x}\Rightarrow du=\frac{1}{2\sqrt{x}}dx$$

$$\int \frac{dx}{\sqrt{x}\cos^2(\sqrt{x})}=2\int \frac{dx}{2\sqrt{x}\cos^2(\sqrt{x})}=2\int \frac{du}{\cos^2(u)}$$
$$=2\int \left(1+\tan^2(u)\right)du=2\tan(u)+C=2\tan\left(\sqrt{x}\right)+C$$

Exercise 10.2 Evaluate the integral $\int \frac{dx}{\sqrt{x}\sin^2(\sqrt{x})}$. (Ans. $-2\cot(\sqrt{x})+C$)

Example 10.45 Evaluate the integral $\int \cot(x)\csc^3(x)dx$.

Solution $\int \cot(x)\csc^3(x)dx=\int \frac{\cos(x)}{\sin(x)}\frac{1}{\sin^3(x)}dx \quad =\int \frac{\cos(x)}{\sin^4(x)}dx=\int \cos(x)\sin^{-4}(x)dx=$ $\frac{\sin^{-4+1}(x)}{-4+1}+C=\frac{1}{-3}\sin^{-3}(x)+C=-\frac{1}{3}\frac{1}{\sin^3(x)}+C=-\frac{1}{3}\csc^3(x)+C$

Example 10.46 Evaluate the integral $\int \frac{\cos(2\sqrt[3]{x})}{\sqrt[3]{x^2}}dx$.

Solution Let's use a substitution and set $u=2\sqrt[3]{x}$. Thus,

$$u=2\sqrt[3]{x}\Rightarrow u=2x^{\frac{1}{3}}\Rightarrow du=\frac{2}{3\sqrt[3]{x^2}}dx$$

$$\int \frac{\cos(2\sqrt[3]{x})}{\sqrt[3]{x^2}}dx=\frac{3}{2}\int \frac{2\cos(2\sqrt[3]{x})}{3\sqrt[3]{x^2}}dx=\frac{3}{2}\int \cos(u)du=\frac{3}{2}\sin(u)+C=\frac{3}{2}\sin\left(2\sqrt[3]{x}\right)+C$$

Example 10.47 Evaluate the integral $\int \frac{\cos(x)}{1-\sin(x)}dx$.

Solution

$$\int \frac{\cos(x)}{1-\sin(x)}dx = \int \cos(x)(1-\sin(x))^{-1}dx$$
$$= -\int -\cos(x)(1-\sin(x))^{-1}dx = -\ln(|1-\sin(x)|) + C$$

Example 10.48 Evaluate the integral $\int \frac{\cos(2x)}{\sqrt{1+\sin(2x)}}dx$.

Solution

$$\int \frac{\cos(2x)}{\sqrt{1+\sin(2x)}}dx = \int \left(\cos(2x)(1+\sin(2x))^{-\frac{1}{2}}\right)dx$$
$$= \frac{1}{2}\int \left(2\cos(2x)(1+\sin(2x))^{-\frac{1}{2}}\right)dx$$
$$= \frac{1}{2}\frac{(1+\sin(2x))^{-\frac{1}{2}+1}}{-\frac{1}{2}+1} + C = (1+\sin(2x))^{\frac{1}{2}} + C = \sqrt{1+\sin(2x)} + C$$

Example 10.49 Evaluate the integral $\int \sec^2(3x)\tan(3x)dx$.

Solution

$$\int \sec^2(3x)\tan(3x)dx = \int \frac{1}{\cos^2(3x)}\tan(3x)dx = \int \left(1+\tan^2(3x)\right)\tan(3x)dx$$
$$= \frac{1}{3}\int 3\left(1+\tan^2(3x)\right)\tan(3x)dx$$
$$= \frac{1}{3}\frac{\tan^{1+1}(3x)}{1+1} + C = \frac{\tan^2(3x)}{6} + C$$

Example 10.50 Evaluate the integral $\int \frac{\cos(x)}{\sin^2(x)+2\sin(x)+1}dx$.

Solution

$$\int \frac{\cos(x)}{\sin^2(x)+2\sin(x)+1}dx = \int \frac{\cos(x)}{(\sin(x)+1)^2}dx = \int \left(\cos(x)(\sin(x)+1)^{-2}\right)dx$$
$$= \frac{(\sin(x)+1)^{-2+1}}{-2+1} + C = -\frac{1}{\sin(x)+1} + C$$

Example 10.51 Evaluate the integral $\int 4\sin(2x)(1+\cos(2x))^7dx$.

Solution

$$\int 4\sin(2x)(1+\cos(2x))^7 dx = 4\int \sin(2x)(1+\cos(2x))^7 dx = \frac{4}{-2}\int$$
$$-2\sin(2x)(1+\cos(2x))^7 dx = \frac{4}{-2}\frac{(1+\cos(2x))^{7+1}}{7+1}+C = \frac{(1+\cos(2x))^8}{-4}+C$$

Example 10.52 Show that $\int \frac{1}{\sqrt{1-x^2}}dx = \sin^{-1}(x)+C$.

Solution Let's assume $y=\sin^{-1}(x)$. Therefore,

$$y=\sin^{-1}(x) \Rightarrow \begin{cases} -1 \le x \le 1 \\ -\frac{\pi}{2} \le y \le \frac{\pi}{2} \end{cases}$$

$$y=\sin^{-1}(x) \Rightarrow x=\sin(y) \Rightarrow \frac{d}{dx}(x)=\frac{d}{dx}(\sin(y)) \Rightarrow 1=y'\cos(y) \Rightarrow y'=\frac{1}{\cos(y)}$$

$$\sin^2(y)+\cos^2(y)=1 \Rightarrow \cos(y)=\pm\sqrt{1-\sin^2(y)} \overset{-\frac{\pi}{2}\le y\le\frac{\pi}{2} \Rightarrow \cos(y)\ge 0}{\Rightarrow} \cos(y)=\sqrt{1-\sin^2(y)}$$

$$y'=\frac{1}{\cos(y)} \Rightarrow y'=\frac{1}{\sqrt{1-\sin^2(y)}} \Rightarrow y'=\frac{1}{\sqrt{1-x^2}} \Rightarrow \int y'dx=\int \frac{1}{\sqrt{1-x^2}}dx \Rightarrow y$$
$$=\int \frac{1}{\sqrt{1-x^2}}dx+C \Rightarrow \sin^{-1}(x)=\int \frac{1}{\sqrt{1-x^2}}dx+$$
$$C \Rightarrow \int \frac{1}{\sqrt{1-x^2}}dx=\sin^{-1}(x)-C \overset{C'=-C}{\Rightarrow} \int \frac{1}{\sqrt{1-x^2}}dx=\sin^{-1}(x)+C'$$

Example 10.53 Show that $\int \frac{1}{\sqrt{1-x^2}}dx = -\cos^{-1}(x)+C$.

Solution Let's assume $y=\cos^{-1}(x)$. Therefore,

$$y=\cos^{-1}(x) \Rightarrow \begin{cases} -1 \le x \le 1 \\ 0 \le y \le \pi \end{cases}$$

$$y=\cos^{-1}(x) \Rightarrow x=\cos(y) \Rightarrow \frac{d}{dx}(x)=\frac{d}{dx}(\cos(y)) \Rightarrow 1=-y'\sin(y) \Rightarrow y'=-\frac{1}{\sin(y)}$$

$$\sin^2(y)+\cos^2(y)=1 \Rightarrow \sin(y)=\pm\sqrt{1-\cos^2(y)} \overset{0\le y\le\pi \Rightarrow \sin(y)\ge 0}{\Rightarrow} \sin(y)=\sqrt{1-\cos^2(y)}$$

$$y'=-\frac{1}{\sin(y)} \Rightarrow y'=-\frac{1}{\sqrt{1-\cos^2(y)}} \Rightarrow y'=-\frac{1}{\sqrt{1-x^2}} \Rightarrow \int y'dx=$$
$$-\int \frac{1}{\sqrt{1-x^2}}dx \Rightarrow y=-\int \frac{1}{\sqrt{1-x^2}}dx+C \Rightarrow \cos^{-1}(x)=-\int \frac{1}{\sqrt{1-x^2}}dx$$
$$+C \Rightarrow \int \frac{1}{\sqrt{1-x^2}}dx=-\cos^{-1}(x)+C$$

Example 10.54 Show that $\int \frac{1}{1+x^2}dx = \tan^{-1}(x)+C$.

Solution Let's assume $y = \tan^{-1}(x)$. Therefore,

$$y = \tan^{-1}(x) \Rightarrow x = \tan(y) \Rightarrow \frac{d}{dx}(x) = \frac{d}{dx}(\tan(y)) \Rightarrow 1 = y'\left(1 + \tan^2(y)\right) \Rightarrow y'$$
$$= \frac{1}{1 + \tan^2(y)} \Rightarrow y' = \frac{1}{1 + x^2} \Rightarrow \int y' dx = \int \frac{1}{1 + x^2} dx \Rightarrow y = \int \frac{1}{1 + x^2} dx$$
$$+ C \Rightarrow \tan^{-1}(x) = \int \frac{1}{1 + x^2} dx + C \Rightarrow \int \frac{1}{1 + x^2} dx = \tan^{-1}(x) - C$$
$$\overset{C' = -C}{\Rightarrow} \int \frac{1}{1 + x^2} dx = \tan^{-1}(x) + C'$$

Example 10.55 Show that $\int \frac{1}{1+x^2} dx = -\cot^{-1}(x) + C$.

Solution Let's assume $y = \cot^{-1}(x)$. Therefore,

$$y = \cot^{-1}(x) \Rightarrow x = \cot(y) \Rightarrow \frac{d}{dx}(x) = \frac{d}{dx}(\cot(y)) \Rightarrow 1 = -y'\left(1 + \cot^2(y)\right) \Rightarrow y'$$
$$= -\frac{1}{1 + \cot^2(y)} \Rightarrow y' = -\frac{1}{1 + x^2} \Rightarrow \int y' dx = -\int \frac{1}{1 + x^2} dx \Rightarrow y =$$
$$-\int \frac{1}{1 + x^2} dx + C \Rightarrow \cot^{-1}(x) = -\int \frac{1}{1 + x^2} dx + C \Rightarrow \int \frac{1}{1 + x^2} dx =$$
$$-\cot^{-1}(x) + C$$

Exercise 10.3 Evaluate $\int \sec(x)dx$. (Hint: $\int \sec(x)dx = \int \frac{1}{\cos(x)} dx = \int \frac{\cos(x)}{\cos^2(x)} dx = \int \frac{\cos(x)}{1 - \sin^2(x)} dx$ then use the substitution $u = \sin(x)$)

Example 10.56 Evaluate the integral $\int \frac{dx}{3x\sqrt{x^2 - 9}}$.

Solution Recall that:

$$\int \frac{1}{|u(x)|\sqrt{u^2(x) - a^2}} du(x) = \frac{1}{a}\sec^{-1}\left(\left|\frac{u(x)}{a}\right|\right) + C \Rightarrow \int \frac{1}{x\sqrt{x^2 - a^2}} dx = \frac{1}{a}\sec^{-1}\left(\left|\frac{x}{a}\right|\right) + C$$

Therefore,

$$\int \frac{dx}{3x\sqrt{x^2 - 9}} = \frac{1}{3}\int \frac{dx}{x\sqrt{x^2 - 3^2}} = \frac{1}{3}\left(\frac{1}{3}\sec^{-1}\left(\left|\frac{x}{3}\right|\right)\right) + C = \frac{1}{9}\sec^{-1}\left(\frac{|x|}{3}\right) + C.$$

Example 10.57 Evaluate the integral $\int \frac{dx}{(x+2)\sqrt{x^2 + 4x - 5}}$.

Solution

$$\int \frac{dx}{(x+2)\sqrt{x^2+4x-5}} = \int \frac{dx}{(x+2)\sqrt{(x+2)^2-9}} = \int \frac{dx}{(x+2)\sqrt{(x+2)^2-3^2}}$$

Using the formula $\int \frac{1}{|u(x)|\sqrt{u^2(x)-a^2}}du(x) = \frac{1}{a}\sec^{-1}\left(\left|\frac{u(x)}{a}\right|\right) + C,$ we obtain:

$$\int \frac{dx}{(x+2)\sqrt{(x+2)^2-3^2}} = \frac{1}{3}\sec^{-1}\left(\left|\frac{x+2}{3}\right|\right) + C = \frac{1}{3}\sec^{-1}\left(\frac{|x+2|}{3}\right) + C$$

Example 10.58 Evaluate the integral $\int \frac{1+\tan^2(x)}{4+\tan^2(x)}dx$.

Solution Let's use a substitution and set $u = \tan(x)$. Thus,

$$u = \tan(x) \Rightarrow du = \left(1 + \tan^2(x)\right)dx$$

$$\int \frac{1+\tan^2(x)}{4+\tan^2(x)}dx = \int \frac{du}{4+u^2} = \int \frac{du}{2^2+u^2} = \frac{1}{2}\tan^{-1}\left(\frac{u}{2}\right) + C = \frac{1}{2}\tan^{-1}\left(\frac{\tan(x)}{2}\right) + C$$

Example 10.59 Evaluate the integral $\int \frac{1}{x^2+2x+5}dx$.

Solution $\int \frac{1}{x^2+2x+5}dx = \int \frac{1}{x^2+2x+1+4}dx = \int \frac{1}{(x+1)^2+4}dx$. Let's use a substitution and set $u = x + 1$. Thus,

$$u = x + 1 \Rightarrow du = dx$$

$$\int \frac{1}{(x+1)^2+4}dx = \int \frac{1}{u^2+2^2}du = \frac{1}{2}\tan^{-1}\left(\frac{u}{2}\right) + C = \frac{1}{2}\tan^{-1}\left(\frac{x+1}{2}\right) + C$$

Example 10.60 Evaluate the integral $\int \frac{\csc^2(x)}{16+\cot^2(x)}dx$.

Solution

$$\int \frac{\csc^2(x)}{16+\cot^2(x)}dx = \int \frac{\frac{1}{\sin^2(x)}}{16+\frac{\cos^2(x)}{\sin^2(x)}}dx = \int \frac{1}{16\sin^2(x)+\cos^2(x)}dx$$

$$= \int \frac{1}{\cos^2(x)\left(16\frac{\sin^2(x)}{\cos^2(x)}+1\right)}dx$$

$$= \int \frac{1}{\cos^2(x)(16\tan^2(x)+1)}dx = \int \frac{1+\tan^2(x)}{16\tan^2(x)+1}dx$$

Let's use a substitution and set $u = \tan(x)$. Thus,

$$u = \tan(x) \Rightarrow du = \left(1 + \tan^2(x)\right)dx$$

$$\int \frac{1+\tan^2(x)}{16\tan^2(x)+1}dx = \int \frac{du}{16u^2+1} = \frac{1}{16}\int \frac{du}{u^2+\left(\frac{1}{4}\right)^2}$$
$$= \frac{1}{16}\times\left(4\tan^{-1}(4u)+C\right) = \frac{1}{4}\tan^{-1}(4u)+\frac{C}{16}$$
$$= \frac{1}{4}\tan^{-1}(4\tan(x))+\frac{C}{16} = \frac{1}{4}\tan^{-1}(4\tan(x))+C'$$

Example 10.61 Evaluate the integral $\int \frac{\sin(x)}{1+\cos^2(x)}dx$.

Solution Let's use a substitution and set $u = \cos(x)$. Thus,

$$u = \cos(x) \Rightarrow du = -\sin(x)dx$$

$$\int \frac{\sin(x)}{1+\cos^2(x)}dx = -\int \frac{-\sin(x)}{1+\cos^2(x)}dx = -\int \frac{du}{1+u^2} = \cot^{-1}(u)+C = \cot^{-1}(\cos(x))+C$$

Example 10.62 Evaluate the integral $\int \frac{\sin(x)}{1+\cos^2(x)}dx$.

Solution Let's use a substitution and set $u = \cos(x)$. Thus,

$$u = \cos(x) \Rightarrow du = -\sin(x)dx$$

$$\int \frac{\sin(x)}{1+\cos^2(x)}dx = -\int \frac{-\sin(x)}{1+\cos^2(x)}dx = -\int \frac{du}{1+u^2} = -\tan^{-1}(u)+C = -\tan^{-1}(\cos(x))+C$$

Exercise 10.4 Show the results of Examples 10.61 and 10.62 are consistent. (Hint: Recall that $\forall x,\ \tan^{-1}(x) + \cot^{-1}(x) = \frac{\pi}{2}$)

Example 10.63 Evaluate the integral $\int \frac{3x}{\sqrt{4-4x^4}}dx$.

Solution Recall that $\int \frac{1}{\sqrt{1-x^2}}du(x) = \sin^{-1}(x) + C$. Let's use a substitution and set $u = x^2$. Thus,

$$u = x^2 \Rightarrow du = 2xdx$$

$$\int \frac{3x}{\sqrt{4-4x^4}}dx = 3\int \frac{x}{2\sqrt{1-x^4}}dx = \frac{3}{2}\int \frac{x}{\sqrt{1-x^4}}dx = \frac{3}{4}\int \frac{2x}{\sqrt{1-(x^2)^2}}dx$$
$$= \frac{3}{4}\int \frac{du}{\sqrt{1-u^2}} = \frac{3}{4}\sin^{-1}(u)+C = \frac{3}{4}\sin^{-1}\left(x^2\right)+C$$

Example 10.64 Evaluate the integral $\int \frac{x}{4+x^4}dx$.

Solution Recall that $\int \frac{1}{a^2+u(x)^2}du(x) = \frac{1}{a}\tan^{-1}\left(\frac{u(x)}{a}\right) + C$. Let's use a substitution and set $u = x^2$. Thus,

$$u = x^2 \Rightarrow du = 2xdx$$

$$\int \frac{x}{4+x^4}dx = \frac{1}{2}\int \frac{2x}{2^2+(x^2)^2}dx = \frac{1}{2}\int \frac{du}{2^2+u^2} = \frac{1}{2}\times\frac{1}{2}\tan^{-1}\left(\frac{u}{2}\right) + C = \frac{1}{4}\tan^{-1}\left(\frac{x^2}{2}\right) + C$$

Example 10.65 Evaluate the integral $\int \frac{(5x-1)^{17}}{(3x+3)^{19}}dx$.

Solution $\int \frac{(5x-1)^{17}}{(3x+3)^{19}}dx = \int \left(\frac{5x-1}{3x+3}\right)^{17} \times \frac{1}{(3x+3)^2}dx$. Let's use a substitution and set $u = \frac{5x-1}{3x+3}$. Thus,

$$u = \frac{5x-1}{3x+3} \Rightarrow du = \frac{5(3x+3)-3(5x-1)}{(3x+3)^2}dx \Rightarrow du = \frac{18}{(3x+3)^2}dx \Rightarrow \frac{1}{(3x+3)^2}dx = \frac{du}{18}$$

$$\int \frac{(5x-1)^{17}}{(3x+3)^{19}}dx = \int \left(\frac{5x-1}{3x+3}\right)^{17} \times \frac{1}{(3x+3)^2}dx = \int u^{17} \times \frac{du}{18}$$
$$= \frac{1}{18}\int u^{17}du = \frac{1}{18}\times\frac{1}{18}u^{18} + C = \frac{1}{324}u^{18} = \frac{1}{324}\left(\frac{5x-1}{3x+3}\right)^{18} + C$$

Example 10.66 Evaluate the integral $\int \frac{x^2}{(4x^2-4x+1)^2}dx$.

Solution $\int \frac{x^2}{(4x^2-4x+1)^2}dx = \int \frac{x^2}{((2x-1)^2)^2}dx = \int \frac{x^2}{(2x-1)^4}dx$. Let's use a substitution and set $u = 2x - 1$. Thus,

$$u = 2x - 1 \Rightarrow du = 2dx \Rightarrow dx = \frac{1}{2}du$$

$$u = 2x - 1 \Rightarrow x = \frac{u+1}{2}$$

$$\int \frac{x^2}{(4x^2-4x+1)^2}dx = \int \frac{x^2}{(2x-1)^4}dx = \int \frac{\left(\frac{u+1}{2}\right)^2}{u^4}\frac{1}{2}du$$
$$= \frac{1}{2}\int \left(\frac{u+1}{2}\right)^2 u^{-4}du = \frac{1}{8}\int (u+1)^2 u^{-4}du$$
$$= \frac{1}{8}\int (u^2+2u+1)u^{-4}du = \frac{1}{8}\int (u^{-2}+2u^{-3}+u^{-4})du$$
$$= \frac{1}{8}\left(\frac{u^{-1}}{-1}+2\frac{u^{-2}}{-2}+\frac{u^{-3}}{-3}\right) = \frac{1}{8}\left(-u^{-1}-u^{-2}-\frac{u^{-3}}{3}\right)$$
$$= -\frac{1}{8}\left(\frac{1}{u}+\frac{1}{u^2}+\frac{1}{3u^3}\right) + C = -\frac{1}{8}\left(\frac{1}{2x-1}+\frac{1}{(2x-1)^2}+\frac{1}{3(2x-1)^3}\right) + C$$

Example 10.67 Evaluate the integral $\int x(x-2)^{50}dx$.

Solution Let's use a substitution and set $u = x - 2$. Thus,

$$u = x - 2 \Rightarrow du = dx$$

$$u = x - 2 \Rightarrow x = u + 2$$

$$\int x(x-2)^{50}dx = \int (u+2)u^{50}du = \int \left(u^{51} + 2u^{50}\right)du = \frac{u^{52}}{52} + 2\frac{u^{51}}{51} + C$$
$$= \frac{1}{52}(x-2)^{52} + \frac{2}{51}(x-2)^{51} + C$$

Example 10.68 Evaluate the integral $\int \frac{x}{x+\sqrt{x^2+1}}dx$.

Solution

$$\int \frac{x}{x+\sqrt{x^2+1}}dx = \int \frac{x\left(x-\sqrt{x^2+1}\right)}{\left(x+\sqrt{x^2+1}\right)\left(x-\sqrt{x^2+1}\right)}dx$$
$$= \int \frac{x\left(x-\sqrt{x^2+1}\right)}{x^2-\left(\sqrt{x^2+1}\right)^2}dx = \int \frac{x\left(x-\sqrt{x^2+1}\right)}{x^2-(x^2+1)}dx$$
$$= \int x\left(\sqrt{x^2+1}-x\right)dx = \int \left(x\sqrt{x^2+1}-x^2\right)dx$$
$$= \int x\sqrt{x^2+1}dx - \int x^2dx = \frac{1}{2}\int 2x(x^2+1)^{\frac{1}{2}}dx - \int x^2dx$$
$$= \frac{1}{2}\frac{(x^2+1)^{\frac{1}{2}+1}}{\frac{1}{2}+1} - \frac{x^3}{3} + C = \frac{1}{3}(x^2+1)^{\frac{3}{2}} - \frac{x^3}{3} + C$$

Example 10.69 Evaluate the integral $\int \frac{1}{\sqrt{x+1}-\sqrt{x}}dx$.

Solution

$$\int \frac{1}{\sqrt{x+1}-\sqrt{x}}dx = \int \frac{\sqrt{x+1}+\sqrt{x}}{\left(\sqrt{x+1}-\sqrt{x}\right)\left(\sqrt{x+1}+\sqrt{x}\right)}dx$$
$$= \int \frac{\sqrt{x+1}+\sqrt{x}}{x+1-x}dx = \int \frac{\sqrt{x+1}+\sqrt{x}}{x+1-x}dx = \int \left(\sqrt{x+1}+\sqrt{x}\right)dx$$
$$= \int \left((x+1)^{\frac{1}{2}} + x^{\frac{1}{2}}\right)dx = \frac{(x+1)^{\frac{1}{2}+1}}{\frac{1}{2}+1} + \frac{x^{\frac{1}{2}+1}}{\frac{1}{2}+1} + C = \frac{2}{3}(x+1)^{\frac{3}{2}} + \frac{2}{3}x^{\frac{3}{2}} + C$$

Example 10.70 Evaluate the integral $\int \frac{7x^2-4}{\sqrt[3]{x^2}}dx$.

Solution

$$\int\frac{7x^2-4}{\sqrt[3]{x^2}}dx=\int\frac{7x^2-4}{x^{\frac{2}{3}}}dx=\int\left(7x^{\frac{4}{3}}-4x^{-\frac{2}{3}}\right)dx=7\frac{x^{\frac{4}{3}+1}}{\frac{4}{3}+1}-4\frac{x^{-\frac{2}{3}+1}}{-\frac{2}{3}+1}+C=3x^{\frac{7}{3}}-12x^{\frac{1}{3}}+C$$

Example 10.71 Evaluate the integral $\int\frac{x^2-1}{\sqrt[3]{x}}dx$.

Solution

$$\int\frac{x^2-1}{\sqrt[3]{x}}dx=\int\left(x^{2-\frac{1}{3}}-x^{-\frac{1}{3}}\right)dx=\int\left(x^{\frac{5}{3}}-x^{-\frac{1}{3}}\right)dx=\frac{x^{\frac{5}{3}+1}}{\frac{5}{3}+1}-\frac{x^{-\frac{1}{3}+1}}{-\frac{1}{3}+1}+C=\frac{3}{8}x^{\frac{8}{3}}-\frac{3}{2}x^{\frac{2}{3}}+C$$

Example 10.72 Evaluate the integral $\int\frac{(1+\sqrt{x})^3-1}{x}dx$.

Solution

$$\begin{aligned}\int\frac{(1+\sqrt{x})^3-1}{x}dx&=\int\frac{1+x^{\frac{3}{2}}+3\sqrt{x}+3x-1}{x}dx\\&=\int\frac{x^{\frac{3}{2}}+3\sqrt{x}+3x}{x}dx=\int\left(x^{\frac{1}{2}}+3x^{-\frac{1}{2}}+3\right)dx\\&=\frac{x^{\frac{1}{2}+1}}{\frac{1}{2}+1}+3\frac{x^{-\frac{1}{2}+1}}{-\frac{1}{2}+1}+3x+C=\frac{2}{3}x^{\frac{3}{2}}+6x^{\frac{1}{2}}+3x+C\end{aligned}$$

Example 10.73 Evaluate the integral $\int\frac{4x-4}{\sqrt[3]{x^2}}dx$.

Solution

$$\begin{aligned}\int\frac{4x-4}{\sqrt[3]{x^2}}dx&=\int\frac{4x-4}{x^{\frac{2}{3}}}dx=\int\left(4x^{\frac{1}{3}}-4x^{-\frac{2}{3}}\right)dx=4\int\left(x^{\frac{1}{3}}-x^{-\frac{2}{3}}\right)dx=4\left(\frac{x^{\frac{4}{3}}}{\frac{4}{3}}-\frac{x^{\frac{1}{3}}}{\frac{1}{3}}\right)+C\\&=4\left(\frac{3}{4}x^{\frac{4}{3}}-3x^{\frac{1}{3}}\right)+C=3x^{\frac{4}{3}}-12x^{\frac{1}{3}}+C\end{aligned}$$

Example 10.74 Evaluate the integral $\int\frac{4}{\sqrt{3x+1}}dx$.

Solution

$$\int\frac{4}{\sqrt{3x+1}}dx=\int4(3x+1)^{-\frac{1}{2}}dx=\frac{4}{3}\int3(3x+1)^{-\frac{1}{2}}dx=\frac{4}{3}\frac{(3x+1)^{-\frac{1}{2}+1}}{-\frac{1}{2}+1}+C=\frac{8}{3}(3x+1)^{\frac{1}{2}}+C$$

Example 10.75 Evaluate the integral $\int\frac{5-3x}{x\sqrt{x}}dx$.

Solution

$$\int \frac{5-3x}{x\sqrt{x}}dx=\int\left(\frac{5}{x\sqrt{x}}-\frac{3}{\sqrt{x}}\right)dx=\int\left(5x^{-\frac{3}{2}}-3x^{-\frac{1}{2}}\right)dx=5\frac{x^{-\frac{3}{2}+1}}{-\frac{3}{2}+1}-3\frac{x^{-\frac{1}{2}+1}}{-\frac{1}{2}+1}+C$$
$$=5\frac{x^{-\frac{1}{2}}}{-\frac{1}{2}}-3\frac{x^{\frac{1}{2}}}{\frac{1}{2}}+C=-10x^{-\frac{1}{2}}-6x^{\frac{1}{2}}+C$$

Example 10.76 Evaluate the integral $\int \frac{(\sqrt{x}-1)(x+\sqrt{x})}{x^2}dx$.

Solution

$$\int \frac{(\sqrt{x}-1)(x+\sqrt{x})}{x^2}dx=\int \frac{x\sqrt{x}+x-x-\sqrt{x}}{x^2}dx=\int \frac{x\sqrt{x}-\sqrt{x}}{x^2}dx=\int\left(\frac{1}{\sqrt{x}}-x^{-\frac{3}{2}}\right)dx$$
$$=\int\left(x^{-\frac{1}{2}}-x^{-\frac{3}{2}}\right)dx=\frac{x^{-\frac{1}{2}+1}}{-\frac{1}{2}+1}-\frac{x^{-\frac{3}{2}+1}}{-\frac{3}{2}+1}+C=2x^{\frac{1}{2}}+2x^{-\frac{1}{2}}+C$$

Example 10.77 Evaluate the integral $\int \frac{\cos(2x)}{\cos(x)-\sin(x)}dx$.

Solution

$$\int \frac{\cos(2x)}{\cos(x)-\sin(x)}dx=\int \frac{\cos^2(x)-\sin^2(x)}{\cos(x)-\sin(x)}dx=\int \frac{(\cos(x)+\sin(x))(\cos(x)-\sin(x))}{\cos(x)-\sin(x)}dx$$
$$=\int(\cos(x)+\sin(x))dx=\sin(x)-\cos(x)+C$$

Example 10.78 Evaluate the integral $\int \frac{\cos(x)-\sin(x)}{1+\sin(2x)}dx$.

Solution

$$\int \frac{\cos(x)-\sin(x)}{1+\sin(2x)}dx=\int \frac{\cos(x)-\sin(x)}{(\sin(x)+\cos(x))^2}dx.$$

Let's use a substitution and set $u=\sin(x)+\cos(x)$. Thus,

$$u=\sin(x)+\cos(x)\Longrightarrow du=(\cos(x)-\sin(x))dx.$$

$$\int \frac{\cos(x)-\sin(x)}{1+\sin(2x)}dx=\int \frac{\cos(x)-\sin(x)}{(\sin(x)+\cos(x))^2}dx=\int \frac{du}{u^2}$$
$$=\int u^{-2}du=\frac{u^{-2+1}}{-2+1}+C=-\frac{1}{u}+C=-\frac{1}{\sin(x)+\cos(x)}+C$$

Example 10.79 Evaluate the integral $\int \frac{1}{\cos^2(x)\sqrt{1+\tan(x)}}dx$.

Solution

$$\begin{aligned}\int \frac{1}{\cos^2(x)\sqrt{1+\tan(x)}}dx &= \int \frac{1+\tan^2(x)}{\sqrt{1+\tan(x)}}dx\\ &= \int \left(1+\tan^2(x)\right)(1+\tan(x))^{-\frac{1}{2}}dx.\end{aligned}$$

Let's use a substitution and set $u = 1 + \tan(x)$. Thus,

$$u = 1 + \tan(x) \Rightarrow du = \left(1+\tan^2(x)\right)dx$$

$$\begin{aligned}&\int \frac{1}{\cos^2(x)\sqrt{1+\tan(x)}}dx = \int \frac{1+\tan^2(x)}{\sqrt{1+\tan(x)}}dx\\ &= \int \left(1+\tan^2(x)\right)(1+\tan(x))^{-\frac{1}{2}}dx = \int u^{-\frac{1}{2}}du = \frac{u^{-\frac{1}{2}+1}}{-\frac{1}{2}+1} + C = 2u^{\frac{1}{2}} + C\\ &= 2\sqrt{1+\tan(x)} + C\end{aligned}$$

Example 10.80 Evaluate the integral $\int \sin(2x)(a\sin^2(x) + b\cos^2(x) + c)^n dx$, with $a, b, c, n \in \mathbb{R}$.

Solution Let $u = a\sin^2(x) + b\cos^2(x) + c$. Therefore,

$$du = (a\sin(2x) - b\sin(2x))dx \Rightarrow du = (a-b)\sin(2x)dx.$$

$$\begin{aligned}&\int \sin(2x)\left(a\sin^2(x) + b\cos^2(x) + c\right)^n dx\\ &= \frac{1}{a-b}\int (a-b)\sin(2x)\left(a\sin^2(x) + b\cos^2(x) + c\right)^n dx = \frac{1}{a-b}\int u^n du\\ &= \frac{1}{a-b}\frac{u^{n+1}}{n+1} + C = \frac{1}{(a-b)(n+1)}\left(a\sin^2(x) + b\cos^2(x) + c\right)^{n+1} + C\end{aligned}$$

Example 10.81 Evaluate the integral $\int 5x^6\cos(x^7)$.

Solution

$$\int 5x^6\cos\left(x^7\right)dx = \frac{5}{7}\int 7x^6\cos\left(x^7\right)dx = \frac{5}{7}\sin\left(x^7\right) + C$$

Example 10.82 Evaluate the integral $\int \cos^2(x)dx$.

Solution

$$\int \cos^2(x)dx = \int \frac{1+\cos(2x)}{2}dx = \frac{1}{2}x + \frac{1}{4}\sin(2x) + C$$

Example 10.83 Evaluate the integral $\int \sin^2(x)dx$.

Solution $\int \sin^2(x)dx = \int \frac{1-\cos(2x)}{2}dx = \frac{1}{2}x - \frac{1}{4}\sin(2x) + C$

Example 10.84 Evaluate the integral $\int \sin^2\left(5x + \frac{\pi}{6}\right)dx$.

Solution As we saw in a previous example, $\int \sin^2(x)dx = \frac{1}{2}x - \frac{1}{4}\sin(2x) + C$. Let's use a substitution and set $u = 5x + \frac{\pi}{6}$. Thus,

$$u = 5x + \frac{\pi}{6} \Rightarrow du = 5dx \Rightarrow dx = \frac{du}{5}$$

$$\int \sin^2\left(5x+\frac{\pi}{6}\right)dx = \int \sin^2(u)\frac{du}{5} = \frac{1}{5}\int \sin^2(u)du = \frac{1}{5}\left(\frac{1}{2}u - \frac{1}{4}\sin(2u) + C\right)$$
$$= \frac{1}{10}u - \frac{1}{20}\sin(2u) + \frac{1}{5}C = \frac{1}{10}\left(5x+\frac{\pi}{6}\right) - \frac{1}{20}\sin\left(2\left(5x+\frac{\pi}{6}\right)\right)$$
$$+ \frac{1}{5}C = \frac{x}{2} + \frac{\pi}{60} - \frac{1}{20}\sin\left(10x+\frac{\pi}{3}\right) + C'$$

Example 10.85 Evaluate the integral $\int \sin^4(x)dx$.

Solution

$$\int \sin^4(x)dx = \int \left(\sin^2(x)\right)^2 dx = \int \left(\frac{1-\cos(2x)}{2}\right)^2 dx$$
$$= \int \left(\frac{1}{4} + \frac{1}{4}\cos^2(2x) - \frac{1}{2}\cos(2x)\right)dx = \int \left(\frac{1}{4} + \frac{1}{4}\left(\frac{1+\cos(4x)}{2}\right) - \frac{1}{2}\cos(2x)\right)dx$$
$$= \int \left(\frac{1}{4} + \frac{1+\cos(4x)}{8} - \frac{1}{2}\cos(2x)\right)dx = \int \left(\frac{3}{8} + \frac{1}{8}\cos(4x) - \frac{1}{2}\cos(2x)\right)dx = \frac{3}{8}x$$
$$+ \frac{1}{32}\sin(4x) - \frac{1}{4}\sin(2x) + C$$

Exercise 10.5 Evaluate the integral $\int \sin^4\left(7x + \frac{\pi}{4}\right)dx$.

Example 10.86 Evaluate the integral $\int \cos^4(x)dx$.

Solution

$$\int \cos^4(x)dx = \int \left(\cos^2(x)\right)^2 dx = \int \left(\frac{1+\cos(2x)}{2}\right)^2 dx$$
$$= \int \frac{1+\cos^2(2x)+2\cos(2x)}{4}dx = \int \left(\frac{1}{4} + \frac{1}{4}\cos^2(2x) + \frac{1}{2}\cos(2x)\right)dx$$
$$= \int \left(\frac{1}{4} + \frac{1}{4}\,\frac{1+\cos(4x)}{2} + \frac{1}{2}\cos(2x)\right)dx$$
$$= \int \left(\frac{1}{4} + \frac{1+\cos(4x)}{8} + \frac{1}{2}\cos(2x)\right)dx$$
$$= \int \left(\frac{3}{8} + \frac{1}{8}\cos(4x) + \frac{1}{2}\cos(2x)\right)dx = \frac{3}{8}x + \frac{1}{32}\sin(4x) + \frac{1}{4}\sin(2x) + C$$

Example 10.87 Evaluate the integral $\int \cos^3(x)dx$.

Solution

$$\int \cos^3(x)dx = \int \cos(x) \times \cos^2(x)dx = \int \cos(x) \times \frac{1+\cos(2x)}{2}dx$$
$$= \frac{1}{2}\int \cos(x)(1+\cos(2x))dx = \frac{1}{2}\int (\cos(x) + \cos(x)\cos(2x))dx$$
$$= \frac{1}{2}\int \left(\cos(x) + \frac{1}{2}(\cos(x+2x) + \cos(x-2x))\right)dx$$
$$= \frac{1}{2}\int \left(\cos(x) + \frac{1}{2}\cos(x+2x) + \frac{1}{2}\cos(x-2x)\right)dx = \frac{1}{2}\int \cos(x)dx$$
$$+\frac{1}{4}\int \cos(3x)dx + \frac{1}{4}\int \cos(-x)dx = \frac{1}{2}\int \cos(x)dx + \frac{1}{4}\int \cos(3x)dx$$
$$+\frac{1}{4}\int \cos(x)dx = \frac{3}{4}\int \cos(x)dx + \frac{1}{4}\int \cos(3x)dx = \frac{3}{4}\sin(x) + \frac{1}{12}\sin(3x)$$
$$+C$$

Example 10.88 Evaluate the integral $\int \cos^3(x)dx$.

Solution This problem was addressed in the previous example. We'll use a more efficient approach here.

$$\int \cos^3(x)dx = \int \cos(x) \times \cos^2(x)dx = \int \cos(x) \times (1 - \sin^2(x))dx.$$

Let's use a substitution and set $u = \sin(x)$. Thus,

$$u = \sin(x) \Rightarrow du = \cos(x)dx$$

$$\int \cos^3(x)dx = \int \cos(x) \times \cos^2(x)dx = \int \cos(x) \times (1 - \sin^2(x))dx = \int (1-u^2)du$$
$$= u - \frac{u^3}{3} + C = \sin(x) - \frac{\sin^3(x)}{3} + C$$

Example 10.89 Use MATLAB to show that the results from Examples 10.87 and 10.88 are equal.

Solution The required code is in Fig. 10.2. This code shows that $\frac{3}{4}\sin(x) + \frac{1}{12}\sin(3x) - \left(\sin(x) - \frac{\sin^3(x)}{3}\right) = 0$, which proves that $\frac{3}{4}\sin(x) + \frac{1}{12}\sin(3x) = \sin(x) - \frac{\sin^3(x)}{3}$.

Fig. 10.2 The MATLAB code

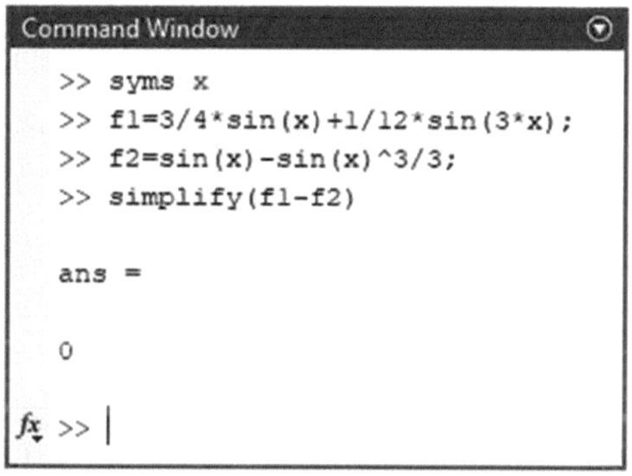

Example 10.90 Evaluate the integral $\int \sin^3(x)dx$.

Solution

$$\int \sin^3(x)dx = \int \sin(x)\sin^2(x)dx = \int \sin(x)\left(1 - \cos^2(x)\right)dx.$$

Let's use a substitution and set $u = \cos(x)$. Thus,

$$u = \cos(x) \Longrightarrow du = -\sin(x)dx$$

$$\int \sin^3(x)dx = \int \sin(x)\sin^2(x)dx = \int \sin(x)\left(1 - \cos^2(x)\right)dx = -\int -\sin(x) \times \left(1 - \cos^2(x)\right)dx = -\int \left(1 - u^2\right)du = -\left(u - \frac{u^3}{3}\right) = \frac{u^3}{3} - \mathrm{u} = \frac{\cos^3(x)}{3} - \cos(x) + C$$

Example 10.91 Evaluate the integral $\int \cos^5(x)dx$.

Solution

$$\int \cos^5(x)dx = \int \cos(x)\cos^4(x)dx = \int \cos(x)\left(1 - \sin^2(x)\right)^2 dx.$$

Let's use a substitution and set $u = \sin(x)$. Thus,

$$u = \sin(x) \Longrightarrow du = \cos(x)dx$$

$$\int \cos^5(x)dx = \int \cos(x)\cos^4(x)dx = \int \cos(x)\left(1 - \sin^2(x)\right)^2 dx = \int \left(1 - u^2\right)^2 du$$
$$= \int \left(1 + u^4 - 2u^2\right)du = u + \frac{u^5}{5} - \frac{2}{3}u^3 + C = \sin(x) + \frac{\sin^5(x)}{5} - \frac{2}{3}\sin^3(x) + C$$

Exercise 10.6 Evaluate the integral $\int \sin^5(x)dx$. (Ans. $-\cos(x)+\frac{2}{3}\cos^3(x)-\frac{1}{5}\cos^5(x)+C$)

Example 10.92 Evaluate the integral $\int \tan^2(x)dx$.

Solution Recall that $\int(1+\tan^2(x))dx = \tan(x)+C$. Thus,

$$\int\left(1+\tan^2(x)\right)dx=\int 1dx+\int \tan^2(x)dx=\tan(x)+C\Rightarrow x+\int \tan^2(x)dx$$
$$=\tan(x)+C\Rightarrow\int \tan^2(x)dx=\tan(x)-x+C$$

Example 10.93 Evaluate the integral $\int \tan^4(x)dx$.

Solution

$$\int \tan^4(x)dx=\int\left(\tan^4(x)+\tan^2(x)-\tan^2(x)+1-1\right)dx$$
$$=\int\left(\tan^4(x)+\tan^2(x)\right)dx+\int\left(-\tan^2(x)-1\right)dx+\int 1dx$$
$$=\int\left(\tan^2(x)\left(\tan^2(x)+1\right)dx-\int\left(1+\tan^2(x)\right)dx+\int 1dx\right.$$

Let's calculate each integral separately. We can calculate the integral $\int(\tan^2(x)(\tan^2(x)+1)dx$ using the substitution $u = \tan(x)$ which gives us $du=(1+\tan^2(x))dx$. Therefore,

$$\int\left(\tan^2(x)\left(\tan^2(x)+1\right)dx=\int u^2du=\frac{u^3}{3}+C_1=\frac{\tan^3(x)}{3}+C_1\right.$$

The remaining two integrals can be evaluated easily as follows:

$$\int\left(1+\tan^2(x)\right)dx=\tan(x)+C_2$$
$$\int 1dx=x+C_3$$

Therefore,

$$\int \tan^4(x)dx=\int\left(\tan^2(x)\left(\tan^2(x)+1\right)dx-\int\left(1+\tan^2(x)\right)\right.$$

$$dx+\int 1dx=\frac{\tan^3(x)}{3}+C_1-(\tan(x)+C_2)+x+C_3$$
$$=\frac{\tan^3(x)}{3}-\tan(x)+x+C_1-C_2+C_3=\frac{\tan^3(x)}{3}-\tan(x)$$
$$+x+C$$

Example 10.94 Evaluate the integral $\int \tan^6(x)dx$.

Solution

$$\int \tan^6(x)dx=\int \left(\tan^6(x)+\tan^4(x)-\tan^4(x)+\tan^2(x)-\tan^2(x)+1-1\right)dx=$$

$$\int \left(\tan^4(x)\left(1+\tan^2(x)\right)dx-\int \left(\tan^2(x)\left(1+\tan^2(x)\right)dx+\right.\right.$$

$$\int \left(1+\tan^2(x)\right)dx-\int 1dx$$

Let's use a substitution and set $u = \tan(x)$. Thus,

$$u=\tan(x)\Rightarrow du=\left(1+\tan^2(x)\right)dx$$

$$\int \tan^6(x)dx=\int \left(\tan^4(x)\left(1+\tan^2(x)\right)dx-\int \left(\tan^2(x)\left(1+\tan^2(x)\right)dx+\right.\right.$$

$$\int \left(1+\tan^2(x)\right)dx-\int 1dx=\int u^4du-\int u^2du+\int du-\int 1dx$$

$$=\frac{u^5}{5}-\frac{u^3}{3}+u-x+C=\frac{\tan^5(x)}{5}-\frac{\tan^3(x)}{3}+\tan(x)-x+C$$

Example 10.95 Evaluate the integral $\int \cot^2(x)dx$.

Solution Recall that $\int (1+\cot^2(x))dx = -\cot(x)+C$. Therefore,

$$\int \cot^2(x)dx=\int \left(\cot^2(x)+1-1\right)dx=\int \left(\cot^2(x)+1\right)dx-\int 1dx=-\cot(x)-x+C$$

Example 10.96 Evaluate the integral $\int \cot^4(x)dx$.

Solution

$$\int \cot^4(x)dx = \int \left(\cot^4(x) + \cot^2(x) - \cot^2(x) + 1 - 1\right)dx = \int \left(\cot^4(x) + \cot^2(x)\right) dx + \int \left(-\cot^2(x) - 1\right)dx + \int 1dx = \int \cot^2(x)\left(1 + \cot^2(x)\right)dx - \int \left(\cot^2(x) + 1\right)dx + \int 1dx$$

Let's calculate each integral separately. We can calculate the integral $\int \cot^2(x)(1 + \cot^2(x))dx$ using the substitution $u = \cot(x)$ which gives us $du = -(1 + \cot^2(x))dx$. Therefore,

$$\int \cot^2(x)\left(1+\cot^2(x)\right)dx = -\int -\cot^2(x)\left(1+\cot^2(x)\right)dx = -\int u^2du = -\frac{u^3}{3} + C_1 = -\frac{\cot^3(x)}{3} + C_1$$

The remaining two integrals can be evaluated easily as follows:

$$\int \left(\cot^2(x) + 1\right)dx = -\int -\left(\cot^2(x) + 1\right)dx = -\cot(x) + C_2$$

$$\int 1dx = x + C_3$$

Therefore,

$$\int \cot^4(x)dx = \int \cot^2(x)\left(1 + \cot^2(x)\right)dx - \int \left(\cot^2(x) + 1\right)dx + \int 1dx = -\frac{\cot^3(x)}{3} + C_1 - \left(-\cot(x) + C_2\right) + x + C_3 = -\frac{\cot^3(x)}{3} + \cot(x) + x + C_1 - C_2 + C_3 = -\frac{\cot^3(x)}{3} + \cot(x) + x + C$$

Example 10.97 Evaluate the integral $\int \tan^3(x)dx$.

Solution

$$\int \tan^3(x)dx = \int \tan(x)\tan^2(x)dx = \int \tan(x)\left(\tan^2(x) + 1 - 1\right)dx = \int \tan(x) \times \left(\tan^2(x) + 1\right)dx - \int \tan(x)dx = \int \tan(x)\left(\tan^2(x) + 1\right)dx + \int -\frac{\sin(x)}{\cos(x)}dx$$
$$= \int \tan(x)\left(\tan^2(x) + 1\right)dx + \int \frac{-\sin(x)}{\cos(x)}dx = \frac{\tan^2(x)}{2} + \ln(|\cos(x)|) + C.$$

Example 10.98 Evaluate the integral $\int \cot^3(x)dx$.

Solution

$$\int \cot^3(x)dx = \int \cot(x)\cot^2(x)dx = \int \cot(x)\left(\cot^2(x)+1-1\right)dx = \int \cot(x) \times \left(\cot^2(x)+1\right)dx - \int \cot(x)dx = -\int -\cot(x)\left(\cot^2(x)+1\right)dx - \int \frac{\cos(x)}{\sin(x)}dx = -\frac{\cot^2(x)}{2} - \ln(|\sin(x)|) + C$$

Exercise 10.7 Evaluate the integral $\int \cot^3\left(4x+\frac{\pi}{3}\right)dx$.

Example 10.99 Evaluate the integral $\int \sqrt{1+\tan^2(x)}\sin(2x)dx$ assuming $\frac{\pi}{2} < x < \frac{3\pi}{2}$.

Solution

$$\int \sqrt{1+\tan^2(x)}\sin(2x)dx = \int \sqrt{\frac{1}{\cos^2(x)}}2\sin(x)\cos(x)dx = 2\int \frac{1}{|\cos(x)|}\sin(x)\cos(x)dx$$

Given that $\frac{\pi}{2} < x < \frac{3\pi}{2}$, it follows that $\cos(x) < 0$. We can therefore write:

$$2\int \frac{1}{|\cos(x)|}\sin(x)\cos(x)dx = 2\int \frac{1}{-\cos(x)}\sin(x)\cos(x)dx = 2\int -\sin(x)dx = 2\cos(x) + C$$

Example 10.100 Evaluate the integral $\int \sqrt{4-x^2}dx$.

Solution The form of the integrand, $\sqrt{a^2-x^2}$, suggests using the substitution $x = a\sin(\theta)$. In this case, $a^2 = 4$, so $a = 2$. Let's use a substitution and set $x = 2\sin(\theta)$. Thus,

$$x = 2\sin(\theta) \Rightarrow \theta = \sin^{-1}\left(\frac{x}{2}\right)$$

$$x = 2\sin(\theta) \Rightarrow dx = 2\cos(\theta)d\theta$$

$$\int \sqrt{4-x^2}dx = \int \sqrt{4-4\sin^2(\theta)}\times 2\cos(\theta)d\theta = \int 2\sqrt{1-\sin^2(\theta)}\times 2\cos(\theta)d\theta$$
$$= \int 2\sqrt{\cos^2(\theta)}\times 2\cos(\theta)d\theta = \int 2\cos(\theta)\times 2\cos(\theta)d\theta$$
$$= 4\int \cos^2(\theta)d\theta = 4\int \frac{1+\cos(2\theta)}{2}d\theta = 2\int 1+\cos(2\theta)d\theta =$$
$$2\left(\theta + \frac{\sin(2\theta)}{2} + C\right) = 2\theta + \sin(2\theta) + 2C = 2\theta + \sin(2\theta) + C'$$

Next, we will simplify the expression. $x = 2\sin(\theta)$ therefore, $\sin(\theta) = \frac{x}{2}$ (Fig. 10.3).

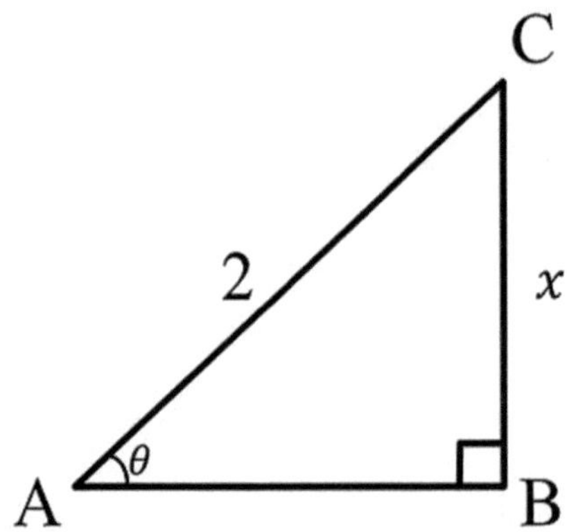

Fig. 10.3 The right triangle for which $\sin(\theta) = \frac{x}{2}$

The Pythagorean theorem can be used to calculate the length of side *AB*(Fig. 10.4).

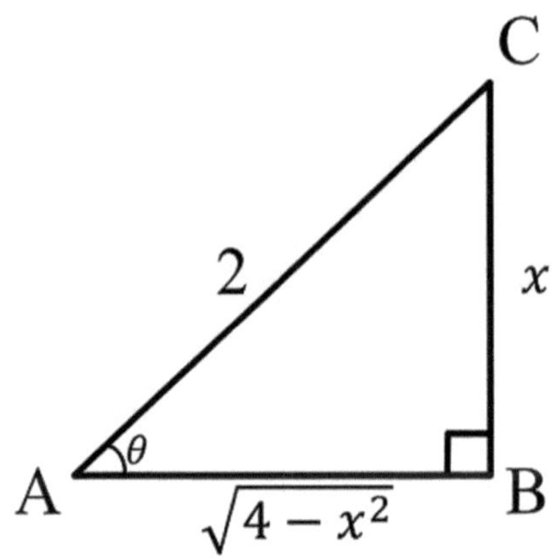

Fig. 10.4 The Pythagorean theorem is used to calculate the length of *AB*

From Fig. 10.4, we get:

$$\sin(\theta) = \frac{x}{2}$$

$$\cos(\theta) = \frac{\sqrt{4 - x^2}}{2}$$

Therefore,

$$\sin(2\theta) = 2\sin(\theta)\cos(\theta) = 2\frac{x}{2}\frac{\sqrt{4 - x^2}}{2} = \frac{x\sqrt{4 - x^2}}{2}$$

$$\int \sqrt{4-x^2}dx = 2\theta + \sin(2\theta) + C' = 2\sin^{-1}\left(\frac{x}{2}\right) + \frac{x\sqrt{4-x^2}}{2} + C'$$

Example 10.101 In the previous example, from a mathematical point of view, $\int 2\sqrt{\cos^2(\theta)} \times 2\cos(\theta)d\theta = \int 2|\cos(\theta)| \times 2\cos(\theta)d\theta$. However, in that example we wrote $\int 2\sqrt{\cos^2(\theta)} \times 2\cos(\theta)d\theta = \int 2\cos(\theta) \times 2\cos(\theta)d\theta$. Let's now explain why this substitution is valid.

When we use the substitution $x = 2\sin(\theta)$ to evaluate the integral $\int \sqrt{4-x^2}dx$, we are implicitly restricting the domain of θ. For this substitution to be a one-to-one function and to ensure the expression under the square root is real, we limit θ to the interval $\left[-\frac{\pi}{2}, \frac{\pi}{2}\right]$. The value of $\cos(\theta)$ is always non-negative (positive or zero).

Therefore, since we are working within a domain where $\cos(\theta) \geq 0$, the simplification $\sqrt{\cos^2(\theta)} = |\cos(\theta)| = \cos(\theta)$ is mathematically sound and the absolute value is not needed.

Example 10.102 Evaluate the integral $\int \sqrt{4+x^2}dx$. (Hint: $\int \sec^3(\theta)d\theta = \frac{1}{2}\sec\theta\tan\theta + \frac{1}{2}\ln(|\sec(\theta) + \tan(\theta)|) + C$)

Solution The form of the integrand, $\sqrt{a^2+x^2}$, suggests using the substitution $x = a\tan(\theta)$. In this case, $a^2 = 4$, so $a = 2$. Let's use a substitution and set $x = 2\tan(\theta)$. Thus,

$$x = 2\tan(\theta) \Rightarrow \theta = \tan^{-1}\left(\frac{x}{2}\right)$$

$$x = 2\tan(\theta) \Rightarrow dx = 2\left(1+\tan^2(\theta)\right)d\theta$$

$$\begin{aligned}\int \sqrt{4+x^2}dx &= \int \sqrt{4+4\tan^2(\theta)} \times 2\left(1+\tan^2(\theta)\right)d\theta \\ &= \int 2\sqrt{1+\tan^2(\theta)} \times 2\left(1+\tan^2(\theta)\right)d\theta \\ &= \int 2\sqrt{\frac{1}{\cos^2(\theta)}} \times \frac{2}{\cos^2(\theta)}d\theta = \int \frac{4}{\cos^3(\theta)}d\theta = 4\int \sec^3(\theta)d\theta \\ &= 4\left(\frac{1}{2}\sec\theta\tan\theta + \frac{1}{2}\ln(|\sec(\theta)+\tan(\theta)|) + C\right) \\ &= 2\sec\theta\tan\theta + 2\ln(|\sec(\theta)+\tan(\theta)|) \\ &+ 4C = 2\sec\theta\tan\theta + 2\ln(|\sec(\theta)+\tan(\theta)|) + C'\end{aligned}$$

Next, we will simplify the expression. $x = 2\tan(\theta)$; therefore, $\tan(\theta) = \frac{x}{2}$ (Fig. 10.5).

Fig. 10.5 The right triangle for which $\tan(\theta) = \frac{x}{2}$

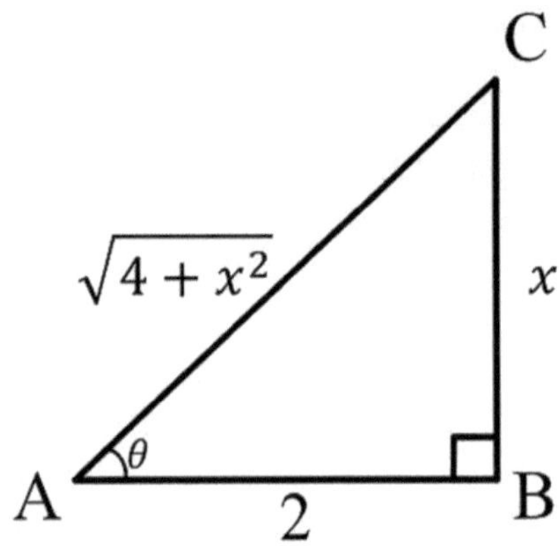

From Fig. 10.5, we get:

$$\sec\theta\tan\theta = \frac{1}{\cos(\theta)}\tan(\theta) = \frac{1}{\frac{2}{\sqrt{4+x^2}}}\frac{x}{2} = \frac{\sqrt{4+x^2}}{2}\frac{x}{2} = \frac{x\sqrt{4+x^2}}{4}$$

$$\sec(\theta) + \tan(\theta) = \frac{1}{\cos(\theta)} + \tan(\theta) = \frac{1}{\frac{2}{\sqrt{4+x^2}}} + \frac{x}{2} = \frac{\sqrt{4+x^2}}{2} + \frac{x}{2} = \frac{x+\sqrt{4+x^2}}{2}$$

Therefore,

$$\int\sqrt{4+x^2}dx = 2\sec\theta\tan\theta + 2\ln(|\sec(\theta)+\tan(\theta)|) + C'$$
$$= 2\frac{x\sqrt{4+x^2}}{4} + 2\ln\left(\left|\frac{x+\sqrt{4+x^2}}{2}\right|\right) + C' = \frac{x\sqrt{4+x^2}}{2}$$
$$+2\ln\left(\left|\frac{x+\sqrt{4+x^2}}{2}\right|\right) + C' = \frac{x\sqrt{4+x^2}}{2} + 2\ln\left(\left|x+\sqrt{4+x^2}\right|\right)$$
$$-2\ln(2) + C' = \frac{x\sqrt{4+x^2}}{2} + 2\ln\left(\left|x+\sqrt{4+x^2}\right|\right) + C''$$

Recall that $\log(ab) = \log(a) + \log(b)$ and $\log\left(\frac{a}{b}\right) = \log(a) - \log(b)$.

Example 10.103 Evaluate the integral $\int\sqrt{x^2-4}dx$. (Hint: $\int\sec(\theta)d\theta = \ln(\sec(\theta) + \tan(\theta)) + C$ and $\int\sec^3(\theta)d\theta = \frac{1}{2}\sec\theta\tan\theta + \frac{1}{2}\ln(|\sec(\theta) + \tan(\theta)|) + C$)

Solution The form of the integrand, $\sqrt{x^2-a^2}$, suggests using the substitution $x = a\sec(\theta)$. In this case, $a^2 = 4$, so $a = 2$. Let's use a substitution and set $x = 2\sec(\theta)$. Thus,

$$x = 2\sec(\theta) \Rightarrow dx = 2\sec(\theta)\tan(\theta)d\theta$$

$$x=2\sec(\theta)\Rightarrow \sec(\theta)=\frac{x}{2}\Rightarrow \frac{1}{\cos(\theta)}=\frac{x}{2}\Rightarrow \cos(\theta)=\frac{2}{x}\Rightarrow\theta=\cos^{-1}\left(\frac{2}{x}\right)$$

$$\int\sqrt{x^2-4}dx=\int\sqrt{(2\sec(\theta))^2-4}\times 2\sec(\theta)\tan(\theta)d\theta$$
$$=4\int\sqrt{\sec^2(\theta)-1}\times\sec(\theta)\tan(\theta)d\theta=4\int\sqrt{\frac{1}{\cos^2(\theta)}-1}\times\sec(\theta)\tan(\theta)d\theta$$
$$=4\int\sqrt{\frac{1-\cos^2(\theta)}{\cos^2(\theta)}}\times\sec(\theta)\tan(\theta)d\theta$$
$$=4\int\sqrt{\frac{\sin^2(\theta)}{\cos^2(\theta)}}\times\sec(\theta)\tan(\theta)d\theta=4\int\frac{\sin(\theta)}{\cos(\theta)}\times\frac{1}{\cos(\theta)}\frac{\sin(\theta)}{\cos(\theta)}d\theta$$
$$=4\int\frac{\sin^2(\theta)}{\cos^3(\theta)}d\theta=4\int\frac{1-\cos^2(\theta)}{\cos^3(\theta)}d\theta=4\int\sec^3(\theta)-\sec(\theta)d\theta$$
$$=4\left(\frac{1}{2}\sec\theta\tan\theta+\frac{1}{2}\ln(|\sec(\theta)+\tan(\theta)|)-\ln(\sec(\theta)+\tan(\theta))+C\right)$$
$$=2\sec\theta\tan\theta+2\ln(|\sec(\theta)+\tan(\theta)|)-4\ln(\sec(\theta)+\tan(\theta))+4C$$
$$=2\sec\theta\tan\theta-2\ln(|\sec(\theta)+\tan(\theta)|)+C'$$

Next, we will simplify the expression. $x=2\sec(\theta)=\frac{2}{\cos(\theta)}$; therefore, $\cos(\theta)=\frac{2}{x}$ (Fig. 10.6).

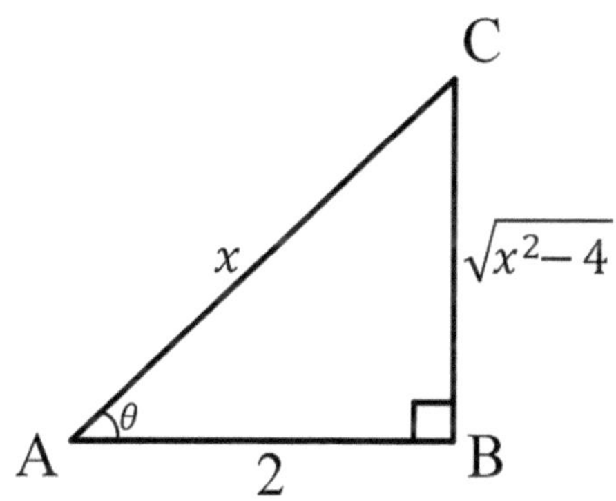

Fig. 10.6 The right triangle for which $\cos(\theta)=\frac{2}{x}$

From Fig. 10.6, we get:

$$\sec\theta\tan\theta=\frac{1}{\cos(\theta)}\times\frac{\sin(\theta)}{\cos(\theta)}=\frac{\sin(\theta)}{\cos^2(\theta)}=\frac{\frac{\sqrt{x^2-4}}{x}}{\left(\frac{2}{x}\right)^2}=\frac{x\sqrt{x^2-4}}{4}$$

$$\sec(\theta)+\tan(\theta)=\frac{1}{\cos(\theta)}+\tan(\theta)=\frac{1}{\frac{2}{x}}+\frac{\sqrt{x^2-4}}{2}=\frac{x}{2}+\frac{\sqrt{x^2-4}}{2}=\frac{x+\sqrt{x^2-4}}{2}$$

Therefore,

$$\begin{aligned}\int\sqrt{x^2-4}dx&=2\sec\theta\tan\theta-2\ln(|\sec(\theta)+\tan(\theta)|)+C'\\&=2\frac{x\sqrt{x^2-4}}{4}-2\ln\left(\left|\frac{x+\sqrt{x^2-4}}{2}\right|\right)+C'=\frac{x\sqrt{x^2-4}}{2}\\&-2\ln\left(\left|\frac{x+\sqrt{x^2-4}}{2}\right|\right)+C'=\frac{x\sqrt{x^2-4}}{2}\\&-2\ln\left(\left|x+\sqrt{x^2-4}\right|\right)+2\ln(2)+C'=\frac{x\sqrt{x^2-4}}{2}-2\ln\left(\left|x+\sqrt{x^2-4}\right|\right)+C''\end{aligned}$$

Example 10.104 Evaluate the integral $\int\frac{1}{\sqrt{4-x^2}}dx$.

Solution Presence of $\sqrt{a^2-x^2}$ suggests using the substitution $x=a\sin(\theta)$. In this case, $a^2=4$, so $a=2$. Let's use a substitution and set $x=2\sin(\theta)$. Thus,

$$x=2\sin(\theta)\Rightarrow\theta=\sin^{-1}\left(\frac{x}{2}\right)$$

$$x=2\sin(\theta)\Rightarrow dx=2\cos(\theta)d\theta$$

$$\begin{aligned}\int\frac{1}{\sqrt{4-x^2}}dx&=\int\frac{1}{\sqrt{4-4\sin^2(\theta)}}\times2\cos(\theta)d\theta=\int\frac{1}{2\sqrt{1-\sin^2(\theta)}}\times2\cos(\theta)d\theta\\&=\int\frac{\cos(\theta)}{\sqrt{\cos^2(\theta)}}d\theta=\int\frac{\cos(\theta)}{\cos(\theta)}d\theta=\int d\theta=\theta+C=\sin^{-1}\left(\frac{x}{2}\right)+C\end{aligned}$$

Example 10.105 Evaluate the integral $\int\frac{1}{\sqrt{4+x^2}}dx$. (Hint: $\int\sec(\theta)d\theta=\ln(\sec(\theta)+\tan(\theta))+C$)

Solution Presence of $\sqrt{a^2+x^2}$ suggests using the substitution $x=a\tan(\theta)$. In this case, $a^2=4$, so $a=2$. Let's use a substitution and set $x=2\tan(\theta)$. Thus,

$$x=2\tan(\theta)\Rightarrow\theta=\tan^{-1}\left(\frac{x}{2}\right)$$

$$x=2\tan(\theta)\Rightarrow dx=2\left(1+\tan^2(\theta)\right)d\theta$$

$$\begin{aligned}\int\frac{1}{\sqrt{4+x^2}}dx&=\int\frac{1}{\sqrt{4+4\tan^2(\theta)}}\times2\left(1+\tan^2(\theta)\right)d\theta\\&=\int\frac{1}{2\sqrt{1+\tan^2(\theta)}}\times2\left(1+\tan^2(\theta)\right)d\theta=\int\sqrt{1+\tan^2(\theta)}d\theta\\&=\int\sqrt{\frac{1}{\cos^2(\theta)}}d\theta=\int\frac{1}{\cos(\theta)}d\theta=\int\sec(\theta)d\theta=\ln(\sec(\theta)+\tan(\theta))+C\end{aligned}$$

Next, we will simplify the expression. $x = 2\tan(\theta)$; therefore, $\tan(\theta) = \frac{x}{2}$ (Fig. 10.7).

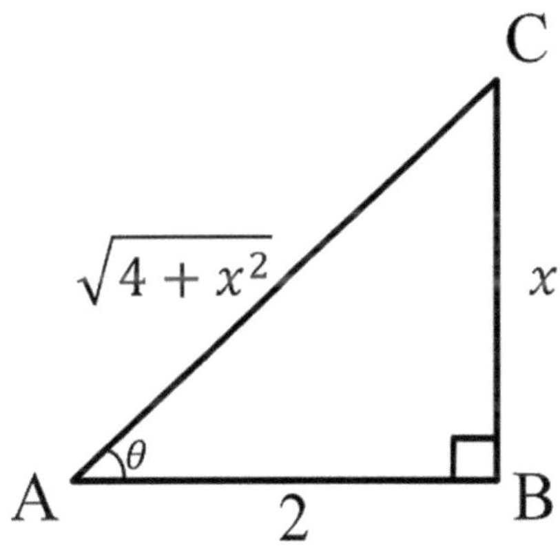

Fig. 10.7 The right triangle for which $\tan(\theta) = \frac{x}{2}$

From Fig. 10.7, we get:

$$\sec(\theta) + \tan(\theta) = \frac{1}{\cos(\theta)} + \tan(\theta) = \frac{1}{\frac{2}{\sqrt{4+x^2}}} + \frac{x}{2} = \frac{\sqrt{4+x^2}}{2} + \frac{x}{2} = \frac{x+\sqrt{4+x^2}}{2}$$

Therefore,

$$\int \frac{1}{\sqrt{4+x^2}} dx = \ln(\sec(\theta) + \tan(\theta)) + C = \ln\left(\frac{x+\sqrt{4+x^2}}{2}\right) + C$$
$$= \ln\left(x+\sqrt{4+x^2}\right) - \ln(2) + C = \ln\left(x+\sqrt{4+x^2}\right) + C'$$

Example 10.106 Evaluate the integral $\int \frac{1}{\sqrt{x^2-4}} dx$. (Hint: $\int \sec(\theta)d\theta = \ln(\sec(\theta) + \tan(\theta)) + C$)

Solution Presence of $\sqrt{x^2-a^2}$, suggests using the substitution $x = a\sec(\theta)$. In this case, $a^2 = 4$, so $a = 2$. Let's use a substitution and set $x = 2\sec(\theta)$. Thus,

$$x = 2\sec(\theta) \Rightarrow dx = 2\sec(\theta)\tan(\theta)d\theta$$

$$x = 2\sec(\theta) \Rightarrow \sec(\theta) = \frac{x}{2} \Rightarrow \frac{1}{\cos(\theta)} = \frac{x}{2} \Rightarrow \cos(\theta) = \frac{2}{x} \Rightarrow \theta = \cos^{-1}\left(\frac{2}{x}\right)$$

$$\int \frac{1}{\sqrt{x^2-4}}dx = \int \frac{1}{\sqrt{(2\sec(\theta))^2-4}} \times 2\sec(\theta)\tan(\theta)d\theta$$
$$= \int \frac{1}{\sqrt{\sec^2(\theta)-1}} \times \sec(\theta)\tan(\theta)d\theta = \int \frac{1}{\sqrt{\frac{1-\cos^2(\theta)}{\cos^2(\theta)}}} \times \sec(\theta)\tan(\theta)d\theta$$
$$= \int \frac{1}{\sqrt{\frac{\sin^2(\theta)}{\cos^2(\theta)}}} \times \sec(\theta)\tan(\theta)d\theta = \int \frac{\cos(\theta)}{\sin(\theta)} \times \sec(\theta)\tan(\theta)d\theta$$
$$= \int \frac{\cos(\theta)}{\sin(\theta)} \times \sec(\theta)\frac{\sin(\theta)}{\cos(\theta)}d\theta = \int \sec(\theta)d\theta = \ln(\sec(\theta)+\tan(\theta))+C$$

Next, we will simplify the expression. $x = 2\sec(\theta) = \frac{2}{\cos(\theta)}$; therefore, $\cos(\theta) = \frac{2}{x}$ (Fig. 10.8).

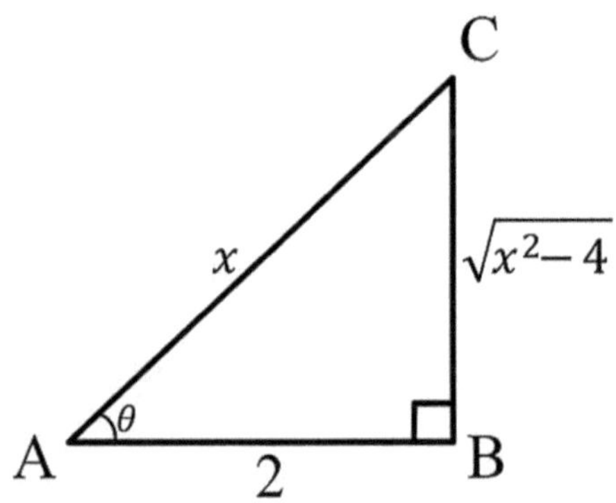

Fig. 10.8 The right triangle for which $cos(\theta) = \frac{2}{x}$

From Fig. 10.8, we get:

$$\sec(\theta)+\tan(\theta) = \frac{1}{\cos(\theta)} + \tan(\theta) = \frac{1}{\frac{2}{x}} + \frac{\sqrt{x^2-4}}{2} = \frac{x}{2} + \frac{\sqrt{x^2-4}}{2} = \frac{x+\sqrt{x^2-4}}{2}$$

Therefore,

$$\int \frac{1}{\sqrt{x^2-4}}dx = \ln(\sec(\theta)+\tan(\theta))+C = \ln\left(\frac{x+\sqrt{x^2-4}}{2}\right)+C$$
$$= \ln\left(x+\sqrt{x^2-4}\right) - \ln(2) + C = \ln\left(x+\sqrt{x^2-4}\right) + C'$$

Example 10.107 Evaluate the integral $\int x^2\sqrt{4-x^2}dx$.

Solution Let's use a substitution and set $x = 2\ sin\ (\theta)$. Thus,

$$x=2\sin(\theta)\Rightarrow\theta=\sin^{-1}\left(\frac{x}{2}\right)$$

$$x=2\sin(\theta)\Rightarrow dx=2\cos(\theta)d\theta$$

$$\begin{aligned}
\int x^2\sqrt{4-x^2}dx&=\int 4\sin^2(\theta)\sqrt{4-4\sin^2(\theta)}\times 2\cos(\theta)d\theta\\
&=\int 4\sin^2(\theta)\times 2\sqrt{1-\sin^2(\theta)}\times 2\cos(\theta)d\theta\\
&=\int 4\sin^2(\theta)\times 2\cos(\theta)\times 2\cos(\theta)d\theta=\int 4\sin^2(\theta)\times 4\cos^2(\theta)d\theta\\
&=4\int(2\sin(\theta)\cos(\theta))^2d\theta=4\int(\sin(2\theta))^2d\theta=4\int\sin^2(2\theta)d\theta\\
&=4\int\frac{1-\cos(4\theta)}{2}d\theta=\int 2-2\cos(4\theta)d\theta\\
&=2\theta-\frac{1}{2}\sin(4\theta)+C=2\theta-\frac{1}{2}(2\sin(2\theta)\cos(2\theta))+C\\
&=2\theta-\sin(2\theta)\cos(2\theta)+C=2\theta-2\sin(\theta)\cos(\theta)\left(\cos^2(\theta)-\sin^2(\theta)\right)+C\\
&=2\theta-2\sin(\theta)\sqrt{1-\sin^2(\theta)}\left(1-\sin^2(\theta)-\sin^2(\theta)\right)+C\\
&=2\theta-2\sin(\theta)\sqrt{1-\sin^2(\theta)}\left(1-2\sin^2(\theta)\right)+C\\
&=2\sin^{-1}\left(\frac{x}{2}\right)-2\frac{x}{2}\sqrt{1-\left(\frac{x}{2}\right)^2}\left(1-2\left(\frac{x}{2}\right)^2\right)+C\\
&=2\sin^{-1}\left(\frac{x}{2}\right)-\frac{x}{2}\sqrt{4-x^2}\left(\frac{2-x^2}{2}\right)+C\\
&=2\sin^{-1}\left(\frac{x}{2}\right)-\frac{x}{4}\sqrt{4-x^2}(2-x^2)+C
\end{aligned}$$

Example 10.108 Evaluate the integral $\int\frac{\sqrt{4-x^2}}{x^2}dx$.

Solution Let's use a substitution and set $x = 2\ \sin\ (\theta)$. Thus,

$$x=2\sin(\theta)\Rightarrow\theta=\sin^{-1}\left(\frac{x}{2}\right)$$

$$x=2\sin(\theta)\Rightarrow dx=2\cos(\theta)d\theta$$

$$\int \frac{\sqrt{4-x^2}}{x^2}dx = \int \frac{\sqrt{4-4\sin^2(\theta)}}{4\sin^2(\theta)} \times 2\cos(\theta)d\theta$$
$$= \int \frac{\sqrt{1-\sin^2(\theta)}}{\sin^2(\theta)} \times \cos(\theta)d\theta = \int \frac{\cos(\theta)}{\sin^2(\theta)}\cos(\theta)d\theta$$
$$= \int \cot^2(\theta)d\theta = \int \left(1+\cot^2(\theta)-1\right)d\theta$$
$$= \int \left(1+\cot^2(\theta)\right)d\theta - \int d\theta$$
$$= -\int -\left(1+\cot^2(\theta)\right)d\theta - \int d\theta = -\cot(\theta) - \theta + C$$

Next, we will simplify the expression. $x = 2\sin(\theta)$; therefore, $\sin(\theta) = \frac{x}{2}$(Fig. 10.9).

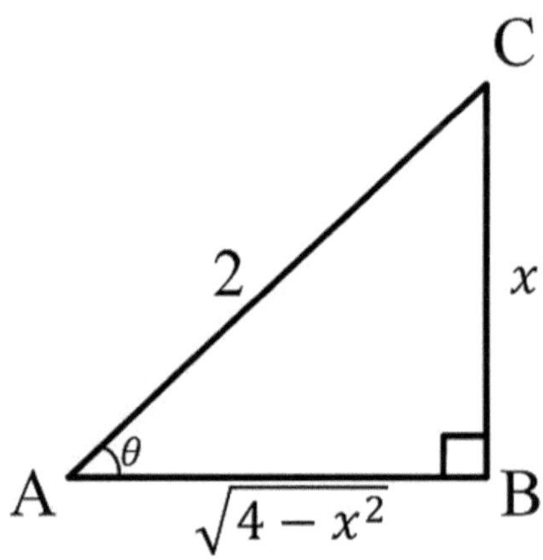

Fig. 10.9 The right triangle for which $\sin(\theta) = \frac{x}{2}$

From Fig. 10.9, we get: $\cot(\theta) = \frac{\sqrt{4-x^2}}{x}$. Therefore, $\int \frac{\sqrt{4-x^2}}{x^2}dx = -\cot(\theta) - \theta + C = -\frac{\sqrt{4-x^2}}{x} - \sin^{-1}\left(\frac{x}{2}\right) + C$.

Example 10.109 Evaluate the integral $\int \frac{1}{x^2\sqrt{x^2+4}}dx$.

Solution Let's use a substitution and set $x = 2\tan(\theta)$. Thus,

$$x = 2\tan(\theta) \Rightarrow \theta = \tan^{-1}\left(\frac{x}{2}\right)$$
$$x = 2\tan(\theta) \Rightarrow dx = 2\left(1+\tan^2(\theta)\right)d\theta$$

$$\int \frac{1}{x^2\sqrt{x^2+4}}dx = \int \frac{2(1+\tan^2(\theta))}{4\tan^2(\theta)\sqrt{4\tan^2(\theta)+4}}d\theta = \int \frac{1+\tan^2(\theta)}{2\tan^2(\theta)\times 2\sqrt{\tan^2(\theta)+1}}d\theta$$
$$= \int \frac{\frac{1}{\cos^2(\theta)}}{4\tan^2(\theta)\sqrt{\frac{1}{\cos^2(\theta)}}}d\theta = \int \frac{\frac{1}{\cos^2(\theta)}}{4\tan^2(\theta)\frac{1}{\cos(\theta)}}d\theta$$
$$= \int \frac{1}{4\tan^2(\theta)\cos(\theta)}d\theta = \frac{1}{4}\int \frac{\cos(\theta)}{\sin^2(\theta)}d\theta$$
$$= \frac{1}{4}\int \cos(\theta)\sin^{-2}(\theta)d\theta = \frac{1}{4}\frac{\sin^{-2+1}(\theta)}{-2+1} + C = -\frac{1}{4\sin(\theta)} + C$$

Next, we will simplify the expression. $x = 2\tan(\theta)$; therefore, $\tan(\theta) = \frac{x}{2}$ (Fig. 10.10).

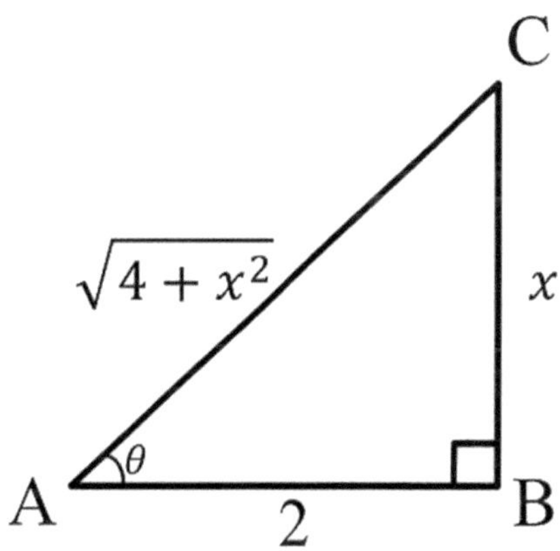

Fig. 10.10 The right triangle for which $\tan(\theta) = \frac{x}{2}$

From Fig. 10.10, we get: $\sin(\theta) = \frac{x}{\sqrt{4+x^2}}$. Therefore, $\int \frac{1}{x^2\sqrt{x^2+4}}dx = -\frac{1}{4\sin(\theta)} + C = -\frac{\sqrt{4+x^2}}{4x} + C$.

Example 10.110 Evaluate the integral $\int \frac{x^3}{\sqrt{x^2+4}}dx$.

Solution Let's use a substitution and set $x = 2\tan(\theta)$. Thus,

$$x = 2\tan(\theta) \Rightarrow \theta = \tan^{-1}\left(\frac{x}{2}\right)$$

$$x = 2\tan(\theta) \Rightarrow dx = 2\left(1 + \tan^2(\theta)\right)d\theta$$

$$\int \frac{x^3}{\sqrt{x^2+4}}dx = \int \frac{(2\tan(\theta))^3}{\sqrt{4\tan^2(\theta)+4}} \times 2\left(1+\tan^2(\theta)\right)d\theta$$
$$= \int \frac{16\tan^3(\theta)(1+\tan^2(\theta))}{2\sqrt{\tan^2(\theta)+1}}d\theta = \int 8\tan^3(\theta)\sqrt{\tan^2(\theta)+1}d\theta = \int 8\tan^3(\theta)$$
$$\times\sqrt{\frac{1}{\cos^2(\theta)}}d\theta = \int 8\tan^3(\theta)\frac{1}{\cos(\theta)}d\theta = 8\int \frac{\sin^3(\theta)}{\cos^4(\theta)}d\theta = 8\int \frac{\sin(\theta)\sin^2(\theta)}{\cos^4(\theta)}d\theta$$
$$= 8\int \frac{\sin(\theta)(1-\cos^2(\theta))}{\cos^4(\theta)}d\theta = 8\int \sin(\theta)\cos^{-4}(\theta)d\theta - 8\int \sin(\theta)\cos^{-2}(\theta) =$$
$$-8\int -\sin(\theta)\cos^{-4}(\theta)d\theta + 8\int -\sin(\theta)\cos^{-2}(\theta) = -8\frac{\cos^{-4+1}(\theta)}{-4+1} + 8\frac{\cos^{-2+1}(\theta)}{-2+1}$$
$$+C = \frac{8}{3\cos^3(\theta)} - \frac{8}{\cos(\theta)} + C$$

Next, we will simplify the expression. $x = 2\tan(\theta)$; therefore, $tan(\theta) = \frac{x}{2}$ (Fig. 10.11).

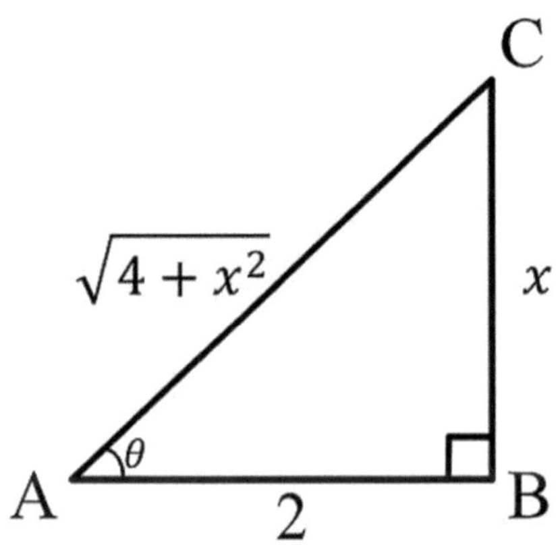

Fig. 10.11 The right triangle for which $\tan(\theta) = \frac{x}{2}$

From Fig. 10.11, we get: $\cos(\theta) = \frac{2}{\sqrt{4+x^2}}$. Therefore,

$$\int \frac{x^3}{\sqrt{x^2+4}}dx = \frac{8}{3\cos^3(\theta)} - \frac{8}{\cos(\theta)} + C = \frac{8}{3\left(\frac{2}{\sqrt{4+x^2}}\right)^3}$$
$$-\frac{8}{\frac{2}{\sqrt{4+x^2}}} + C = \frac{(4+x^2)\sqrt{4+x^2}}{3} - 4\sqrt{4+x^2} + C$$
$$= \left(\frac{4+x^2}{3} - 4\right)\sqrt{4+x^2} + C = \frac{(x^2-8)\sqrt{4+x^2}}{3} + C.$$

Example 10.111 Evaluate the integral $\int \frac{x^3+x}{x-1}dx$.

Solution According to Fig. 10.12, $x^3 + x = (x^2 + x + 2)(x - 1) + 2$.

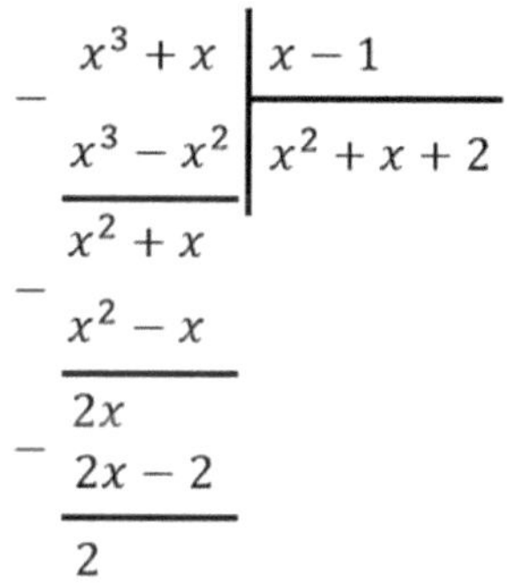

Fig. 10.12 Dividing $x^3 + x$ by $x - 1$

$$\int \frac{x^3+x}{x-1}dx = \int x^2+x+2+\frac{2}{x-1}dx = \frac{x^3}{3}+\frac{x^2}{2}+2x+2\ln(|x-1|) + C = \frac{x^3}{3}+\frac{x^2}{2}+2x+\ln\left(|x-1|^2\right)+C = \frac{x^3}{3}+\frac{x^2}{2}+2x+\ln\left(x^2-2x+1\right)+C$$

Example 10.112 Evaluate the integral $\int \frac{x^2+2x-1}{2x^3+3x^2-2x}dx$.

Solution The code in Fig. 10.13 is used to calculate the roots of the denominator.

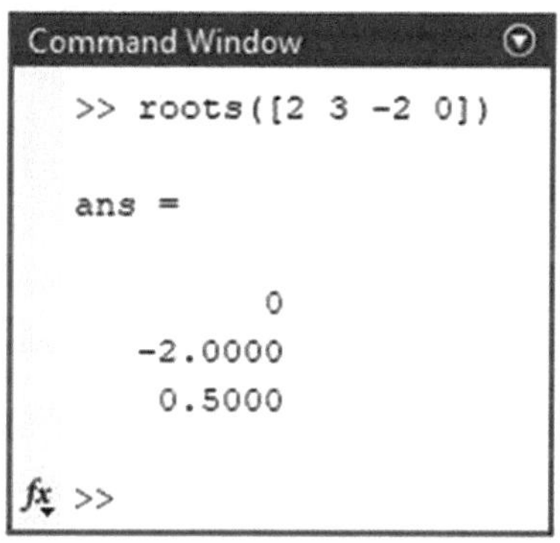

Fig. 10.13 The MATLAB code

$2x^3 + 3x^2 \quad 2x$ can be factored as: $2x^3 + 3x^2 - 2x = 2(x - 0)(x - (-2))(x - 0.5) = x(x + 2)(2x - 1)$. Therefore,

$$\frac{x^2+2x-1}{2x^3+3x^2-2x} = \frac{A}{x}+\frac{B}{2x-1}+\frac{C}{x+2} \overset{\text{use partial fraction expansion}}{\Rightarrow} A=\frac{1}{2}, B=\frac{1}{5}, C=-\frac{1}{10}$$

$$\int \frac{\frac{1}{2}}{x} + \frac{\frac{1}{5}}{2x-1} - \frac{\frac{1}{10}}{x+2} dx = \frac{1}{2} \ln|x| + \frac{1}{10} \ln|2x-1| - \frac{1}{10} \ln|x+2| + C$$

Example 10.113 Evaluate the integral $\int \frac{4}{(2x-7)^2} dx$.

Solution

$$\int \frac{4}{(2x-7)^2} dx = \int 4(2x-7)^{-2} dx = 2\int 2(2x-7)^{-2} dx = 2\frac{(2x-7)^{-2+1}}{-2+1} = \frac{-2}{2x-7}$$

Exercise 10.8 Show that for $a \neq 0$, $\int \frac{dx}{x^2-a^2} = \frac{1}{2a} \ln\left|\frac{x-a}{x+a}\right| + C$.

Example 10.114 Evaluate the integral $\int \frac{x^4-2x^2+4x+1}{x^3-x^2-x+1} dx$.

Solution Using polynomial division and partial fraction decomposition, we get:

$$\frac{x^4-2x^2+4x+1}{x^3-x^2-x+1} = x+1+\frac{4x}{x^3-x^2-x+1} = x+1+\frac{1}{x-1}+\frac{2}{(x-1)^2}+\frac{-1}{x+1}$$

$$\int \frac{x^4-2x^2+4x+1}{x^3-x^2-x+1} dx = \int \left(x+1+\frac{1}{x-1}+\frac{2}{(x-1)^2}+\frac{-1}{x+1}\right) dx = \frac{x^2}{2}+x$$
$$+\ln(|x-1|)-\ln(|x+1|)+2\frac{(x-1)^{-2+1}}{-2+1}$$
$$+C = \frac{x^2}{2}+x+\ln\left(\frac{|x-1|}{|x+1|}\right)-\frac{2}{x-1}$$
$$+C = \frac{x^2}{2}+x-\frac{2}{x-1}+\ln\left(\left|\frac{x-1}{x+1}\right|\right)+C$$

Example 10.115 Evaluate the integral $\int \frac{x+4}{x^2+5x+6} dx$.

Solution The quadratic equation $x^2 + 5x + 6 = 0$ has roots at $x = -2$ and $x = -3$, which means the expression can be factored into $(x + 2)(x + 3)$. Thus,

$$\int \frac{x+4}{x^2+5x+6} dx = \int \frac{x+4}{(x+2)(x+3)} dx = \int \left(\frac{2}{x+2} + \frac{-1}{x+3}\right) dx$$
$$= 2\ln(|x+2|) - \ln(|x+3|) + C$$

Example 10.116 Evaluate the integral $\int \frac{x+4}{x^2+4x+13} dx$.

Solution The quadratic denominator has complex roots at $x = -2 + 3i$ and $x = -2 - 3i$, which would make partial fraction decomposition a lengthy and complex process. Instead, let's use a simpler method.

First, we need to rewrite the numerator, $x + 4$, to include the derivative of the denominator, $x^2 + 4x + 13$, which is $2x + 4$. We can do this by splitting the fraction: $x+4=\frac{1}{2}(2x+4)+2$. Now, substitute this back into the integral:

$$\int \frac{x+4}{x^2+4x+13}dx=\int \frac{\frac{1}{2}(2x+4)+2}{x^2+4x+13}dx=\frac{1}{2}\int \frac{2x+4}{x^2+4x+13}dx$$
$$+\int \frac{2}{x^2+4x+13}dx=\frac{1}{2}\int \frac{2x+4}{x^2+4x+13}dx+\int \frac{2}{(x+2)^2+3^2}dx$$
$$=\frac{1}{2}\ln\left(\left|x^2+4x+13\right|\right)+2\times\frac{1}{3}\tan^{-1}\left(\frac{x+2}{3}\right)$$
$$+C=\frac{1}{2}\ln\left(\left|x^2+4x+13\right|\right)+\frac{2}{3}\tan^{-1}\left(\frac{x+2}{3}\right)+C$$

It is worth mentioning that this example required using the formula $\int \frac{du}{u^2+a^2}=\frac{1}{a}\tan^{-1}\left(\frac{u}{a}\right)+C$.

Example 10.117 Evaluate the integral $\int \frac{x^2+4}{x^2+4x+13}dx$.

Solution

$$\int \frac{x^2+4}{x^2+4x+13}dx=\int \frac{x^2+4x-4x+9-9+4}{x^2+4x+13}dx=\int \frac{x^2+4x+13-4x-9}{x^2+4x+13}dx$$
$$=\int \left(1-\frac{4x+9}{x^2+4x+13}\right)dx=\int \left(1-\frac{2(2x+4)+1}{x^2+4x+13}\right)dx$$
$$=\int 1dx-2\int \frac{2x+4}{x^2+4x+13}dx-\int \frac{1}{x^2+4x+13}dx$$
$$=\int 1dx-2\int \frac{2x+4}{x^2+4x+13}dx-\int \frac{1}{(x+2)^2+3^2}dx=$$
$$x-2\ln\left(\left|x^2+4x+13\right|\right)-\frac{1}{3}\tan^{-1}\left(\frac{x+2}{3}\right)+C$$

It is worth mentioning that this example required using the formula $\int \frac{du}{u^2+a^2}=\frac{1}{a}\tan^{-1}\left(\frac{u}{a}\right)+C$.

Example 10.118 Evaluate the integral $\int \frac{x^3+4}{x^2+4x+13}dx$.

Solution

$$\begin{aligned}\int \frac{x^3+4}{x^2+4x+13}dx &= \int \frac{x^3+4x^2+13x-4x^2-13x+4}{x^2+4x+13}dx \\ &= \int x-\frac{4x^2+13x-4}{x^2+4x+13}dx \\ &= \int x-\frac{4x^2+13x+3x-3x+52-52-4}{x^2+4x+13}dx \\ &= \int x-\frac{4x^2+16x+52-3x-56}{x^2+4x+13}dx \\ &= \int x-4+\frac{3x+56}{x^2+4x+13}dx \\ &= \int x-4+\frac{\frac{3}{2}(2x+4)+50}{x^2+4x+13}dx \\ &= \int x-4+\frac{\frac{3}{2}(2x+4)+50}{x^2+4x+13}dx \\ &= \int x-4+\frac{\frac{3}{2}(2x+4)}{x^2+4x+13}+\frac{50}{(x+2)^2+3^2}dx \\ &= \frac{x^2}{2}-4x+\frac{3}{2}\ln\left(\left|x^2+4x+13\right|\right)+\frac{50}{3}\tan^{-1}\left(\frac{x+2}{3}\right)+C\end{aligned}$$

It is worth mentioning that this example required using the formula $\int \frac{du}{u^2+a^2} = \frac{1}{a}\tan^{-1}\left(\frac{u}{a}\right) + C$.

Example 10.119 Evaluate the integral $\int \frac{2x^2-x+4}{x^3+4x}dx$.

Solution

$$\frac{2x^2-x+4}{x^3+4x} = \frac{2x^2-x+4}{x(x^2+4)} = \frac{A}{x} + \frac{Bx+C}{x^2+4}$$

$$A = \lim_{x\to 0} \frac{2x^2-x+4}{x^2+4} = 1$$

$$\frac{Bx+C}{x^2+4} = \frac{2x^2-x+4}{x(x^2+4)} - \frac{1}{x} \Rightarrow Bx+C = \frac{2x^2-x+4}{x} - \frac{x^2+4}{x} \Rightarrow Bx^2 + Cx = 2x^2-x+4-x^2-4 \Rightarrow Bx^2+Cx = x^2-x \Rightarrow B=1, C=-1$$

$$\frac{2x^2-x+4}{x^3+4x} = \frac{2x^2-x+4}{x(x^2+4)} = \frac{1}{x} + \frac{x-1}{x^2+4}$$

$$\int \frac{2x^2-x+4}{x^3+4x}dx = \int \frac{1}{x} + \frac{x-1}{x^2+4}dx = \int \frac{1}{x}dx + \frac{1}{2}\int \frac{2x}{x^2+4}dx - \int \frac{1}{x^2+4}dx$$

We will calculate each integral separately:

$$\int \frac{1}{x} dx = ln|x|$$

$$\frac{1}{2}\int \frac{2x}{x^2+4} dx = \frac{1}{2}\ln|x^2+4| = \frac{1}{2}\ln(x^2+4)$$

$$-\int \frac{1}{x^2+4} dx = -\frac{1}{2}\tan^{-1}\frac{x}{2}$$

Therefore,

$$\int \frac{2x^2-x+4}{x^3+4x} dx = \int \frac{1}{x} dx + \frac{1}{2}\int \frac{2x}{x^2+4} dx - \int \frac{1}{x^2+4} dx = \ln|x| + \frac{1}{2}\ln(x^2+4) - \frac{1}{2}\tan^{-1}\frac{x}{2} + C$$

Example 10.120 Evaluate the integral $\int \frac{2x^2-x+4}{x^3+4x} dx$.

Solution A solution to this problem was provided in the previous example. An alternative approach for the partial fraction decomposition is used here.

$$\frac{2x^2-x+4}{x^3+4x} = \frac{2x^2-x+4}{x(x^2+4)} = \frac{2x^2-x+4}{x(x+2i)(x-2i)} = \frac{A}{x} + \frac{B}{x+2i} + \frac{C}{x-2i}$$

$$A = \lim_{x\to 0} \frac{2x^2-x+4}{x^2+4} = 1$$

$$B = \lim_{x\to -2i} \frac{2x^2-x+4}{x(x-2i)} = \frac{2(-2i)^2-(-2i)+4}{-2i(-2i-2i)} = \frac{-8+2i+4}{-8} = \frac{1}{2} - \frac{1}{4}i$$

$$C = \lim_{x\to 2i} \frac{2x^2-x+4}{x(x+2i)} = \frac{2(2i)^2-2i+4}{2i(2i+2i)} = \frac{-8-2i+4}{-8} = \frac{1}{2} + \frac{1}{4}i$$

$$\frac{2x^2-x+4}{x^3+4x} = \frac{A}{x} + \frac{B}{x+2i} + \frac{C}{x-2i} = \frac{1}{x} + \frac{\frac{1}{2}-\frac{1}{4}i}{x+2i} + \frac{\frac{1}{2}+\frac{1}{4}i}{x-2i} = \frac{1}{x} + \frac{x-1}{(x+2i)(x-2i)} = \frac{1}{x} + \frac{x-1}{x^2+4}$$

$$\int \frac{2x^2-x+4}{x^3+4x} dx = \int \frac{1}{x} + \frac{x-1}{x^2+4} dx = \int \frac{1}{x} dx + \frac{1}{2}\int \frac{2x}{x^2+4} dx - \int \frac{1}{x^2+4} dx$$

We will calculate each integral separately:

$$\int \frac{1}{x} dx = \ln|x|$$

$$\frac{1}{2}\int \frac{2x}{x^2+4} dx = \frac{1}{2}\ln|x^2+4| = \frac{1}{2}\ln(x^2+4)$$

$$-\int \frac{1}{x^2+4} dx = -\frac{1}{2}\tan^{-1}\frac{x}{2}$$

Therefore,

$$\int \frac{2x^2-x+4}{x^3+4x}dx=\int \frac{1}{x}dx+\frac{1}{2}\int \frac{2x}{x^2+4}dx-\int \frac{1}{x^2+4}dx=\ln|x|+\frac{1}{2}\ln\left(x^2+4\right)-\frac{1}{2}\tan^{-1}\frac{x}{2}+C$$

Example 10.121 Evaluate the integral $\int \frac{dx}{x(1+x^4)}$.

Solution Let's use a substitution and set $u = x^4$. Thus,

$$u=x^4 \Rightarrow du=4x^3dx$$

$$\int \frac{dx}{x(1+x^4)}=\frac{1}{4}\int \frac{4x^3dx}{x^4(1+x^4)}=\frac{1}{4}\int \frac{du}{u(1+u)}$$
$$=\frac{1}{4}\int \left(\frac{1}{u}-\frac{1}{u+1}\right)du=\ln(|u|)-\ln(|u+1|)+C$$
$$=\ln\left(\left|\frac{u}{u+1}\right|\right)+C=\ln\left(\left|\frac{x^4}{x^4+1}\right|\right)+C$$

Example 10.122 Evaluate the integral $\int \frac{x^2+1}{x^4+1}dx$.

Solution

$$\int \frac{x^2+1}{x^4+1}dx=\int \frac{x^2\left(1+\frac{1}{x^2}\right)}{x^2\left(x^2+\frac{1}{x^2}\right)}dx=\int \frac{1+\frac{1}{x^2}}{x^2+\frac{1}{x^2}}dx=\int \frac{1+\frac{1}{x^2}}{\left(x-\frac{1}{x}\right)^2+2}dx$$

Let's use a substitution and set $u=x-\frac{1}{x}$. Thus,

$$u=x-\frac{1}{x}\Rightarrow du=1+\frac{1}{x^2}$$

$$\int \frac{x^2+1}{x^4+1}dx=\int \frac{1+\frac{1}{x^2}}{\left(x-\frac{1}{x}\right)^2+2}dx=\int \frac{1}{u^2+2}du=\frac{1}{\sqrt{2}}\tan^{-1}\left(\frac{u}{\sqrt{2}}\right)+C=\frac{1}{\sqrt{2}}\tan^{-1}\left(\frac{x-\frac{1}{x}}{\sqrt{2}}\right)+C$$

Example 10.123 Evaluate the integral $\int \frac{x^2-1}{x^4+1}dx$.

Solution

$$\int \frac{x^2-1}{x^4+1}dx=\int \frac{x^2\left(1-\frac{1}{x^2}\right)}{x^2\left(x^2+\frac{1}{x^2}\right)}dx=\int \frac{1-\frac{1}{x^2}}{x^2+\frac{1}{x^2}}dx=\int \frac{1-\frac{1}{x^2}}{\left(x+\frac{1}{x}\right)^2-2}dx$$

Let's use a substitution and set $u=x+\frac{1}{x}$. Thus,

$$u=x+\frac{1}{x}\Rightarrow du=1-\frac{1}{x^2}$$

$$\int \frac{x^2-1}{x^4+1}dx = \int \frac{1-\frac{1}{x^2}}{\left(x+\frac{1}{x}\right)^2-2}dx = \int \frac{1}{u^2-2}du$$

$$= \int \frac{1}{\left(u-\sqrt{2}\right)\left(u+\sqrt{2}\right)}du = \int \left(\frac{\frac{1}{2\sqrt{2}}}{u-\sqrt{2}} - \frac{\frac{1}{2\sqrt{2}}}{u+\sqrt{2}}\right)du$$

$$= \frac{1}{2\sqrt{2}}\int \frac{1}{u-\sqrt{2}}du - \frac{1}{2\sqrt{2}}\int \frac{1}{u+\sqrt{2}}du$$

$$= \frac{1}{2\sqrt{2}}\ln\left(\left|u-\sqrt{2}\right|\right) - \frac{1}{2\sqrt{2}}\ln\left(\left|u+\sqrt{2}\right|\right) + C$$

$$= \frac{1}{2\sqrt{2}}\ln\left(\left|\frac{u-\sqrt{2}}{u+\sqrt{2}}\right|\right) + C$$

$$= \frac{1}{2\sqrt{2}}\ln\left(\left|\frac{x+\frac{1}{x}-\sqrt{2}}{x+\frac{1}{x}+\sqrt{2}}\right|\right) + C = \frac{1}{2\sqrt{2}}\ln\left(\left|\frac{x^2-\sqrt{2}x+1}{x^2+\sqrt{2}x+1}\right|\right) + C$$

Example 10.124 Evaluate the integral $\int x^3 \cos(x)dx$.

Solution The tabular method, also known as the DI method, is particularly well-suited for integrals that involve products of polynomial functions with exponential, trigonometric, or logarithmic functions. For instance, $\int x^2 e^x dx$, $\int x^3 \sin(x)dx$ or $\int x^4 \ln(x)dx$ can be calculated easily with tabular integration technique.

Tabular integration involves creating a table with two columns: one for repeated differentiation of a function and the other for repeated integration of another function. The product of the diagonal entries, with alternating signs, is then summed to obtain the integral.

Figures 10.14 and 10.15 illustrate how to perform tabular integration on the function provided in this example.

Fig. 10.14 Calculation of the $\int x^3 \cos(x)dx$ (First step)

Derivatives	Integrals
x^3	$\cos(x)$
$3x^2$	$\sin(x)$
$6x$	$-\cos(x)$
6	$-\sin(x)$
0	$\cos(x)$

Fig. 10.15 Calculation of the $\int x^3 \cos(x)dx$ (Second step)

Derivatives		Integrals
x^3	+	$\cos(x)$
$3x^2$	−	$\sin(x)$
$6x$	+	$-\cos(x)$
6	−	$-\sin(x)$
0		$\cos(x)$

According to Fig. 10.15, $\int x^3 \cos(x)dx = x^3 \sin(x) + 3x^2 \cos(x) - 6x \sin(x) - 6 \cos(x) + C$

Example 10.125 Table of derivatives and integrals for $\int x^3 \cos(x)dx$ is reshown in Fig. 10.16.

Fig. 10.16 Tabular integration of $\int x^3 \cos(x)dx$

Derivatives		Integrals
x^3	+	$\cos(x)$
$3x^2$	−	$\sin(x)$
$6x$	+	$-\cos(x)$
6	−	$-\sin(x)$
0		$\cos(x)$

By considering only the first two rows (Fig. 10.17), we get:

$$\int x^3 \cos(x)dx = x^3 \sin(x) - \int 3x^2 \sin(x)dx.$$

Fig. 10.17 The first two rows of Fig. 10.16

Derivatives		Integrals
x^3	+	$\cos(x)$
$3x^2$	−	$\sin(x)$

By considering only the first three rows (Fig. 10.18), we get:

$$\int x^3 \cos(x)dx = x^3 \sin(x) + 3x^2 \cos(x) - \int 6x \cos(x)dx.$$

Fig. 10.18 The first three rows of Fig. 10.6

Derivatives		Integrals
x^3	+	$\cos(x)$
$3x^2$	−	$\sin(x)$
$6x$	+	$-\cos(x)$

By considering only the first four rows (Fig. 10.19), we get:

$$\int x^3 \cos(x)dx = x^3 \sin(x) + 3x^2 \cos(x) - 6x \sin(x) + \int 6 \sin(x)dx.$$

Fig. 10.19 The first four rows of Fig. 10.16

Derivatives		Integrals
x^3	+	$\cos(x)$
$3x^2$	−	$\sin(x)$
$6x$	+	$-\cos(x)$
6	−	$-\sin(x)$

Example 10.126 Evaluate the integral $\int (x^3 - 2x)e^{4x}dx$.

Solution The table of derivatives and integrals is shown in Fig. 10.20.

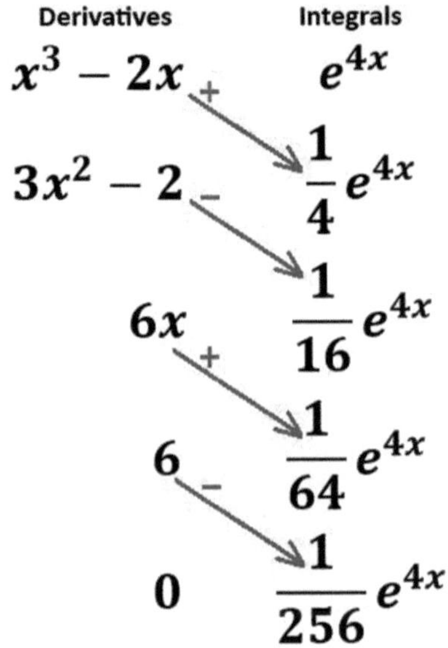

Fig. 10.20 Tabular integration of $\int(x^3 - 2x)e^{4x}dx$

From Fig. 10.20, we get:

$$\int (x^3 - 2x)e^{4x}dx = \frac{1}{4}(x^3 - 2x)e^{4x} - \frac{(3x^2 - 2)}{16}e^{4x} + \frac{6x}{64}e^{4x} - \frac{6}{256}e^{4x} + C.$$

Example 10.127 Table of derivatives and integrals for $\int(x^3 - 2x)e^{4x}dx$ is reshown in Fig. 10.21.

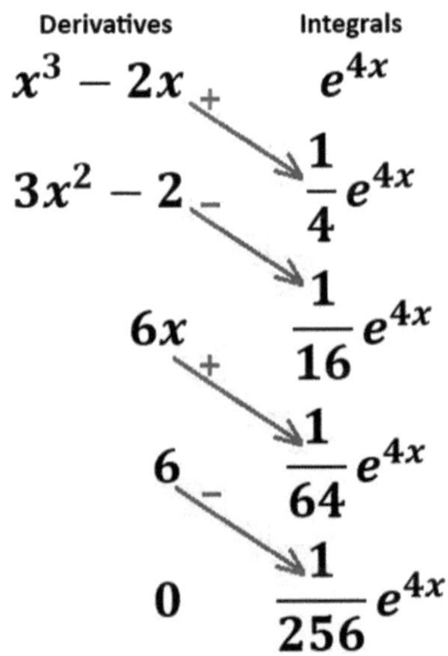

Fig. 10.21 Tabular integration of $\int(x^3 - 2x)e^{4x}dx$

By considering only the first two rows (Fig. 10.22), we get:

$$\int (x^3 - 2x)e^{4x}dx = \frac{1}{4}(x^3 - 2x)e^{4x} - \int \frac{(3x^2 - 2)}{4}e^{4x}dx.$$

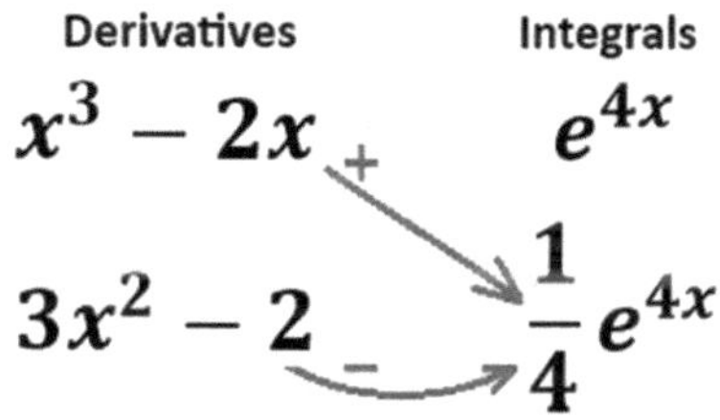

Fig. 10.22 First two rows of Fig. 10.21

By considering only the first three rows (Fig. 10.23), we get:

$$\int (x^3 - 2x)e^{4x}dx = \frac{1}{4}(x^3 - 2x)e^{4x} - \frac{(3x^2 - 2)}{16}e^{4x} + \int \frac{6x}{16}e^{4x}dx.$$

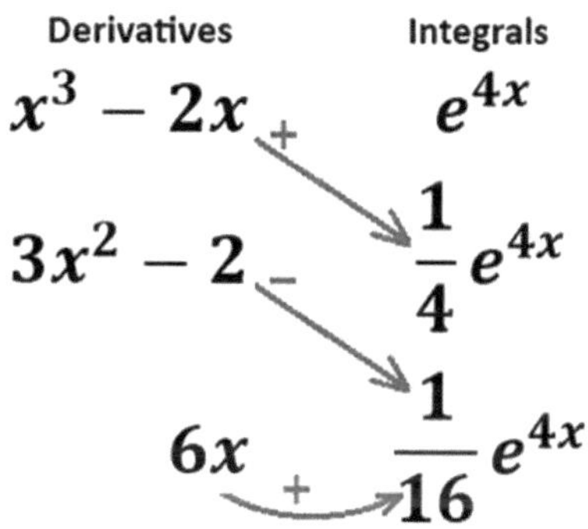

Fig. 10.23 First three rows of Fig. 10.21

Example 10.128 Evaluate the integral $\int e^x \cos(x)dx$.

Solution In this example, successive differentiation of e^x does not result in a value of zero. The first three rows of the table of derivatives and integrals for $\int e^x \cos(x)dx$ is shown in Fig. 10.24.

Fig. 10.24 Calculation of the $\int e^x \cos(x)dx$

Derivatives	Integrals
e^x +	$\cos(x)$
e^x −	$\sin(x)$
e^x +	$-\cos(x)$

From Fig. 10.24, we get:

$$\int e^x \cos(x)dx = e^x \times \sin(x) - e^x \times -\cos(x) + \int -e^x \cos(x)dx$$
$$+C \Rightarrow 2\int e^x \cos(x)dx = e^x \sin(x) + e^x \cos(x)$$
$$+C \Rightarrow \int e^x \cos(x)dx = \frac{e^x \sin(x)}{2} + \frac{e^x \cos(x)}{2}$$
$$+\frac{C}{2} \Rightarrow \int e^x \cos(x)dx = \frac{e^x}{2}(\sin(x) + \cos(x)) + C'.$$

Exercise 10.9 Evaluate the integral $\int e^x \sin(x)dx$. (Ans. $-\frac{e^x}{2}(\cos(x) - \sin(x)) + C$)

Example 10.129 Evaluate the integral $\int \ln(x)dx$.

Solution Successive differentiation of ln(x) does not result in a value of zero. The first two rows of the table of derivatives and integrals for $\int \ln(x)dx$ is shown in Fig. 10.25.

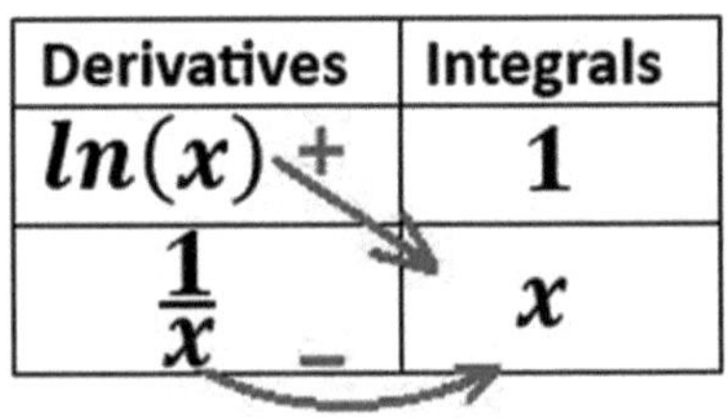

Fig. 10.25 Calculation of the $\int \ln(x)dx$

From Fig. 10.25, we get:

$$\int \ln(x)dx = x\ln(x) - \int \frac{1}{x} \times xdx = x\ln(x) - \int dx = x\ln(x) - x + C.$$

Example 10.130 Evaluate the integral $\int \ln(\ln(x))dx$.

Solution Let's use a substitution and set $u = \ln(x)$. Thus,

$$u = \ln(x) \Rightarrow du = \frac{1}{x}$$

$$\int \frac{\ln(\ln(x))}{x}dx = \int \ln(u)du = u\ln(u) - u + C = \ln(x)\ln(\ln(x)) - \ln(x) + C$$

Example 10.131 Evaluate the integral $\int \frac{\ln(x)}{x^2}dx$.

Solution The first two rows of the table of derivatives and integrals for $\int \frac{\ln(x)}{x^2}dx$ is shown in Fig. 10.26.

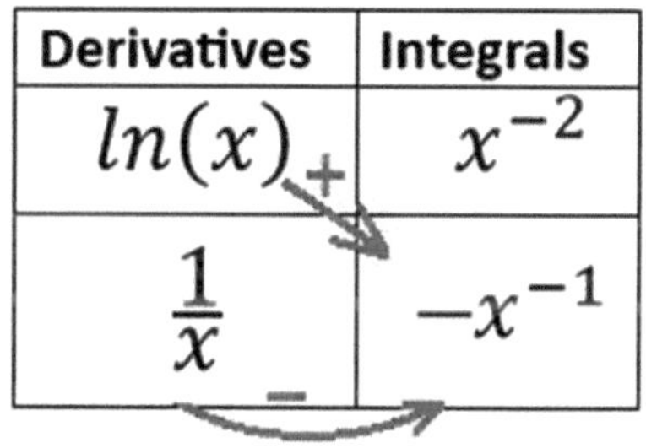

Fig. 10.26 Calculation of the $\int \frac{\ln(x)}{x^2}dx$

From Fig. 10.26, we get:

$$\int \frac{\ln(x)}{x^2}dx = -x^{-1}\ln(x) + \int \frac{1}{x}\times x^{-1}dx = -x^{-1}\ln(x) + \int x^{-2}dx = -\frac{\ln(x)}{x} - x^{-1} + C = -\frac{\ln(x)}{x} - \frac{1}{x} + C.$$

Example 10.132 Evaluate the integral $\int \sin^{-1}(x)dx$.

Solution The first two rows of the table of derivatives and integrals for $\int \sin^{-1}(x)dx$ is shown in Fig. 10.27.

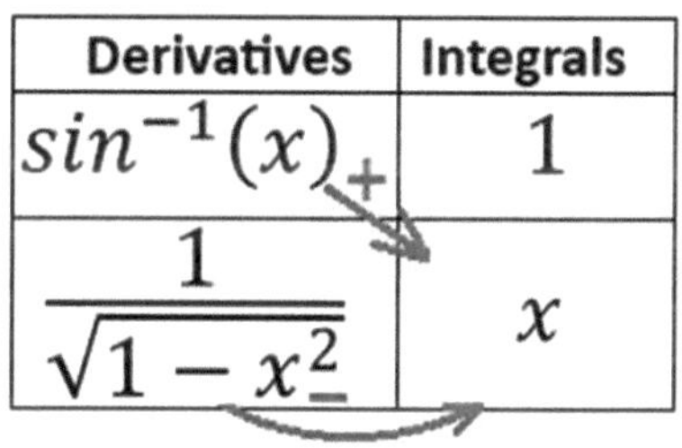

Fig. 10.27 Calculation of the $\int sin^{-1}(x)dx$

From Fig. 10.27, we get:

$$\int \sin^{-1}(x)dx = x\sin^{-1}(x) - \int \frac{x}{\sqrt{1-x^2}}dx = x\sin^{-1}(x) - \int x\left(1-x^2\right)^{-\frac{1}{2}}dx$$

$$= x\sin^{-1}(x) - \frac{1}{-2}\int -2x\left(1-x^2\right)^{-\frac{1}{2}}dx = x\sin^{-1}(x) + \frac{1}{2}\frac{\left(1-x^2\right)^{-\frac{1}{2}+1}}{-\frac{1}{2}+1} + C$$

$$= x\sin^{-1}(x) + \sqrt{1-x^2} + C$$

Example 10.133 Evaluate the integral $\int \tan^{-1}(x)dx$.

Solution The first two rows of the table of derivatives and integrals for $\int \tan^{-1}(x)dx$ is shown in Fig. 10.28.

Fig. 10.28 Calculation of the $\int \tan^{-1}(x)dx$

Derivatives	Integrals
$tan^{-1}(x)$ +	1
$\frac{1}{1+x^2}$	x

From Fig. 10.28, we get:

$$\int \tan^{-1}(x)dx = x\tan^{-1}(x) - \int \frac{x}{1+x^2}dx = x\tan^{-1}(x) - \frac{1}{2}\int \frac{2x}{1+x^2}dx$$
$$= x\tan^{-1}(x) - \frac{1}{2}\ln(|1+x^2|) + C = x\tan^{-1}(x) - \frac{1}{2}\ln(1+x^2) + C$$

Exercise 10.10 Evaluate the integral $\int \cos^{-1}(x)dx$. (Ans. $\int \cos^{-1}(x)dx = x\cos^{-1}(x) - \sqrt{1-x^2} + C$)

Exercise 10.11 Evaluate the integral $\int \cot^{-1}(x)dx$. (Ans. $\int \cot^{-1}(x)dx = x\cot^{-1}(x) + \frac{1}{2}\ln(1+x^2) + C$)

Exercise 10.12 Evaluate the following integrals. Correct answers are provided for verification.

$$\int x^2 e^x dx = e^x(x^2 - 2x + 2) + C$$

$$\int x^4 \ln(x)dx = \frac{x^5}{5}\left(\ln(x) - \frac{1}{5}\right) + C$$

$$\int x^2 \sin(x)dx = 2x\sin(x) - (x^2 - 2)\cos(x) + C$$

Example 10.134 The code shown in Fig. 10.29 computes $f(\pi)$ for the function $f(x) = e^{3x}\cos(2x)$.

Fig. 10.29 Calculation of the $f(\pi)$ for $f(x) = e^{3x} \cos(2x)$

```
Command Window
>> syms x
>> f=exp(3*x)*cos(2*x);
>> subs(f,x,pi)

ans =

exp(3*pi)

>> eval(ans)

ans =

   1.2392e+04

fx >>
```

Example 10.135 The code in Fig. 10.30 calculates $\int e^{3x} \cos(2x)dx$.

Fig. 10.30 Calculation of the $\int e^{3x} \cos(2x)dx$

```
Command Window
>> syms x
>> f=exp(3*x)*cos(2*x);
>> F=simplify(int(f,x))

F =

(exp(3*x)*(3*cos(2*x) + 2*sin(2*x)))/13

>> pretty(F)
exp(3 x) (cos(2 x) 3 + sin(2 x) 2)
----------------------------------
               13

fx >>
```

The symbolic integration function int returns the antiderivative without the constant of integration. You must manually add this constant when writing the final mathematical answer. Therefore,

$$\int e^{3x} \cos(2x)dx = \frac{e^{3x}(3\cos(2x) + 2\sin(2x))}{13} + C$$

Example 10.136 The code in Fig. 10.31 calculates $\int \frac{1}{x^2+4} dx$.

Fig. 10.31 Calculation of the $\int \frac{1}{x^2+4}dx$

```
Command Window
>> syms x
>> f=1/(x^2+4);
>> F=simplify(int(f,x))

F =

atan(x/2)/2

>> pretty(F)
    / x \
atan| - |
    \ 2 /
---------
    2

fx >>
```

The symbolic integration function int returns the antiderivative without the constant of integration. You must manually add this constant when writing the final mathematical answer. Therefore,

$$\int \frac{1}{x^2+4}dx = \frac{\tan^{-1}\left(\frac{x}{2}\right)}{2} + C$$

Example 10.137 The code in Fig. 10.32 calculates $\int |x^2 - 5x + 6| dx$ assuming that $2 < x < 3$.

Fig. 10.32 Calculation of the $\int \frac{1}{x^2+4}dx$ with the assumption $2 < x < 3$

```
Command Window
>> syms x
>> assume(2<x & x<3)
>> f=abs(x^2-5*x+6);
>> F=simplify(int(f,x))

F =

-(x*(36 - 15*x + 2*x^2))/6

>> pretty(F)
                       2
  x (36 - 15 x + 2 x )
- --------------------
           6

fx >> |
```

We will now confirm the result presented in Fig. 8.3 through hand calculation.

$$f(x) = |x^2 - 5x + 6| \overset{2<x<3}{\Rightarrow} f(x) = |x^2 - 5x + 6| = -(x^2 - 5x + 6)$$

$$\int |x^2 - 5x + 6| dx = \int -(x^2 - 5x + 6) dx = \int (-x^2 + 5x - 6) dx = -\frac{x^3}{3}$$
$$+\frac{5}{2}x^2 - 6x + C = -\frac{(2x^3 - 15x^2 + 36x)}{6} + C = -\frac{x(2x^2 - 15x + 36)}{6} + C$$

The obtained result agrees with the one shown in Fig. 10.32.

References for Further Study

1. Engineering Mathematics with MATLAB® and Simulink®, Farzin Asadi, Springer, 2025.
2. Applied Numerical Analysis with MATLAB®/Simulink®: For Engineers and Scientists, Farzin Asadi, Springer, 2023.
3. Advanced Engineering Mathematics, Erwin Kreyszig, Wiley, 2020.
4. A Problem-Solving Approach to Single-Variable Calculus (Volume II), Farzin Asadi, Springer, 2026.
5. A Problem-Solving Approach to Multi-Variable Calculus (Volume I), Farzin Asadi, Springer, 2026.
6. A Problem-Solving Approach to Multi-Variable Calculus (Volume II), Farzin Asadi, Springer, 2026.
7. A Problem-Solving Approach to Multi-Variable Calculus (Volume III), Farzin Asadi, Springer, 2026.
8. A Problem-Solving Approach to Ordinary Differential Equations (Volume I), Farzin Asadi, Springer, 2026.
9. A Problem-Solving Approach to Ordinary Differential Equations (Volume II), Farzin Asadi, Springer, 2026.
10. A Problem-Solving Approach to Ordinary Differential Equations (Volume III), Farzin Asadi, Springer, 2026.
11. A Problem-Solving Approach to Ordinary Differential Equations (Volume IV), Farzin Asadi, Springer, 2026.

Chapter 11
Essentials of MATLAB®

11.1 Introduction

MATLAB® (short for MATrix LABoratory) is a programming and numerical computing platform created specifically for engineers and scientists. It is used to analyze and design the systems and products that are transforming our world.

At the core of MATLAB is the MATLAB language, a matrix-based language that provides the most natural and expressive way to write computational mathematics.

The keywords of the MATLAB language are listed in Table 11.1.

Table 11.1 Keywords of the MATLAB language

Break	Case	Catch	Classdef
continue	else	elseif	end
for	function	global	if
otherwise	parfor	persistent	return
spmd	switch	try	while

Millions of engineers and scientists worldwide use MATLAB for a wide range of applications in both industry and academia. MATLAB serves as both an analysis and a design tool. This chapter will introduce the essentials of MATLAB and its common daily uses.

11.2 MATLAB Environment

The MATLAB environment is shown in Fig. 11.1. It is composed of four parts. First part is a collection of icons that are required frequently. The second section (current folder browser) enables you to interactively manage files and folders in MATLAB.

F. Asadi, *A Problem-Solving Approach to Single-Variable Calculus*,
https://doi.org/10.1007/978-3-032-13505-6_11

Use the current folder browser to view, create, open, move, and rename files and folders in the current folder. Use the icon shown in Fig. 11.2 to open the desired folder. The third section of MATLAB is command window. The MATLAB commands are entered here and their results are shown here. You need to press Enter key of your key board to run the commands written in command window. The fourth section of MATLAB (the workspace) contains variables that you create within or import into MATLAB from data files or other programs.

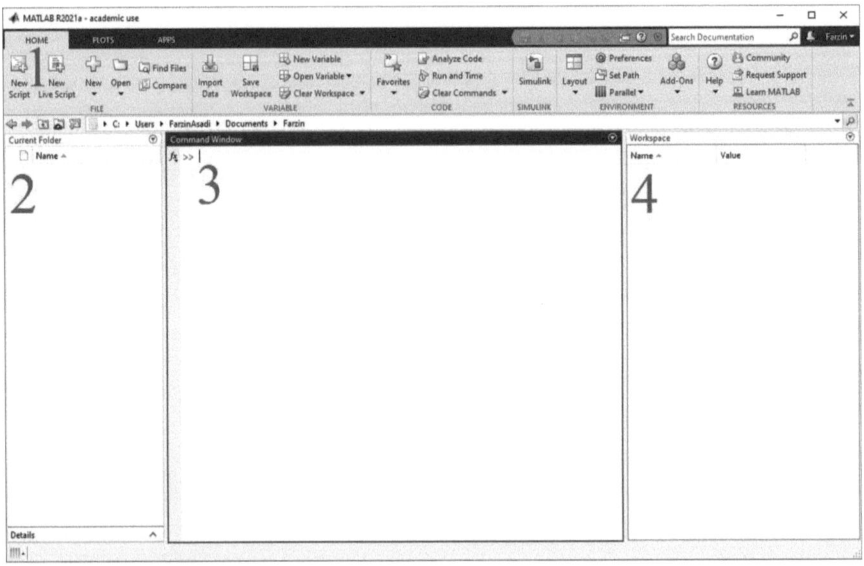

Fig. 11.1 MATLAB environment

Fig. 11.2 Brows for folder icon

The layout of MATLAB can be changed using the layout icon (Fig. 11.3). This book uses the default layout.

Fig. 11.3 Layout icon

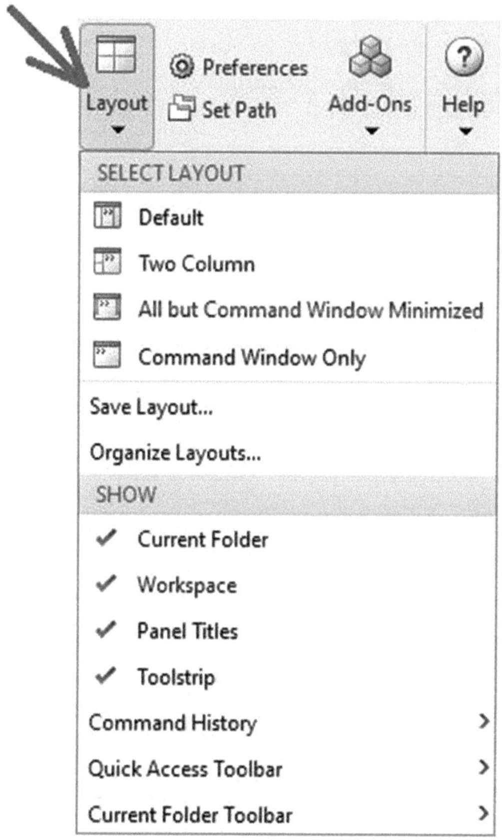

11.3 Basic Operations with MATLAB

Let's define a complex variable with a value of $1 + 2i$. This can be done with the aid of commands shown in Figs. 11.4, 11.5, or 11.6. In MATLAB, both of i and j can be used to enter the imaginary part of a complex number.

Fig. 11.4 Defining variable $a = 1 + 2i$

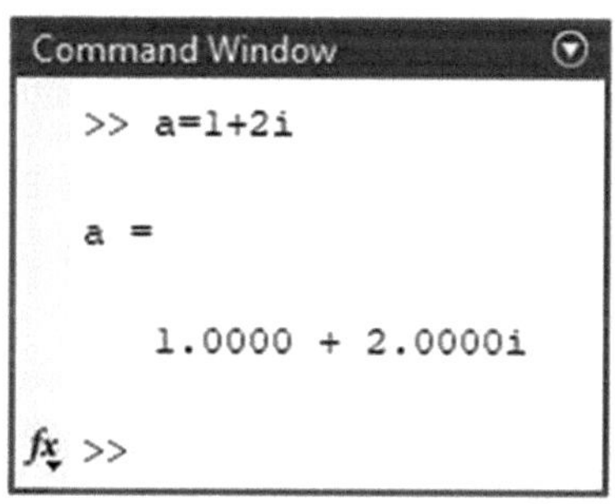

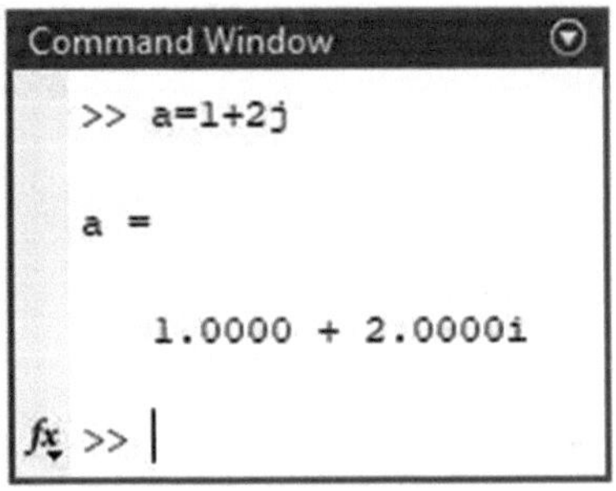

Fig. 11.5 Defining variable a = 1 + 2*i*

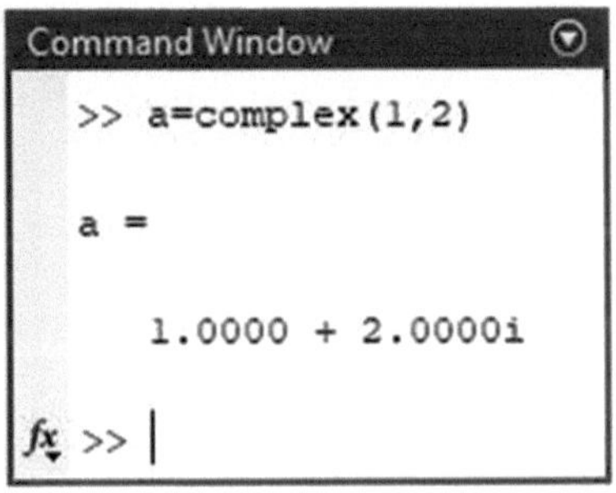

Fig. 11.6 Defining variable a = 1 + 2*i*

After running the commands of Figs. 11.4, 11.5, or 11.6, a new variable will be added to the workspace window (Fig. 11.7).

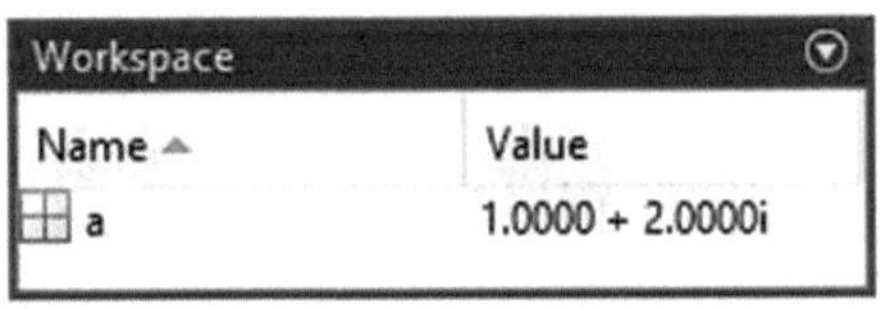

Fig. 11.7 Variable a is added to the workspace

To suppress the output of a command (i.e., prevent the result from being displayed in the command window), simply end the line with a semicolon (Fig. 11.8).

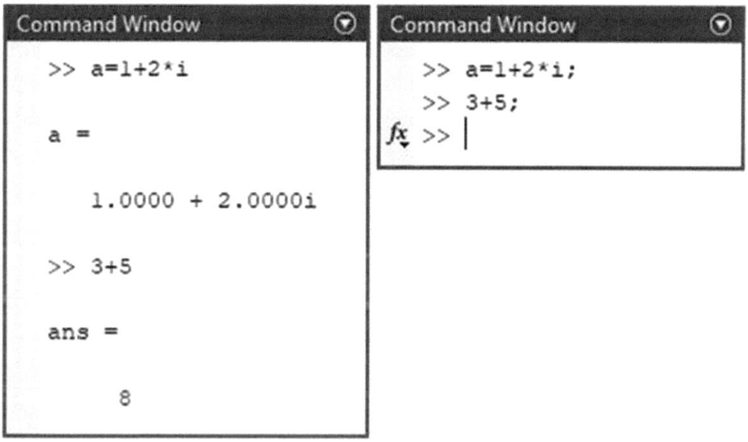

Fig. 11.8 The semicolon suppresses the output of a command

You can use the built-in functions real and imag to extract the real and imaginary parts of any complex number, respectively (Fig. 11.9).

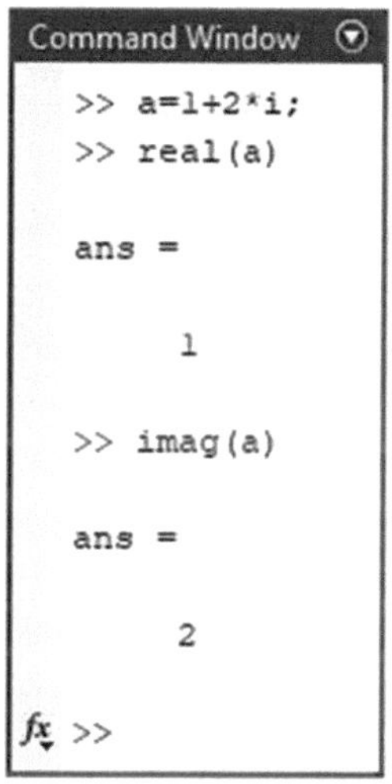

Fig. 11.9 Calculation of real and imaginary parts with real and imag commands

You can use the conj function to calculate the complex conjugate of a number (Fig. 11.10).

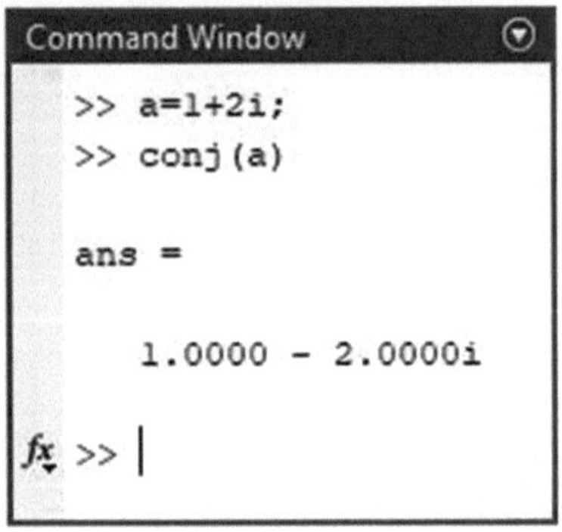

Fig. 11.10 Calculation of the complex conjugate with conj command

The point $a = 1 + 2i$ is shown in Fig. 11.11. The polar form of this point is $\sqrt{1^2+2^2}e^{i\tan^{-1}\left(\frac{2}{1}\right)}$. Value of $\sqrt{1^2+2^2}$ and $\tan^{-1}\left(\frac{2}{1}\right)$ is calculated with the aid commands shown in Fig. 11.12. The magnitude and principal argument of a complex number can be calculated using the commands shown in Fig. 11.13.

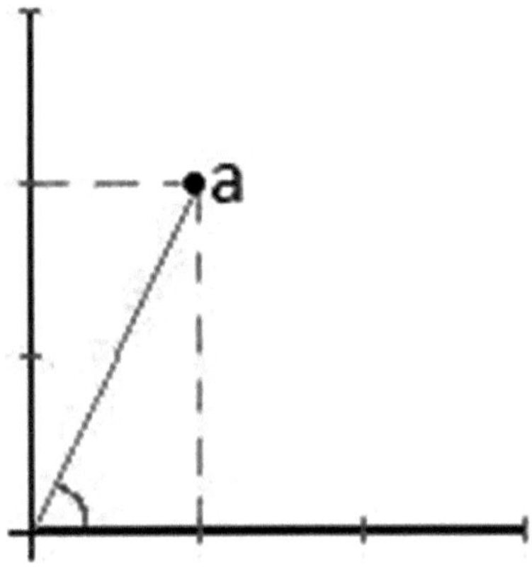

Fig. 11.11 Representation of 1 + 2*i* in the complex plane

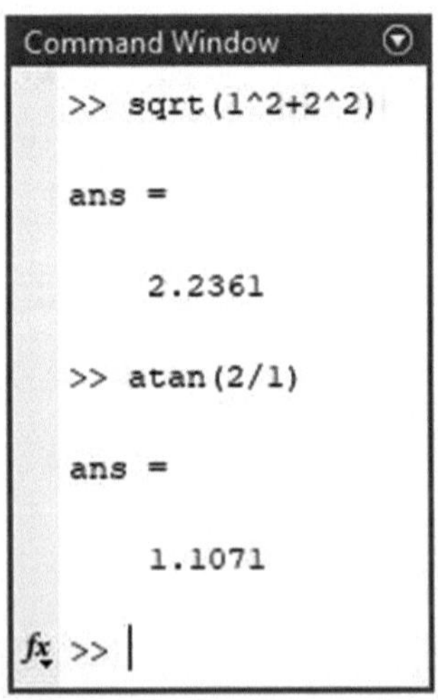

Fig. 11.12 Calculation of magnitude and angle of 1 + 2*i*

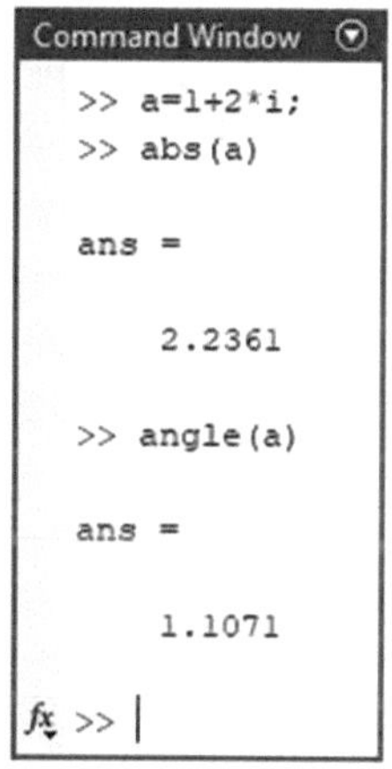

Fig. 11.13 Calculation of magnitude and angle of $1 + 2i$

Case sensitivity dictates whether a programming language treats uppercase and lowercase letters as distinct (case-sensitive) or equivalent (case-insensitive). MATLAB is a case-sensitive language; this characteristic is demonstrated in Figs. 11.14 and 11.15.

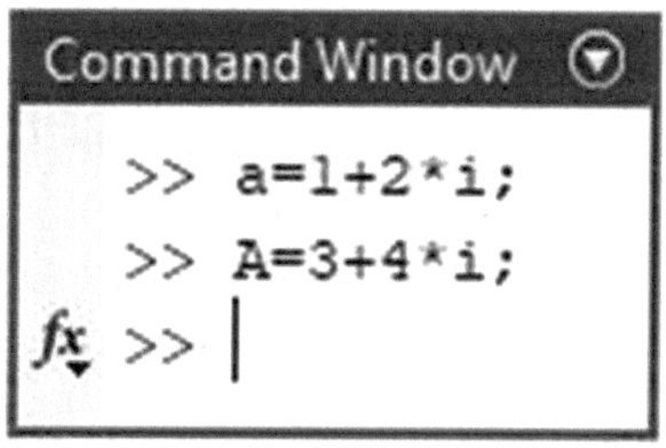

Fig. 11.14 MATLAB is a case-sensitive language

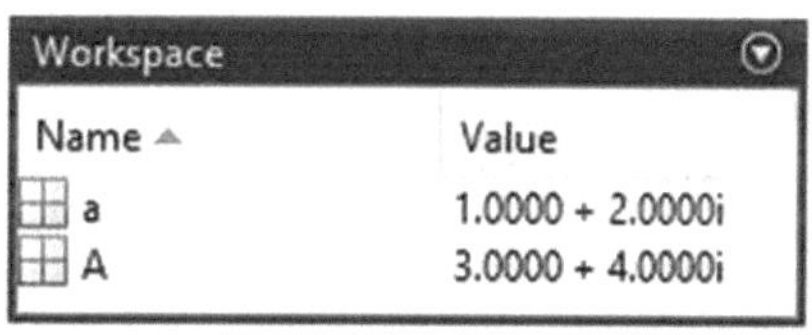

Fig. 11.15 a and A are two different variables

MATLAB automatically uses the default variable ans (short for answer) to store the result of a command when no output variable is explicitly defined. For example, in Fig. 11.16, the result of the multiplication is stored in the user-defined variable c. Conversely, in Fig. 11.17, the result is saved in ans because the user did not define any variable.

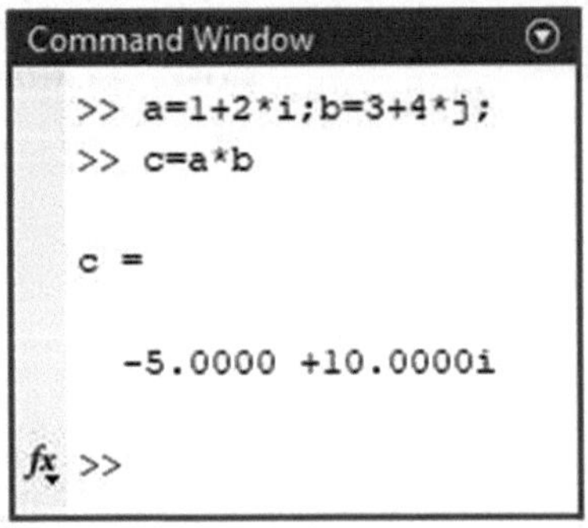

Fig. 11.16 Variable c equals to product of a and b

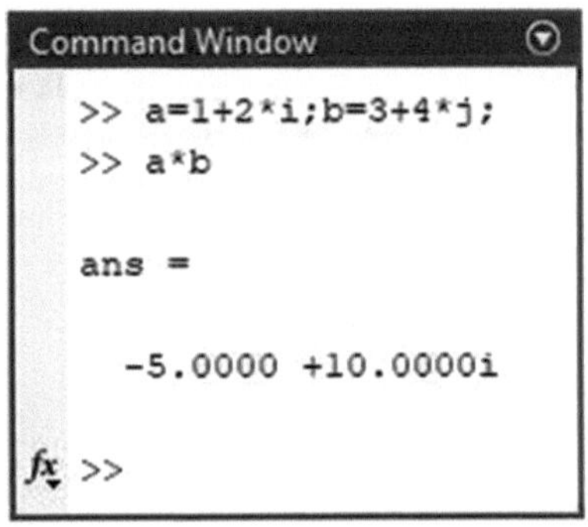

Fig. 11.17 Default variable ans saves the calculation result

Commands in Fig. 11.18 do some basic operations on two complex numbers. The percent symbol (%) is used for indicating a comment line.

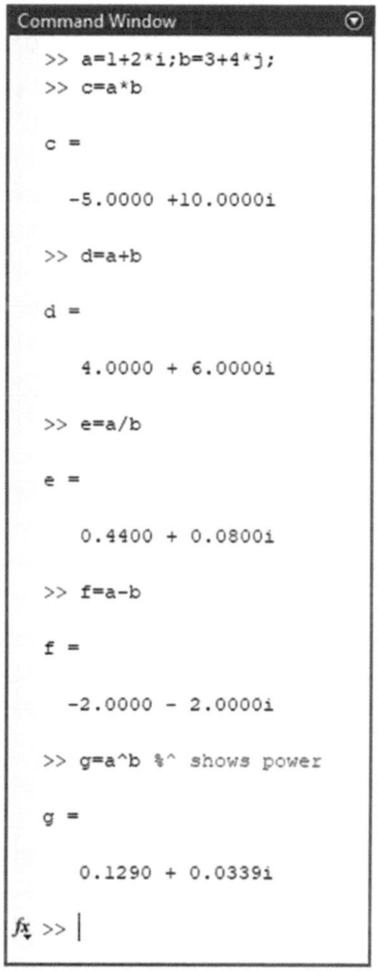

Fig. 11.18 Basic operations in MATLAB

The reminder of division of two real numbers can be found using the mod or rem functions (Fig. 11.19).

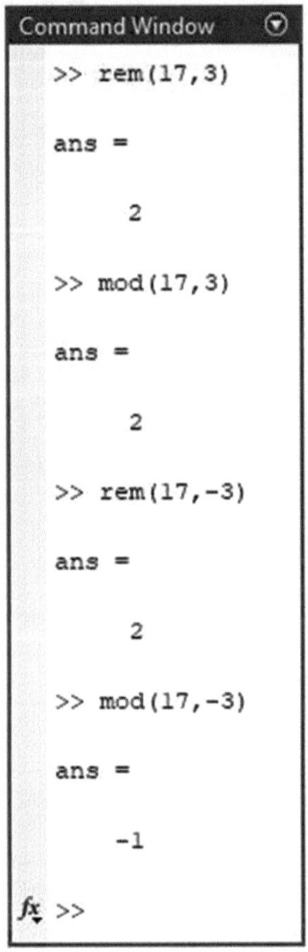

Fig. 11.19 rem and mod commands

The concept of remainder after division is not uniquely defined, and the two functions mod and rem each compute a different variation. The mod function produces a result that is either zero or has the same sign as the divisor. The rem function produces a result that is either zero or has the same sign as the dividend.

Another difference is the convention when the divisor is zero. The mod function follows the convention that mod(a,0) returns a, whereas the rem function follows the convention that rem(a,0) returns NaN. NaN stands for "Not a Number" and is used to represent values that are not real or complex numbers. Expressions like 0/0 and inf/inf result in NaN.

Both mod and rem functions have their uses. For example, in signal processing, the mod function is useful in the context of periodic signals because its output is periodic (with period equal to the divisor).

11.4 Clearing the Command Window and Workspace

To clear the contents of the command window, use the clc command (Fig. 11.20). Figure 11.21 shows the appearance of the command window after clc has been executed.

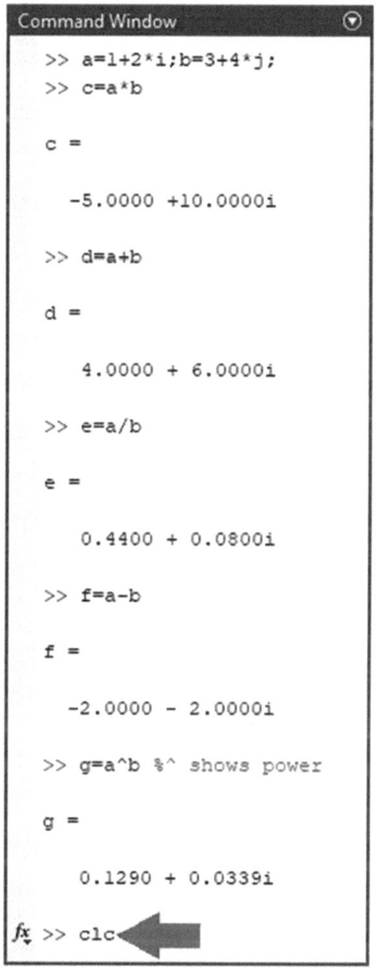

Fig. 11.20 clc command clears the Command Window

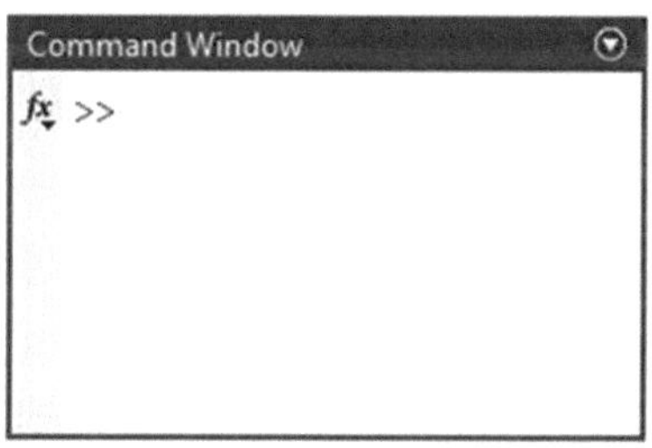

Fig. 11.21 The Command Window is cleared using the clc command

Alternatively, you can clear the command window by right-clicking anywhere within it and selecting Clear Command Window (Fig. 11.22) from the context menu that appears.

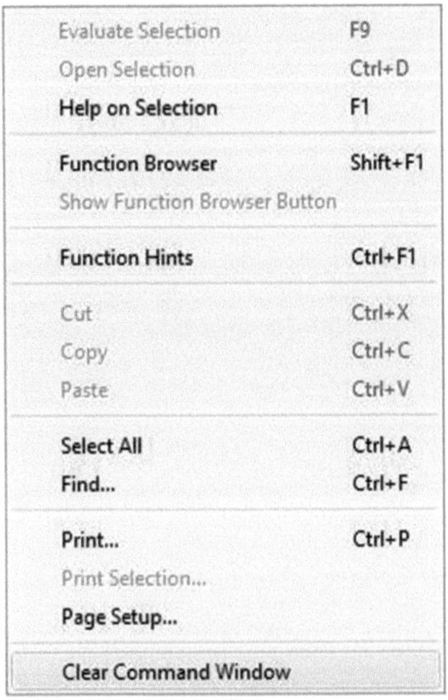

Fig. 11.22 Clicking the clear command window is another way to clear the command window

Clearing the command window does not affect the variables (Fig. 11.23). If you want to clear all the variables, you need to use the command shown in Fig. 11.24 or 11.25. If you want to clear a specific variable, then you need to write its name after the clear command. For instance, the command shown in Fig. 11.26, clears the variable g only. There is another way to remove the variable g as well. You can right-click on it and select delete from the appeared menu (Fig. 11.27).

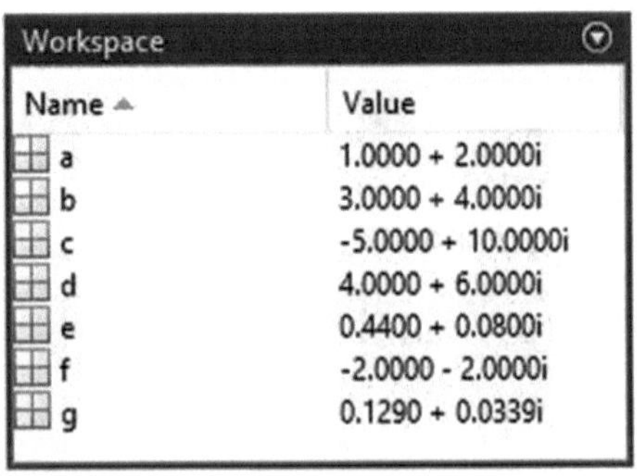

Fig. 11.23 Clearing the command window has no effect on the variables

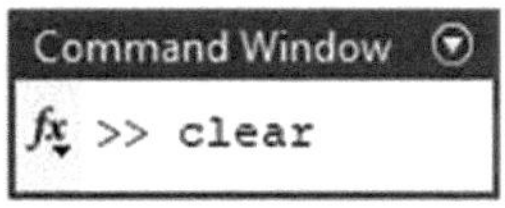

Fig. 11.24 Deleting all the workspace variables

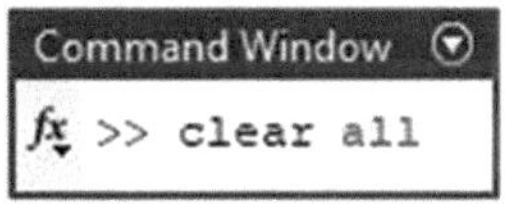

Fig. 11.25 Deleting all the workspace variables

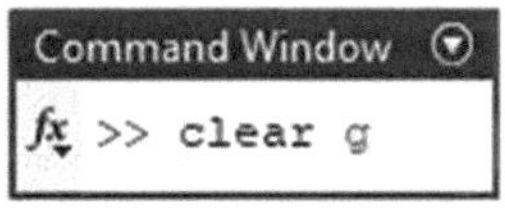

Fig. 11.26 Deleting the variable g of Workspace

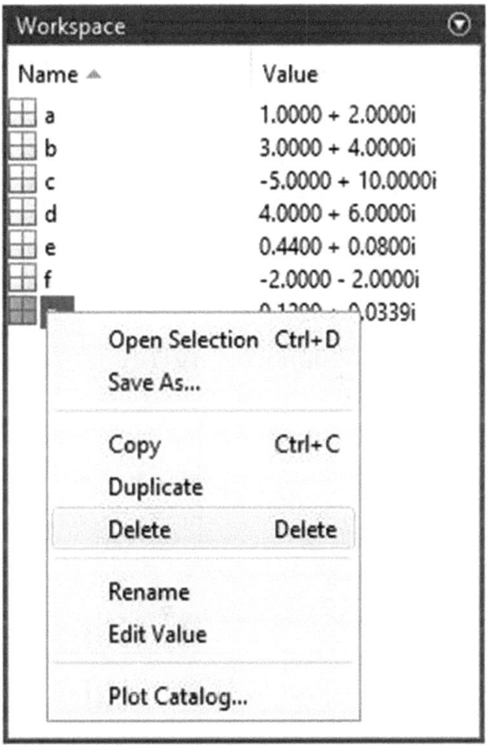

Fig. 11.27 Deleting the variable g of Workspace

You can use the up and down arrows of the keyboard to access the previous entered commands (Fig. 11.28).

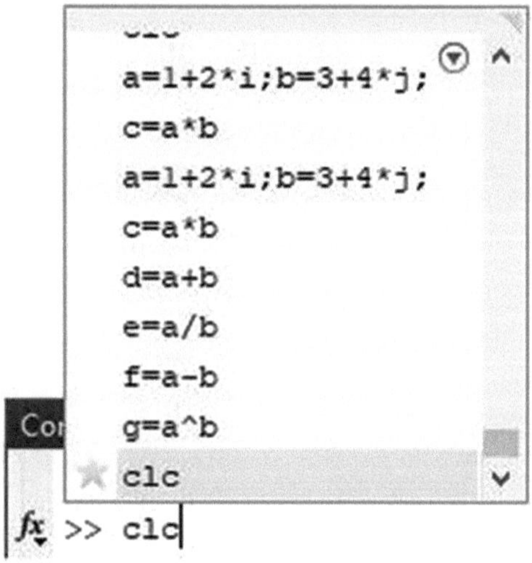

Fig. 11.28 Accessing to the previously entered commands

11.5 Basic Matrix Operations

Anything in MATLAB is a matrix. Even a number is considered as a 1×1 matrix. Assume that you want to enter $A = [1\ \ 2\ \ 3], B = \begin{bmatrix} 4 \\ 5 \\ 6 \end{bmatrix}$ and $C = \begin{bmatrix} 7 & 8 & 9 \\ 10 & 11 & 12+2i \\ 13 & 14 & 16 \end{bmatrix}$ to MATLAB. The commands shown in Figs. 11.29 or 11.30 do this job.

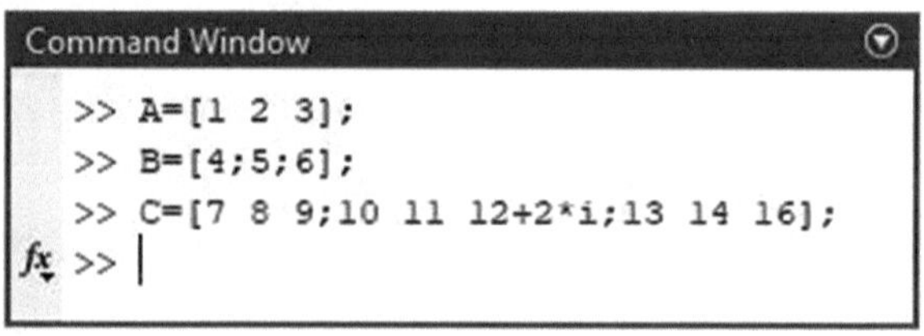

Fig. 11.29 Defining the matrices A, B, and C

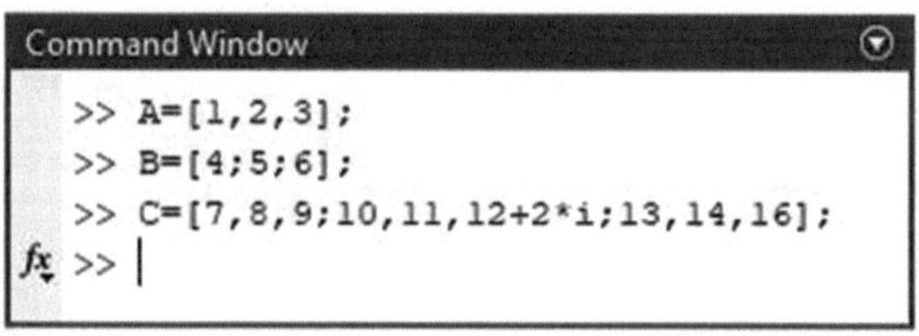

Fig. 11.30 Defining the matrices A, B, and C

You can access the element in the *i*th row and *j*th column of a matrix x using the syntax $x(i,j)$. For matrices that are strictly one-dimensional (vectors), a shorthand notation is available: if x is a $1 \times n$ (row) vector, you can use $x(i)$ instead of $x(1,i)$. Similarly, if x is an $n \times 1$ (column) vector, you can use $x(i)$ instead of $x(i,1)$. The term vector is generally applied to both $1 \times n$ and $n \times 1$ matrices.

Figure 11.31 presents examples of commands used to read the elements of the matrices defined in Figs. 11.29 and 11.30.

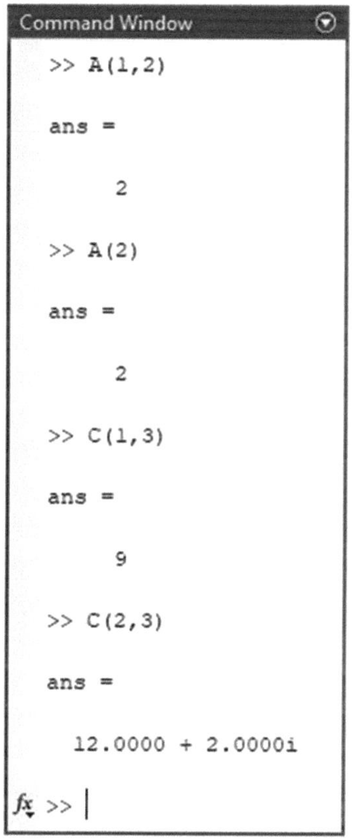

Fig. 11.31 Accessing the elements of the matrices (Matrices A, B, and C are defined in Fig. 11.29)

The command shown in Fig. 11.32, changes the element in the second row and third column of matrix C to $12 + 12i$.

```
Command Window
>> C(2,3)=12+12*i;
>> C

C =

   7.0000 + 0.0000i   8.0000 + 0.0000i   9.0000 + 0.0000i
  10.0000 + 0.0000i  11.0000 + 0.0000i  12.0000 +12.0000i
  13.0000 + 0.0000i  14.0000 + 0.0000i  16.0000 + 0.0000i

>>
```

Fig. 11.32 Changing the element on the second row and third column of C (Matrix C is defined in Fig. 11.29)

To obtain the dimensions (number of rows and columns) of a matrix, use the size command (Fig. 11.33).

Fig. 11.33 Calculation of number of rows and columns of a matrix (Matrices A, B, and C are defined in Fig. 11.29)

```
Command Window
>> size(A)

ans =

     1     3

>> size(B)

ans =

     3     1

>> size(C)

ans =

     3     3

>>
```

You can save the number of rows and the number of columns in separate variables. The command shown in Fig. 11.34 saves the number of rows and the number of columns of matrix A in the variables RowA and ColumnA, respectively.

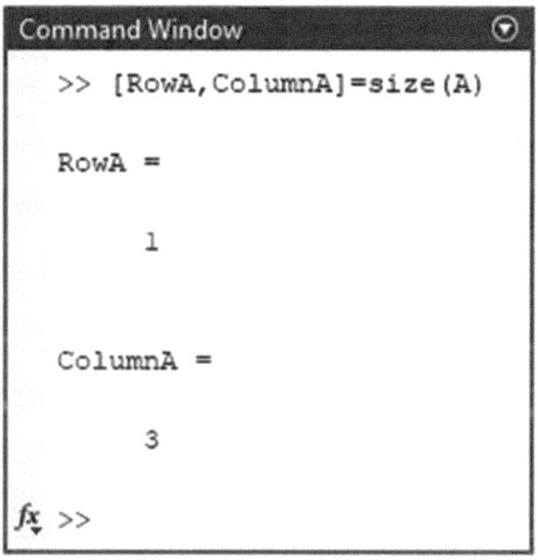

Fig. 11.34 Number of rows and columns of matrix A

The multiplication, summation, and subtraction of two matrices can be done with the aid of *, + , and - operators, respectively. Note that the size of two matrices must be the same for summation and subtraction and the number of columns of the first matrix must be the same as the number of rows of the second matrix; otherwise, the operation will not be done.

MATLAB has elementwise operators as well. The elementwise operator does the operation on the corresponding elements. For instance, in Fig. 11.35, [1 2 3].^[4 5 6] = [1^4 2^5 3^6] = [1 32 729] or [1 2 3].*[4 5 6] = [1*4 2*5 3*6] = [4 10 18].

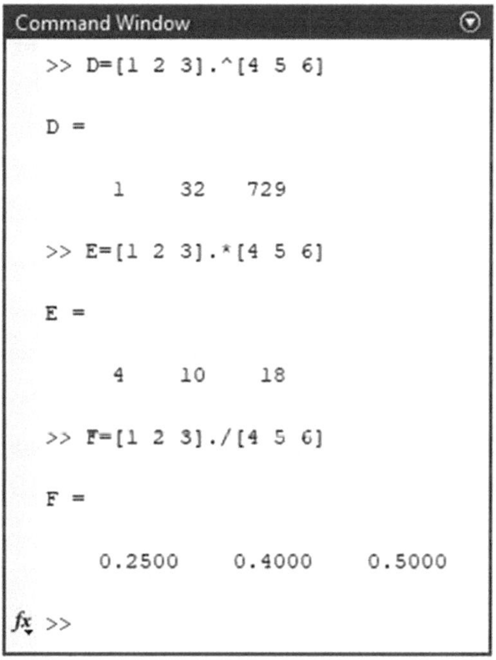

Fig. 11.35 Elementwise commands

You can find the minimum and maximum of a vector by using the min and max commands, respectively (Fig. 11.36).

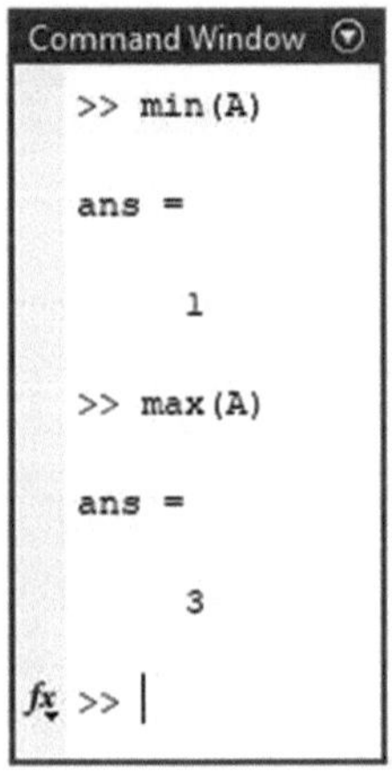

Fig. 11.36 Calculation of minimum and maximum of vector A

The min and max functions can be applied to matrices as well. If the input of these functions is an m×n matrix, they return an n × 1 vector containing the minimum/ maximum element from each column (Fig. 11.37).

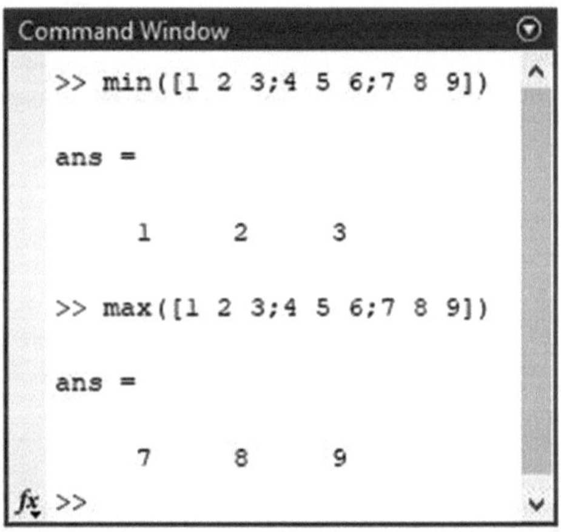

Fig. 11.37 The min and max commands can also be applied to matrices

When a matrix contains complex elements, the min and max functions operate based on the magnitude of each element. Recall that the magnitude of a complex number $a + bi$ is $\sqrt{a^2 + b^2}$. In Fig. 11.38, the corresponding magnitude matrix is $\begin{bmatrix} 7 & 8 & 9 \\ 10 & 11 & 16.971 \\ 13 & 14 & 16 \end{bmatrix}$. Therefore, the max function returns 12 + 12i as the maximum of the third column since 12+12i has the largest magnitude among the elements in that column.

```
Command Window
>> C

C =

    7.0000 + 0.0000i   8.0000 + 0.0000i   9.0000 + 0.0000i
   10.0000 + 0.0000i  11.0000 + 0.0000i  12.0000 +12.0000i
   13.0000 + 0.0000i  14.0000 + 0.0000i  16.0000 + 0.0000i

>> min(C)

ans =

     7     8     9

>> max(C)

ans =

   13.0000 + 0.0000i  14.0000 + 0.0000i  12.0000 +12.0000i

fx >>
```

Fig. 11.38 Calculation of minimum and maximum element of matrix C

The transpose of a matrix can be calculated using the single quote operator or transpose function (Fig. 11.39). If the matrix is real, then two methods produce the same results. However, when the matrix is complex, the results are not the same. The transpose function simply flips the matrix over its diagonal (i.e., switches the row and column indices of the matrix), whereas the single quote function calculates the complex conjugate of elements and then changes the row and column indices.

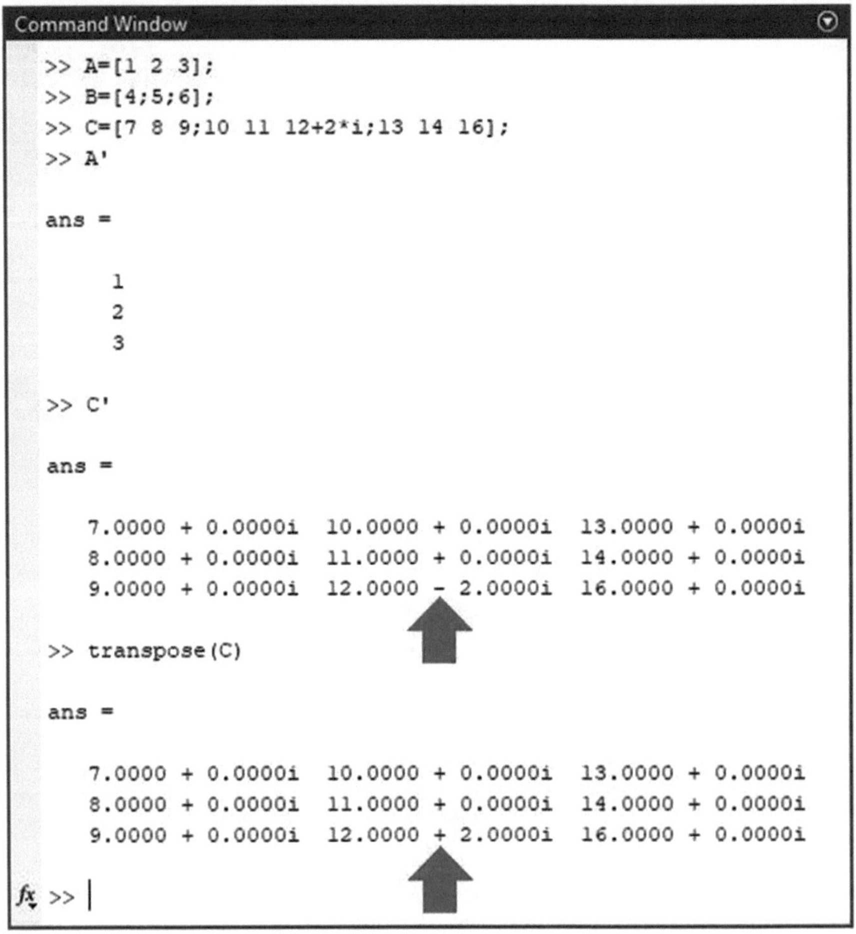

Fig. 11.39 Different methods of calculation of transpose of a matrix

You can use the det function to calculate the determinant of square matrix. The inverse of a matrix X can be calculated using the inv(X) or X^-1 commands (Fig. 11.40).

```
Command Window
>> det(C)

ans =

  -3.0000 +12.0000i

>> inv(C)

ans =

  -2.3529 - 0.0784i   0.0392 + 0.1569i   1.3137 - 0.0784i
   2.1176 - 0.1961i   0.0980 + 0.3922i  -1.2157 - 0.1961i
   0.0588 + 0.2353i  -0.1176 - 0.4706i   0.0588 + 0.2353i

>> C^-1

ans =

  -2.3529 - 0.0784i   0.0392 + 0.1569i   1.3137 - 0.0784i
   2.1176 - 0.1961i   0.0980 + 0.3922i  -1.2157 - 0.1961i
   0.0588 + 0.2353i  -0.1176 - 0.4706i   0.0588 + 0.2353i

fx >>
```

Fig. 11.40 Calculation of the determinant and inverse of matrix C

Assume you want to ensure the two matrices calculated in Fig. 11.40—using inv (C) and C^-1—are identical. You can verify this by computing the difference between them. As shown in Fig. 11.41, the difference is zero, which confirms that the results obtained from both commands are the same.

Fig. 11.41 Calculation of the difference between the C^-1 and inv(C) results

```
Command Window
>> C^-1-inv(C)

ans =

     0     0     0
     0     0     0
     0     0     0

fx >>
```

Assume that you want to solve the linear system $\begin{bmatrix} 7 & 8 & 9 \\ 10 & 11 & 12+2i \\ 13 & 14 & 16 \end{bmatrix} x = \begin{bmatrix} 4 \\ 5 \\ 6 \end{bmatrix}$. The commands shown in Fig. 11.42 can be used to calculate the vector x.

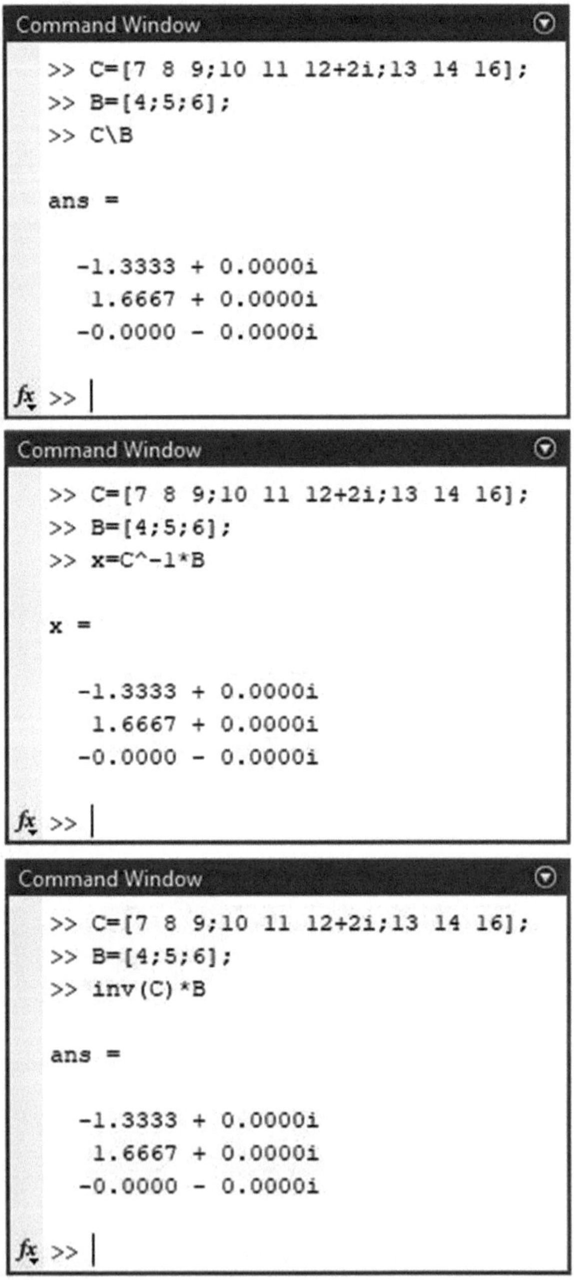

Fig. 11.42 Calculation of vector x

11.6 Trigonometric Functions

Trigonometric functions of MATLAB are listed in Table 11.2. The listed functions can also take complex arguments.

Table 11.2 Trigonometric functions of MATLAB

MATLAB function	Description	Example
`sin`	sin(x). Takes the input in radians	`>>sin(pi/2)` `ans = 1`
`sind`	sin(x). Takes the input in degrees	`>>sind(30)` `ans = 0.5000`
`asin`	$\sin^{-1}(x)$. Returns the output in radians	`>>asin(0.5)` `ans = 0.5236`
`asind`	$\sin^{-1}(x)$. Returns the output in degrees	`>>asind(.5)` `ans = 30.0000`
`cos`	cos(x). Takes the input in radians	`>>cos(1)` `ans = 0.5403`
`cosd`	cos(x). Takes the input in degrees	`>>cosd(60)` `ans = 0.5000`
`acos`	$\cos^{-1}(x)$. Returns the output in radians	`>>acos(.5)` `ans = 1.0472`
`acosd`	$\cos^{-1}(x)$. Returns the output in degrees	`>>acosd(.5)` `ans = 60.0000`
`tan`	tan(x). Takes the input in radians	`>>tan(1)` `ans = 1.5574`
`tand`	tan(x). Takes the input in degrees	`>>tand(60)` `ans = 1.7321`
`atan`	$\tan^{-1}(x)$. Returns the output in radians	`>>atan(1)` `ans = 0.7854`
`atand`	$\tan^{-1}(x)$. Returns the output in degrees	`>>atand(1)` `ans = 45`
`sec`	sec(x). Takes the input in radians	`>>sec(1)` `ans = 1.8508`
`secd`	sec(x). Takes the input in degrees	`>>secd(10)` `ans = 1.0154`
`asec`	$\sec^{-1}(x)$. Returns the output in radians	`>>asec(30)` `ans = 1.5375`
`asecd`	$\sec^{-1}(x)$. Returns the output in degrees	`>>asecd(30)` `ans = 88.0898`
`csc`	csc(x). Takes the input in radians	`>>csc(1)` `ans = 1.1884`
`cscd`	csc(x). Takes the input in degrees	`>>cscd(1)` `ans = 57.2987`
`acsc`	$\csc^{-1}(x)$. Returns the output in radians	`>>acsc(1)` `ans = 1.5708`
`acscd`	$\csc^{-1}(x)$. Returns the output in degrees	`>>acscd(1)` `ans = 90`

(continued)

Table 11.2 (continued)

MATLAB function	Description	Example
`cot`	cot(x). Takes the input in radians	`>>cot(1)` `ans = 0.6421`
`cotd`	cot(x). Takes the input in degrees	`>>cotd(30)` `ans = 1.7321`
`acot`	$\cot^{-1}(x)$. Returns the output in radians	`>>acot(1)` `ans = 0.7854`
`acotd`	$\cot^{-1}(x)$. Returns the output in degrees	`>>acotd(1)` `ans = 45`

For instance, assume we want to calculate $sin(x)^4 - 3\ cos\ (x)$ for $x = 30°$. The commands shown in Figs. 11.43 or 11.44 perform this calculation.

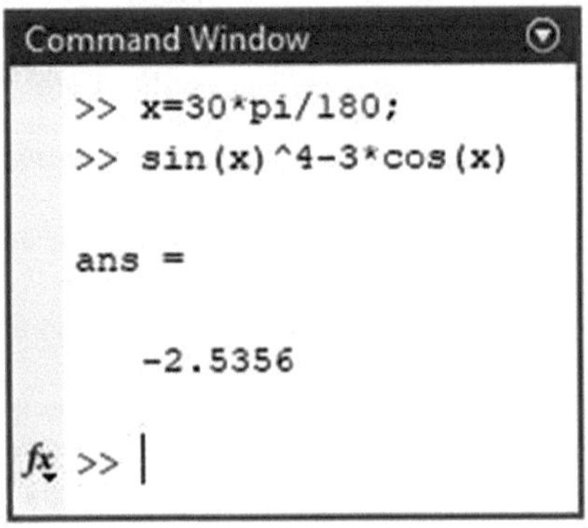

Fig. 11.43 Calculation of $\sin(x)^4 - 3\cos(x)$ for $x = 30°$

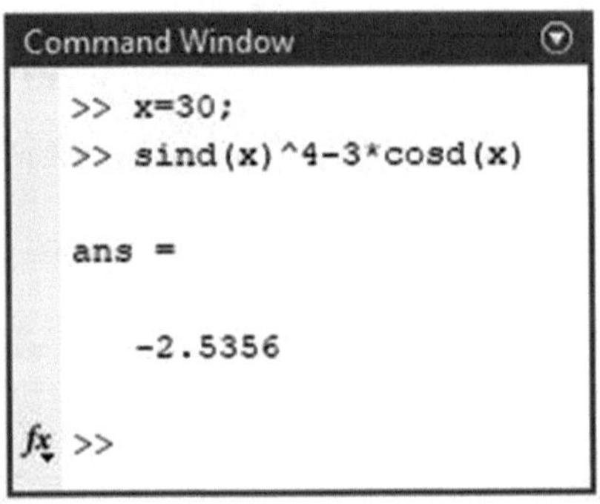

Fig. 11.44 Calculation of $\sin(x)^4 - 3\cos(x)$ for $x = 30°$

11.7 Hyperbolic Functions

Hyperbolic functions of MATLAB are listed in Table 11.3. The listed functions can also take complex arguments.

Table 11.3 Hyperbolic functions of MATLAB

Function	MATLAB Representation
$\sinh(x)$	`sinh(x)`
$\cosh(x)$	`cosh(x)`
$\tanh(x)$	`tanh(x)`
$\coth(x)$	`coth(x)`
$\mathrm{sech}(x)$	`sech(x)`
$\mathrm{csch}^{-1}(x)$	`csch(x)`
$\sinh^{-1}(x)$	`asinh(x)`
$\cosh^{-1}(x)$	`acosh(x)`
$\tanh^{-1}(x)$	`atanh(x)`
$\coth^{-1}(x)$	`acoth(x)`
$\mathrm{sech}^{-1}(x)$	`asech(x)`
$\mathrm{csch}^{-1}(x)$	`acsch(x)`

11.8 Logarithmic and Exponential Function

Logarithmic and exponential functions of MATLAB are listed in Table 11.4. The listed functions can also take complex arguments.

Table 11.4 Logarithmic and exponential functions of MATLAB

MATLAB Function	Description	Example
`log`	Calculates the natural logarithm (with base of $e = 2.7182$)	`>>log (2.7182)` `ans = 1.0000`
`log10`	Calculates the common logarithm (with base of 10)	`>>log10 (100)` `ans = 2`
`log2`	Calculates the binary logarithm with (with base of 2)	`>>log2(4)` `ans = 2`
`exp`	Calculates e^x	`>>exp(2)` `ans = 7.3891`
`sqrt`	Calculates the square root	`>>sqrt(16)` `ans = 4`
`power`	Calculates the power function	`>>power (2,3)` `ans = 8`
`^`	Calculates the power function	`>>2^3` `ans = 8`

11.9 Rounding Functions

Rounding functions of MATLAB are shown in Table 11.5.

Table 11.5 Rounding function of MATLAB

MATLAB function	Expression	Example
fix	It truncates the decimal part	`fix([-4.6 4.6])` `ans = [-4 4]`
floor	It rounds down to the nearest integer	`>>floor(8.4687)` `ans = 8`
`ceil`	It rounds up to the nearest integer	`>>ceil(4.4)` `ans = 5`
`round`	Rounds to the nearest integer	`>>round(4.55)` `ans = 5`

11.10 Colon Operator

The colon operator can be used for creating regularly spaced vectors, index into arrays, and define the bounds of a `for` loop. For instance, assume that you want to make a vector t from 0 to 1 with 0.1 steps. The command shown in Fig. 11.45 makes this vector.

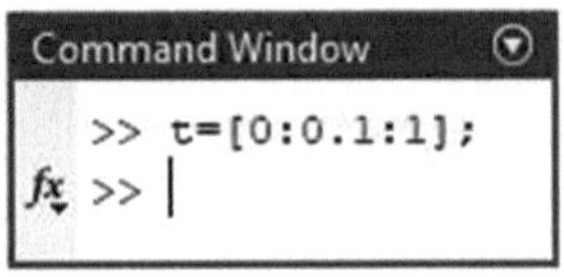

Fig. 11.45 Defining the vector t with colon operator

The values of vector t are shown in Fig. 11.46.

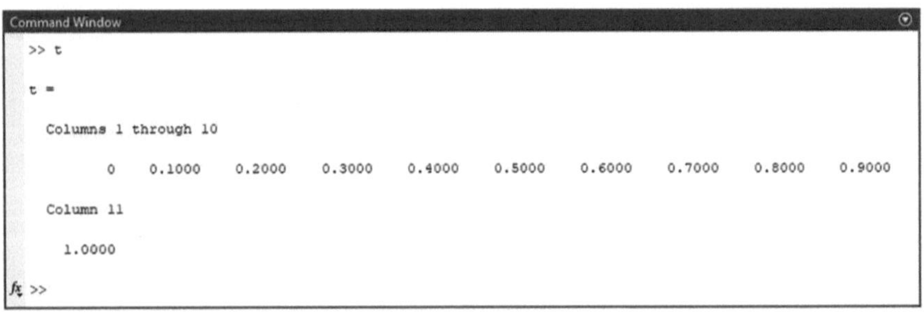

Fig. 11.46 Values of vector t defined in Fig. 11.45

The default step of colon operator equals to one. So, x = [3:10] and x = [3:1:10] are the same. Both of these commands define x = [3 4 5 6 7 8 9 10]. The command shown in Fig. 11.47 defines a random matrix with 8 rows and 6 columns. The A(5: 8:) command in Fig. 11.47, selects the elements in the 5th, 6th, 7th, and 8th rows.

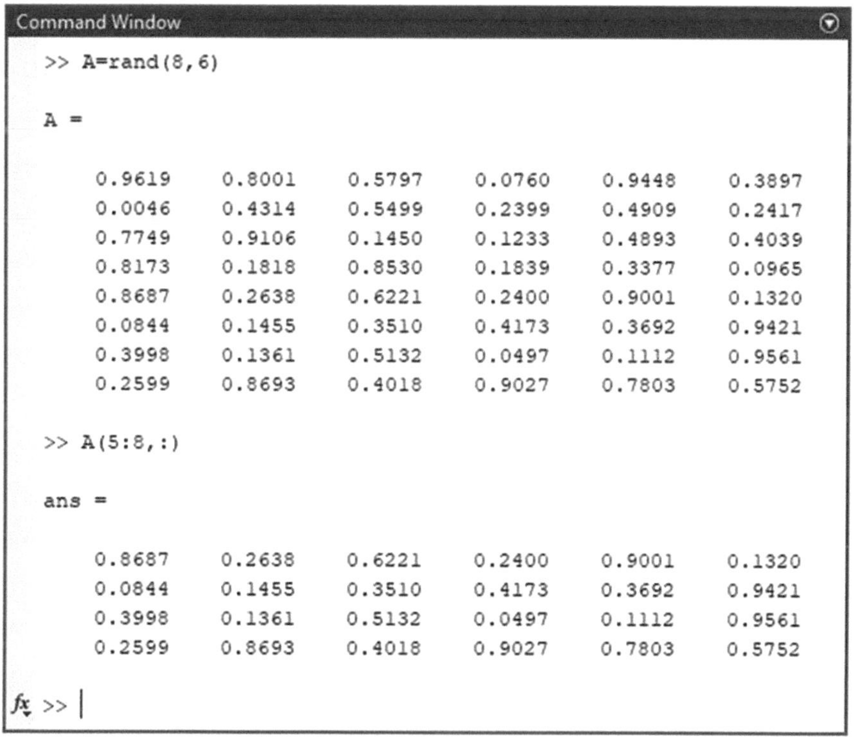

Fig. 11.47 Defining a random matrix and selecting rows 5, 6, 7, and 8

The command A(:,4:6) in Fig. 11.48 selects all the elements in the 4th, 5th, 6th columns.

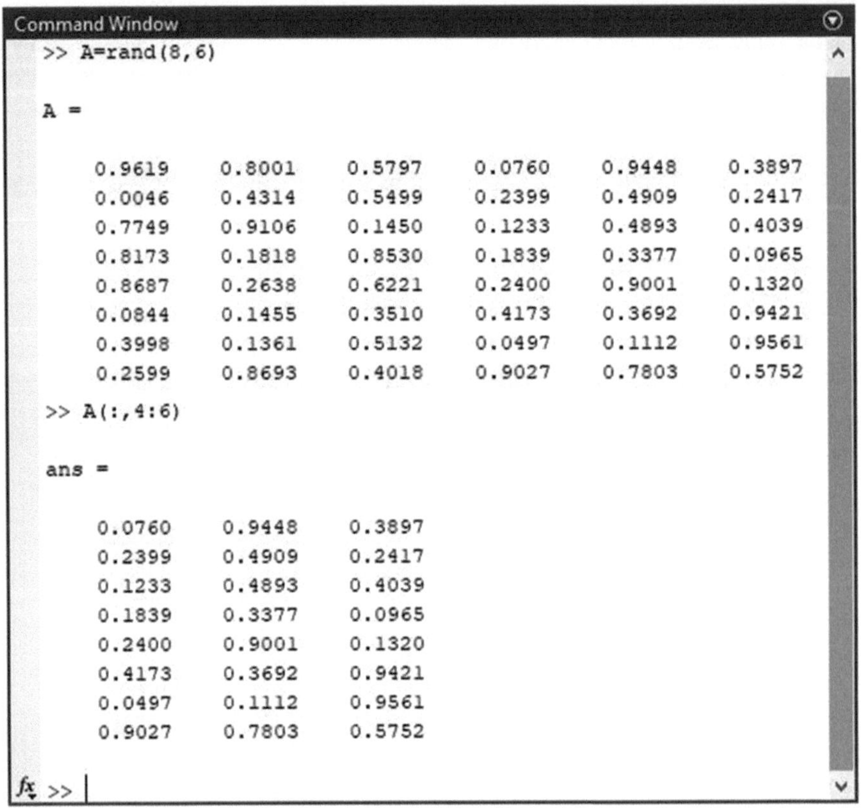

Fig. 11.48 Defining a random matrix and selecting columns 4, 5, and 6

The command A(2:5,3:4) in Fig. 11.49 selects the elements A_{ij} where $2 \leq i \leq 5$ and $3 \leq j \leq 4$.

```
Command Window
>> A=rand(8,6)

A =

    0.9619    0.8001    0.5797    0.0760    0.9448    0.3897
    0.0046    0.4314    0.5499    0.2399    0.4909    0.2417
    0.7749    0.9106    0.1450    0.1233    0.4893    0.4039
    0.8173    0.1818    0.8530    0.1839    0.3377    0.0965
    0.8687    0.2638    0.6221    0.2400    0.9001    0.1320
    0.0844    0.1455    0.3510    0.4173    0.3692    0.9421
    0.3998    0.1361    0.5132    0.0497    0.1112    0.9561
    0.2599    0.8693    0.4018    0.9027    0.7803    0.5752

>> A(2:5,3:4)

ans =

    0.5499    0.2399
    0.1450    0.1233
    0.8530    0.1839
    0.6221    0.2400

>>
```

Fig. 11.49 Defining a random matrix and selecting the elements A_{ij} where $2 \le i \le 5$ and $3 \le j \le 4$

The colon operator is used in for loops as well. For instance, the code in Fig. 11.50 counts the number of elements that are greater than or equal to 0.5.

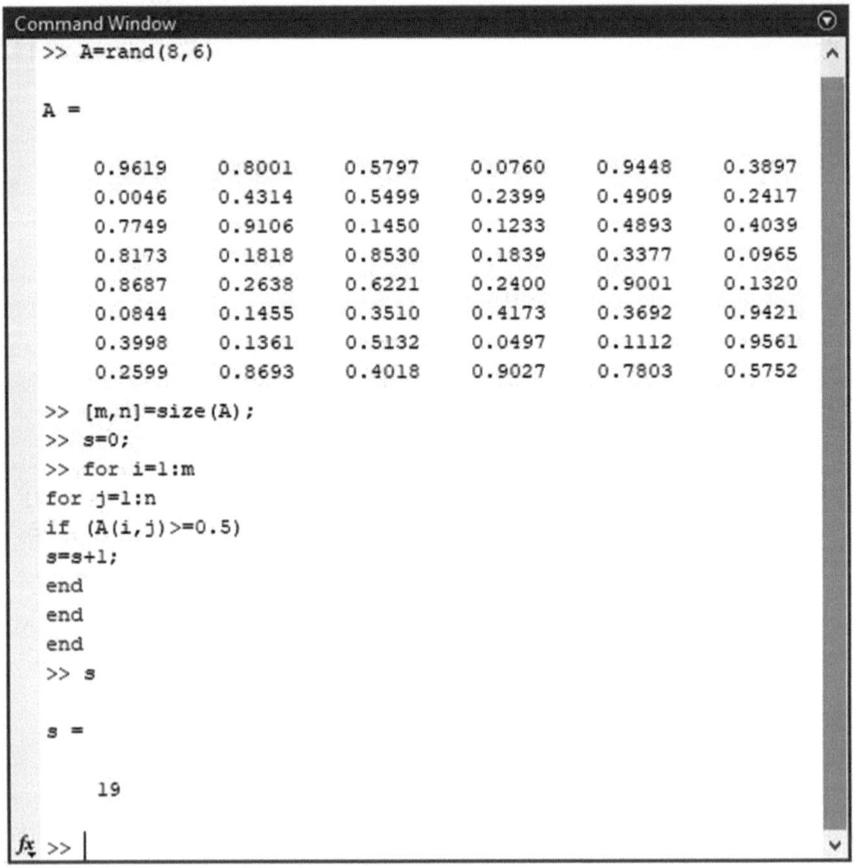

Fig. 11.50 Counting the number of elements which are equal or bigger than 0.5

11.11 Linspace and Logspace Commands

The linspace(X1, X2, N) generates N linearly spaced points between X1 and X2. MATLAB assumes N = 100 when it is not entered. For instance, the command in Fig. 11.51 generates a linear space from 1 to 10 with 6 members. We expect the difference between two consecutive elements to be $\frac{10-1}{6-1} = 1.8$.

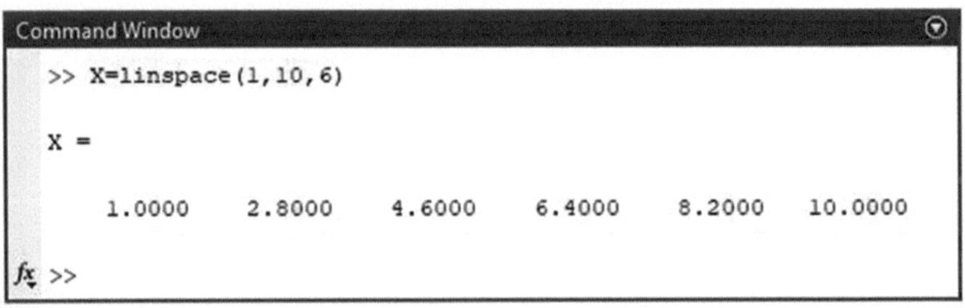

Fig. 11.51 Defining the variable X with linspace command

The difference between two consecutive elements can be calculated using the diff command (diff command can be used to calculate the derivative of a function as well). According to Fig. 11.52, the difference between two consecutive elements is 1.8.

```
Command Window
>> X=linspace(1,10,6)

X =

    1.0000    2.8000    4.6000    6.4000    8.2000   10.0000

>> diff(X)

ans =

    1.8000    1.8000    1.8000    1.8000    1.8000

fx >>
```

Fig. 11.52 Difference between the consecutive elements of vector X

logspace(X1, X2,N) generates a row vector of N logarithmically equally spaced points between 10^{X1} and 10^{X2}. MATLAB assumes N=100 when it is not entered. For instance, the command in Fig. 11.53, makes a logarithmically equally spaced from 100 to 1000 with 5 members.

```
Command Window
>> Y=logspace(2,3,5)

Y =

   1.0e+03 *

    0.1000    0.1778    0.3162    0.5623    1.0000

fx >>
```

Fig. 11.53 Defining the variable Y with logspace command

The ratio between any two consecutive elements is constant (Fig. 11.54). This ratio can be calculated from the $10^2 \times x^{5-1} = 10^3$ equation. Figure 11.55 solves this equation. According to Fig. 11.55, the only positive answer of this equation is 1.7783. This value matches the one shown in Fig. 11.54.

```
Command Window
>> Y=logspace(2,3,5)

Y =

   1.0e+03 *

    0.1000    0.1778    0.3162    0.5623    1.0000

>> Y(2)/Y(1)

ans =

    1.7783

>> Y(3)/Y(2)

ans =

    1.7783

>> Y(4)/Y(3)

ans =

    1.7783

>> Y(5)/Y(4)

ans =

    1.7783

fx >>
```

Fig. 11.54 Ratio between the consecutive elements of vector Y

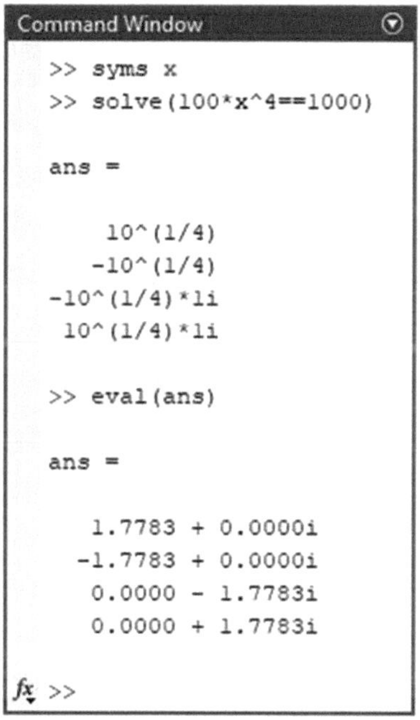

Fig. 11.55 Solution of $10^2 \times x^{5-1} = 10^3$

11.12 Ones, Zeros, and Eye Commands

The ones(m,n) command makes an m × n matrix which all of its elements are one. The zeros (m,n) makes an m × n matrix which all of its elements are zero. The eye (n) command makes an n × n identity matrix. The eye(m,n) command makes an m×n matrix with 1's on the diagonal and 0 elsewhere (Fig. 11.56).

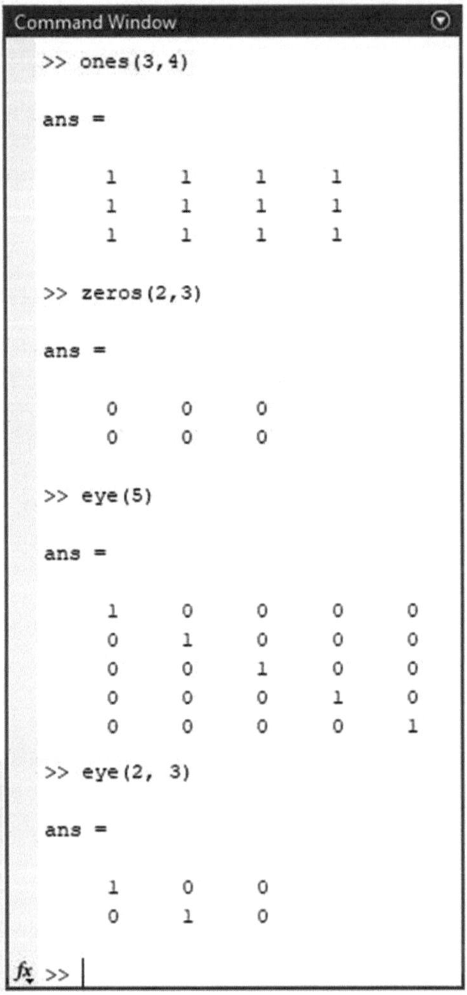

Fig. 11.56 Example for ones, zeros, and eye commands

11.13 Format Command

The format command permits you to set the output display format for command window. The long format displays results with greater precision in the decimal places (Fig. 11.57).

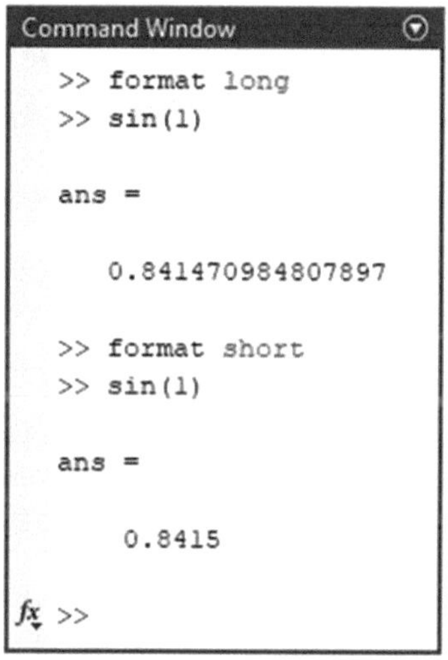

Fig. 11.57 The format command

11.14 Polynomial Functions

Assume that you want to solve the $x^2 - 4x + 3 = 0$. The roots command takes the coefficients and returns the roots (Fig. 11.58).

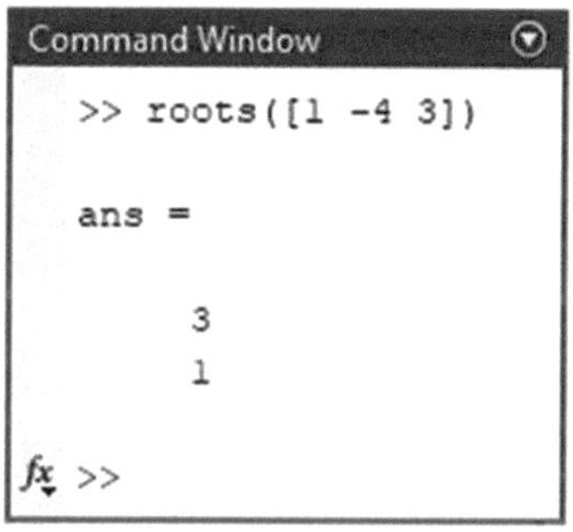

Fig. 11.58 The roots command calculates the roots of a polynomial

Assume that you want to find the polynomial whose roots are -1+2i, -1-2i, and 6. The poly function gives the coefficients of the polynomial which has those roots. According to Fig. 11.59, the polynomial is $p(x) = x^3 - 4x^2 - 7x - 30$.

Fig. 11.59 Poly command takes the roots of a polynomial function and gives its coefficients

```
>> poly([-1+2i -1-2i 6])

ans =

     1    -4    -7   -30

>> |
```

Assume that we want to calculate the value of $p(x) = x^3 - 4x^2 - 7x - 30$ at $x = 5$. The polyval function can be used for this purpose (Fig. 11.60).

Fig. 11.60 Calculation of a polynomial function's value at a point with polyval command

```
>> polyval([1 -4 -7 -30],5)

ans =

   -40

>> |
```

Assume that you want to calculate the product of $p_1(x) = x + 3$ and $p_2(x) = x^2 + 5x + 6$. The commands shown in Fig. 11.61 calculate the product of these two polynomials. According to Fig. 11.61, the product is $p(x) = x^3 + 8x^2 + 21x + 18$.

Fig. 11.61 Calculation of the product of two polynomials with the conv command

```
>> conv([1 3],[1 5 6])

ans =

     1     8    21    18

>>
```

11.15 Partial Fraction Expansion

You can use the residue command to calculate the partial fraction expansions. For instance, assume that we want to find the partial fraction expansion of $\frac{s+1}{s^2+5s+6}$. The commands shown in Fig. 11.62 calculate the residues at the poles.

The poles of a rational function are the values of the variable that make the denominator equal to zero. Residues are the coefficients of the terms in a partial fraction expansion that correspond to the poles of the function.

According to Fig. 11.62, the partial fraction expansion of $\frac{s+1}{s^2+5s+6}$ is equal to $\frac{2}{s+3} + \frac{-1}{s+2}$. The residue at pole -3 is 2, and the residue at pole -2 is -1.

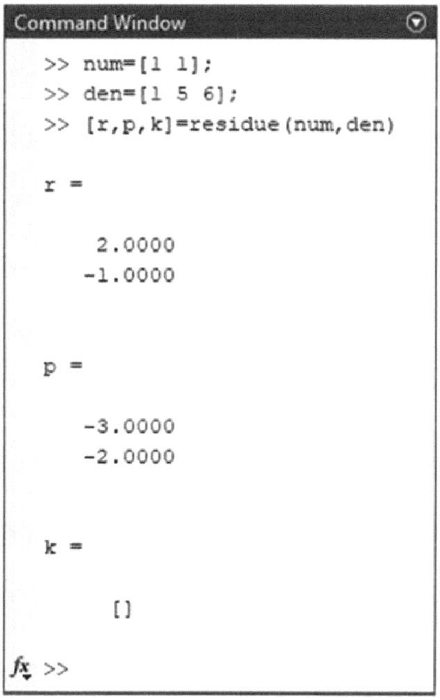

Fig. 11.62 residue command output for the function $\frac{s+1}{s^2+5s+6}$

Let's consider another example: Finding the partial fraction expansion of the expression $\frac{1}{s^3(s-0.5)}$. By applying the partial fraction expansion technique, it is found that $\frac{1}{s^3(s-0.5)} = \frac{-2}{s^3} + \frac{-4}{s^2} + \frac{-8}{s} + \frac{8}{s-0.5}$. Figure 11.63 demonstrates that this result is correct.

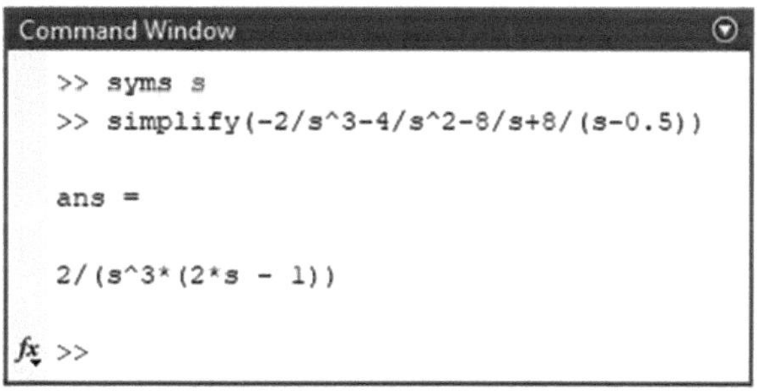

Fig. 11.63 Verification of $\frac{-2}{s^3} + \frac{-4}{s^2} + \frac{-8}{s} + \frac{8}{s-0.5} = \frac{1}{s^3(s-0.5)}$

The commands shown in Fig. 11.64 calculate the residues. The calculated residues agree with the hand calculations.

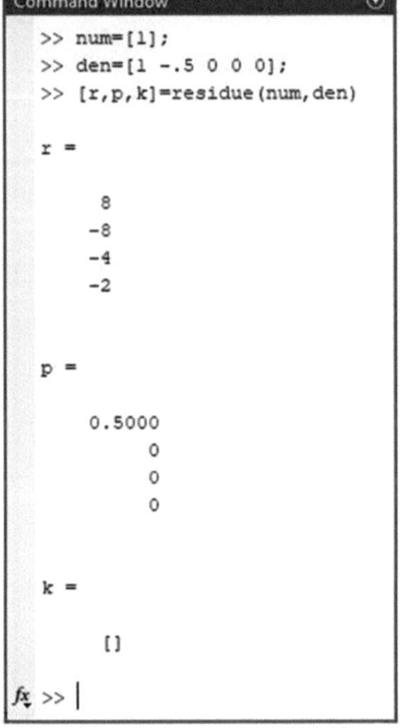

Fig. 11.64 residue command output for the function $\frac{1}{s^3(s-0.5)}$

In both of the above examples, the matrix **k** was a null matrix. The matrix **k** is always null when the degree of the numerator is less than the degree of the denominator. However, when the degree of the numerator is equal to or greater than the degree of the denominator, it will not be null.

For instance, consider the expression $\frac{s^3+3s^2+7s+4}{s^2+2s+1}$. The commands shown in Fig. 11.73 calculate the partial fraction of this function. Note that matrix k is not null in this case since the degree of the numerator is greater than that of denominator. According to Fig. 11.65, $\frac{s^3+3s^2+7s+4}{s^2+2s+1} = s + 1 + \frac{4}{(s+1)} + \frac{-1}{(s+1)^2}$.

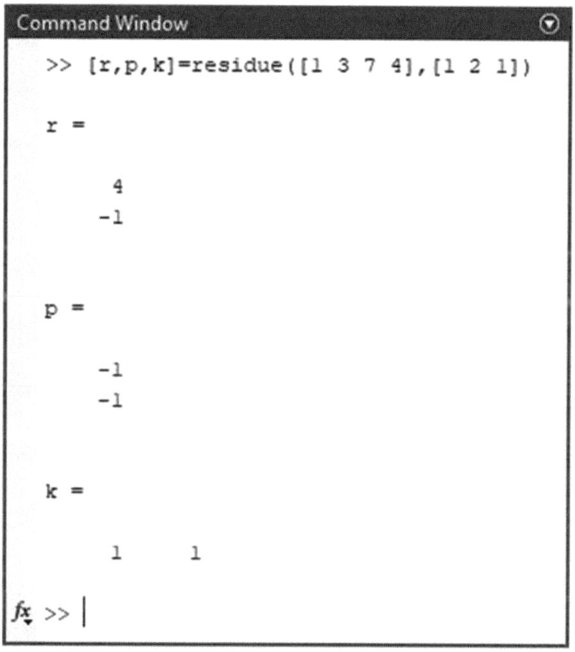

Fig. 11.65 residue command output for the function $\frac{s^3+3s^2+7s+4}{s^2+2s+1}$

11.16 Calculation of Limit, Derivative, and Integral

Table 11.6 shows the calculation of limits, derivatives, and integrals using MATLAB. The syms command is used to create the necessary symbolic variables for these operations.

Table 11.6 Calculating limits, derivatives, and integrals in MATLAB

Mathematical expression	MATLAB representation
$\lim_{x\to 0} \frac{\sin(x)}{x} = ?$	`>>syms x` `>> limit(sin(x)/x,0)`
$\frac{d}{dx}\left(\frac{x^3-x}{x^2+8}\right) = ?$	`>>syms x` `>>diff((x^3-x)/(x^2+8))`
$\int \frac{x^3-x}{x^2+8} dx = ?$	`>>syms x` `>>int((x^3-x)/(x^2+8))`
$\int_0^1 \frac{x^3-x}{x^2+8} dx = ?$	`>>syms x` `>>int((x^3-x)/(x^2+8),x,0,1)`
$\int_0^1 \frac{x^3-x}{x^2+8} dx = ?$	`>>f=@(x)(x.^3-x)./(x.^2+8);` `>>integral(f,0,1)`

The command in Fig. 11.66 calculates $\frac{d^2f(x)}{dx^2}$ for the function $f(x) = \sin(x) + \cos(x^2)$.

```
Command Window
>> syms x
>> f=sin(x)+cos(x^2);
>> diff(f,x,2)

ans =

- 2*sin(x^2) - sin(x) - 4*x^2*cos(x^2)

>>
```

Fig. 11.66 Calculation of $\frac{d^2}{dx^2}(\sin(x) + cos(x^2))$

For calculating indefinite integrals in MATLAB, the primary tool is the int command from the symbolic math toolbox. This command is used to find the general antiderivative of a symbolic expression. To use it, one must first define the variable (s) as symbolic using the syms command. For example, Fig. 11.67 shows the code that calculates the integral $\int \sin(3x + 7)dx$.

```
Command Window
>> syms x
>> int(sin(3*x+7)^2)

ans =

x/2 - sin(14 + 6*x)/12

>> pretty(ans)
x   sin(14 + 6 x)
- - -------------
2        12

>>
```

Fig. 11.67 Calculation of $\int \sin^2(3x + 7)dx$ using the int command

The MATLAB int and integral commands both calculate definite integrals, but they differ fundamentally in their approach and the type of result they return.

The int command, part of the symbolic math toolbox, performs symbolic integration, meaning it first finds the exact mathematical antiderivative of the function and then evaluates it at the specified limits; this method yields the exact mathematical value of the definite integral.

In contrast, the integral command performs numerical integration, employing adaptive algorithms to approximate the area under the curve; this method returns a double-precision floating-point number that is a high-accuracy approximation of the definite integral.

Consequently, int is used when an exact, formulaic solution is required, while integral is used for faster computation or when integrating a function for which no simple antiderivative exists, or when the function is defined by data points rather than a symbolic expression.

The code in Figs. 11.68 or 11.69 calculates the definite integral $\int_0^1 \frac{x^3-x}{x^2+8}dx$. Both the int and integral commands successfully calculate the value of this integral.

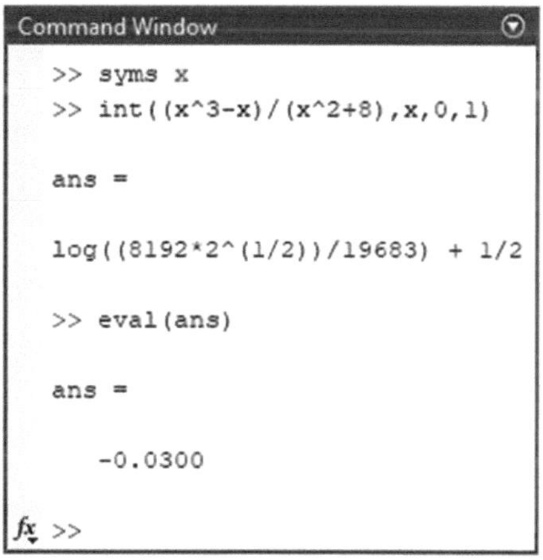

Fig. 11.68 Calculation of $\int_0^1 \frac{x^3-x}{x^2+8}dx$ using the int command

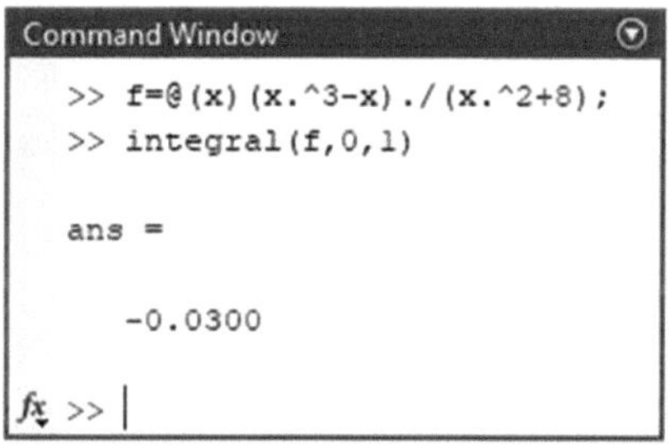

Fig. 11.69 Calculation of $\int_0^1 \frac{x^3-x}{x^2+8}dx$ using the integral command

The code in Figs. 11.70 or 11.71 calculates the definite integral $\int_0^1 e^{-x^2}dx$. Both the int and integral commands successfully calculate the value of this integral.

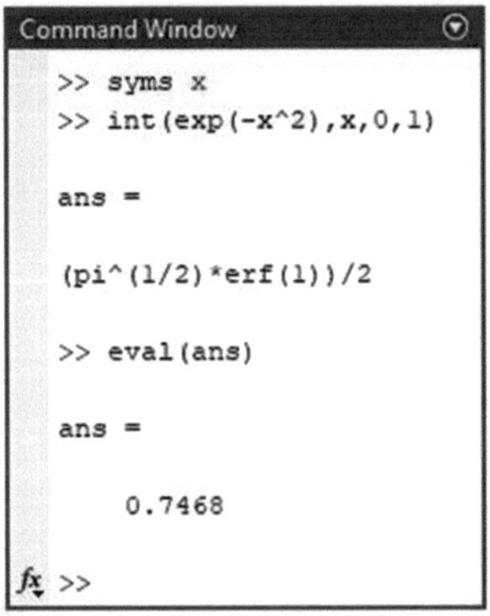

Fig. 11.70 Calculation of $\int_0^1 e^{-x^2}dx$ using the int command

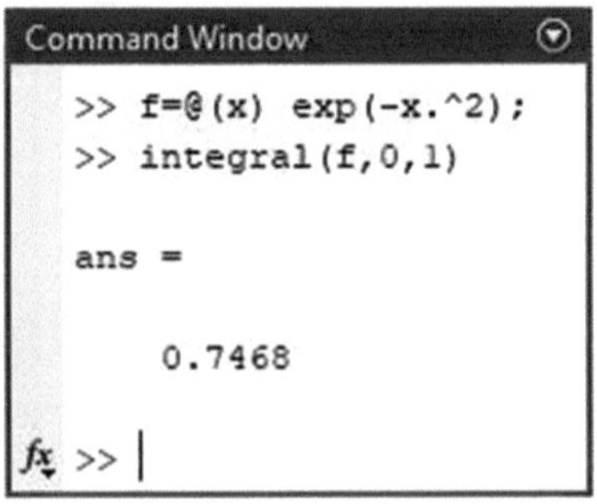

Fig. 11.71 Calculation of $\int_0^1 e^{-x^2}dx$ using the integral command

The code in Fig. 11.72 used the int command and failed to calculate the integral $\int_0^{\pi} \sin(\sin(x))dx$, while the code in Fig. 11.73 used the integral command to successfully calculate its value.

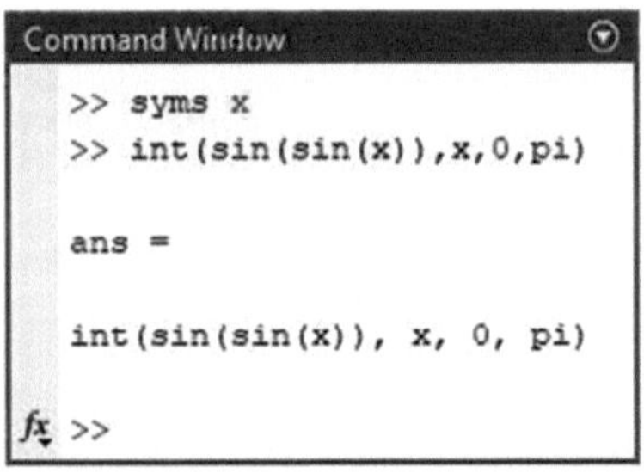

Fig. 11.72 The int command cannot calculate $\int_0^{\pi} \sin(\sin(x))dx$

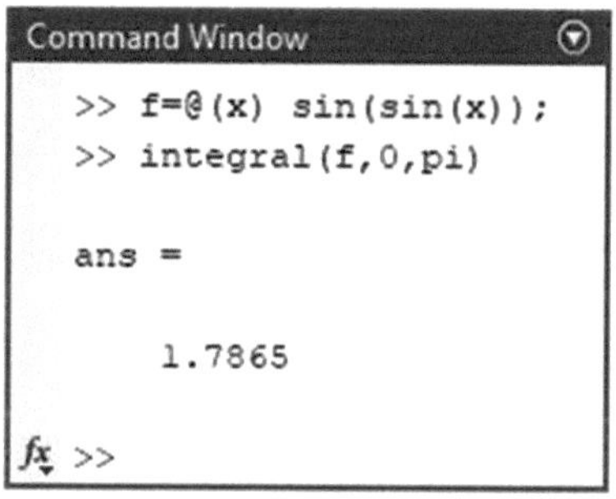

Fig. 11.73 The integral command successfully calculates $\int_0^{\pi} \sin(\sin(x))dx$

The trapz function is specifically designed to perform numerical integration on a set of discrete numeric data points, where the mathematical function that generated the data is unknown or unavailable. Let's examine an example using the data provided in Table 11.7.

Table 11.7 Sample data points

x	y
0	0.0000
0.1	0.0998
0.2	0.1987
0.3	0.2955
0.4	0.3894
0.5	0.4794

To compute the numerical integral of the data presented in Table 11.7, you would use the commands illustrated in Fig. 11.74.

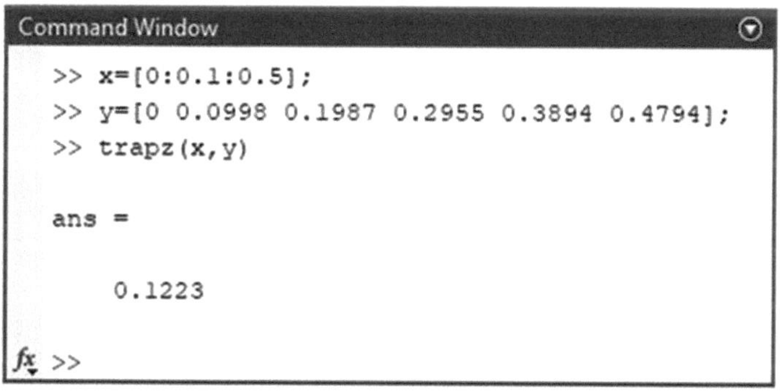

Fig. 11.74 Integrating Table 11.7's data using the quad command

MATLAB results can sometimes be difficult to read when displayed on a single line. In these instances, you can use the pretty function to display the result in a more readable format, as shown in Fig. 11.75.

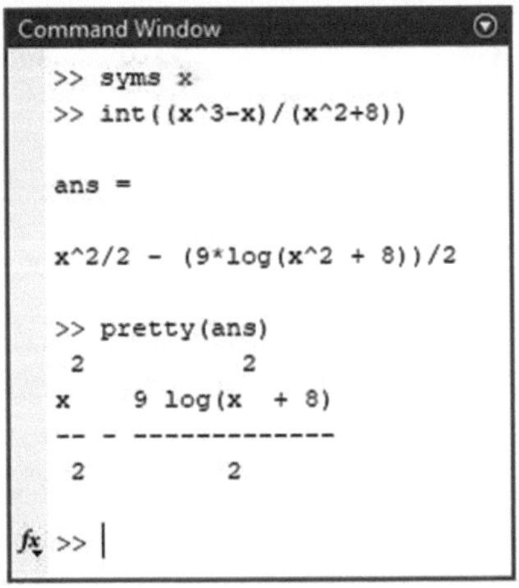

Fig. 11.75 The pretty command improves the output's readability

The simplify command, can be used to simplify the result of symbolic calculation (Fig. 11.76).

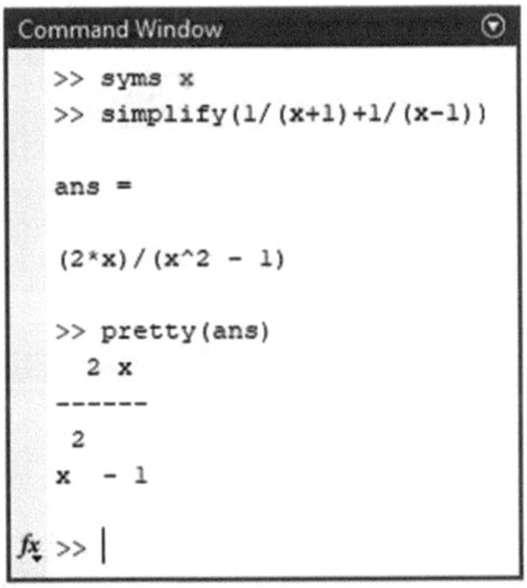

Fig. 11.76 Simplification of $\frac{1}{x+1} + \frac{1}{x-1}$

11.17 Plotting the Graph of Data

Plotting a function's graph in MATLAB is easily achieved using the plot command. The commands illustrated in Fig. 11.77 generate a plot of the function $y = \sin(x)$ over the interval $0 \le x \le 2\pi$, with the resulting graphic output shown in Fig. 11.78.

Fig. 11.77 Plotting the function *sin*(x) over the interval $[0, 2\pi]$

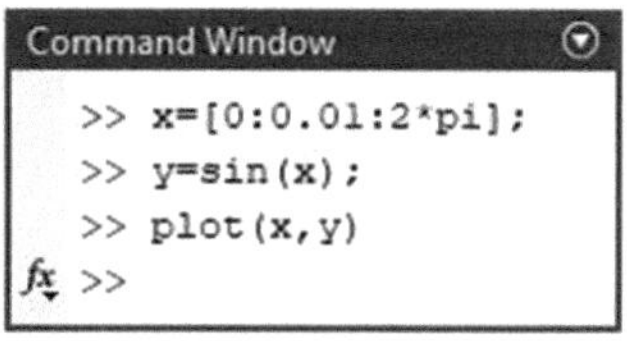

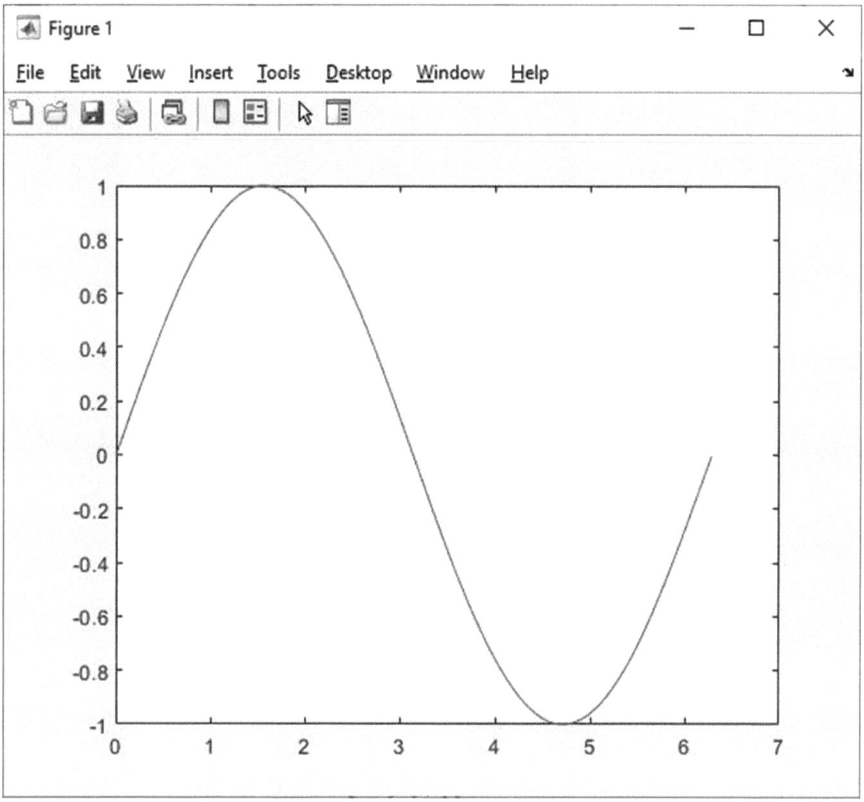

Fig. 11.78 The output generated by the code in Fig. 11.77

To create more customized plots, you can utilize the operators detailed in Tables 11.8, 11.9, and 11.10.

Table 11.8 Line types

MATLAB command	Line type
–	Solid
:	Dotted
–	Dashdot
–	Dashed

Table 11.9 Colors

MATLAB command	Color
R	Red
g	Green
b	Blue
c	Cyan
m	Magenta
y	Yellow
k	Black
w	White

Table 11.10 Plot symbols

MATLAB command	Plot symbol
.	Point
+	Plus
*	Star
o	Circle
x	x-mark
s	Square
d	Diamond
v	Triangle (down)
^	Triangle (up)
<	Triangle (left)
>	Triangle (right)

Let's study an example. Table 11.11 presents voltage and current values for a resistor.

Table 11.11 Voltage current measurements for resistor R1

V (Volt)	I (Amper)
0.499	0.10
0.985	0.20
1.508	0.31
1.969	0.41
2.528	0.53
2.935	0.61
3.481	0.73
3.971	0.83
4.486	0.94
4.960	1.04
5.502	1.15
6.007	1.26
6.60	1.38

We aim to plot this data, using circles to mark the data points and connecting them with a black dashed line. The plot must be appropriately labeled, using "Voltage (V)" for the horizontal axis, "Current (A)" for the vertical axis, and "I-V for a resistor" as the title. The necessary MATLAB commands are shown in Fig. 11.79, and the resulting graph is displayed in Fig. 11.80.

```
Command Window
>> V=[0.499 0.985 1.508 1.969 2.528 2.935 3.481 3.971 4.486 4.960 5.502 6.007 6.60];
>> I=[0.1 0.2 0.31 0.41 0.53 0.61 0.73 0.83 0.94 1.04 1.15 1.26 1.38];
>> plot(V,I,'--ko')
>> title('I-V for a resistor')
>> xlabel('Voltage(V)')
>> ylabel('Current(A)')
>> grid on
fx >>
```

Fig. 11.79 The commands for generating the plot of Table 11.11

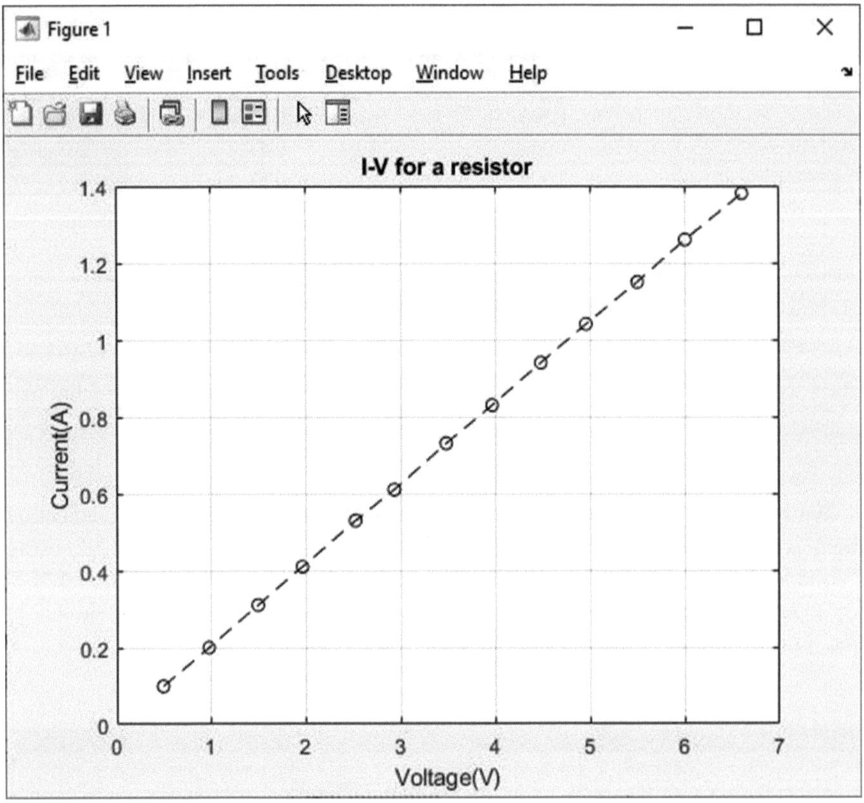

Fig. 11.80 The resulting plot generated by the code shown in Fig. 11.79

Figure 11.81 shows the commands to plot the data contained in Table 11.11. Unlike the previous plot, this version uses red stars for data points and connects them with a solid black line (the output is shown in Fig. 11.82).

Command Window

```
>> V=[0.499 0.985 1.508 1.969 2.528 2.935 3.481 3.971 4.486 4.960 5.502 6.007 6.60];
>> I=[0.1 0.2 0.31 0.41 0.53 0.61 0.73 0.83 0.94 1.04 1.15 1.26 1.38];
>> plot(V,I,'k',V,I,'r*'),grid minor
>>
```

Fig. 11.81 The commands for generating the plot of Table 11.11

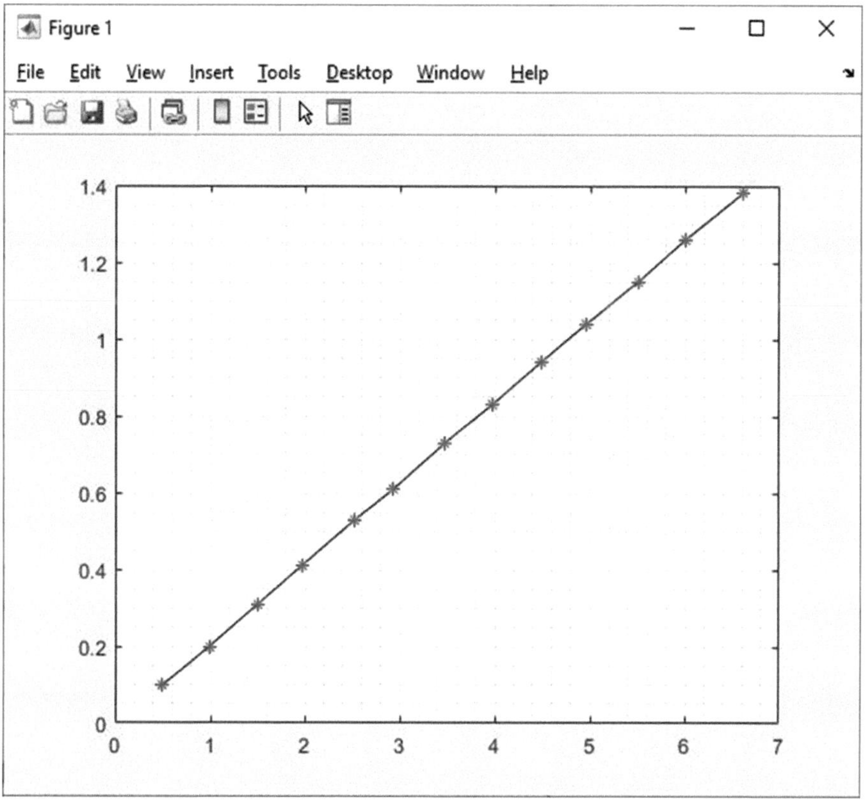

Fig. 11.82 The resulting plot generated by the code shown in Fig. 11.81

Figure 11.82 currently lacks labels and a title. To add the necessary labels and title, you can use the Insert menu (Fig. 11.83).

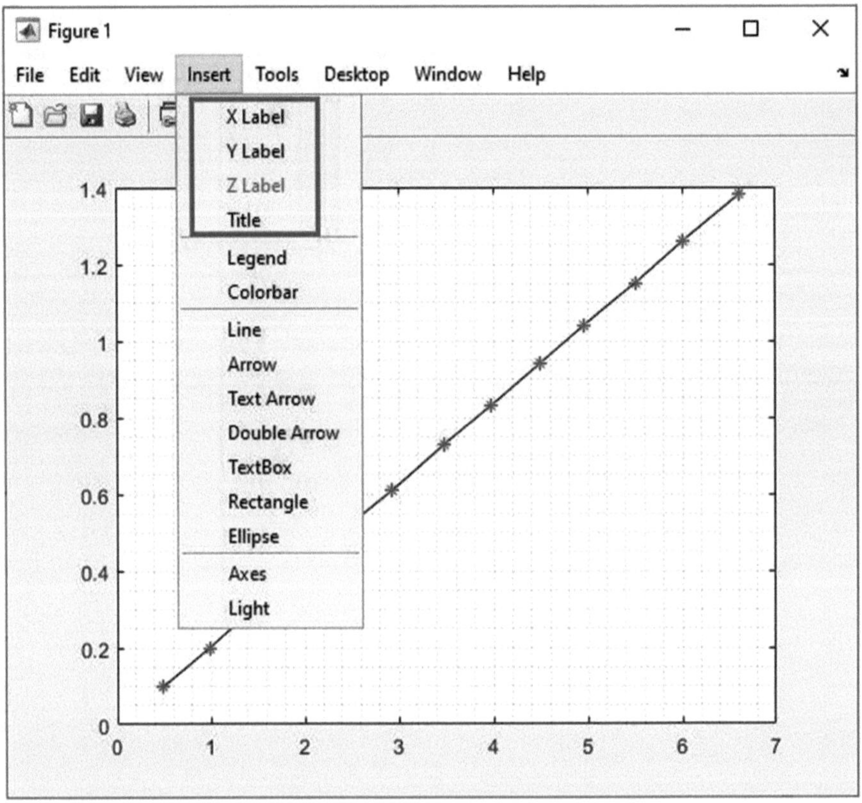

Fig. 11.83 The insert menu can be used for adding a title and axis labels

To transfer the resulting graph to the clipboard, select the Edit > Copy Figure command (Fig. 11.84). Once on the clipboard, the image can be pasted into other applications, such as MS Word®, by using the Ctrl+V keyboard shortcut.

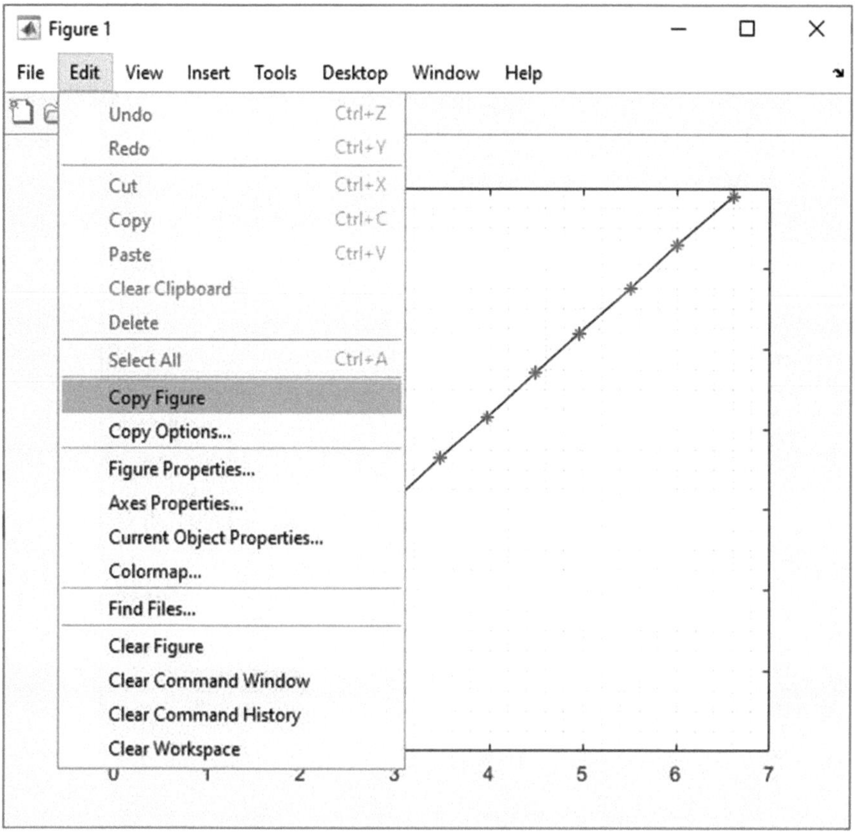

Fig. 11.84 Copying the figure to the clipboard

You can also save the generated graph as a graphical file. To do this, use File> Save as (Fig. 11.85). When the Save As window appears, select the desired output file format from the Save as type drop-down list (Fig. 11.86).

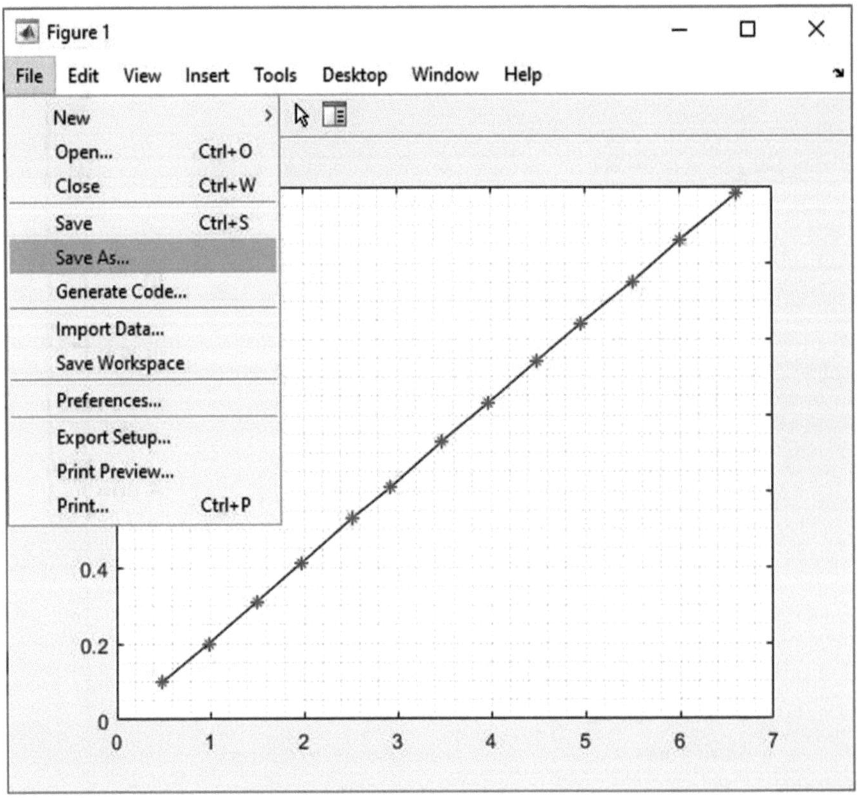

Fig. 11.85 Exporting the figure to a graphical file format

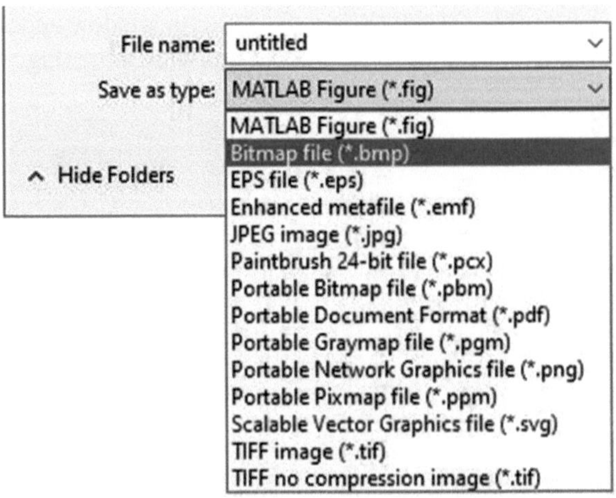

Fig. 11.86 Choice of the desired file type

To show multiple data sets concurrently on a single set of axes, the hold on command is required. Consider an exercise where we plot the data from Table 11.11 and a second data set in Table 11.12 on the same figure.

Table 11.12 Voltage current measurements for resistor R2

V (Volt)	I (Amper)
0.579	0.10
0.978	0.17
1.598	0.28
1.976	0.34
2.496	0.43
2.953	0.51
3.458	0.60
4.068	0.71
4.450	0.78
4.917	0.86
5.35	0.93
5.75	1.01
6.37	1.11
6.60	1.15

The commands shown in Fig. 11.87 generate a combined plot of the data from both tables. The resulting graph is displayed in Fig. 11.88.

```
Command Window
>> V1=[0.499 0.985 1.508 1.969 2.528 2.935 3.481 3.971 4.486 4.960 5.502 6.007 6.60];
>> I1=[0.1 0.2 0.31 0.41 0.53 0.61 0.73 0.83 0.94 1.04 1.15 1.26 1.38];
>> V2=[0.579 0.978 1.598 1.976 2.496 2.953 3.458 4.068 4.450 4.917 5.35 5.75 6.37 6.60];
>> I2=[0.1 0.17 0.28 0.34 0.43 0.51 0.60 0.71 0.78 0.86 0.93 1.01 1.11 1.15];
>> plot(V1,I1,'b',V1,I1,'r*')
>> hold on
>> plot(V2,I2,'k',V2,I2,'r+')
>> grid minor
>> xlabel('Voltage (V)')
>> ylabel('Current (A)')
>> title('Comparison of I-V plot of R1 and R2')
fx >>
```

Fig. 11.87 Plotting data from Tables 11.11 and 11.12 on a single graph

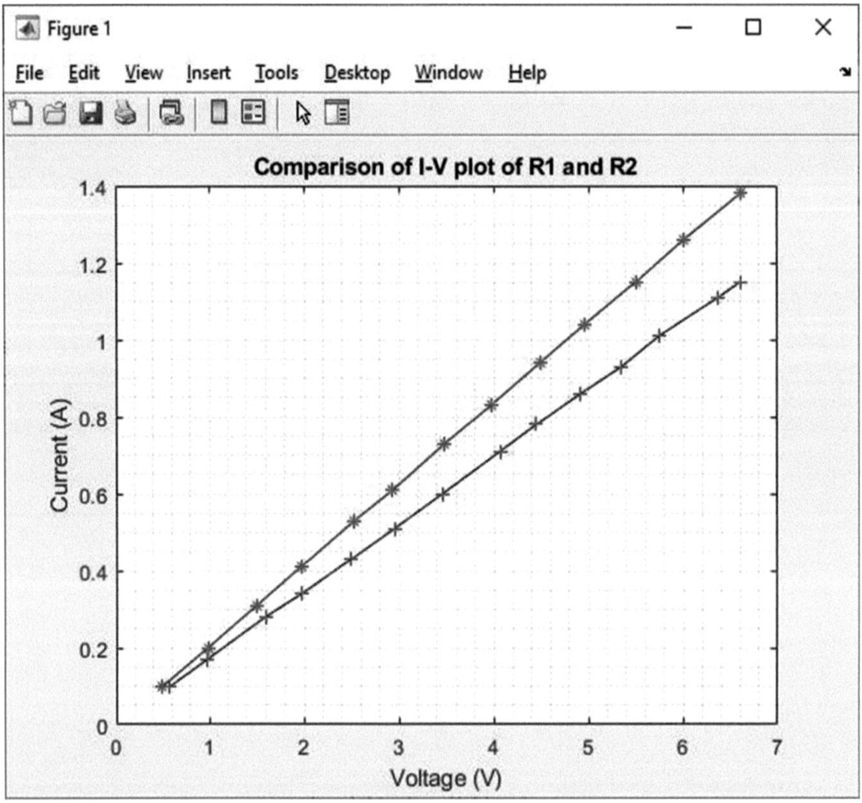

Fig. 11.88 The combined plot generated by the code shown in Fig. 11.87

You can add a legend to the graph by selecting Insert > Legend (Fig. 11.89).

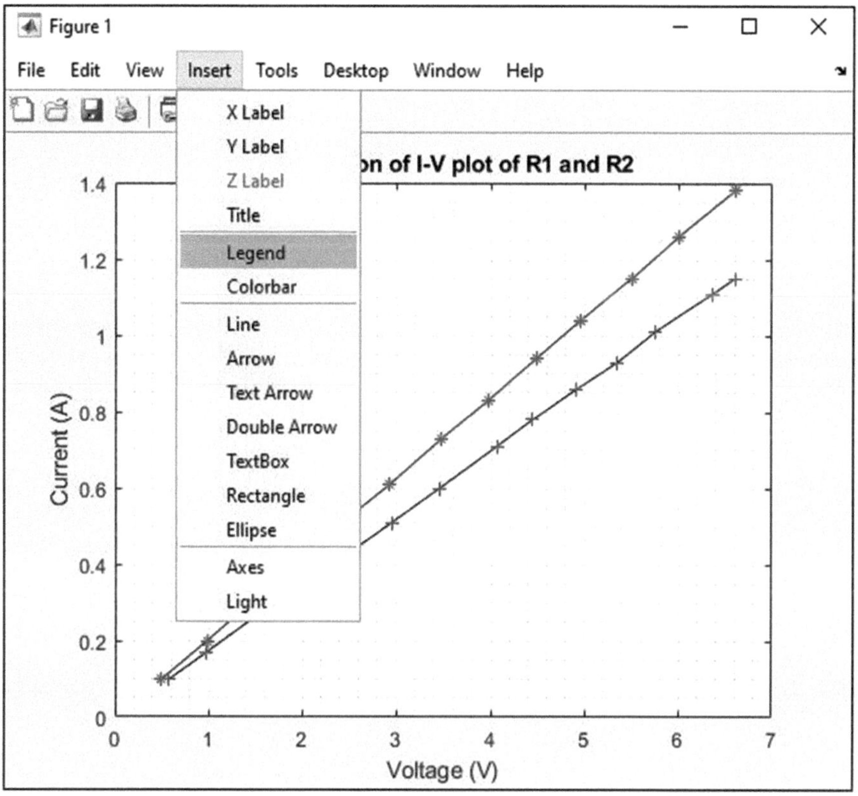

Fig. 11.89 Adding a legend to the graph

The legend shown in Fig. 11.90 is added to the graph immediately after executing the Insert> Legend command.

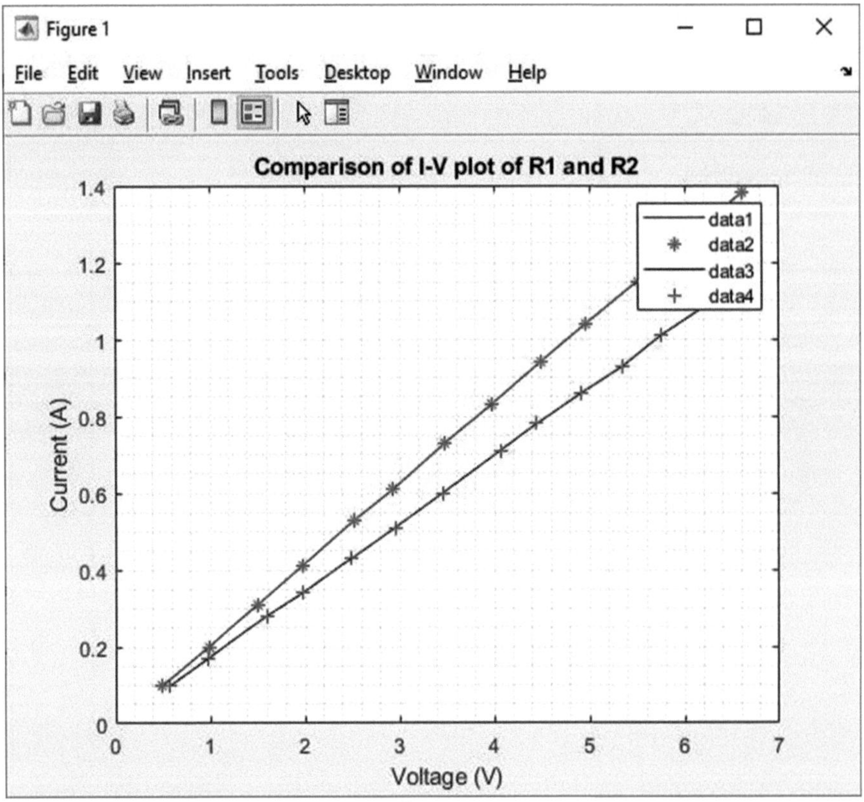

Fig. 11.90 The legend added to the graph

To customize the legend labels shown in Fig. 11.90, double-click the default text (e.g., "data1") and enter the appropriate description. Apply this editing process to data2, data3, and data4. the legend's position can be adjusted manually by dragging it (Fig. 11.91), or you may right-click the box to choose from a list of predefined placement options (Fig. 11.92).

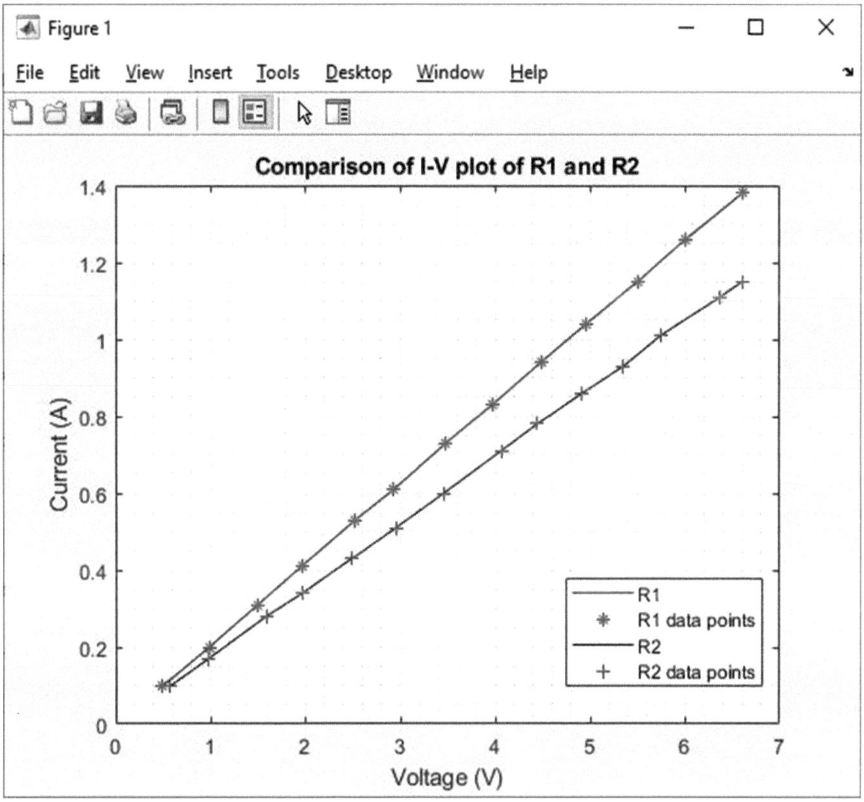

Fig. 11.91 Customizing the plot legend

Fig. 11.92 Predefined locations for legend insertion

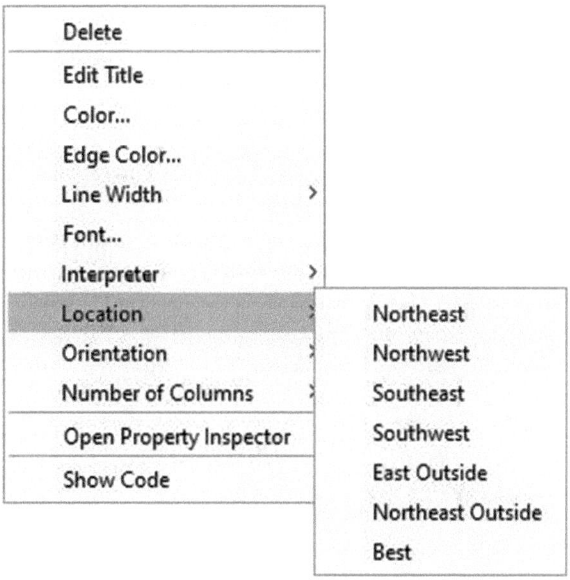

11.18 Accessing Help in MATLAB

To retrieve more information about the commands covered in this chapter, use the help command. For instance, entering help roots will display the documentation for the roots command (Fig. 11.93).

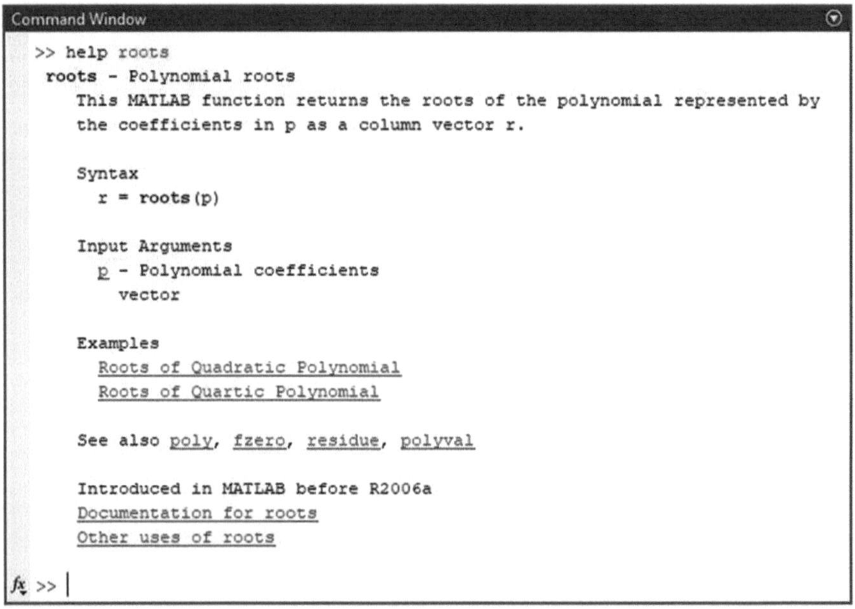

Fig. 11.93 Documentation for the roots command

Use the doc command (Fig. 11.94) to access the MATLAB Help Center.

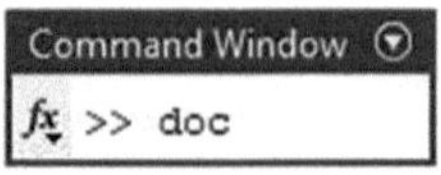

Fig. 11.94 doc command

Let's examine an alternative method for seeking assistance. On the left side of the command window, there is a small f_x icon (Fig. 11.95). Clicking this icon (or pressing Shift+F1) opens a drop-down menu (Fig. 11.96). This menu lists your installed MATLAB products. You can then click any product name to view a complete list of its related functions (Fig. 11.97). Clicking a function will display its specific help documentation (Fig. 11.98).

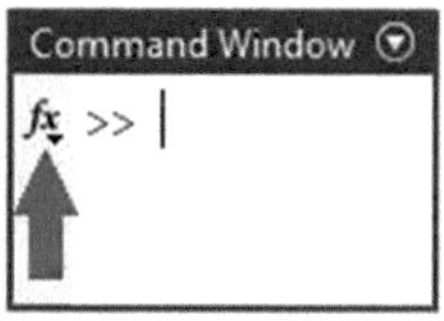

Fig. 11.95 The f_x button

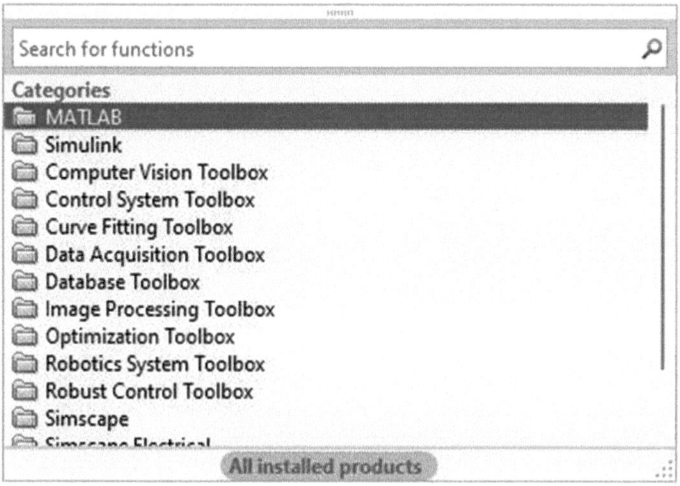

Fig. 11.96 The window displayed after clicking the f_x button

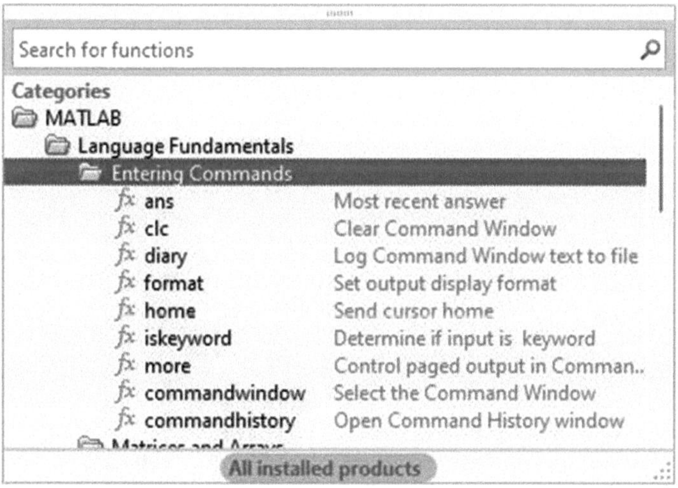

Fig. 11.97 List of commands under the MATLAB Language Fundamentals category

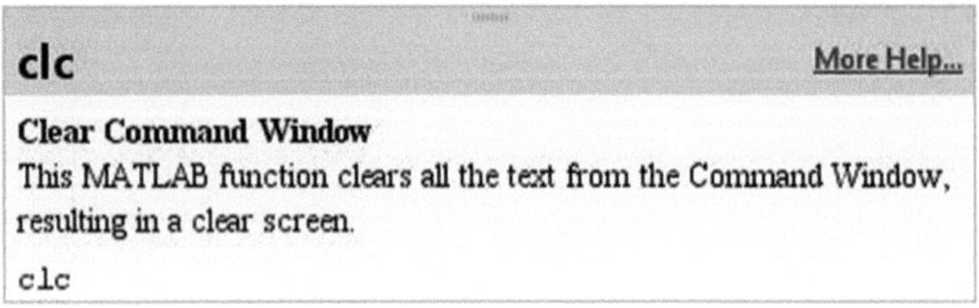

Fig. 11.98 Documentation for the clc command

References for Further Study

1. Mathematical Functions of MATLAB: https://www.mathworks.com/help/symbolic/mathematical-functions.html
2. Engineering Mathematics with MATLAB® and Simulink®, Farzin Asadi, Springer, 2025.
3. Applied Numerical Analysis with MATLAB®/Simulink®: For Engineers and Scientists, Farzin Asadi, Springer, 2023.
4. A Problem-Solving Approach to Single-Variable Calculus (Volume II), Farzin Asadi, Springer, 2026.
5. A Problem-Solving Approach to Multi-Variable Calculus (Volume I), Farzin Asadi, Springer, 2026.
6. A Problem-Solving Approach to Multi-Variable Calculus (Volume II), Farzin Asadi, Springer, 2026.
7. A Problem-Solving Approach to Multi-Variable Calculus (Volume III), Farzin Asadi, Springer, 2026.
8. A Problem-Solving Approach to Ordinary Differential Equations (Volume I), Farzin Asadi, Springer, 2026.
9. A Problem-Solving Approach to Ordinary Differential Equations (Volume II), Farzin Asadi, Springer, 2026.
10. A Problem-Solving Approach to Ordinary Differential Equations (Volume III), Farzin Asadi, Springer, 2026.
11. A Problem-Solving Approach to Ordinary Differential Equations (Volume IV), Farzin Asadi, Springer, 2026.

Chapter 12
Curve Fitting in MATLAB

12.1 Introduction

Curve fitting is the process of creating a mathematical function that best represents a series of data points. This chapter will guide you through solving curve fitting problems using MATLAB's Curve Fitting Toolbox.

12.2 Fitting Data to a Straight Line

Consider the data presented in Table 12.1. We aim to establish the linear relationship ($y = mx + b$) that optimally fits this data set. This optimal fit is achieved by minimizing the sum of squared errors, $\sum_{i=1}^{14} (mx_i + b - y_i)^2$, through the selection of appropriate parameters m and b.

Table 12.1 Data for linear curve fitting

x	y
0.579	0.100
0.978	0.175
1.598	0.283
1.976	0.345
2.496	0.433
2.953	0.515
3.458	0.606
4.068	0.717
4.450	0.788
4.917	0.866
5.355	0.935
5.753	1.016
6.378	1.118
6.601	1.152

F. Asadi, *A Problem-Solving Approach to Single-Variable Calculus*,
https://doi.org/10.1007/978-3-032-13505-6_12

The data is entered into MATLAB using the commands shown in Fig. 12.1.

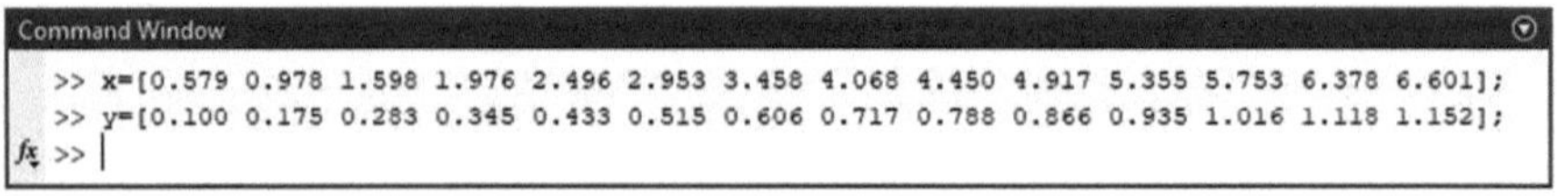

Fig. 12.1 Entering the given data into MATLAB

Use the cftool command (Fig. 12.2) to open the Curve Fitting Toolbox (Fig. 12.3).

Fig. 12.2 Running the Curve Fitting Toolbox

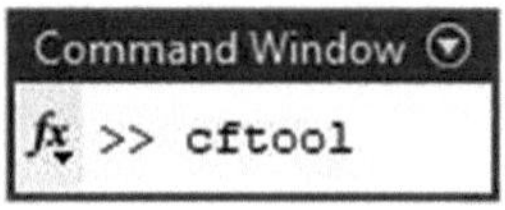

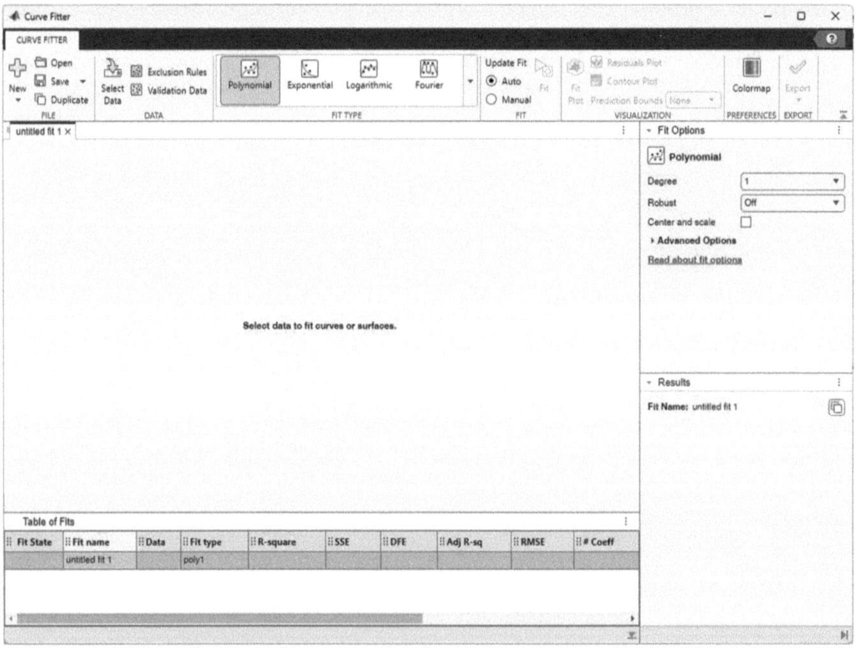

Fig. 12.3 Curve Fitting Toolbox

The Curve Fitting Toolbox offers the ability to fit the given data using various models. If you click the small arrow shown in Fig. 12.4, a list of all supported models will be displayed (Fig. 12.5).

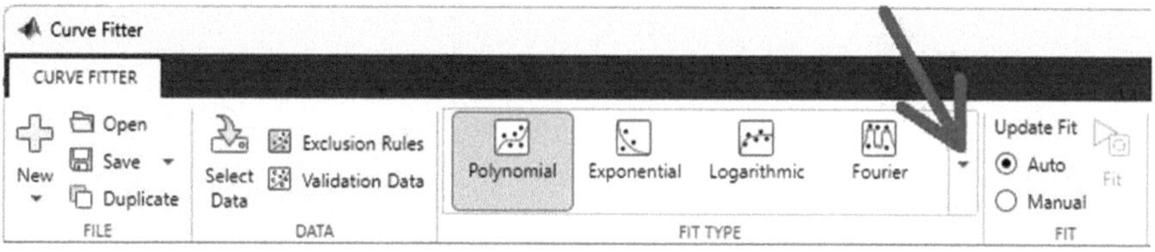

Fig. 12.4 Click the small triangle to view all supported models

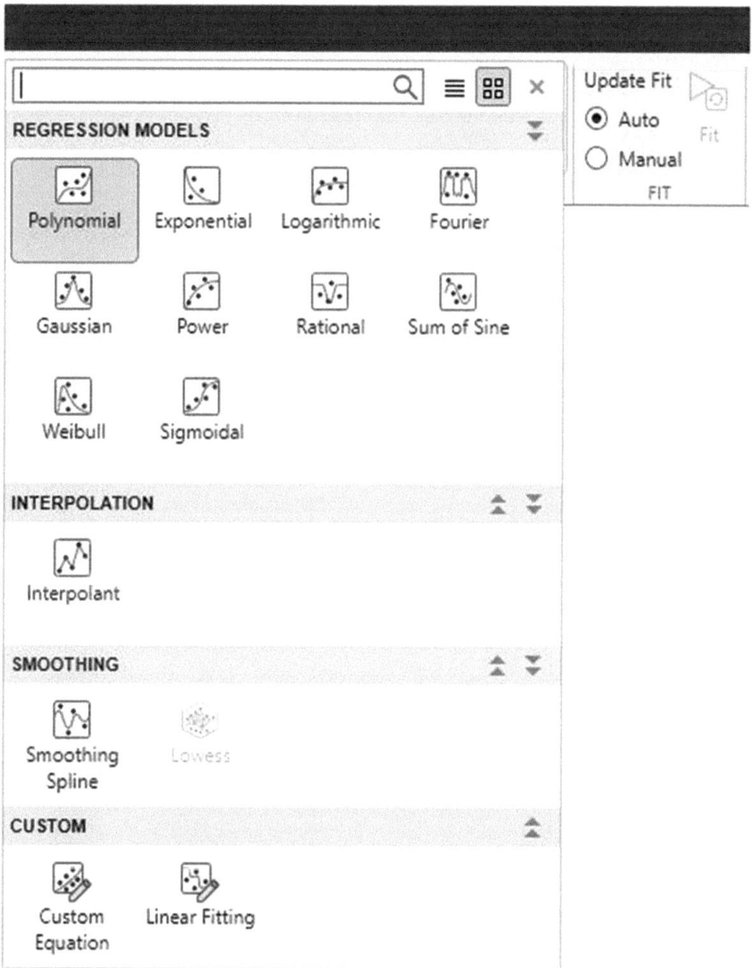

Fig. 12.5 Supported models

In this example, we aim to find a linear model (a first-degree polynomial) for the data shown in Table 12.1. To do this, ensure the following options are selected in the Curve Fitting Toolbox: Polynomial, Update Fit: Auto, and Degree: 1 (Fig. 12.6).

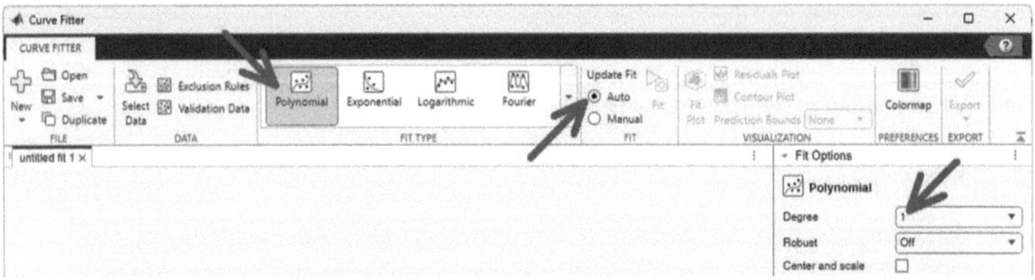

Fig. 12.6 Settings required for fitting a linear model on the data set

Click the Select Data button (Fig. 12.7). This action will open the Select Fitting Data window (Fig. 12.8).

Fig. 12.7 Select Data button

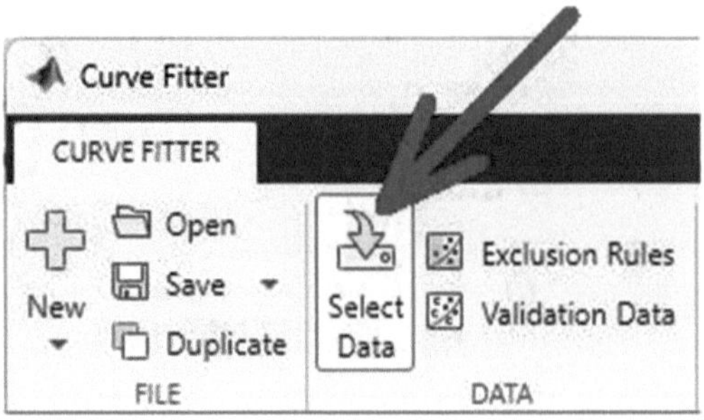

Fig. 12.8 Select Fitting Data window

In the Select Fitting Data window, choose the variable "x" for the X data field and the variable "y" for the Y data field (Fig. 12.9).

Fig. 12.9 Select x for the X data field and y for the Y data field

Once the Close button in Fig. 12.9 is clicked, the Curve Fitting Toolbox automatically computes the coefficients for the best-fit model (Fig. 12.10).

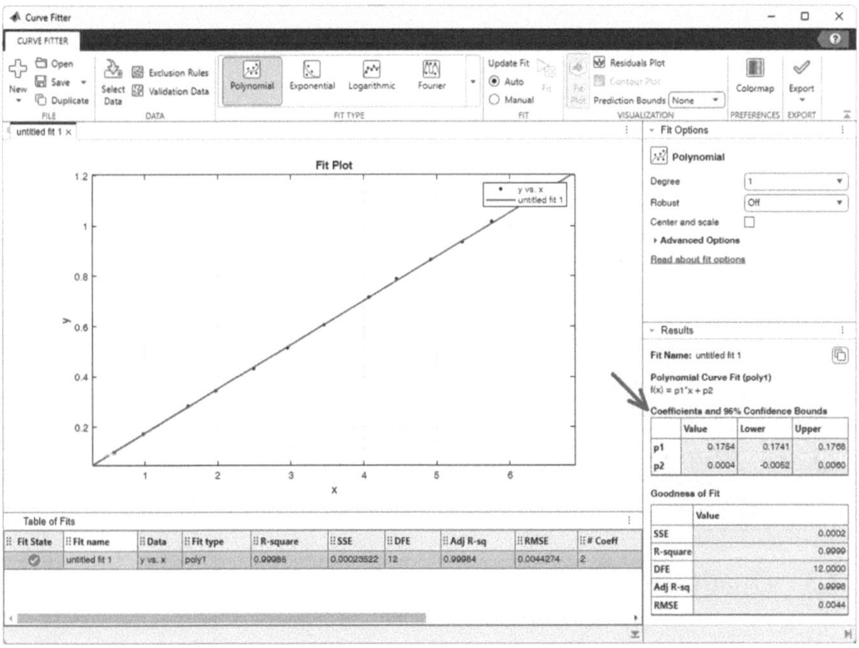

Fig. 12.10 The coefficients for the best-fit model are calculated

The Results section (Fig. 12.11) indicates that the best linear fit for the data in Table 12.1 is defined by the equation $y = 0.1754x + 0.0004$.

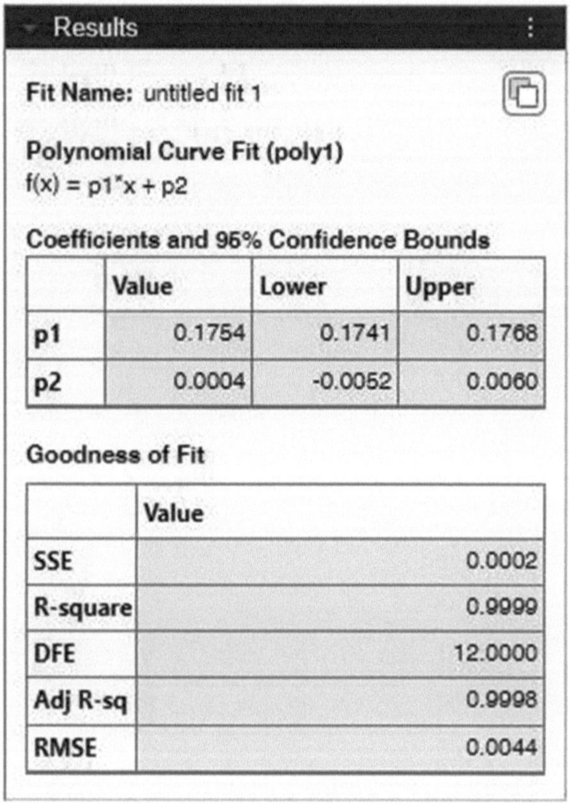

Fig. 12.11 The Results window

R-square (Fig. 12.11), also called the coefficient of determination, is a statistical measure that tells you how well your fitted curve represents the actual data points. It is a single number between 0 and 1. If your R-square is close to 1 (e.g., 0.95 or higher), it means your model is an excellent fit. Conversely, an R-square close to 0 means the model does a very poor job of fitting the data.

Therefore, when using the Curve Fitting Toolbox, you should look for an R-square value as close to 1 as possible to confirm that you have found a reliable mathematical relationship for your data set.

12.3 Quadratic Curve Fitting

The objective is to determine the coefficients a, b, and c for the second-order polynomial ($y = ax^2 + bx + c$) that best models the data in Table 12.2.

Table 12.2 Data for quadratic curve fitting

x	y
2	11
3	15
5	31
8	73
9	84
10	105
13	170

The data is entered into MATLAB using the commands shown in Fig. 12.12.

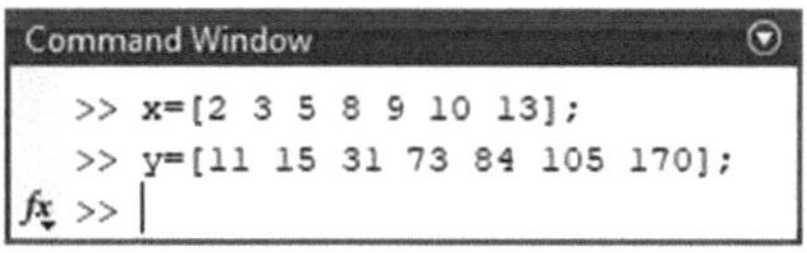

Fig. 12.12 Entering the given data into MATLAB

Execute the cftool command to launch the Curve Fitting Toolbox (Fig. 12.13).

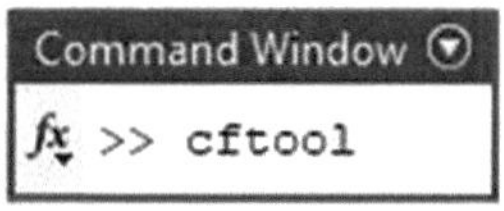

Fig. 12.13 Running the Curve Fitting Toolbox

In this example, we aim to find a quadratic model (a second-degree polynomial) for the data shown in Table 12.2. To do this, ensure the following options are selected in the Curve Fitting Toolbox: Polynomial, Update Fit: Auto, and Degree: 2 (Fig. 12.14).

Fig. 12.14 Settings required for fitting a second-order model on the data set

The Results window (Fig. 12.15) indicates that the best second-order polynomial fit for the data presented in Table 12.2 is the equation $y = 0.9110x^2 + 0.8603x + 4.8883$.

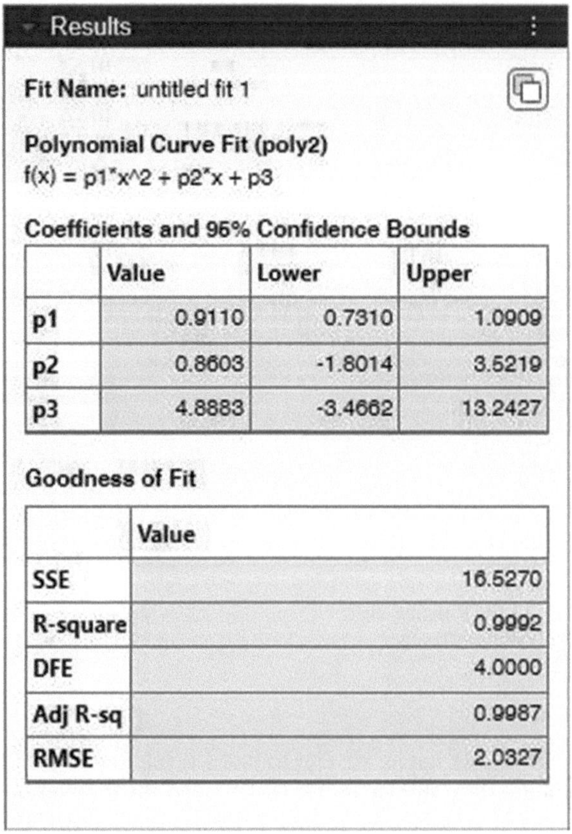

Fig. 12.15 The Results window

The Sum of Squared Errors (SSE in Fig. 12.15) is a measure of the total discrepancy or error between the actual observed data points and the values predicted by your fitted curve (the model). The primary goal of the curve fitting process is to minimize the sum of squared errors.

When you have a data set and fit multiple models to it (for instance, a quadratic and an exponential model), the model that achieves the lowest SSE is the one that best represents the data.

The calculation of SSE for this example is illustrated in Table 12.3.

Table 12.3 Calculation of the SSE

x	y	$f(x) = 0.9110x^2 + 0.8603x + 4.8883$	$(y_i - f(x_i))^2$	$\sum_{i=1}^{7} (y_i - f(x_i))^2$
2	11	10.2529	0.5582	16.5270
3	15	15.6682	0.4465	
5	31	31.9648	0.9308	
8	73	70.0747	8.5574	
9	84	86.4220	5.8661	
10	105	104.5913	0.1670	
13	170	170.0312	0.0010	

12.4 Fitting Data to a Custom Mathematical Equation

This section focuses on fitting the data in Table 12.4 to the custom exponential equation $y = a(e^{bx} - 1)$. Our goal is to determine the best parameters a and b by minimizing the SSE.

Table 12.4 Data for fitting $y = a(e^{bx} - 1)$

x	y
0.50	0.50
0.53	0.97
0.55	1.44
0.57	1.93
0.58	2.42
0.60	3.40
0.61	4.38
0.62	5.37
0.63	6.36
0.64	7.34
0.68	17.00
0.70	24.00
0.72	34.00
0.73	39.00

The data is entered into MATLAB using the commands shown in Fig. 12.16.

```
Command Window
>> x=[0.50 0.53 0.55 0.57 0.58 0.60 0.61 0.62 0.63 0.64 0.68 0.70 0.72 0.73];
>> y=[0.50 0.97 1.44 1.93 2.42 3.40 4.38 5.37 6.36 7.34 17.00 24.00 34.00 39.00];
>>
```

Fig. 12.16 Entering the given data into MATLAB

Execute the cftool command to launch the Curve Fitting Toolbox (Fig. 12.17).

Fig. 12.17 Running the Curve Fitting Toolbox

```
Command Window
>> cftool
```

Click the small triangle (Fig. 12.18) and choose the Custom Equation option (Fig. 12.19). This action opens the Fit Options window (Fig. 12.20).

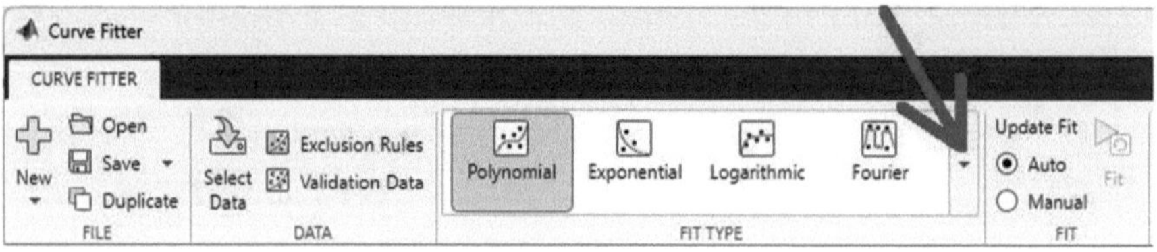

Fig. 12.18 Click the small triangle to view all supported models

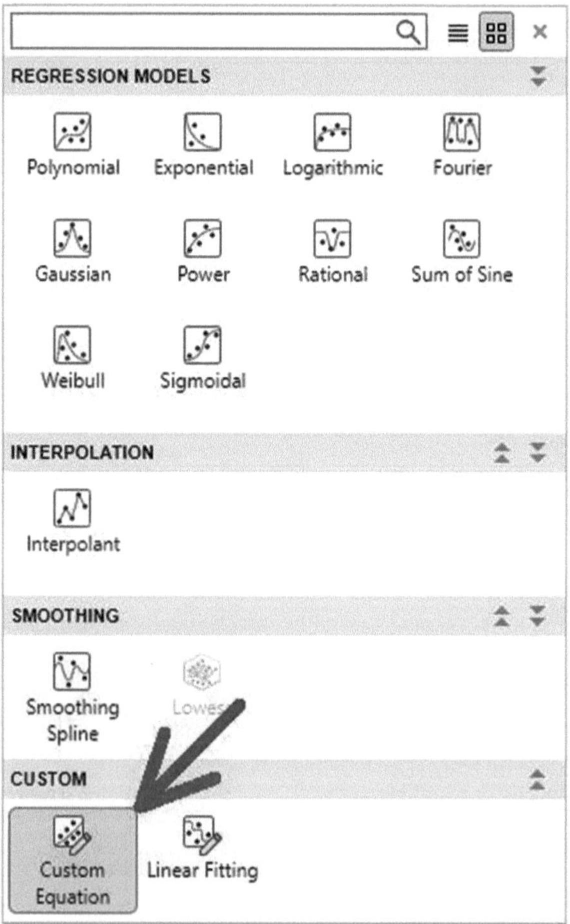

Fig. 12.19 Select the Custom Equation option

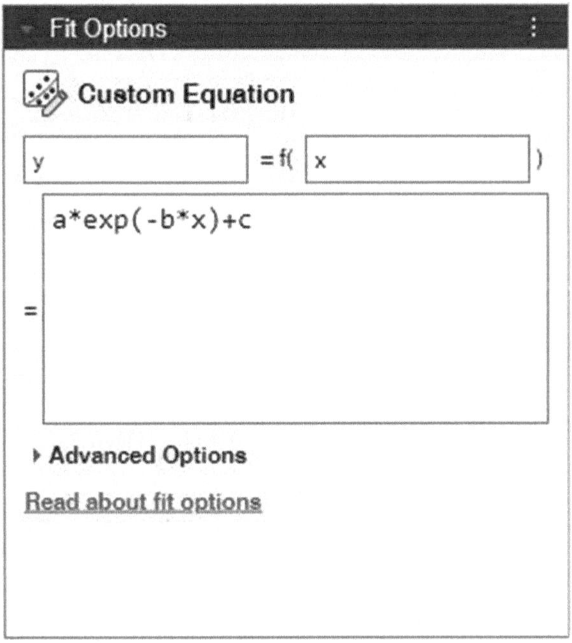

Fig. 12.20 Fit Options window

Enter the custom equation $y = a(e^{bx} - 1)$ into the Fits Options window, as shown in Fig. 12.21.

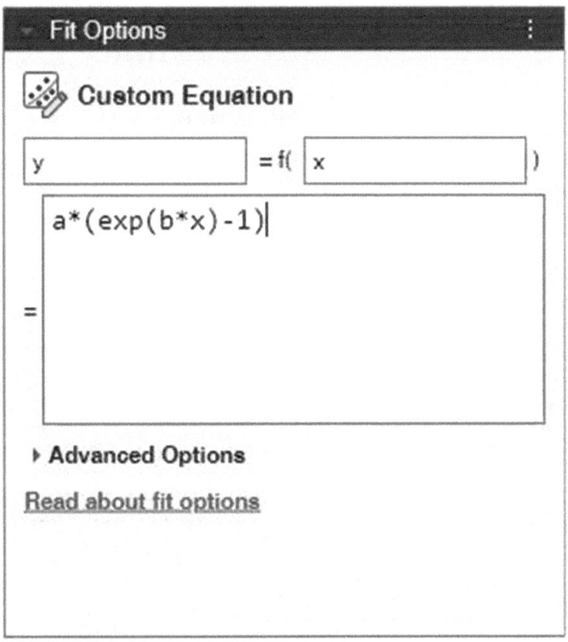

Fig. 12.21 Entering the custom equation into the Fits Options window

The Results window (Fig. 12.22) indicates that the fit for the data presented in Table 12.4 is the equation $y = 0.0001(e^{17.9944x} - 1)$.

Fig. 12.22 The Results window

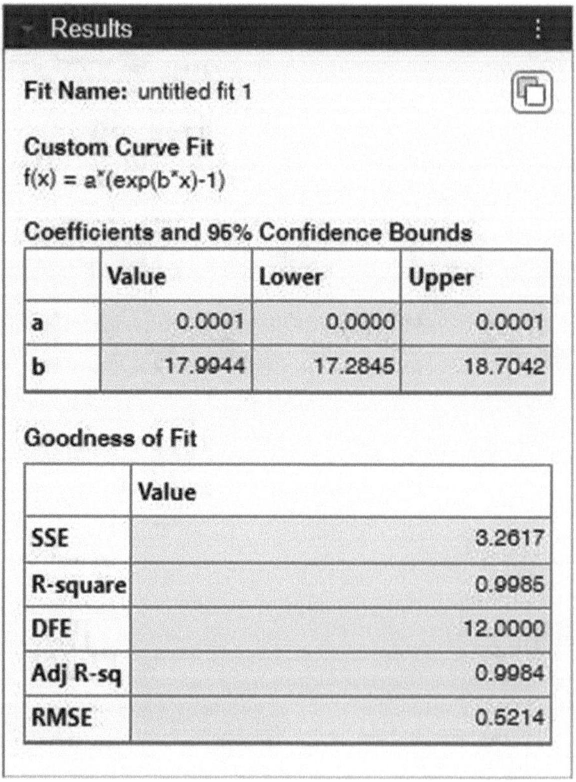

References for Further Study

1. Engineering Mathematics with MATLAB® and Simulink®, Farzin Asadi, Springer, 2025.
2. Applied Numerical Analysis with MATLAB®/Simulink®: For Engineers and Scientists, Farzin Asadi, Springer, 2023.
3. Advanced Engineering Mathematics, Erwin Kreyszig, Wiley, 2020.
4. A Problem-Solving Approach to Single-Variable Calculus (Volume II), Farzin Asadi, Springer, 2026.
5. A Problem-Solving Approach to Multi-Variable Calculus (Volume I), Farzin Asadi, Springer, 2026.
6. A Problem-Solving Approach to Multi-Variable Calculus (Volume II), Farzin Asadi, Springer, 2026.
7. A Problem-Solving Approach to Multi-Variable Calculus (Volume III), Farzin Asadi, Springer, 2026.

8. A Problem-Solving Approach to Ordinary Differential Equations (Volume I), Farzin Asadi, Springer, 2026.
9. A Problem-Solving Approach to Ordinary Differential Equations (Volume II), Farzin Asadi, Springer, 2026.
10. A Problem-Solving Approach to Ordinary Differential Equations (Volume III), Farzin Asadi, Springer, 2026.
11. A Problem-Solving Approach to Ordinary Differential Equations (Volume IV), Farzin Asadi, Springer, 2026.

Chapter 13
Calculator Skills and Formula Writing in Microsoft® Word®

13.1 Introduction

Calculators have become an essential tool for engineers, significantly enhancing their efficiency and accuracy in a wide range of tasks. Calculators play an important role in university exams, as well. So, it is important to learn how to work with your calculator correctly. By mastering your calculator, you can significantly improve your problem-solving skills and achieve better results in your engineering studies.

A common mistake in calculator usage is setting the wrong angle mode. If your calculator is in radian mode but you input an angle in degrees, the result will be incorrect. Always double-check your calculator's mode setting to ensure it's in degrees or radians, as needed, before performing calculations.

First part of this chapter offers reference operations to aid in your calculator usage. Endeavor to replicate the provided correct results. Depending on your calculator's model, some operations may not be possible. Second part of this chapter shows how to write a formula in Microsoft® Word®.

13.2 Basic Operations

The following are some basic operations and their values.

1. $12345679 \times 8 = 98765432$
2. $54.63 \times 632.58 = 34557.8454$
3. $12.653 \times 98.74 = 1249.3572$
4. $\frac{542.68}{6.24} = 86.9679$
5. $\frac{364.56}{5.25} = 69.44$
6. $(1 + 2j) + (3 - 8j) = 4 - 6j$
7. $(1 + 2j) - (3 - 8j) = -2 + 10j$

F. Asadi, *A Problem-Solving Approach to Single-Variable Calculus*,
https://doi.org/10.1007/978-3-032-13505-6_13

8. $(1 + 2j) \times (3 - 8j) = 19 - 2j$
9. $\frac{(1+2j)}{3-8j} = -0.1781 + 0.1918j$

13.3 Exponents and Radicals

The following are some example calculations for exponents and radicals.

1. $14^{2.5} = 733.3648$
2. $3.2^{4.3} = 148.6432$
3. $12.96^{2.24} = 310.6241$
4. $\sqrt{17.63} = 4.1988$
5. $\sqrt{965.22} = 31.0680$
6. $\sqrt[3]{752.53} = 9.0958$
7. $\sqrt[5]{12036} = 6.5478$

13.4 Logarithm

The following are some example calculations for logarithms.

1. $\log_{10}56.3 = 1.7505$
2. $\log_{10}1200.698 = 3.0794$
3. $\ln(56.3) = 4.0307$
4. $\ln(1200.698) = 7.0907$
5. $\ln(0.33) = -1.1087$

13.5 Conversion of Complex Numbers to Polar Form

The following are some examples of converting complex numbers between Cartesian and polar forms.

1. $1 + 6j = 6.0828e^{1.4056j}$
2. $-3 + 5j = 5.8310e^{j2.1112}$
3. $-16 - 3j = 16.2788e^{-2.9562j}$
4. $10 - 9j = 13.4536e^{-j0.7328}$
5. $12.56e^{-j1.8} = -2.8537 - 12.2315j$
6. $7.23e^{j.8} = 5.0372 + 5.1865j$

13.6 Roots of Polynomial and Nonlinear Equations

The following are some examples of finding the roots of polynomials and nonlinear equations.

1. $x^2 - 3x + 2 = 0 \Rightarrow x_1 = 1, x_2 = 2$
2. $x^3 - 11.77x^2 + 41.985x - 47.025 = 0 \Rightarrow x_1 = 2.5, x_2 = 3, x_3 = 6.27$
3. $x^2 + 2x + 10 = 0 \Rightarrow x_1 = -1 + 3j, x_2 = -1 - 3j$
4. $x^3 + 2x^2 - 51x - 232 = 0 \Rightarrow x_1 = -5 - 2j, x_2 = -5 + 2j, x_3 = 8$
5. $x^2 - 2\cos(x) = 0 \Rightarrow x_1 = -1.02162, x_2 = 1.02162$
6. $x + e^{-x} - 2 \Rightarrow x_1 = -1.146189, x_2 = +1.841402$

13.7 Derivative and Integrals

The following are some examples of evaluating integrals and derivatives.

1. $\int_0^3 \sin(x)dx = 1.9900$
2. $\int_0^1 x^2 dx = 0.3333$
3. $\int_0^2 x^2 \sin(2x)dx = 0.1371$
4. $\int_0^{0.5} e^{-6x} dx = 0.1584$
5. $\int_0^2 \frac{3}{x+1} dx = 3.2958$
6. $\frac{d}{dx}(\cos(x))\big|_{\frac{\pi}{3}} = -0.8660$
7. $\frac{d}{dx}(1 + \tan^2(x))\big|_{0.25} = 0.5440$
8. $\frac{d}{dx}\left(\frac{1+x}{\sin(x)}\right)\Big|_{0.5} = -3.6413$
9. $\frac{d}{dx}\left(\frac{x^2+9}{(x^3+6x-4)}\right)\Big|_{1.5} = -1.6868$

13.8 Trigonometric Ratios

Table 13.1 presents trigonometric ratios for several angles.

Table 13.1 Trigonometric ratios for some sample angles

Radian	Degrees
$\sin(0.2563) = 0.2535$	$\sin(10°) = 0.1736$
$\sin(0.5236) = 0.5000$	$\sin(30°) = 0.5000$
$\sin(0.7238) = 0.6622$	$\sin(50°) = 0.7660$
$\sin(1.5) = 0.9975$	$\sin(70°) = 0.9397$
$\sin(2) = 0.9093$	$\sin(123°) = 0.8387$
$\cos(0.2563) = 0.9673$	$\cos(10°) = 0.9848$
$\cos(0.5236) = 0.8660$	$\cos(30°) = 0.866$
$\cos(0.7238) = 0.7493$	$\cos(50°) = 0.6428$
$\cos(1.5) = 0.0707$	$\cos(70°) = 0.3420$
$\cos(2) = -0.4161$	$\cos(123°) = -0.5446$
$\tan(0.2563) = 0.2621$	$\tan(10°) = 0.1763$
$\tan(0.5236) = 0.5774$	$\tan(30°) = 0.5774$
$\tan(0.7238) = 0.8838$	$\tan(50°) = 1.1918$
$\tan(1.5) = 14.1014$	$\tan(70°) = 2.7475$
$\tan(2) = -2.1850$	$\tan(123°) = -1.5399$

13.9 Inverse Trigonometric Ratios

Table 13.2 shows the inverse trigonometric functions for a sample of trigonometric ratios.

Table 13.2 Inverse trigonometric functions for a sample of trigonometric ratios

$\sin^{-1}(0.310) = 0.3152 \text{ Rad} = 18.0596°$	$\cos^{-1}(0.52) = 1.0239 \text{ Rad} = 58.6677°$
$\sin^{-1}(0.652) = 0.7102 \text{ Rad} = 40.6915°$	$\cos^{-1}(0.73) = 0.7525 \text{ Rad} = 43.1136°$
$\sin^{-1}(0.711) = 0.7909 \text{ Rad} = 45.3163°$	$\cos^{-1}(0.95) = 0.3176 \text{ Rad} = 18.1949°$
$\tan^{-1}(1.45) = 0.967 \text{ Rad} = 55.4077°$	$\cot^{-1}(0.65) = 0.9944 \text{ Rad} = 56.9761°$
$\tan^{-1}(3.35) = 1.2807 \text{ Rad} = 73.3792°$	$\cot^{-1}(7.35) = 0.1352 \text{ Rad} = 7.7477°$
$\tan^{-1}(8.62) = 1.4553 \text{ Rad} = 83.3827°$	$\cot^{-1}(10.87) = 0.0917 \text{ Rad} = 5.2562°$

13.10 Microsoft Word Equation Editor

Click the Insert button (Fig. 13.1) to open the Insert ribbon (Fig. 13.2).

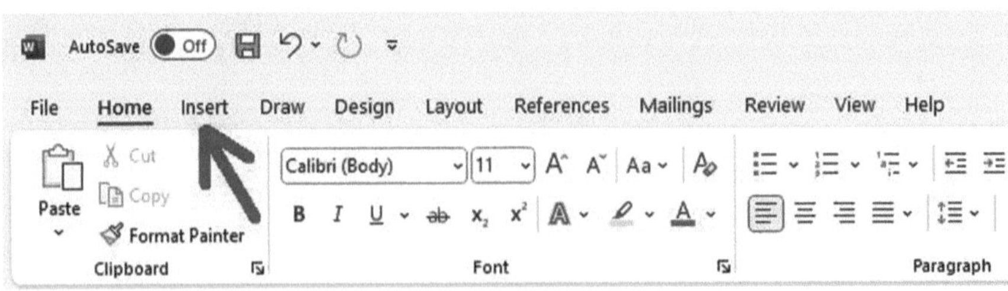

Fig. 13.1 Insert button

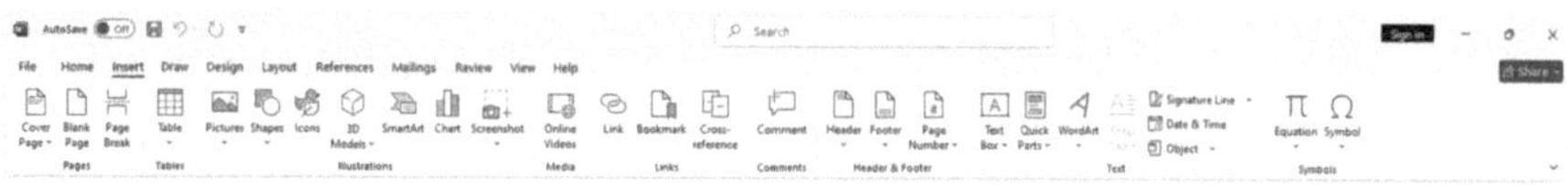

Fig. 13.2 Insert ribbon

Click on the Equations button (Fig. 13.3).

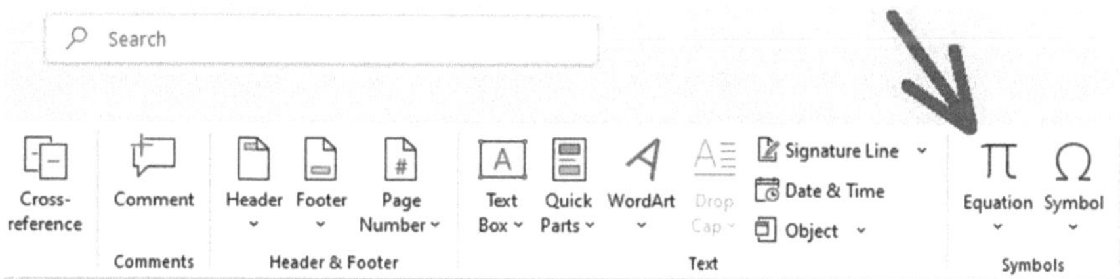

Fig. 13.3 Equations button

Clicking the Equation button will insert an equation box (Fig. 13.4) into your Word document. You can then use the tools provided in the toolbar to input your mathematical expressions (Fig. 13.5).

Fig. 13.4 Equation box

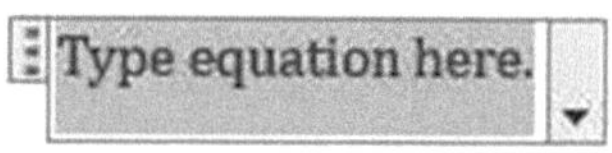

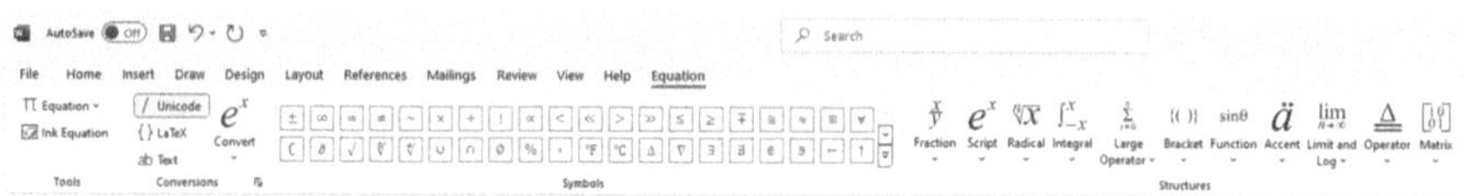

Fig. 13.5 Tools for writing mathematical expressions

Clicking the "Other" button (Fig. 13.6) will open a new window (Fig. 13.7), providing access to additional tools. From there, expand the "Basic Math" drop-down menu (Fig. 13.8) to explore a wider range of mathematical symbols and functions.

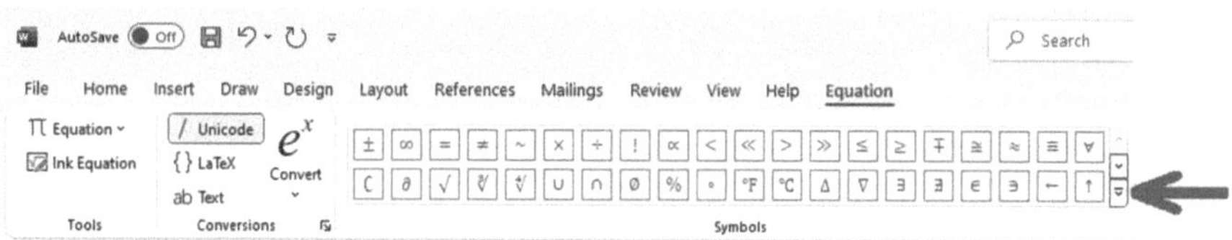

Fig. 13.6 Other button

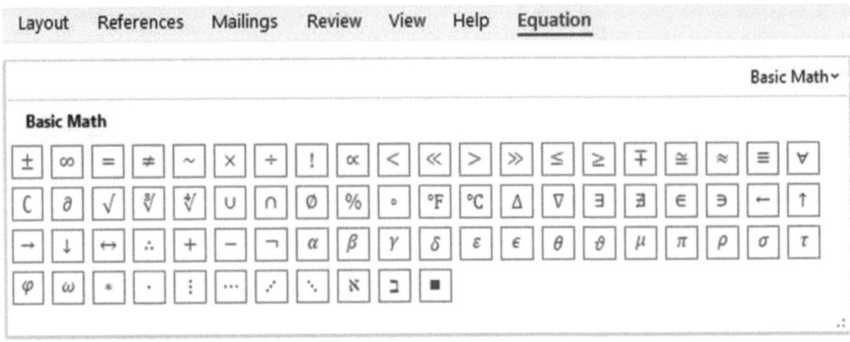

Fig. 13.7 Opened window

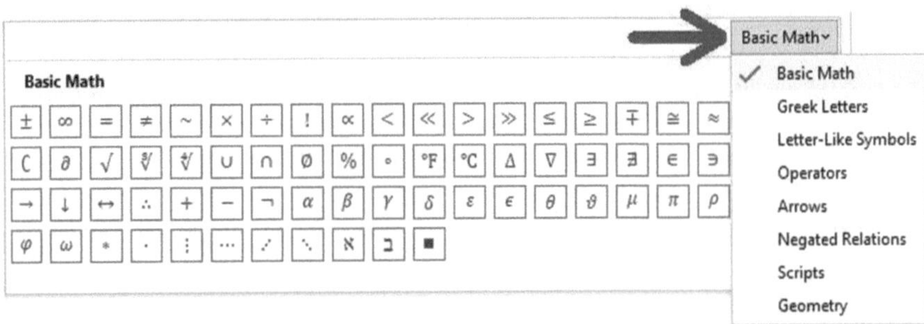

Fig. 13.8 Provided mathematical symbols and functions

References for Further Study

1. Engineering Mathematics with MATLAB® and Simulink®, Farzin Asadi, Springer, 2025.
2. Applied Numerical Analysis with MATLAB®/Simulink®: For Engineers and Scientists, Farzin Asadi, Springer, 2023.
3. Advanced Engineering Mathematics, Erwin Kreyszig, Wiley, 2020.
4. A Problem-Solving Approach to Single-Variable Calculus (Volume II), Farzin Asadi, Springer, 2026.
5. A Problem-Solving Approach to Multi-Variable Calculus (Volume I), Farzin Asadi, Springer, 2026.
6. A Problem-Solving Approach to Multi-Variable Calculus (Volume II), Farzin Asadi, Springer, 2026.
7. A Problem-Solving Approach to Multi-Variable Calculus (Volume III), Farzin Asadi, Springer, 2026.
8. A Problem-Solving Approach to Ordinary Differential Equations (Volume I), Farzin Asadi, Springer, 2026.
9. A Problem-Solving Approach to Ordinary Differential Equations (Volume II), Farzin Asadi, Springer, 2026.
10. A Problem-Solving Approach to Ordinary Differential Equations (Volume III), Farzin Asadi, Springer, 2026.
11. A Problem-Solving Approach to Ordinary Differential Equations (Volume IV), Farzin Asadi, Springer, 2026.

MIX
Papier aus verantwortungsvollen Quellen
Paper from responsible sources
FSC® C105338

If you have any concerns about our products,
you can contact us on
ProductSafety@springernature.com

In case Publisher is established outside the EU,
the EU authorized representative is:
Springer Nature Customer Service Center GmbH
Europaplatz 3, 69115 Heidelberg, Germany

Printed by Libri Plureos GmbH
in Hamburg, Germany